MAJOR INDUSTRIAL RESEARCH UNIT STUDIES

No. 62

OPEN SHOP CONSTRUCTION REVISITED

AF777467

by

Herbert R. Northrup

with the assistance of

David O. Northrup

Industrial Research Unit
The Wharton School, Vance Hall / CS
University of Pennsylvania
Philadelphia, Pennsylvania 19104
U.S.A.

© 1984 by the Trustees of the University of Pennsylvania
Library of Congress Catalog Number 84–048502
MANUFACTURED IN THE UNITED STATES OF AMERICA
ISBN: 0–89546–047–5 (Hardcover)
ISBN: 0–89546–049–1 (Softcover)

Foreword

In 1975, the Wharton Industrial Research Unit published *Open Shop Construction,* coauthored by myself and Professor Howard G. Foster from the State University of New York at Buffalo. This was the first thorough analysis of the nonunion segment of construction, which has undergone a spectacular expansion since the mid-1960s and accounted for a clear majority of the industry's volume by 1975. At the time of the original study, it was believed that the open shop share of construction volume would tend to level off in future years. This, however, proved not to be the case. There continued to be a clear indication that open shop contractors were gaining in most areas of the country, both increasing their share where they were already dominant and winning jobs in former union strongholds. Moreover, open shop contractors were clearly increasing their capability to handle larger and larger jobs. By January 1981, it was decided, therefore, to reexamine the field in order to determine whether a revised study should be issued. What this reexamination revealed, however, was that the extent of open shop growth had increased so dramatically that an entire new field study and questionnaire were required, in addition to an examination of considerably more government, industry, association, and union data than even extant for the original study.

With some misgiving, it was decided to proceed. Fund raising to support the research proved very difficult because of the recession and because this book is a revision and expansion of the original work. We did, however, receive three small grants from the Houston Chapter of the Associated General Contractors, the Merit Shop Foundation, and H. B. Zachry. These covered only about 10 percent of the research costs, and the remainder were met, first, by having a very limited staff and, second, by drawing on the general funds of the Wharton Industrial Research Unit, which depend heavily on the membership contributions of the unit's Industry Research Advisory Group. All funds from the sale of the book will revert to the unit, as is the case with all books published by the unit.

Unfortunately, there were no funds available to utilize Professor Foster's expertise, and he was already committed in any case. The field work, except for a few interviews which I conducted, was done by David O. Northrup. He holds a Master of Architecture degree from the University of Texas at Austin and worked as an architectural consultant before coming to Wharton for his Master in Business Administration degree, which he received in December 1982. While at Wharton, he did field work in forty-five states to collect information for the study and wrote memoranda used in the study. He is now in a managerial position in a major construction company. I wrote the book between the spring of 1983 and the summer of 1984. Of course, my work with Professor Foster on the earlier edition was of critical value in accomplishing this task.

Unless one has done extensive research concerning open shop construction, it is hard to comprehend the difficulties inherent in the subject. Each open shop contractor operates somewhat differently and employs, deploys, trains, and compensates his labor uniquely. Therefore, the only way to find out about the subject is to collect every scrap of information that is available, then spend considerable time, money, and energy interviewing numerous persons throughout the country, and finally to organize all this information into a comprehensible whole. This is what was done for the original study, and this procedure was followed once again. The methodology is described in more detail in Chapter II; the organization of the book is explained in Chapter I. It is hoped that readers will find the results as fascinating as doing the job was to the author.

In addition to those persons already mentioned, the following individuals helped with the research of this book or preparation of the manuscript. Michael V. Nguyen, an M.B.A. candidate, did the data processing of the 1982 questionnaire sent to 22,000 construction contractors and of the National Labor Relations Board data; Erik B. Halverson, a student in the University of Pennsylvania's Moore School of Engineering, data processed the Equal Employment Opportunity Commission data utilized in Chapter XII. Susan A. Torelli, the Industrial Research Unit's librarian, processed the U.S. Census data and searched out numerous sources of information. O. P. Suri word processed all chapters but one, which was done by Mrs. Cynthia K. Smith; my secretary, Mrs. Mary Jane Welsh, and our now retired office manager, Mrs. Margaret E. Doyle, corrected chapters and inserted new information into the manuscript. Besides these duties, Mrs. Welsh spent many hours helping to telephone persons all over the country to gather information for the study, and Mrs. Doyle handled all administrative matters associated with the project prior to her December 31, 1983 retirement. After that date the fiscal aspects were administered by Mrs. Marthenia A. Perrin, the unit's fiscal coordinator. The work was fact checked by Ms. Amie D. Thornton and edited by Ms. Bette R. Haitsch, the unit's chief editor, assisted by Ms. Kate C. Bradford and Ms. Thornton. Both Ms. Haitsch and Ms. Thornton prepared the index.

Numerous contractors, contractor association officials, and some local union and government personnel, submitted to tiresome field interviews, answered bothersome telephone calls, and supplied information whenever requested. Certainly, this work could not have been completed without their unselfish assistance. Since many of these contributors prefer anonymity and space does not permit naming all of them, only a few are mentioned: Victoria L. Tanner, Director of Communications, Associated Builders and Contractors; Dennis M. Bradshaw, Executive Director, Manpower Services, Associated General Contractors; Dick Lewis, Executive Director, Houston Chapter of the Associated General Contractors; Robert M. Gasperow, Executive Director, Construction Labor Research Council; Jay M. Kraker, Labor Editor, *Engineering News-Record;* and Peter A. Cockshaw, Editor and Publisher, *Cockshaw's Construction Labor & Opinion.*

As in all works published by the Wharton Industrial Research Unit, the author is solely responsible for the research and for all opinions expressed, which should not be attributed to the University of Pennsylvania or to any grantors involved.

Philadelphia
July 1984

HERBERT R. NORTHRUP, *Director*
Industrial Research Unit
University of Pennsylvania

TABLE OF CONTENTS

LIST OF TABLES

TABLE PAGE

TABLE PAGE

CHAPTER V

TABLE PAGE

TABLE PAGE

TABLE PAGE

TABLE PAGE

TABLE PAGE

LIST OF FIGURES

PART ONE

Introduction and Overview

CHAPTER I

The Industry and Its Organizations

Since 1975, when the first edition of *Open Shop Construction* was published, a tremendous change has occurred in the construction industry's unionization pattern. The trend toward open shop, or nonunion construction, has accelerated. Whereas ten years ago open shop construction accounted for an estimated 55 percent of construction volume, today it accounts for close to 70 percent.

This dramatic decline of union power and influence has resulted in considerable literature concerning open shop construction, in contrast to the relative absence of such information when the research for the original work was done. Nevertheless, this literature lacks comprehensive nationwide information and analysis of the issues which have caused this union decline; an examination of the actual division between the unionized and open shop segments of the industry on a local, state, regional, and national basis for each sector; and an analysis of how open shop contractors hire, train, deploy, and compensate labor; and how these methods differ from those of unionized contractors. This book is designed to fill that gap and to update and expand the 1975 study.

It therefore, focuses on that portion of the construction industry which operates nonunion or, more accurately, "open shop." (We use the term "open shop" as synonymous with nonunion because it is truly descriptive of much of nonunion construction, where the general contractor is nonunion but some subcontractors operate union, or vice versa.) Prior to our first volume, this sector was the neglected stepchild of the industry and, to a large extent, still is today. Not only are unions frequently seen as coincident with the industry, but even when union behavior is itself the subject of analysis, it is seldom explicitly approached in terms of response to nonunion competition. The situation is especially anomalous when construction unions have lost significant ground since the mid-1970s and, as a result, today do not constitute even one-third of the industry's workforce. This book, then, addresses the employee relations situation applicable to the substantial majority of the industry.

In this introductory chapter, we shall briefly describe the construction industry and the organizations which operate within its boundaries and affect its destiny. The following chapter will outline the major questions which are addressed, summarize the factors which have propelled open shop construction to the forefront, and conclude with an account of the information sources and research methodology on which our study is based.

THE CONSTRUCTION INDUSTRY

In any given year, construction may account for approximately 7 to 15 percent of the nation's economic activity. Table I-1 shows that the annual value of construction put into place in 1983, a recession year, was in excess of $265 billion, or in constant 1979 dollars, $168.6 billion, down $10 billion from 1979. Privately financed construction, as shown in Table I-1, was approximately four times larger than government-financed work. We shall see throughout this book that the Davis-Bacon Act and state prevailing wage legislation (discussed in Chapter II) have a profound impact on wages, fringes, training, and deployment of labor, and that these laws have heavily supported union growth and power. Although they apply only to public construction projects, their influence has been felt in the private construction segment as well.

As shown in Figures I-1 and I-2, construction is a volatile industry characterized by wide fluctuations in product demand. Private construction in particular is likely to vary widely as a result of business conditions, fluctuations in interest rates, demand for the users' products, or as in the case of nuclear energy, public fear or opposition. The demand for publicly financed construction is more likely to fluctuate at a slower pace as the political decision-making process takes longer to evolve. Figure I-2 shows, however, a secular decline in public construction since 1977. Because the union-open shop role in construction varies significantly from one industry sector to another, a decline, for example, in public utility or industrial construction will have a quite different effect on the unionized and open shop segments of the industry than a similar decline in residential or small commercial construction.

Contract construction generates substantial employment, but the number of workers varies considerably because of seasonal factors and the number of persons who work short periods in the industry. For example, in February 1984, local construction employment, not seasonally adjusted, was 6.1 million.[1] Seasonal factors can add or substract two million or more from these totals. Moreover, the total

[1] *Employment and Earnings,* Vol. 31 (March 1984), Table A-25, p. 35.

TABLE I-1

Annual Value of New Construction Put in Place in the United States, 1979-1983

(millions of dollars)

Type of Construction	Current Dollars					Constant (1977) Dollars				
	1979	1980	1981	1982	1983	1979	1980	1981	1982	1983
Total new construction	230,413	230,712	239,418	232,049	265,090	178,951	161,098	157,587	150,581	168,620
Private construction	181,621	175,699	186,069	180,979	214,802	141,354	123,559	122,446	116,913	135,637
Residential buildings	99,030	87,261	86,566	74,810	112,867	75,958	60,911	55,893	47,230	69,647
New housing units	78,587	63,139	62,664	51,916	85,184	60,268	44,080	40,484	32,776	52,558
One unit	60,790	45,662	44,401	35,663	62,133	46,630	31,858	28,686	22,524	38,337
Two or more units	17,798	17,476	18,263	16,253	23,051	13,638	12,223	11,798	10,253	14,221
Nonhousekeeping	2,206	2,997	3,876	4,221	5,259	1,687	2,092	2,497	2,664	3,244
Additions and alterations	18,236	21,125	20,025	18,673	n.a.	14,003	14,739	12,911	11,790	n.a.
Nonresidential buildings	47,298	52,434	60,818	65,134	61,117	36,064	35,444	39,356	41,867	38,880
Industrial	14,950	13,837	17,030	17,343	13,144	11,425	9,360	11,018	11,147	8,363
Office	9,461	13,318	17,473	23,049	21,699	7,205	8,995	11,305	14,816	13,806
Other commercial	15,463	16,627	16,775	14,235	14,570	11,766	11,242	10,858	9,149	9,267
Religious	1,548	1,637	1,665	1,543	1,801	1,180	1,106	1,078	992	1,145
Educational	806	1,175	1,171	1,356	1,411	614	793	758	872	898
Hospital and institutional	3,530	4,046	4,907	5,875	6,579	2,702	2,734	3,176	3,776	4,185
Miscellaneous	1,540	1,794	1,798	1,733	1,912	1,172	1,214	1,164	1,114	1,216
Farm nonresidential	5,588	5,274	4,809	4,987	n.a.	4,268	3,562	3,111	3,205	n.a.
Public utilities	28,253	29,480	32,609	34,796	n.a.	23,943	22,792	23,248	23,775	n.a.
Telephone and telegraph	6,343	6,733	7,074	7,110	6,430	5,436	5,243	5,117	4,908	4,324
Other public utilities	21,910	22,747	25,534	27,686	n.a.	18,508	17,549	18,131	18,867	n.a.
Railroads	1,189	1,256	1,334	1,314	n.a.	917	849	882	877	n.a.
Electric light and power	17,480	17,647	19,827	22,067	n.a.	14,913	13,830	14,247	15,066	n.a.
Gas	2,650	3,035	3,631	3,928	n.a.	2,189	2,266	2,491	2,668	n.a.
Petroleum pipelines	591	809	742	377	n.a.	489	604	511	256	n.a.
All other private	1,452	1,250	1,268	1,252	1,529	1,121	849	838	836	1,021

TABLE I-1 (Continued)

Type of Construction	Current Dollars					Constant (1977) Dollars				
	1979	1980	1981	1982	1983	1979	1980	1981	1982	1983
Public construction	48,792	55,013	53,349	51,070	50,287	37,597	37,539	35,141	33,668	32,983
Buildings	15,558	18,517	17,792	16,997	17,203	11,895	12,540	11,515	10,908	10,915
Housing and redevelopment	1,211	1,648	1,722	1,658	1,672	929	1,150	1,112	1,047	1,033
Industrial	1,112	1,441	1,655	1,632	1,797	852	974	1,071	1,049	1,144
Educational	6,903	8,050	6,737	5,927	5,282	5,273	5,432	4,361	3,810	3,362
Hospital	1,648	1,785	2,083	1,991	2,133	1,261	1,205	1,348	1,280	1,357
Other	4,684	5,593	5,595	5,789	6,319	3,580	3,779	3,621	3,722	4,020
Highways and streets	11,997	13,770	13,599	13,521	14,177	8,585	8,462	8,633	9,127	9,640
Military facilities	1,647	1,880	1,964	2,205	2,470	1,248	1,251	1,271	1,438	1,588
Conservation and development	4,587	5,090	5,300	5,027	4,825	3,821	3,821	3,666	3,329	3,169
Sewer systems	7,298	7,171	5,935	5,529	5,253	5,971	5,409	4,163	3,659	3,448
Water supply facilities	2,490	3,266	3,004	2,902	2,016	2,048	2,456	2,090	1,947	1,322
Miscellaneous public	5,215	5,318	5,754	4,889	4,344	4,029	3,601	3,804	3,262	2,902

Source: U.S. Department of Commerce, Bureau of Census, *Value of New Construction Put in Place,* Construction Reports C30-84-1 (Washington, D.C.: Government Printing Office, 1984), Table 1, p. 3.

n.a. = not available

FIGURE I-1
New Private Construction Put in Place in the United States
1964–1982
(billions of 1977 dollars)

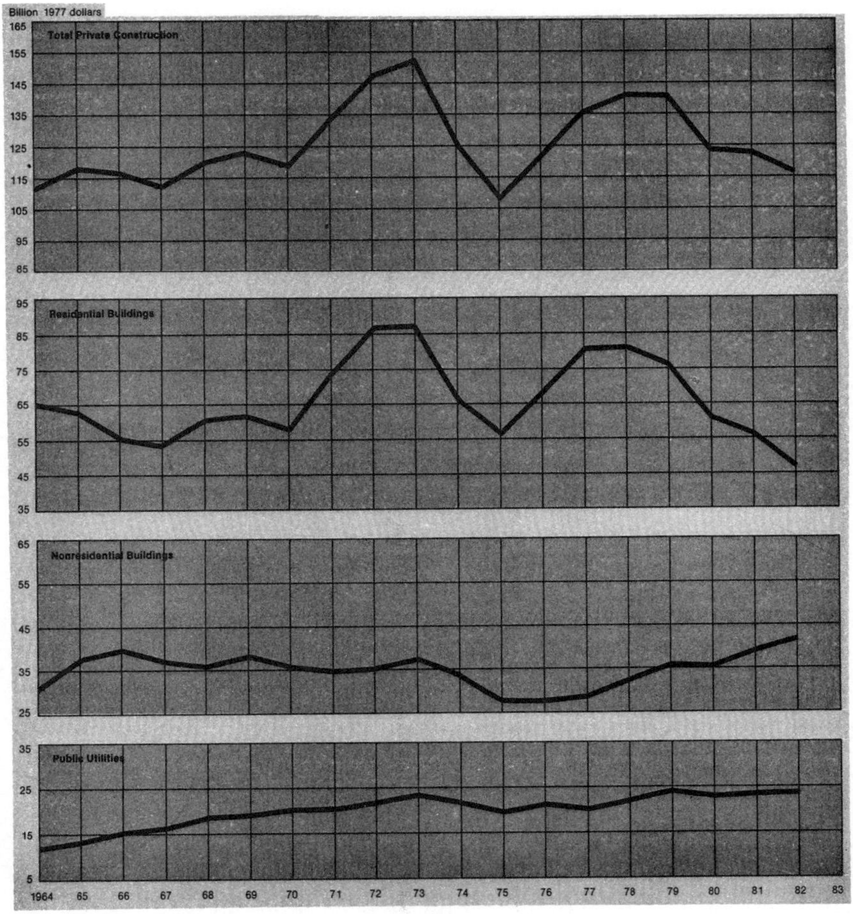

Source: *Construction Review,* Vol. 29 (July-August 1983), p. 3.

employment figures can remain constant, but a million or more persons may be in and out of the industry working for short periods of time.

Unemployment can be very high in construction because of the severe fluctuations in product demand. Thus in February 1983, while unemployment in the economy stood at 11.3 percent, construction unemployment hit 27.3 percent; the comparable figures

FIGURE I-2
New Public Construction Put in Place in the United States, 1964–1982
(Billions of 1977 dollars)

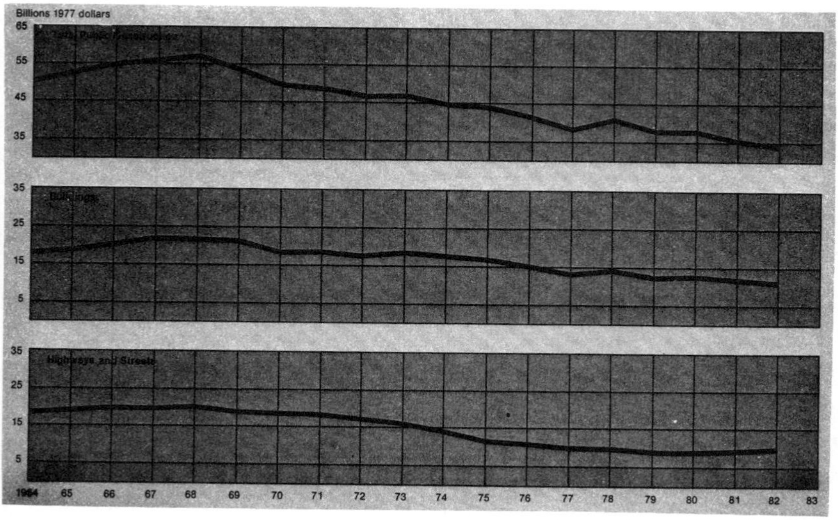

Source: *Construction Review,* Vol. 29 (July-August 1983), p. 4.

for February 1984 for the economy were 8.4, and for construction, 21.1 percent.[2]

Construction workers are also very mobile. As a result, construction in general has a very high turnover. Every full-time construction job is filled, on the average, by nearly two workers.

As in most industries, the salient characteristics of the construction industry, its labor market, and employee relations are reflections of the nature of the construction product. Although there is great variety in the kinds of facilities produced by the industry, all construction products exhibit a number of common properties: immobility, complexity, durability, and often a speculative value.

In one sense, there is no "market" for most kinds of construction products. With the significant exception of single-unit residences and a few types of small factory buildings, virtually all construction is "made to order." Work is started at the request of the buyer, not the producer, and the determination of what and where to build is made by those outside the industry. The result is an extreme heterogeneity of demand, a condition necessitating a kind of industry

[2] *Ibid.,* Table A-13, p. 24.

organization which is capable of adjusting rapidly to share fluctuations in organizational and manpower requirements.

The need for flexibility, together with the vast heterogeneity and complexity of construction products, has led to a system of functional specialization. In the typical case, a buyer engages a general contractor who has overall responsibility for the project. The general contractor, in turn, may engage any number of specialized contractors who perform certain operations such as wiring, plumbing, or painting. As we shall note in several chapters, open shop general contractors are much more likely to perform a substantial share of the specialty work, such as plumbing or electrical work, than are unionized contractors, but even the most self-sufficient general contractors will contract out specialties like brick or tile work.

The need for specialty contractors in the industry is very great. It would simply be too costly for general contractors to supply these different needs on all occasions, and to maintain the engineering and work capability required by construction projects of different sizes and in different locations. Subcontractors specializing in particular trades and skills provide the capability to perform their work because of their concentration and experience in their trades.

Because of the immobility of the construction product, most building structures obviously cannot be erected in one place and transported to another; hence, the mass production technique found in other goods-producing industries are largely absent. This characteristic has been tempered by the growing use of prefabricated materials which allow certain portions of buildings to be produced on what approaches an "assembly-line" basis. In addition, large-scale housing tracts utilizing one basic plan may permit a kind of "factory" approach. Nevertheless, the construction market is predominantly local in character.[3] To be sure, there are nationwide construction firms whose gross receipts run into the billions of dollars, but these typically undertake only the most extensive and costly projects, leaving small and medium-sized jobs (the bulk of all construction work) to local contractors. Moreover, these large national contractors will often recruit much of their workforce from the area where the projects are located.

The durability and costliness of construction products render the demand for them highly unstable. The replacement or renovation of most existing structures can be deferred to a time when the economic outlook is most favorable. As a result, the demand for

[3] See Daniel Quinn Mills, *Industrial Relations and Manpower in Construction* (Cambridge, Mass.: MIT Press, 1972), p. 8.

construction tends to fluctuate much more dramatically than for other products. This imposes on the industry a need for much greater flexibility than that required for other industries. It also creates an element of uncertainty which magnifies the problems otherwise inherent in developing and maintaining a skilled labor force.

The composition of the workforce on different projects will vary widely. For example, a brick building and a frame building will obviously require different mixes of carpenters and bricklayers. It is also common for the crew of an individual project to change markedly in composition and size over a relatively short time. In some cases the necessary workforce adjustments are very abrupt. In the union sector of the industry, these adjustments are effected by the use of a labor pool situated in part in the union hall. In open shop construction, the pool is more amorphous, the mechanics of the adjustment process are less structured, and the adjustments themselves are smaller because the open shop contractor has, as we shall see, very practical reasons to go to greater lengths to keep his employment patterns steady. Moreover, the deployment of labor by open shop contractors is, as described in Chapter IX, both quite different and more economical than that by union contractors.

Types of Construction

Construction embraces a wide variety of building activities. The product may range from a small, single-family home erected in a few months by a three-man work crew and some specialty subcontractors to a gigantic dam or oil refinery worth billions of dollars and requiring the labor of thousands over several years. The industry, unfortunately, has no consistently defined breakdown of its constituent parts. Most employment statistics, for example, use only these categories: general building contractors, highway and heavy contractors, and special trades contractors.[4] It is thus impossible to discern the number of workers employed, for example, in residential construction at a given time.

Since we shall have occasion to distinguish among various types of construction throughout this book, it will be useful to describe briefly four industrial subdivisions to which we shall frequently refer. *Residential construction* includes predominantly the building of single-family homes and apartments. (Government statistics also

---

[4] In addition, however, there are firms engaged principally in land development and construction which are officially classified as "subdividers and developers" and "operative builders" under the general umbrella of "real estate."

include such structures as dormitories and hotels.) The residential portion of the industry may be further divided into homes, town houses and low-rise apartment complexes (under four stories), on the one hand, and high-rise apartment buildings, on the other. *Commercial construction* includes such projects as stores, office buildings, warehouses, small factories, hospitals, nursing homes, and service stations. *Industrial construction* refers primarily to large factories, power plants, refineries, and other large structures built for establishments engaged in secondary economic activity. *Heavy and highway construction* includes streets, roads, bridges, dams, pipelines, airports, and subways. Government data do differentiate between private and public construction, but public projects may be in any of the aforementioned classes: residential construction (low-income housing), commercial construction (schools, office buildings), industrial construction (power plants), and heavy and highway construction.

Firms operating in the various industrial subdivisions tend to differ in a number of ways. Most firms throughout the industry are small,[5] particularly in residential construction and, to a lesser extent, in commercial construction. Industrial, heavy, and major highway work, largely because of the size of the projects involved, is more likely to be concentrated in the hands of large nationwide contractors whose yearly volume is well into the eight- or nine-figure range. Furthermore, there tends to be more specialization in residential construction among both general contractors and subcontractors; a firm that builds homes and small apartments is likely to build only homes and small apartments. Commercial and industrial contractors, on the other hand, are usually more diversified. Heavy and highway construction is more seasonal than the other divisions and also utilizes a larger proportion of unskilled labor. Finally, there are differences in the amount of work under the direct control of the general contractor. Although this characteristic is to a great extent a feature of the specified contractor involved, the amount of subcontracting is generally greatest in residential construction (many homebuilders subcontract all their work) and least in heavy and highway work.

In short, we are dealing with an industry of prodigious variety that makes generalizations hazardous. For that reason, we shall examine in Part Two the extent of open shop activity separately by industry sector before bringing our findings together in state and regional summaries.

---

[5] Most studies show that the average contractor has under ten employees. Of course, this number can be greatly expanded for specific projects.

ORGANIZATIONAL STRUCTURE

Basic to an understanding of the extent of open shop construction is an explanation of the complex organizational structure of the industry. There are literally hundreds of thousands of contracting firms with payrolls and a like number of one-man operations. Many contractors are organized into an intricate and sometimes overlapping network of trade associations, of which there are currently about sixty operating nationally and an unknown number operating locally.[6] While an undetermined, but apparently substantial, number of firms, especially the smaller ones, do not belong to any contractors' association, some belong to several such associations. These may be classified into four broad categories: those comprising general contractors; those comprising contractors who specialize in a particular type of project; those comprising contractors who specialize in a particular phase of construction work; and organizations which attempt to coordinate all or some of the first three groups.

Estimates of union membership in the construction industry are difficult to make for several reasons. Unions in all industries are likely to inflate their numbers in order to give an impression of greater strength, and unfortunately, the U.S. Bureau of Labor Statistics has never made a critical examination of union membership data, despite severe criticisms from scholars in the field.[7] In construction, the problem is magnified by the fact that thousands of workers maintain union membership in order to retain eligibility for union-sponsored welfare and pension plans, return to work in unionized construction even though they may be not working at all, may be employed as managers in construction, may be working on an open shop job, or may be working outside the industry. Furthermore, many construction unions represent workers in other industries. Thus, any estimate of union membership in construction is likely to have a considerable margin of error.

With these caveats, we estimate on the basis of evidence presented throughout this study that approximately 30 to 40 percent of all construction workers are unionized, although we believe that the actual number of workers employed in unionized construction is generally closer to the lower figure. Nearly all these unionized

---

[6] A directory of national trade associations, professional societies, and labor unions involved in construction and building material industries is found in *Construction Review,* Vol. 27 (April 1981), pp. 4–28.

[7] For an analysis of the problems of estimating union membership and citation of sources, see Gordon F. Bloom and Herbert R. Northrup, *Economics of Labor Relations,* 9th ed. (Homewood, Ill.: Richard D. Irwin, 1981), pp. 69–71.

workers belong to one of the craft unions which are collectively referred to as the "building trades." The union structure is highly fragmented; there are currently sixteen national organizations and thousands of locals across the country. Fifteen of these unions are affiliated with the Building and Construction Trades Department (BCTD) of the AFL-CIO. The sixteenth, the Teamsters Union, is not now a member of the AFL-CIO, but in many areas Teamsters locals representing construction truck drivers still work closely with the other trades.

The multiplicity of employer and employee organizations naturally makes for a highly complex collective bargaining structure in the industry. Annually, the number of labor agreements negotiated can exceed 3,500. Most of these agreements are executed at the local level and, in many cases, cover only a single metropolitan area and sometimes only a portion of a given labor market. Although there has been some tendency in recent years toward a more centralized structure with contracts covering wider areas and several local unions, the trend has not been headlong. There is, moreover, some bargaining at the national level, particularly on certain very specialized projects (e.g., elevator or pipeline construction) or on large projects (e.g., dams) in remote areas where no local agreement adequately covers the work.

General contractors' associations that negotiate usually do so directly with each of the basic trades: Carpenters, Bricklayers, Ironworkers, Cement Masons, Laborers, and Teamsters. Some of the basic trades, however, may also have contracts with associations of specialty contractors, such as the Mason Contractors Association (Bricklayers) or the Drywall Contractors Association (Carpenters). Specialty contractors' associations, whose members characteristically perform work that is subcontracted by general contractors, deal primarily with a single union representing the craft which they predominantly employ. Examples are the Mechanical Contractors Association (Plumbers and Steamfitters), the National Electrical Contractors Association (Electricians), and the Painting and Decorating Contractors Association (Painters and Paperhangers).

Contractors' associations often expend much of their energies dealing with labor relations matters. In the unionized sector, they assume such key functions as negotiating contracts and assisting member contractors with the grievance process. They also invariably become involved in such issues as jurisdictional disputes and the administration of apprenticeship programs. Associations representing open shop contractors handle different kinds of labor relations problems: they are principally concerned with assisting members who are objects of union organizing campaigns. In both cases, however, the building trades constitute major preoccupations for staff members of these associations.

From this discussion it should not be inferred, that labor issues comprise the entire raison d'être for contractor groups. The broad range of their activities has been well summarized by Mills:

> The local associations perform a wide variety of functions for their members, including public relations, lobbying, legal advice, labor relations activities, and members' benefits (such as group life insurance for contractors or types of liability insurance) and they deal with architects, owners, suppliers, and others. The national office of the association also conducts lobbying and public relations and provides legal and industrial relations advice. It often publishes periodicals carrying trade news, innovations, legislative reports, and analyses of the national scene as it affects members' concerns. Each national association normally holds a national convention and may sponsor trade shows as well.[8]

The following sections are devoted to a more detailed examination of the various organizations which are important components of the construction industry. We begin with the building trades unions, since an understanding of the history, role, and function of contractors' associations—both union and open shop—depends in large measure on an apppreciation of the character of the labor organizations with which they deal. We then turn to the associations themselves, with special emphasis on those which have had significant roles in the growth of open shop activity.

The Building Trades Unions

The building trades claim to represent about 4.1 million workers in a variety of industries, crafts, and subcrafts. Table I-2 summarizes pertinent data about the fifteen organizations which comprise the BCTD. It gives the official version of the union membership based upon figures supplied by the unions. (For reasons already stated, we believe that these data overstate the actual membership—perhaps by as much as 10 to 20 percent.) The table also indicates that some of these unions have a substantial proportion of their memberships outside the contract construction industry. The Carpenters, Laborers, and Sheet Metal Workers, for example, also represent workers in various manufacturing industries, and the Electrical Workers have a large number of members both in manufacturing and electric utility companies. Thus, even if the data in Table I-2

[8] Mills, *Industrial Relations and Manpower,* p. 11.

TABLE I-2
The AFL-CIO Building Trades Unions, 1983

Union[a]	Claimed Membership	Estimated Percent in Construction	No. of Affiliated Local Unions
Asbestos Workers, International Association of Heat & Frost Insulators and	20,000	100	130
Boilermakers, Iron Shipbuilders, Blacksmiths, Forgers & Helpers, International Brotherhood of	145,000	22	400
Bricklayers and Allied Trades, International Union of	145,000	100	618
Carpenters and Joiners of America, United Brotherhood of	750,000	75	2,682
Electrical Workers, International Brotherhood of	1,100,000	2.0	1,467
Elevator Constructors, International Union of	18,000	100	109
Iron Workers, International Association of Bridge, Structural & Ornamental	174,000	100	323
Laborers' International Union of North America	610,000	75	805
Operating Engineers, International Union of	420,000	80	210
Painters and Allied Trades, International Brotherhood of	170,000	85	830
Plasterers' and Cement Masons' International Association of the United States and Canada, Operative	57,000	99	365
Plumbing and Pipe Fitting Industry of the United States and Canada, *United Association* of Journeymen and Apprentices of the	325,000	100	533
Roofers, Damp and Waterproof Workers Association, United Slate, Tile and Composition	31,000	100	177
Sheet Metal Workers' International Association	160,000	65	283
Marble, Slate and Stone Polishers, Rubbers and Sawyers, *Tile and Marble Setters,* Helpers and Marble Mosaic and Terrazzo Workers' Helpers, International Association of	9,000	97	117

Source: *National Trade and Professional Associations of the United States,* Nineteenth Edition (Washington, D.C.: Columbia Books, 1984), as modified by new information.

[a] Italicized portion is abbreviated or commonly used name.

are correct, the representation of the building trades unions in construction is about 2.6 million persons. Since this would indicate that the unions represent about 32.8 percent of the total construction employees, we believe tht the data and analysis presented in Part Two of this book will clearly indicate that this representation is certainly not underestimated.

The BCTD itself, like the AFL-CIO, is a federation of national unions whose primary function is to advance the general interests of unionized construction labor. There are also state and local branches of the department (sometimes called "councils"). In many instances, these interunion bodies have no collective bargaining responsibilities, but rather perform various political, public relations, and research activities designed to benefit all the member unions. In other instances, the local branches actually do some or all of the bargaining.

The issue of union jurisdiction has long been endemic to the labor relations scene in construction. The preservation of job opportunities is a vital responsibility of the union leadership, especially in an industry marked by intermittent and uncertain employment conditions. A loss of jobs to the members of a sister union is only slightly less distasteful than losses to technology, recession, or nonunion firms. The craft unions guard their jurisdictional terrain with great vigor, often striking to protest job assignments to a different craft despite the illegality of such actions under section 8(b)(4)(D) of the National Labor Relations (Taft-Hartley) Act. The building trades have sought, with varying degrees of success over the years, to develop and maintain private adjudication mechanisms for the resolution of jurisdictional disputes.[9] Nevertheless, the problem remains, with "intraunion or interunion matters" often constituting the basis for approximately one-third or more of the strikes in the construction industry in any one year.[10] It is significant that many of the contractors and association executives whom we interviewed in 1982 and in 1984—both union and nonunion—cited jurisdictional strikes as a major factor in the recent growth of the open shop sector.

Jurisdictional rivalries may pose serious problems for contractors in several ways. First and most obvious, of course, are the delays caused by stoppages which grow out of conflicting jurisdictional claims. Although jurisdictional strikes tend to be shorter than those

[9] As this is being written, the unions and various associations are once again attempting to reconstruct a system of jurisdictional settlement after its break down.

[10] Data from U.S. Bureau of Labor Statistics.

for other reasons, the loss of a few days can wreak havoc with a construction schedule. Second, even if the dispute is settled without a stoppage, the resolution may not be the one preferred by the contractor. Employers are not always indifferent with respect to who wins a job assignment, especially if there is an appreciable wage differential between the competing crafts, or if the settlement requires additional costly labor and/or manhours. Finally, jurisdictional stipulations may require that assignments be given to a craft whose union is less able to provide the number or quality of workers needed to man the job, again adding to expenses.

The locus of authority within the building trades is generally centered at the local level. Local officials have the major responsibility for executing and policing collective bargaining agreements, administering apprenticeship programs, and operating hiring halls or job referral services. The national organizations promulgate general guidelines and provide technical staff assistance to the locals through international representatives, but their actual power over the affairs of the locals is severely limited. Since local autonomy is a principle of long historical tradition in the building trades, national officers who attempt to exert authority (sometimes even authority given to them by the union constitution) over important locals may do so only at a considerable political risk.

The key person in a building trades local is the business agent. He is an elected, salaried official who, in larger locals, may appoint several full-time representatives to help him run the union. The business agent is often the union's chief spokesman at the bargaining table, and between contracts his major responsibility is policing the trade. Through a network of union stewards and other officials, he ensures that the provisions of the agreement are followed, that the union's jurisdiction is not violated, and that individual grievances at the site are addressed. Because of his various functions, he is highly visible to the membership, and thus his job tenure depends largely upon his ability to respond to the needs of his constituency. The other officers of the local—president, treasurer, and so forth—deal principally with the internal affairs of the organization and have relatively little influence over the relationship between the union and employers.

Partly because of their decentralized political structure, most local unions in the building trades tend to be workably democratic organizations. But the relative weakness of the national officers and the need for local officials to respond aggressively to the demands of the membership make it difficult to correct some of the policies for which the unions have been roundly criticized. National leaders and a few brave local officers may speak passionately about the dangers of excessive wage increases and about the contribution that restrictive union practices have made to the growth of open shop competition, but the political realities of their positions usually preclude them from going significantly beyond speechmaking until, as in the early 1980s, loss of work to the open shop and the recession made it clear beyond any doubt that in many localities the survival of unionized construction was at stake. Even then, as recounted in Chapter VIII, wage reductions and productivity improvements may not have been sufficient to stem the tide.

There is also some evidence that the higher the official in the union hierarchy, the more likely he is to be sensitive to the claims of minority groups that they have been systematically and invidiously excluded from job opportunities,[11] as will be discussed in Chapter XII.

Since construction workers, like most people, tend to define their interests narrowly, the challenges which the building trades now face are attributable more to policies dictated by their own organizational structure than to any particular human failing of their leadership. Their failure to meet these challenges have resulted, to a major extent, in the increasing growth of open shop construction.

Contractors' Associations

As noted above, there are about sixty national associations of construction contractors in the United States. Some contractors hold membership in two or more of these associations, while others do not belong to any of them. Some associations are comprised predominantly or exclusively of unionized contractors; others have a substantial number of open shop contractors among their membership; still others are entirely composed of open shop contractors. In this section we shall introduce some of the larger and more important associations.

National Association of Home Builders (NAHB). This is the largest contractor organization; about 37,000 of its 120,000 members are builders engaged predominantly in residential construction. Homebuilders comprise the vast majority of the membership, with a relatively small number of specialty contractors who service the builders. Most NAHB members do little work outside the residential sector of the industry, although a fair number engage in such related activities as land development, remodeling, and real estate broker-

[11] Derek C. Bok and John T. Dunlop, *Labor and the American Community* (New York: Simon and Schuster, 1970), pp. 134–135; Herbert R. Northup, *Organized Labor and the Negro* (New York: Kraus Reprint Company, 1971), pp. 232, 236–238.

age. Much of the homebuilders' work tends to be contracted out to specialty firms. An earlier Industrial Research Unit study found that two-thirds of the homebuilders in Erie County, New York, performed no building work with their own labor.[12] This condition was found to be largely true elsewhere, as revealed by our field studies both in 1974 and 1982.

As we shall see in Chapter III, residential construction is the most unorganized sector of the industry. In prior years, when residential construction unionism was more prevalent, a number of NAHB chapters bargained for their members. Today only a few NAHB chapters have labor agreements with some or all of the building trades unions, and even in these jurisdictions, an increasing number of builders are conducting open shop operations outside the coverage of the contracts. There are also a few cases of individual builders with union contracts, but these instances are infrequent and dwindling. On the whole, the NAHB is an organization comprised predominantly of open shop employers, a direction in which the association has steadily moved for the past quarter century.

NAHB chapters are found in most labor market areas of any appreciable size. In a few localities where the Associated Builders and Contractors (ABC) has not yet established itself, the NAHB constitutes the principal organizational base for residential contractors. On the whole, however, the association has not filled the need of accommodating the large number of subcontractors who specialize in residential work, a constituency to which the ABC has had a special appeal. Since the ABC has also made significant inroads among contractors specializing in apartment building, we can expect to find a growing overlap in the kinds of firms served by the two organizations.

The NAHB has a sizable national staff headquartered in Washington, D.C., which deals with a wide range of issues of general concern to the homebuilding industry. Its Department of Labor Relations is principally a staff advisory and research service; it engages in no collective bargaining activities directly. Although some building trades unions have, from time to time, announced new efforts to organize the residential sector, our interviews with the NAHB and its local chapter personnel confirm that these efforts have not resulted in any dramatic breakthrough. Since residential builders tend to perform relatively little work outside that sector, the NAHB does not occupy a prominent position on the unions'

[12] Howard G. Foster, *Manpower in Homebuilding: A Preliminary Analysis,* Manpower and Human Resources Studies No. 3 (Philadelphia: Industrial Research Unit, The Wharton School, University of Pennsylvania, 1974), pp. 31–36.

enemies list, a status largely reserved for the ABC. The NAHB has historically neither promoted nor fought open shop construction and, as a consequence, has not had a conspicuous role in open vs. union shop controversies.

Associated Builders and Contractors (ABC). Since our original study in 1975, the ABC has grown from 8,000 to more than 16,000 contractor members, plus 2,000 suppliers. In some parts of the country, this fast-paced growth has since slowed considerably. But in areas just opening up, an increase in membership every year or two is being anticipated. There are also highly unionized areas, such as St. Louis and New York City, that still have no local chapters.

Originally founded in the early 1950s in Maryland, the ABC has expanded to include almost every state in the country. (The most notable exception is the Carolinas, where the Associated General Contractors (AGC) has been aggressively open shop for so long that the ABC has avoided the area). ABC activity has been concentrated in areas where open shops have not traditionally been conspicuous. There are now active chapters in the Northeast, the upper Midwest, and the Far West. The ABC's membership is, of course, open shop, and its growth, examined in detail in Chapter IV, is testimony to the expansion of the open shop segment. A few ABC members do have union contracts or sometimes work union under exceptional circumstances. In some cases, formerly all-union contractors have joined the ABC through open shop branches which are operated on a doublebreasted basis (a term designating common ownership by one party or company of two firms, one operating union, the other open shop). A number of firms are members both of the AGC and ABC.

The ABC is an association both of general and specialty contractors, with the latter actually in the majority. The latter found that most specialty contractor organizations were dominated, as discussed below, by unionized firms, and the ABC offered support, relations with general contractors, and many other needed services. As noted in Chapter IV, the ABC has more members in certain key specialty trades than do some traditional specialty associations.

The ABC is headquartered in Washington, D.C., where a large staff functions in such specialties as labor law, public relations, training, employee benefits, and other relevant matters. It has led the fight against illegal picketing and construction union violence and played a significant role in congressional and state legislative affairs. Its staff operates and advises on benefit and apprentice programs, and its home office is considered the central headquarters of open shop interests.

Associated General Contractors (AGC). The AGC is an organization composed of approximately 8,400 general contractors in local and state chapters throughout the country. In many states, there are separate chapters for building contractors and heavy and highway contractors, with a few affiliates comprised of municipal utility contractors. Heavy and highway chapters are often organized on a statewide basis, while in building construction there are usually separate chapters in several metropolitan areas within a state. In smaller states both building and highway contractors are included in a single chapter. The Carolinas Chapter is unique in that it covers two states and all branches of construction within the AGC's jurisdiction. It has long been a powerful open shop force in the industry.

Although open shop contractors now constitute more than one-third of the AGC's membership, their numbers vary greatly from chapter to chapter. Virtually all of the highway chapters have open shop representation, with open shops comprising a majority in many states, as discussed in Chapter VI. Building chapters are less likely to serve both union and nonunion firms. The bylaws of some building chapters require the contractor to give power of attorney to the association, which, in practical terms, means that the contractor is bound by any labor agreements that are negotiated by the chapter. The effect, of course, is to preclude open shop membership. In other chapters this power of attorney is optional, which therefore permits open shops to take advantage of the other services offered by the association. In recent years, many building chapters have amended their bylaws to allow nonunion contractors to join; the Massachusetts, New York State, and New Orleans chapters are cases in point.

Although labor relations matters occupy a substantial portion of the energies of association staff members at the local level, there are other valuable services which make membership attractive to the nonunion contractor. Where AGC chapters have effectively barred open shops from membership, other associations, most notably the ABC, have stepped in to fill the vacuum. The current receptiveness displayed by many AGC chapters toward the interests of nonunion contractors has doubtless, in large measure, been occasioned by the successful organizational challenge mounted by the ABC.

At the national level, the AGC participates in a wide range of activities, most of which hold no relevance to the union status of the contractor. There are, for example, some forty national committees dealing with such varied topics as the environment, fuel and material supply, metric conversion, and tax and fiscal affairs. Most of the committees have full-time AGC staff members serving on them. On labor-related matters, there are committees on equal employment opportunity, labor, manpower and training, and open shop. The very existence of a national open shop committee, which has an able, well-rounded staff of its own and which, among other things, sponsors conferences across the country dealing with the distinct problems of open shop contractors, attests to the interest in open shop operations within the AGC. Open shop contractors also exert greater influence in the councils of the national AGC, with open shop contractors often (as in 1984) serving as the association's president.

National Constructors Association (NCA). The NCA is an organization of about fifty of the largest unionized contractors who specialize in large industrial and heavy projects, such as the construction of utility generating facilities, large industrial complexes, dams, and other megaprojects. The NCA negotiates labor agreements with the construction unions, represents its members' political interests, and makes studies of such relevant matters as open shop penetration in the industrial sector. In recent years, the NCA members have been rather divided concerning the role of the organization and its performance.[13]

Specialty Trades Associations. A large number of the contractor associations in construction are composed of specialty contractors. Indeed, in many trades, there is more than one association, sometimes split between union and open shop, in others between larger and smaller contractors, and occasionally divided along other lines. In Chapter IV, we shall examine in detail the union-open shop division within the specialty trades, particularly the highly skilled electrical and mechanical trades. Here we have merely listed the major specialty associations in Table I-3 with a few pertinent comments.

Umbrella Organizations. There have been numerous attempts in recent years to bring this wide array of contractors' associations under a coordinating body that could act on some matters, for example, questions before Congress, with one voice. For the most part, success has eluded the organizers of such a movement. Three such organizations, however, continue to exist: the Associated Specialty Contractors (ACA), the American Subcontractors Association (ASA), and the National Construction Employers Council (NCEC). The ACA's membership is composed of eight national specialty contractors associations. It coordinates lobbying and legislative activities of member associations, works with the AGC to obtain uniform

[13] "NCA Drops Its Industrial Focus" *Engineering News-Record,* July 14, 1983, pp. 28–29.

TABLE I-3
Principal Specialty Trade Associations in Construction

Associations	Unions which Bargain with Association	Comments on Membership
American Road and Transportation Builders	Operating Engineers, Teamsters, and others	Approximately, 60 percent open shop; about 2,300 contractor members.
Elevator Constructors Employers Association	Elevator Constructors	All Unionized contractors; about 60 contractor members[a]
Independent Electrical Contractors	—	All open shop; about 2,000 contractor members.
International Association of Wall & Ceiling Industries	Plasterers	About 80 percent unionized contractors among approximately 500–1,000 members.
Mason Contractors Association of America	Bricklayers	All unionized contractors; about 1,700 contractor members.
Mechanical Contractors Association of America	Plumbers and Pipefitters	All unionized contractors; about 1,600 members.
National Association of Heating-Cooling Contractors	Plumbers and Pipefitters	Approximately 60 percent open shop; about 5,900 contractor members.
National Electrical Contractors Association	Electrical Workers (IBEW)	Approximate 95 percent unionized among, 6,100 contractor members.
National Erectors; Association	Iron Workers	All unionized contractors; about 175 contractor members.
National Insulation Contractors Association	Asbestos Workers	All unionized contractors. In 1984 decided to admit and to service open shop contractors; about 450
National Roofing Contractors Association	Roofers	Approximately 60 percent union; about 2,000 contractor members.
Painting and Decorating Contractors Association	Painters	Split approximately 52 percent unionized, 48 percent open shop; about 4,000 contractor members.
Sheet Metal and Air Conditioning Contractors National Association	Sheet Metal Workers	Approximately 96 percent unionized contractors with about 2,400 contractor members.

Source: C. B. Pitcher, "A Directory of National Trade Associations, Professional Societies, and Labor Unions of the Construction and Building Materials Industries," *Construction Review,* Vol. 27 (April 1981), pp. 4–28; updated by Industrial Research Unit research and field interviews.

[a] With the exception of its New York City chapter, this is the only association that bargains on a national basis.

labor contracts, and does some collective bargaining training, among other work. The ACA also has a special membership in the NCEC.[14] The ASA is largely a lobbying organization which seeks support for issues of importance to subcontractors (who are in the specialty trades) in such matters as timely payment of charges and state laws pertinent to its members.[15]

The NCEC was established as another attempt to coordinate contractor association activities on a broad front. Its aim was to become the voice of unionized contractors, but it has never been successful in obtaining the support of all key associations, particularly the specialty associations. It has, however, succeeded in establishing relations with the BCTD, with which it set up broad bargaining guidelines and a National Market Recovery Program,[16] which is discussed in Chapter VIII. It has worked with large users on matters of local interest, and it has published various studies and "how-to-do-it" materials, on such topics as contract costing. At the end of 1983 it had thirteen association members, but lost one in January 1984.

Users' Organization

Concerned about high construction costs, executives of major corporations formed the Construction Users' Anti-Inflation Roundtable in 1969 to bring together large users in an effort to control the burgeoning construction costs of plant and large facility construction. In 1972, this group merged with the Business Roundtable, which maintains a separate staff devoted to construction problems. The purpose of the Business Roundtable's construction division is to give support to contractors in order to decrease the upward spiral of construction costs. It has succeeded in obtaining users' agreements to limit scheduled overtime and in shuting down construction under national agreements when local strikes occur to avoid giving work to striking unionists. In addition, it has lent talent to local groups engaged in bargaining. To effectuate its work, the Roundtable has also established chapters in various parts of the country. Recently, the Roundtable sponsored "The Construction Industry

---

[14] Interview, Associated Specialty Contractors executive, Washington, D.C., February 26, 1982.

[15] "Subs Seek New Policy-Making Role," *Engineering News-Record,* December 17, 1981, p. 73; "Subs Push for State Reforms," *Engineering News-Record,* July 1, 1982, p. 89.

[16] See, e.g., "Building Trades-NCEC Sign Market Recovery Agreement," *Construction Labor Report,* Vol. 28 (November 17, 1982), p. 1257; "NCEC Regrouping After Losses" *Engineering News-Record,* December 17, 1981, p. 170; and "Subcontractor Pact Eludes NCEC," *Engineering News-Record,* December 15, 1983, p. 58.

Cost Effectiveness Project," a major study issued in the form of twenty-three pamphlets plus a summary report entitled *More Construction for the Money.*[17] Reference to, and discussion of, various parts of this project are found in several later chapters of this book, particularly in Part Three.

The Roundtable does not openly promote open shop construction. It does, however, encourage alternatives to the union shop where it is feasible, and this, of course, involves opening up bidding opportunities to open shop contractors. Moreover, by furthering an exchange of information among users, the problems of union construction are aired and opportunities to overcome them by special arrangements, such as project agreements, wider area agreements, and the use of open shop contractors for industrial construction projects are discussed.

CONCLUSION

The construction industry thus has a diverse organizational structure of many components each created because of a particular need. The interplay of these organizations in an industry which is itself highly complicated, both affects and is affected by, the extent of union organization. In the following chapter, we provide an overview of the extent of organization; examine the economics which has pushed the industry toward a larger open shop component; and describe the methodology upon which the detailed information in the chapters of Parts Two and Three are based.

---

[17] *More Construction for the Money,* Summary Report of The Construction Industry Cost Effectiveness Project (New York: The Business Roundtable, 1983).

CHAPTER II

Overview

The wide scope and atomistic structure of the construction industry make estimates of any kind difficult, particularly for such an emotion-laden subject as the extent of union organization, for which, furthermore, there has been no systematic collection of data. Later chapters review the limited data available by industry sector, and region. This section gives an overview of the subject.

EXTENT OF UNIONIZATION AND UNIONIZED CONSTRUCTION

Using government surveys, Professor Daniel Quinn Mills has estimated that union membership in the construction industry in 1967 included 44 to 45 percent of the industry's blue-collar employees.[1] A U.S. Bureau of Labor Statistics (BLS) report, based on the March 1971 *Current Population Survey,* found that by 1970, 54.7 percent of all craftsmen and only 29.9 percent of all laborers in the industry were union members.[2] More whites were union members than nonwhites (42.2 percent and 33.7 percent, respectively), although the opposite was true in the manufacturing industry. Regionally, percentages of union membership ranged from 51.0 in the West to 22.2 in the South. The findings of this study, although it involved a rather small sample, strongly suggest that the open shop dominated the construction industry in 1970. The data on union membership which will be presented in this book leave little doubt about the continuing erosion of union membership and the decline of the unionized segment of the construction industry. Union membership can now claim no more than 30 percent of the industry's employees. We have also estimated that the dollar volume of construction produced by unionized craftsmen is not likely to exceed 30 percent of the total.

[1] David Quinn Mills, *Industrial Relations and Manpower in Construction* (Cambridge: MIT Press, 1972), pp. 16–17 (footnotes omitted).

[2] U.S., Bureau of Labor Statistics, *Selected Earnings and Demographic Characteristics of Union Members, 1970,* Report No. 417 (Washington, D.C.: Government Printing Office, 1972), Table 1, p. 6; Table 8, p. 21.

Traditionally, open shop construction has been concentrated in certain industry sectors and in certain regions and locations. Until recently, the general configurations of these concentrations followed historic patterns. For that reason, this book devotes considerable attention to early developments which give perspective to recent changes. During the years since 1970, open shop construction has increased in the sectors and regions in which it has historically dominated. At the same time, sectors and regions which traditionally have been union strongholds have been significantly penetrated by the open shop.

Industry Sector

Residential construction always has been, and remains, the least unionized sector of the industry. Probably more than 95 percent of single-family and low-rise (under four- or five-story) housing is built open shop. Except in some parts of California and in a few large cities of the Northeast and Midwest,[3] unions have lost position in this sector since the early 1950s.

In the early 1970s, unions held a much stronger position in high-rise apartment construction, but this too has been eroded. As a result, unions now control less than 20 percent of high-rise construction, and their control is declining. This is significant not only in itself but also because of the capability which large apartment construction gives open shop builders. The contractor who can build a high-rise apartment can build a high-rise commercial building.

Open shop contractors' success in residential construction, which by 1983 was more than 90 percent of combined low-rise and high-rise construction, is no small factor in the total construction picture. Residential construction accounts for between one-third and two-fifths of all construction work. With control of more than 90 percent of residential work, the open shop would be in a clear majority in the industry if it controlled only 25 percent of nonresidential construction. The open shop has won far more than this percentage in commercial, industrial, highway, and heavy construction.

The residential sector is heavily open shop for several reasons. Among these are the small size and geographical dispersion of jobs in this sector. The erosion of union influence is probably linked to the decline of residential construction in central cities and the centrifugal movement of building away from their cores, creating a much broader territory for the unions to police. When this burden is coupled with the low numbers of potential members to be found

[3] Regional definitions are found later in this chapter.

among residential contractors, many unions have simply abandoned the residential market. Furthermore, the inflation of union wages and union craft demarcations result in costs that homebuilders cannot bear.

The size of projects is significant in all industry sectors. In the commercial sector, jobs under $1 million in costs such as stores, professional buildings, service stations, and moderate-sized shopping plazas have long been predominantly open shop. Large projects, like major shopping centers, high-rise office buildings, and university complexes, although once more likely to be built by union labor, have more and more often been won by open shop contractors since 1975. Since much new commercial building (especially smaller structures) has tended to follow residential movement, the smaller commercial work is as much open shop as is residential building; as the open shop contractors have built larger and larger residential structures, they have gained the capability to win over small commercial work. By 1984, the unionized segment was strongest in large office construction within major cities, but was losing commercial work of most types outside the cities.

Traditionally, the industrial sector, including industrial plants, public utility generating facilities, and other megaprojects, has been a union preserve. Except for a few large open shop contractors, most open shop contractors lacked the capability to compete for such jobs and were reluctant to challenge the construction unions in the heart of their domain. In the years since 1975 a significant change has occurred. A dozen or more open shop contractors have developed the capability to handle major projects and are actively competing for them. In addition, many unionized contractors, including most of the largest ones, have established or purchased open shop companies and operate them on a dual, or doublebreasted, basis. These open shop concerns have separate managements and personnel and are maintained as separate operations. Doublebreasted companies are more competitive; they also reduce the number of union jobs and the power of unions. Despite the gains and the rising importance of doublebreasted companies, open shop contractors are least powerful in the industrial sector of construction, although they do control the majority of small industrial jobs.

Highway and heavy construction unionization has been hard hit in recent years, despite having been insulated from open shop competition by the prevailing wage laws, especially the Davis-Bacon Act and similar state laws. Nonetheless, large segments of these sectors are dominated by open shop contractors, and considerable open shop gains have been made in recent years.

Public construction generally follows the same patterns as the private sector construction: discernibly greater nonunion work on smaller, less complex projects. As populations move to the suburbs, public building work follows them: schools, sewer lines, streets, sidewalks, municipal buildings, and the like. Much of this small work has been done nonunion, despite the operation of prevailing wage laws which effectively remove the advantage of lower wage rates enjoyed by open shop contractors. Larger projects, however, many of which tend to be located either in the major cities or in remote areas with no local labor supply, are more likely to be built union, but less so than formerly.

Another factor which has until recent years accounted for the varying levels of unionization by industry division is relative skill requirements. Large-scale commercial and industrial projects tend to employ more craftsmen and fewer laborers than either highway or residential building work. Unions have historically been stronger among skilled workers, since these workers are more difficult to recruit or replace. Thus, it is interesting that the spurt in open shop construction during the last ten years has coincided with two related developments: a sharp increase in the higher skilled electrical and mechanical work done by open shop contractors and the increased use of formal training programs by open shop contractors.

Geographic Area

Geography influences the extent of open shop construction activity in several ways. Within a market, union strength wanes as work moves farther from the city. More building is open shop in exurban and rural areas than in cities and suburbs. The size and nature of projects, of course, are still important. A power plant, even in a rural area, may be built with union labor because of long-standing relationships between the utility and a union contractor, or a union contractor may obtain the job because of political pressure or the desire of the utility to gain union support for customer rate increases. Residential and small commercial jobs within cities are often open shop. Other things being equal, however, in most market areas there is a direct relationship between unionization and the distance of construction projects from major metropolitan centers.

There are also discernible variations in unionization among different parts of the country. In the South, open shop contractors predominate, and in some southern states the union project is a rarity. The situation in the South is related to population concentration. Rural areas generally are not congenial to unions; northern

New England, for example, has little union building. But that is not the whole story; Chapter VII examines the factors which distinguish different regions.

Finally, the size of a metropolitan area probably has a substantial relationship to the extent of unionization. Open shop penetration in such giant population centers as New York, Chicago, Detroit, and San Francisco has been substantially less marked than in cities like Columbus, Denver, and Rochester. The reasons for this relate to factors mentioned earlier—the size of projects and the centralization of building activity. Another factor, however, may be the extent of unionization in other industries in an area. The fact that the workforces of larger cities are often heavily unionized, affects purchasers' attitudes and local politics.

Other Changes

The union workforce is organized along jealously guarded craft lines, and a metropolitan area of any size will usually have headquarters for local organizations of ten to fifteen different national unions. Major cities may have scores of building trades locals. These locals generally go about their organizing and bargaining activities independent of one another. Consequently, some trades in a given locality may be more strongly unionized than others.

Numerous significant changes have occurred in the open shop portion of the construction industry since the first edition of this book was published in 1975. The Associated Builders and Contractors (ABC)—the voice of the open shop—has become the largest organization of contractors outside the residential sector. Open shop firms have also gained increasing influence in other associations, such as the Associated General Contractors (AGC), which were once dominated by unionized contractors. All these factors have historical antecedents. This volume is the second major inquiry—and the only one with up-to-date data—to explore the changes which have occurred, the reasons for these changes, and the prospects for the future.

Union Response to Open Shop Competition

Perhaps the best evidence of the increasing penetration of nonunion construction into the previously unchallenged domain of the building trades is found in the rhetorical and substantive reaction of unions. This reaction has taken various forms, ranging from impassioned speeches by national leaders urging their members to improve their productivity to outbursts of violence against open shop contractors. Unions in many parts of the country have become defensive in their contract negotiations and have made significant concessions both within and without the confines of their labor agreements, including considerable wage reductions and productivity improvement changes.

THE LABOR MARKET IN OPEN SHOP CONSTRUCTION

Construction unions are central to the labor market processes of the industry, more so than in the case with most other labor organizations. Wages and other forms of compensation are set through negotiations in which unions have generally enjoyed a preponderance of bargaining power. A significant amount of worker training (although by no means all of it, as popularly imagined) is carried on through apprenticeship programs in which decisions on matters such as the number of new applicants to be admitted are controlled by the unions. Many unemployed workers and employers seeking help rely on union-administered hiring halls. Perhaps most important, the allocation of the workforce within the employing unit is constrained by various contractual rules relating to work assignments, crew sizes, use of nonworking foremen, manning requirements, and job classifications. Foremen and lower supervisors must belong to the appropriate craft unions, thus restraining management direction at those levels. The strong influence of unions over the construction labor market and the widespread, but increasingly inaccurate, notion that unionized construction composes almost the entire industry have served to direct scholarly attention away from the nonunion labor market.

Employment

Turnover is an important feature of the construction labor market. Some of it is occasioned by influences endemic to the industry: the demand for construction labor tends to be cyclical. Employment patterns in the individual firm will fluctuate according to the size and number of contracts which it is able to obtain. When work falls off, employees are released, and they are not always available when activity resumes. This condition characterizes all industries to some extent, but the large and often abrupt swings in demand, along with the atomistic and competitive structure, exacerbate it in the construction industry. Furthermore, the relatively large component of skilled labor makes recruitment even more difficult.

Union hiring halls, despite their abuses and restrictions, provide a central pool of labor for employers. In the open shop, much recruitment is informal, consisting of random solicitations or a grapevine of workers, employers, and suppliers. The newspaper, the public employment service, and (rarely) the placement service of vocational schools may be used. Recently, open shop associations have begun to develop their own hiring halls or job referral systems.

Deployment and Utilization

Open shop contractors and their spokesmen contend that their major advantage over union builders is the greater flexibility which they enjoy in deploying their workforce. In broadest terms, this flexibility has three components: (1) the contractor's use of unskilled and semiskilled workers to perform work of which they are capable, rather than assigning such work to higher paid craftsmen; (2) the use of a skilled worker for a variety of tasks, some of which may be part of the job description of a different skilled occupation; and (3) dispensing with unnecessary and/or unproductive labor.

Occupational distinctions in the open shop sector are often blurred. For example, cement may be poured and finished by mixed crews consisting of laborers, carpenters, ironworkers, and cement masons. Similarly the helper (although he may be called by another name), a worker who labors alongside a craftsman and may perform the more routine aspects of the trade, is an integral part of the open shop workforce. This classification had become almost extinct in the union sector, until somewhat revived by the givebacks of the early 1980s.

Another element of human resource utilization is manning requirements: the stipulation of crew sizes; the required use of nonworking foremen; and limitations on the jobs a worker may perform or the number of times he may be shifted from one job to another in the course of a day. The open shop contractor, of course, is free of these restrictions. Nor does he necessarily have to provide workers with pay for time not on the job, showup pay, travel pay, and a guaranteed forty-hour workweek.

Seasonal Employment Patterns

Seasonal fluctuations in employment impose an uncommonly heavy burden on the construction labor market. In a typical year, summer employment in the industry will exceed the level of the previous winter by 30 to 35 percent. In part, these swings are accomplished by releasing workers during the off season and recalling them when activity resumes. But other forms of labor force adjustment are at play.

Although seasonal employment is a problem throughout the industry, spokesmen for the open shop sector usually assert that employment patterns in the open shop are steadier than those among union employers. The proposition is a plausible one. As the open shop contractor's sources of new skilled workers may well be less reliable than those of his unionized counterparts, he has a greater incentive to retain his regular employees. What does the open shop builder do to avoid layoffs in the off season? The peak season, of course, presents a very different problem: how are manpower needs to be met? How is the industry's workforce supplemented by temporary workers and overtime schedules? These questions will be addressed in Chapter IX.

Development and Training

More than 50 percent of workers in the construction industry are officially classified as "craftsmen, foremen and kindred workers." The industry depends heavily upon skill development, both internal and external. Yet little research has been done on skill development in the open shop sector. Apprenticeship programs in the open shop sector are comparatively new, in part because of the difficulty open shop associations have had in obtaining government approval. But many programs are now fully registered, most notably those sponsored through the ABC, the AGC, the Independent Electrical Contractors (IEC), the National Association of Home Builders (NAHB), and various specialty associations.

Wages and Benefits

Wages in construction are commonly identified with the rates negotiated by the building trades unions. Thus, while explosive wage settlements since the late 1960s focused popular attention on wages and their contribution to inflationary pressures, little attention has been directed toward pay scales in the open shop sector. But as it becomes increasingly clear that the wages of large numbers of construction workers are not set through collective bargaining, this omission represents an ever more consequential gap in our knowledge of the industry. On average, hourly rates of nonunion workers are lower than those fixed by union agreements. But beyond such generalizations, very little has heretofore been understood about wage relationships within the open shop sector and between the union and nonunion sectors.

Wage-related areas of interest include: the factors which influence the setting of nonunion wages; the extent to which union-nonunion differentials in hourly wages are reflected in relative annual earnings, or whether more steady employment in the open shop sector offsets lower rates; relative wages within the open shop sector, or whether rates vary with individuals as compared with single job rates in the union sector; and the extent to which fringe benefits are paid in open shop construction. Union agreements, although they continue to devote a smaller proportion of compensation to wages than is the economywide average, have placed increasing emphasis on such benefits as health insurance, pensions, vacations, and supplemental unemployment benefits. This practice has been followed only to a limited degree in the open shop sector. Finally, differences in regard to scheduled hours and overtime pay give the open shop another clear economic advantage.

An important and related issue involving wages is the effect of public policy, particularly prevailing wage laws, on wage rates. These laws are the federal Davis-Bacon Act and similar state laws, known as "little Davis-Bacon Acts." As these laws have been important in determining the extent of unionized construction control in the various industry sectors, the deployment of labor, and employment compensation and hours, they are briefly summarized and analyzed in this chapter.[4]

Minorities and Women

Members of minority groups, especially blacks, have long been underrepresented in the skilled occupations of the construction industry. Indeed, the industry is often regarded as a center of racial discrimination with the building trades unions as the prime offenders. Government programs have been developed to remedy such discrimination, but they have often ignored the bulk of the industry—the open shop.

The government, through equal opportunity legislation, has also aided women in gaining employment in construction. Although women made some gains during the 1970s, they still compose a relatively small proportion of the industry.

---

[4] Our analysis of prevailing wage laws is relatively brief because they are the subject of a forthcoming Wharton Industrial Research Unit study, and as set forth in note 5 below, the portion dealing with the state laws had already been prepublished as a long article.

PREVAILING WAGE LEGISLATION

The earliest and best known of the federal prevailing wage laws, the Davis-Bacon Act of 1931, is only one of scores of federal laws which employ the prevailing wage principle.[5] It requires that workmen in each job category employed by contractors doing federal public works costing $2,000 or more, or work in which the federal government is a sponsor or is financially involved as participant, guarantor, or other interested party, must receive wages and fringe benefits no less than those prescribed by the U.S. Department of Labor as "prevailing" for similar types of construction in a particular locality.

The Davis-Bacon Act should properly be called a "minimum wage" act rather than a "prevailing wage" act, for it specifies the lowest wage rates that can be paid to workmen employed in the construction of all federal public works in a particular area. The rates required by the law, to which separate mandatory fringe benefits must be added, are different for each craft and subcraft (there are frequently fifty or more categories of equipment operators, for example, for each type of construction, and for each locality where the construction work is to be done.

As a federal wage law, the Davis-Bacon Act is unusual in the narrowness of the groups to which its benefits apply. Unlike the Fair Labor Standards Act (FLSA) or most other federal labor laws, the Davis-Bacon Act does not apply universally to all work or to all workers. Its wage benefits are limited to a select group of employees engaged by private contractors to perform construction, repair, or maintenance activities on federal public works. Because the Tenth Amendment to the Constitution limits the ability of the

---

[5] This summary analysis closely follows the works of the leading scholar on the subject, Professor Armand J. Thieblot, Jr. See his *The Davis-Bacon Act,* Labor Relations and Public Policy Series, Report No. 10 (Philadelphia: Industrial Research Unit, The Wharton School, University of Pennsylvania, 1975); and his "Prevailing Wage Laws of the States," *Government Union Review,* Vol. 4 (Fall 1983), pp. 1–65. Dr. Thieblot is revising the Davis-Bacon study for a publication by the Wharton Industrial Research unit. It will also include the study of state prevailing wage laws and an analysis of the Service Contract Act, by Beverly H. Burns, Esq.

For other studies of the Davis-Bacon Act, see also U.S. Comptroller General, *The Davis-Bacon Act Should Be Repealed,* Report HRD-79-18 (Washington, D.C.: General Accounting Office, 1979); Robert S. Goldfarb and John F. Morrall III. "The Davis-Bacon Act: An Appraisal of Recent Studies." *Industrial & Labor Relations Review,* Vol. 34 (January 1981), pp. 191-206; Morgan O. Reynolds, "Understanding Political Pricing of Labor Services: The Davis-Bacon Act," *Journal of Labor Research,* Vol. III (Summer 1982), pp. 295–310; and Michael R. Metzger and Robert S. Goldfarb, "Do Davis-Bacon Minimum Wages Raise Product Quality?," *Journal of Labor Research,* Vol. IV (Summer 1983), pp. 265–272.

federal government to dictate contract terms selectively to the states, contracts for public works entered into by states or by local entities, such as cities or school districts, are not subject to the Davis-Bacon provisions. States that wanted benefits similar to those of the Davis-Bacon Act for construction workers on nonfederal projects had to enact their own prevailing wage laws.

Not all states have enacted such laws, and not all of those that have done so have followed the example set by the federal Davis-Bacon Act, except in that they all establish super-minimum wage rates for construction workers on governmentally sponsored public works. Even the states that have followed the federal example have incorporated changes in purpose or in administration which have caused them to differ substantially in concepts, definitions, procedures, and rules from the federal parent. Thirty-seven states and the District of Columbia now have prevailing wage laws; they are all effectively different, tied together by little more than a common name. (See Table II-1.)

Because of the Davis-Bacon Act and other prevailing wage laws, there are few public works on which wage rates can be freely set by the contractor. The ostensible purpose of prevailing wage laws is to neutralize the effect of public construction on local wage standards. Much of the legislative debate about the Davis-Bacon Act focused on the "itinerant"[6] contractor able to undermine local standards by importing cheaper labor, a practice made easy by the depressed economic conditions of the time. By requiring that workers be paid according to locally prevailing wages, the act sought to keep federal construction projects from lowering wage rates through "unfair" competition. Although there was no explicit intention to exert an upward pressure on wages, students of the act agree that this has occurred.

Federal prevailing wage laws are administered by the Employment Standards Administration of the U.S. Department of Labor. The Davis-Bacon Act requires that those contracts it covers "contain a provision stating the minimum wages to be paid various classes of laborers and mechanics which shall be based upon the wages that will be determined by the Secretary of Labor to be prevailing for the corresponding classes of laborers and mechanics employed on projects of a character similar to the contract work in the city, town, village, or other civil subdivision of the state in which the

---

[6] "... the word 'itinerant' is a pejorative one. There seems to be no reason why local contractors should be no protected against competition any more than local manufactures, local storekeepers or local professors are." Thieblot, *The Davis-Bacon Act,* note 1, pp. 1–2.

TABLE II-1
Existence of Prevailing Wage Laws by State

States Having Prevailing Wage Laws	Date of Law	States Without Prevailing Wage Laws
Alaska	1931	Georgia
Arkansas	1955	Iowa
California	1931	Mississippi
Colorado	1933	North Carolina
Connecticut	1935	North Dakota
Delaware	1962	South Carolina
District of Columbia	1931	South Dakota
Hawaii	1955	Vermont
Idaho	1911	Virginia
Illinois	1931	
Indiana	1935	
Kansas	1891	
Kentucky	1982	
Louisiana	1968	
Maine	1933	
Maryland	1945	
Massachusetts	1914	
Michigan	1965	
Minnesota	1973	
Missouri	1957	
Montana	1931	
Nebraska	1923	
Nevada	1937	
New Hampshire	1941	
New Jersey	1913	
New Mexico	1937	
New York	1897	
Ohio	1931	
Oklahoma	1965	
Oregon	1959	
Pennsylvania	1961	
Rhode Island	1935	
Tennessee	1953	
Texas	1933	
Washington	1945	
West Virginia	1933	
Wisconsin	1931	
Wyoming	1967	

States with Repealed Laws	Date of Law	Date of Repeal
Alabama	1969	1981
Florida	1933	1979
Utah	1933	1981

States With Ineffective Laws	Date of Law	Date
Arizona Unconstitutional	1912	1979

Source: Individual state laws and Wharton Industrial Research Unit survey, state departments of labor, 1982.
Reproduced from Armand J. Thieblot, Jr. "Prevailing Wage Laws of the States," *Government Union Review,* Vol. 4 (Fall 1983), p. 7.

work is to be performed. . . ."[7] A 1964 amendment to the Act redefined "wages" to include the costs of such supplemental benefits as pensions, medical insurance, holidays, and vacation days. When a planned project nears the bidding stage, the contracting federal agency must secure a prevailing wage determination from the Department of Labor. Prevailing wage determinations are published in the *Federal Register* and, unless amended or challenged, apply to all similar projects in the area. Because of the administrative burden of making wage surveys, area determinations may cover a wide territory, including urban, suburban, and rural localities, sometimes encompassing an entire state, and often including areas with quite different wage patterns. Officially, the county is the geographic district used wherever possible, but practice has been something different.

The Davis-Bacon Act does not define the concept of "prevailing," but leaves that task, along with other administrative details, to the Secretary of Labor. In 1935, the Secretry of Labor adopted the survey as his means of identifying wage rates, using (usually) the county as the geographic boundary within which to identify wage rates paid to workmen in various classes of labor. He then adopted a three-step procedure for identifying the prevailing rates: if, of the rates identified by the survey, any one rate to the penny was paid to a majority of workmen, that rate was taken as prevailing; but if there were no such rate, then the rate to the penny paid to 30 percent of workmen was selected; and, if there were no such rate, then the average of all the rates in the survey was adopted as prevailing. This method has been most commonly called the "30 percent rule."

Since 1964, the Department has provided a three-member Wage Appeals Board to consider objections to the determinations fixed by the Employment Standards Administration. In these instances, the appellant must show that the original determinations were in error. As such an appeal will normally demand a private survey of local wages, requiring substantial investigative resources, it is not surprising that the caseload of the Wage Appeals Board has not been large,[8] especially since the appellant has no guarantee of winning a contract even if he wins his case.

At least four aspects of the statutory language and of past Labor Department administrative policies have, in the views of many contractors and of most academic observers, as well as of the United

---

[7] Title 40, Sec. 276a, U.S. Code.

[8] Fifteen or so cases per year reach the Board.

States General Accounting Office, served to distort the original intent of the Davis-Bacon Act. In many cases the result has been merely to extend the standards established by collective bargaining agreements to cover an entire industry or area. In others it has been to promulgate arbitrary wage rates which relate neither to any logical concept of "prevailing" wages nor to collectively bargained standards. Ironically, a law initially directed at "itinerant" contractors may have had its greatest adverse impact on indigenous open shops with strong ties to their localities.

Prevailing Wage Impact

The effect of the 30 percent rule has been to give disproportionate weight to collectively bargained wage rates. As we shall show, open shop wages tend to vary both within and among employing units, whereas union wages are typically established at a single rate for each craft. To take the most extreme example, in a locality which is 30 percent unionized, 70 percent of the workers in a given craft may earn a wide range of wage rates all below the union rate; yet the "prevailing" rate would be fixed at the higher union scale. In theory, of course, the rule can act to fix wages below or above the union scale; in practice the former has been very rare, unless open shop contractors have made their own surveys, because the Department of Labor historically has used union contracts much more than actual surveys to establish prevailing wages.[9]

Definition of Locality

Although the Department of Labor selected the county as its key geographic area, other areas have been used. Use of the county sometimes seems to have stretched the intent of Congress, when the county involved was partially urban and partially rural, or otherwise differentiated by diverse wage patterns. Although the Act makes explicit reference to "city, town, or village," the Department of Labor has frequently made liberal use of the catchall phrase "other civil subdivision of the state." On numerous occasions, wage rates have been imported from outside the county involved or applied statewide. Although some of these cases undoubtedly involved projects in remote areas where the very concept of a prevailing

[9] In an earlier study of the Davis-Bacon Act, D. Gujurati found that over 80 percent of a sample of Department of Labor determinations were pegged at the union scale, and only 8 of 372 determinations were based on actual wage surveys. D. Gujurati, "The Economics of the Davis-Bacon Act," *Journal of Business*, Vol. 40 (July, 1967), pp. 303–316.

wage was ambiguous, there seems little justification for importing standards set in distant metropolitan communities. If a wage is set lower than that necessary to attract sufficient numbers of workers, then the market will provide the adjustment. But if the wage is fixed at a rate determined by a union many miles away, it may well be substantially higher than necessary to man the job. In either case, there is no apparent need to protect locally established standards, as no such standards exist.

Worker Classification

Historically, the Department of Labor has generally recognized only three classifications of workers: journeymen, laborers, and certified apprentices. An exception was made where open shop contractors predominated; here a separate determination existed for the "helper" category, particularly if the open shop group provided extensive data. Otherwise, open shop contractors faced the Hobson's choice of paying journeyman's wages to their helpers or not using them on the job at all. Surveys by the General Accounting Office over the years have uncovered numerous misclassifications of workers which resulted in the application of rates higher than those which actually prevailed.

Identification of "Similar" Projects

The Davis-Bacon Act mandates that prevailing wages be determined according to wage rates paid "on projects of a character similar to the contract work," but the definition of "similar" is left to the discretion of the Department of Labor. Perhaps the most controversial of the Department's decisions in this regard have involved public housing projects, where prevailing wage determinations were based on commercial building scales rather than on those of the largely nonunion homebuilding sector. After severe criticism from the General Accounting Office, the Labor Department in 1972 agreed to let the Department of Housing and Urban Development conduct its own wage surveys on federally funded housing construction. In some instances involving heavy or highway construction, determinations have been based on substantially dissimilar projects. On other occasions, the definition of a similar project has been drawn so narrowly that it was necessary to import wage standards from very distant regions. It must be acknowledged that the task of specifying a similar project in the immediate locality is not always easy. In such instances it might be argued that the very notion of a "prevailing" wage is meaningless.

The 1983–1984 Regulations

There appears little doubt that the Davis-Bacon Act and other prevailing wage laws have inflated the cost of public construction; numerous studies have come to this conclusion.[10] Certainly, any measure which required open shop contractors to raise their wages to union levels and to deploy labor according to union standards would have this effect. Nevertheless, the political power of the construction unions enabled the act to survive this criticism until 1983, when Secretary of Labor Raymond J. Donovan issued new regulations altering the manner in which prevailing wage laws were established and their conditions determined. The Donovan regulations were bitterly fought and litigated by the construction unions,[11] but by mid-1984 it appeared that these new regulations would go into effect. They will effect the following changes:

1. The new regulations eliminate the 30 percent rule and substitute a requirement that the prevailing wage for a classification be set at the rate which is being paid to a majority of the workers involved; if a majority of workers in a single class do not earn a single wage, then a weighted average will be used. This provision will reduce the impact of collectively bargained wages paid to minority workers in areas in which open shop wages vary widely.
2. Where there has not been sufficient similar construction in the county at a prevailing wage, wages paid on similar construction will be used—except that data from projects in metropolitan counties may not be used for setting prevailing wages in rural counties, and vice versa. This meets one of the major criticisms of the act, that it brings higher urban wage rates into lower-wage rural areas. As union strength is concentrated in urban areas, the net effect of the new regulation will be to reduce government support of union power and expansion.
3. The Donovan regulations make a rather dramatic break with the past by excluding from the prevailing wage calculation prior federal or federally assisted projects under the Davis-

---

[10] *Ibid.*, p. 309. See U.S. Comptroller General, note 5 above.

[11] Building & Construction Trades Department, AFL-CIO v. Raymond J. Donovan, Secretary of Labor, 553 F. Supp. 352 (D.C. Cir. 1982), *affirmed in part, and reversed in part,* 712 F.2d 611 (D.C. Cir. 1983), *cert. den.,* 104 S. Ct. 975 (1984). The district court enjoined all the proposed regulatory changes except the elimination of the 30 percent rule; the appellate court affirmed the right of the Secretary of Labor to change the 30 percent rule and reversed the district court on all other matters except a reporting method, but modified the Secretary's provision for the helper classification.

Bacon Act. These new regulations, which exclude highway and heavy projects where there is little nonfederal construction, provide that prior federal projects shall not be considered in wage determinations for building and residential construction projects "unless it is determined that there is insufficient wage data to determine the prevailing wages in absence of such data." Again, the new rule addresses a widespread criticism of the act: Once prevailing wages were determined under the 30 percent rule for a large federal project, that project bound all future ones in the area and skewed rates upward at the union rate, although the original intent of the act was merely to prevent contractors from paying less on government projects than was being paid on private projects.[12]

4. The new regulations encourage the use of helpers by permitting the helper classification on jobs where helpers were prohibited by union rules. The Donovan proposals were somewhat narrowed by the Court of Appeals,[13] so that the helper classification must prevail in an area before it can be utilized, but this new regulation is significant advance toward effective use of labor on government projects. It increases the opportunities of open shop contractors to use their more efficient methods, possibly winning more government-sponsored work.
5. The Donovan regulations will also create new and expanded wage determinations. The Department of Labor has predicted that it will double its former average of 2,000 wage surveys per year, and outside contractors have been employed to assist in this work.[14] The results could provide the most complete information ever on open shop contractor wages, benefits, and classifications.

Without doubt, the new regulations should make it easier for open shop contractors to win government jobs, further eroding union power and control. The extent of possible open shop gains under the new rules should not, however, be overestimated, nor should the success of open shop contractors under the old rules be underestimated. In the past, despite the substantial reduction or elimination of the wage advantages they otherwise enjoy, many open shop firms have been eminently successful in competing for public

---

[12] The appellate court decision contains a good summary of these congressional reports.

[13] See the decision on this reasoning.

[14] "School to Survey Wages," *Engineering News-Record,* September 8, 1983, p. 61; and "DOL Wage Surveys to Double," *Engineering News-Record,* September 1, 1983, p. 50.

contracts. Because of their greater flexibility in deploying manpower and their freedom from costly jurisdictional disputes and such requirements as show-up pay, guaranteed workweek, and strict overtime hours, some open shop contractors experience significantly lower labor costs even when paying collectively bargained wages and supplemental benefits. Many field interviews with highway and public works contractors in 1974 and in 1982 revealed this situation.

Prevailing wage standards, moreover, often eliminate one nagging problem faced by nonunion firms in periods of higher employment: the recruitment of labor. Employers paying wage rates of $10 to $20 per hour are not obliged to beat the bushes for help; those we interviewed talked rather of long waiting lists of applicants.

On the other hand, open shop contractors may face an employee relations problem not encountered by their unionized counterparts. Firms which work in both the public and private sectors may find themselves paying grossly disparate wage rates on different projects. Some contractors expressed dismay at the difficulty of maintaining morale and satisfaction among workers who knew that other employees were making several dollars more an hour because they were employed on a prevailing wage job. Some of them try to rotate opportunities on the higher paying projects. Another alternative is to pay the higher scale on all work, but few employers will surrender their competitive wage advantage.

The extent to which prevailing wage laws have limited open shop construction is not entirely clear. The unions have used such laws to their maximum advantage in highway and heavy construction, but the open shop has made substantial gains there in recent years. The new regulations for Davis-Bacon Act administration should promote more open shop penetration into this and other industry sectors where state laws with concurrent jurisdiction are not administered in support of union hegemony. Although these regulations are subject to change whenever a new federal administration, or even a new Secretary of Labor, comes into power, they should add to the open shop segment's capability and ability to compete.

THE COMPETITIVE FACTORS

Since the mid-1960s, union control has been reduced to less than one-third of construction dollar volume in the United States. Competitive factors have contributed heavily to the current situation.

In interviews in 1974 and in 1982 with scores of union and nonunion contractors and with spokesmen for their associations, we looked for reasons for the increase in open shop activity. They most often cited these three factors: wages, productivity, and strikes—although not necessarily in that order. The unions' success in winning prodigious wage increases, instituting inefficient and restrictive work rules and practices, and interrupting production has given concern not only to contractors but also to their customers, as shown by the formation of the Construction Users Roundtable, the predecessor of the Business Roundtable, in the early 1970s, and the continued interest and activity of the Roundtable.

Between 1970 and 1980, according to the BLS, construction union wage rates rose from an average of $6.18 per hour to an average of $12.21 per hour, an increase of about 50 percent. Employer benefit payments in this decade advanced from $0.80 per hour to $2.83 per hour, an increase of more than 300 percent.[15] BLS has not published more recent data, but the Construction Labor Research Council reports that the average increase in wage and fringe benefits from 1974 to 1984 was close to 125 percent, bringing the average combined wage-fringe average hourly rate to $19.35 per hour. Moreover, despite the lowest wage increases in a decade during 1983, wages and fringes in union construction rose 47.5 percent in the five years prior to January 1, 1984.[16]

These union wage increases, plus costly fringe benefits, have given a substantial cost advantage to open shop contractors. Even when open shop contractors pay their top mechanics the equivalent of the union scale, their freedom to pay a range of rates to individuals with varying capabilities results in a significantly lower average wage. But wages are far from the whole story; even in public construction, where prevailing wage laws have often required the universal payment of union scales, open shop contractors have also won traditional union domains. The cost impact of union / nonunion wage differentials, moreover, is ambiguous. It cannot easily be determined whether higher wages allow union contractors to choose among the best qualified and most productive workers, assuming union rules permit such selection.

The effect of union work rules is less ambiguous. Jurisdictional demarcations require that unskilled work be performed by highly paid journeymen, or lead to delays because work cannot be assigned to employees who are immediately available at the site. Restrictions on working hours mean that interruptions because of weather or

---

[15] U.S., Bureau of Labor Statistics, *Union Wages and Benefits: Building Trades, July 1, 1980,* Bulletin 2091 (Washington, D.C.: Government Printing Office, 1981), p. 3.

[16] "1984 Construction Labor Rate Trends and Outlook," *Construction Labor Research Council,* February 1984, p. 2.

materials shortages must be made up at overtime rates. There may be limitations on the number of machines a worker can operate in a day, or requirements that automatic equipment be manned. Although it is impossible to say precisely how much these work rules add to labor costs, their effect appears substantial. Construction unions in many parts of the country admitted as much during the 1982 to 1984 "giveback" period, by giving up some of these restrictions as well as reducing wages or waiving wage increases in attempts to counter substantial unemployment and increased market gains by open shop contractors. It remains unclear whether these union concessions are sufficient to regain lost work, although in some areas they have undoubtedly helped union workers obtain or hold work.

Unionized and Open Shop Cost Comparisons

Cost comparisons between different jobs are fraught with potential error because no two jobs are exactly alike. We have, however, acquired a number of comparisons, made under varying circumstances, which emphasize the cost differences between unionized and open shop construction. Some studies attempt to refute the economic advantages of open shop building.

Report of a Major User. A large *Fortune* 500 company determined in the late 1960s that its construction costs were far too high. After a careful study, it decided to permit open shop contractors to bid on jobs where possible (that is, where political factors permitted and violence would be controlled). Because this company has been building similar facilities, both union and open shop, since at least 1972, it has made excellent cost comparisons covering the periods from 1972 to 1977 and from 1978 to 1982. The earlier study involved 154 projects costing a total of $3.4 billion and showed that union labor accounted, on average, for 30 percent of total project cost; open shop labor represented only 22 percent. This 8-percentage-point difference resulted in savings of more than 25 percent in labor costs and 7 percent in total project costs.[17] At this time, open shop construction represented about 50 percent of the company's total construction dollar volume.

The later study yielded similar results. By then, the company was using open shop contractors for at least 60 percent of its new construction volume. Examining sixty-two projects, and using 1977

---

[17] These studies are reported in *Cockshaw's Construction Labor News + Opinion,* Vol. 13 (July 1983), pp. 1–3.

dollars to ensure comparison with the original study, the company found that open shop tradesmen installed $1.85 worth of materials per labor dollar, but union tradesmen installed only $1.25 worth. Union labor represented 27 percent of total project costs; open shop labor, 21 percent. The net difference involved a 6 percent savings in total project costs. Over the ten-year period, the company saved $400 million by using open shop contractors. According to the company:

> The major reason for these savings is that the merit shop [open shop] has the freedom to manage its jobs. A perfect example of this is the determination of crew mix which makes up a significant dollar difference between open and closed [union] shop.
>
> On our union projects, the average is ten journeymen to one apprentice whereas our merit shop jobs have an average ratio of three journeymen to one apprentice. . . .open shoppers don't have to pay skilled craftsmen to do unskilled tasks or "bull work." [18]

Large Metal Manufacturing Facility. In 1980, a major metal products manufacturer built a facility in a southern state using an open shop contractor. The company received cost comparisons from a number of contractors and decided to use an open shop contractor on the basis of the information summarized in Figure II-1 and Table II-2, which shows a 44 percent labor premium as the cost of building union shop. The premium is based on: 1) wage differentials; 2) deployment differentials, again involving use of journeymen for subjourneymen work; 3) daily overtime, scheduling, and report-in premium pay required by the union contracts and not paid by open shop contractors; and 4) pay for time not worked by union stewards. The premium does not take into account the possible costs of strikes, jurisdictional disputes, and other unforeseen interruptions of production caused by union relations problems. It is estimated that 50 percent of the union stewards' worktime will be nonproductive—that is, spent on grievances and union matters—and that in a union situation journeymen would unload materials, whereas laborers would do most of this essentially unskilled labor in open shop conditions.

A number of factors other than wage and manning differences drive the union contractor's labor cost higher. The company's analysis, based on actual practice and contractual provisions in the area involved, summarizes these factors as follows:

1. On union projects, each craft has its own "work claims" resulting in possible work stoppages due to jurisdictional disputes. In open

[18] *Ibid.* The figures regarding journeymen and apprentice ratios under open shop conditions may include helpers as well as apprentices.

FIGURE II-1
Construction Labor Cost Comparison
Large Factory, 1980

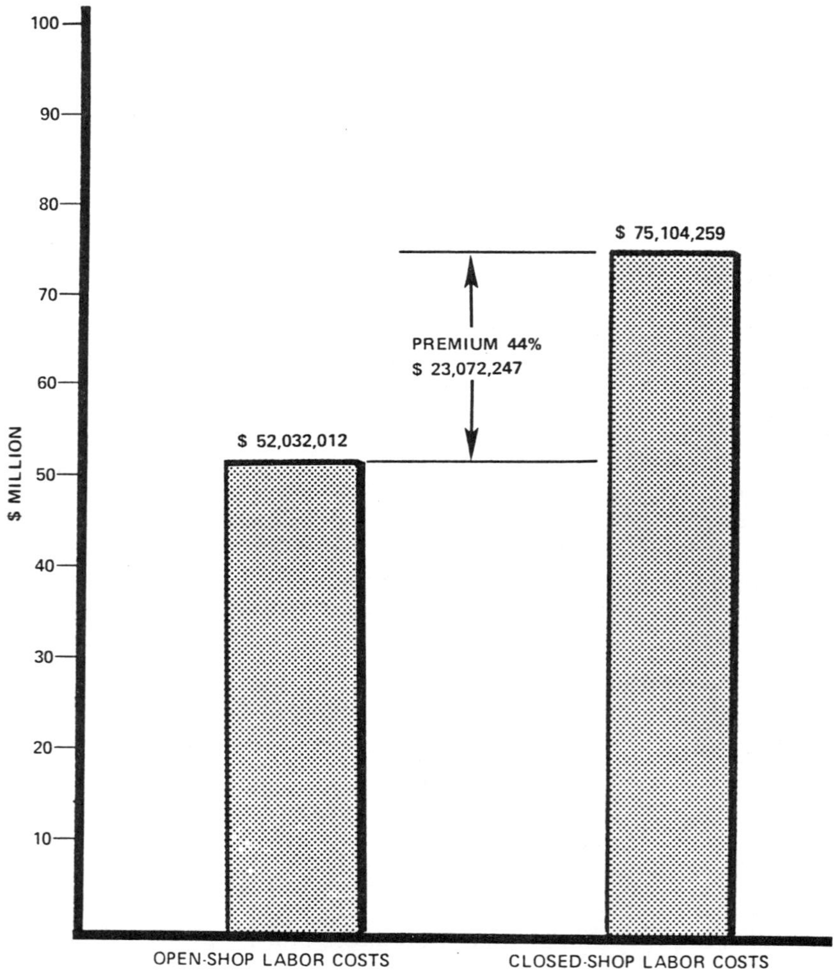

Source: Large metal manufacturing facility, southern state, 1980.

TABLE II-2
Union Wage Premium[a]
Large Factory, 1980

Area	Premium / Building Trades
Union Stewards	$ 234,497
Warehousing	660,250
Operating Engineers	792,000
Carpenters	1,646,400
Ironworkers	2,149,200
Pipe Fitters, Welders, Instrumentation	4,788,000
Electrical	2,526,000
Millwrights	1,146,000
Masons	1,090,800
Painters	1,476,000
Insulators	1,897,200
Cement Finishers	498,000
Laborers	1,852,200
Truck Drivers	855,000
Overtime	1,460,700
	$23,072,247

Source: See Figure II-1

[a] PREMIUM $= \dfrac{\text{Premium } \$23{,}072{,}247}{\text{Total Open-Shop Labor } \$52{,}032{,}012} = 44\%$

shop work a worker may be used temporarily in another craft, if expedient.

2. On union projects, any work in excess of eight hours per day is on an overtime basis. On open shop projects, overtime is calculated in excess of forty hours per week. Normally our open shop projects operate ten hours Monday through Thursday. This allows us to make up on Friday and Saturday any time lost due to weather at straight time.
3. Second-shift work is normally at a premium rate on a union project, i.e., seven and one-half hours' work for eight hours' pay. There is no premium pay for shift work on an open shop project.
4. On a union project, work assignments between crafts cannot be varied in case of manpower problems or workloads.
5. Union weekend work (Saturday afternoon and Sunday) is at double time. Open shop work allows incorporation of Saturday and Sunday as part of a four-day, ten-hour shift if expedient. This is generally done for painting crews in areas where they would be unable to operate during normal working hours and it is impractical to utilize a second shift.
6. On union work each individual craft has a steward to provide representation. This means that a certain percentage of the steward's time is spent in nonproductive work. Depending upon the specific situation, up to 25 percent of his time can be expected to be lost.
7. Most union crafts receive vacation and holiday fringes. Open shop workers are paid only for hours actually worked.

8. The insulators, boilermakers, and pipefitter crafts receive a travel allowance and travel time depending on the distance to the worker's home.[19]

Although these comparisons may exaggerate the advantages of the open shop, they point up the basic economic fact that, both in direct compensation and in labor use, open shop operators have tremendous economic advantages, which the union wage escalation since the late 1960s has exacerbated.

The Speed Study. Two studies have attempted a realistic comparison of costs and productivity under open shop and union shop conditions. The first, by William S. Speed, presented "an extensive financial analysis of the labor costs associated with each type of construction." [20] His comparisons are summarized in Table II-3, which also compares costs under a union project agreement [21] and under an open shop contractor who is nevertheless paying union rates and benefit costs of $0.40 per hour (an appropriate figure at the time of Speed's analysis).

Table II-3 shows that under all conditions Speed found that open shop construction was more economical, even though he included 15 percent more supervisors for open shop construction on the grounds that a supervisor "on an open shop contract has to be more responsible for construction methods and workmanship because the workers in his crew may not be as skilled as their ... counterparts." [22] (This assumption is becoming less correct as open shop contractors and their crews become more experienced.) In any case, these added open shop costs were greatly outweighed by lower labor rates, superior deployment, lower premiums, and other factors set forth in Table II-3.

The Heizelman Study.[23] This study analyzed productivity and costs in a chemical plant, refinery, or similar facility. Following

---

[19] The data herein were received on a confidential basis and are utilized with the understanding that the contractor and user will not be identified.

[20] William S. Speed, "Construction Labor Cost Comparison of Open and Closed Shop Construction Projects," *Transactions of the American Association of Cost Engineers,* Twenty-Second Annual Meeting, San Francisco, 1978, p. 140. Mr. Speed is a certified construction cost engineer who has managed engineering departments in both open shop and unionized companies.

[21] Project agreements are special contract arrangements that usually apply to major jobs and include concessions by unions to improve productivity. They are discussed in Chapter VIII, as well as more briefly in earlier chapters.

[22] Speed, "Construction Labor Cost Comparison" p. 144.

[23] John R. Heizelman, "Productivity: Union v. Open Shop," *Transactions of the American Association of Cost Engineers,* Twenty-Sixth Annual Meeting, Toronto, reproduced in *The Builder and Contractor,* August 1982, pp. 18–25. Mr. Heizelman is Senior Estimator, ARCO Oil and Gas Company.

TABLE II-3
Comparison of Open and Closed Shop Construction Costs

	Closed Shop Project with Project Labor Agreement	Closed Shop Project without Project Labor Agreement	Open Shop with Open Shop Rates	Open Shop with Union Rates and $.40/hr. Benefits
Direct	$ 8,843,789	$ 8,843,789	$ 7,384,665	$ 9,192,111
Benefits	+ 953,493	+ 1,113,743	+ 150,000	+ 371,640
Stewards	+ 65,458	+ 87,277	—	—
Helpers	− 539,471	—	− 420,926	− 523,950
Strikes and Jurisdictional Disputes	—	+ 221,094	—	—
Check In-Out, Breaks	518,395	+ 1,139,275	392,313	498,076
Tools and Material Warehousing	—	+ 96,016	—	—
Added Supervision	—	—	+ 53,063	+ 53,063
Total	$ 9,841,664	$ 11,501,194	$ 7,559,115	$ 9,590,940
Cost Per Manhour Worked	$ 9.84	$ 11.50	$ 7.55	$ 9.59

Source: William S. Speed, "Construction Labor Cost Comparison of Open and Closed Shop Construction Projects," *Transactions of the American Association of Cost Accountants,* Twenty-Second Annual Meeting, San Francisco, 1978, p. 150.

much the same procedure as Speed, Heizelman concluded that savings amounted to more than $2 million on a project requiring 700,000 direct manhours and a fifteen-month construction period. Table II-4 summarizes this study's conclusion, which includes no extra charges to unions for work stoppages. Overtime premiums and items such as call-in pay are included under direct labor costs. This study shows a 20 percent cost premium for building union.

Studies Claiming Union Construction Is Equally or More Productive

Many studies, including a number pertaining to public utilities (both fossil and nuclear-fired), paper manufacturing plants, and coal mine preparation, have demonstrated the cost advantages of open shop construction over unionized construction under varied circumstances. Two studies dispute this conclusion; one of them in several versions purports to show that union construction is actually less costly and more productive. Although all other evidence points to the contrary, including the evidence presented in Chapters III through VII of this book of the expansion of open shop construction because of its economic advantages, these two studies deserve some attention.

TABLE II-4
Union Shop-Open Shop Cost Comparison, Refinery Project
700,000 Manhours, 15-Month Construction Period, 1982

	Union Shop	Open Shop
Direct labor cost	$ 9,247,840	$ 8,004,584
Cost of stewards	83,230	—
Fringe benefits	1,296,370	800,456
	$10,627,440	$ 8,805,040
Insurance and tax on labor 17%	1,586,280	1,360,780
Total labor cost	$12,213,720	$10,165,820
Savings using open shop	$ 2,047,900	
Cost of money at 12%	153,600	
	$ 2,201,500	

Source: John R. Heizelman, "Productivity: Union v. Open Shop," *Transactions of the American Association of Cost Engineers,* Twenty-Sixth Annual Meeting, Toronto, reproduced in *Builder and Contractor,* August 1982, pp. 18-25.

The Bourdon-Levitt Study. In 1980, Clinton C. Bourdon of Harvard and Raymond E. Levitt published *Union and Open-Shop Construction,*[24] which was based upon a survey conducted for the U.S. Department of Housing and Urban Development in 1976. Various aspects of their findings will be examined and questioned in later chapters, particularly their assertion that open shop construction has won a smaller share of the market than claimed by other studies, including this work. Here we are concerned with their Chapter 5, "Relative Costs of Union and Open-Shop Construction."

They note that open shop contractors have proved themselves in the marketplace but "it is important to determine exactly the extent to which this success is due directly to lower hourly compensation costs in contrast to different skill mixes, the lack of union restrictions, disputes, and the like." They declare that merely examining relative compensation is insufficient "without any allowance for offsetting skills and productivity, and ignores many other important components of total labor costs." [25] They then quote a 1965 study of residential construction by Mandelstamm,[26] which adduced data to demonstrate that union productivity overcame lower open shop wage rates; the initial work of Steven Allen, discussed below, to the

[24] Clinton C. Bourdon and Raymond E. Levitt, *Union and Open Shop Construction* (Lexington, Mass.: D.C. Heath and Company, Lexington Books, 1980).

[25] *Ibid.,* p. 81.

[26] Allan B. Mandelstamm, "The Effects of Unions on Efficiency in the Residential Construction Industry: A Case Study," *Industrial and Labor Relations Review,* Vol. 18 (July 1965), pp. 503-521.

same effect; and the works of Freeman, Medoff, and associates at Harvard who claim that the union impact in other industries enhances productivity and thus lowers costs.[27] The authors feel that these studies offset the work of Speed, which is criticized because, unlike those who conclude that union construction is more economical, Speed includes savings to the open shop contractor for the use of helpers, for avoiding unproductive expenditures on union stewards, and for time lost on strikes and work breaks. Bourdon and Levitt conclude that no estimates are definitive and that whether open shop or unionized construction is more economical depends on a variety of factors.[28]

The problems with their analysis are, first, the manner in which they state the problem and, second, their disregard of market forces. By stating that "different skill mixes, lack of union restrictions and the like" are not part of labor costs, they ignore a major part of labor costs and the advantages of the open shop. Mandelstamm's study of residential construction cannot be credited, for since he wrote, open shop construction, already holding a majority in that industry sector in 1965, has increased its share of residential construction to more than 90 percent—something that would be inconceivable if unionized construction were sufficiently productive to overcome the open shop contractors' lower labor costs. This is not the place to analyze the Freeman-Medoff thesis in detail.[29] Their claim is that when the union enters, the "shock effect" produces higher productivity and lower costs. Again the market is skeptical, but in any case reference to their work is irrelevant, not only because they deal with other industries but because, in construction, open shop contractors are taking work away from unionized ones because of the latter's high costs, not vice versa. Thus, the Bourdon-Levitt thesis that a host of factors ultimately determine whether open shop or unionized construction is more economical may be true in a very limited situation, but barring any unusual factors, the market has found that open shop construction costs considerably less.

The Allen Thesis. Professor Steven G. Allen of North Carolina State University has written a number of papers since 1979 pur-

---

[27] For a thorough statement of the views of Richard B. Freeman and James L. Medoff, see their recent book *What Do Unions Do?* (New York: Basic Books, 1984). This work includes the working papers and articles which were earlier available to Bourdon and Levitt.

[28] Bourdon and Levitt, *Union and Open Shop Construction,* p. 85.

[29] For a strong critique of Freeman and Medoff see Morgan O. Reynolds, *Power and Privilege: Labor Unions in America* (New York: Universe Books, 1984).

porting to prove that union construction is more efficient.[30] His work is closely associated with the Freeman-Medoff studies. Allen's first work in 1979 was underwritten by the Center to Protect Workers' Rights, described by *Engineering News-Record* as "a research offshoot of the Building and Construction Trades Department, AFL-CIO."[31] Since then, he has enjoyed support from the U.S. Department of Labor, the Freeman-Medoff group at Harvard University, and the National Bureau of Economic Research.

In his first paper, Allen found that union construction was more productive both in large office building construction and in school construction. His data then applied to 1972. In each case, he claims to have kept capital, firm size, regions, and other variables constant. Essentially, he claims that higher union pay is justified because union workers are more productive, but he never explains in detail why this is so.

Allen's first paper was severely and rightly criticized for combining Census of Construction Industries and Current Population Survey data, which, in effect, was "combining different universes."[32] He has since used data from the Federal Housing Authority and the U.S. Department of Labor. He has retreated from the view that union construction is more efficient in school construction and smaller office buildings. He claims, however, that union construction has many more economies of scale, and that unionized construction demonstrates greater productivity in larger office buildings (defined by him as buildings with more than 66,185 square feet of space).

Allen's analysis includes an impressive array of calculations and econometric modeling. Since his data base was not available, comment on the extent to which his sampling is representative and his methodology and calculations proper cannot be definitive. We can, however, comment on the appropriateness of the data, their pertinence, and the conclusions drawn therefrom.

The basic data were gathered by the U.S. Department of Labor, Bureau of Labor Statistics as part of its Construction Labor and Material Requirement Series: the school data were gathered in 1972; the commercial office building data in 1974. More than three times

---

[30] Professor Allen kindly gave the author a set of his papers which are being published in various journals. The most current ones are entitled "Can Union Labor Cost Less," April 1983 and "Union Work Rates and Efficiency in the Building Trades," August 1983.

[31] "Contractors Dispute Union Productivity Study," *Engineering News-Record,* December 13, 1979, p. 65.

[32] *Ibid.* This was the reported comment of Aaron Sabghir, Director, Construction and Building Products Division, U.S. Department of Commerce.

as many union firms as open shop ones for office buildings, and five times as many union firms than open shop ones for schools, responded to the surveys. Allen believes he corrected for the data imbalances, but the number of open shop sources is very small.

There is some institutional support for Allen's conclusions. School construction has become heavily open shop in most areas, and large office building construction in urban areas has been one of the last major strongholds of unionized construction. The more economical form has won out—the market has triumphed—if Allen is correct. Several considerations cast doubt on Allen's attempt (and that of Freeman and Medoff) to generalize from these data. In 1972, open shop contractors were already expanding into school construction and they have continued to do so since then. Even though Allen's union sample was five times as large as his nonunion one (fifty-seven to eleven), his results conform to market realities. In other words, the sample is deficient but the results appear correct in terms of the market. Nevertheless, there have been so many changes in the open shop / union construction market mix since 1972 that cost comparisons for that year, even with a more representative sample, should be considered inappropriate for 1984.

In 1974, open shop constructors were building few high-rise office buildings anywhere. The nineteen nonunion buildings, as compared with the sixty-four union ones in this sample, were undoubtedly small buildings; if Allen found high-rise ones, they were in the South and in small communities or in suburbs. Open shop contractors then had little experience with such construction. Even if the data are accurate and representative, they have little bearing on the situation ten years later.

Allen attempts to rationalize his conclusion by pointing to a number of factors which he takes as given: greater union training and practical experience on the part of union workers, the greater efficiency of the hiring hall as an employment mechanism, less supervision required on union construction, more labor-saving machinery, and more professional and efficient management in the union sector.[33] He claims all these advantages for union construction by citing sympathetic publications and ignoring those which present contrary facts.[34]

---

[33] According to a number of executive search firms, open shop firms have been offering "attractive compensation packages" to recruit experienced construction managers and are paying salaries at least equal to, and in some cases 10 to 20 percent higher than, union firms in order to acquire the most talented managers. See "Execs Lured to Open Shop," *Engineering News-Record,* March 31, 1983, p. 53.

[34] He cites particularly Mills and Bourdon and Levitt. He even cites an article by Howard G. Foster, coauthor of our first edition, but carefully refrains from citing

Since the Allen studies were made, a University of Michigan study of unionized construction workers, based upon lengthy structured questionnaires, cast further doubts upon some of his assumptions. The workers interviewed were "confident that they can perform their assigned tasks." They reported, however, that they received "irritating setbacks from poor supervision and management of the work site . . . being delayed an average of 1.2 hours each day while they waited for the resources necessary to do their jobs." Moreover, the survey found that "performance and satisfaction were not strongly related, suggesting that high performers do not reap higher rewards for their relatively higher contributions. . . . Workers express satisfaction with the levels of pay and fringe benefits but also report that these rewards offer little or no incentive to perform. As long as a worker performs at the level necessary to keep his job, he enjoys the full schedule of monetary benefits." Hence the study concludes that the "motivational picture for unionized construction work is not particularly bright."[35]

In Chapter XI, data are presented which show that unionized contractors overwhelmingly pay a single rate within a craft whereas open shop contractors pay a wide range of rates. Open shop contractors frequently pay their outstanding performers the union rate or more, and then pay other workers a range of considerably lower pay based upon worker knowledge, ability, and productivity. It would, therefore, not be surprising if the as yet unfinished study of open shop construction workers by the same University of Michigan group found that motivation and rewards are much better correlated by open shop than by unionized pay practices. Certainly, there is little support in these findings for Allen's claim of managerial superiority or greater worker productivity in unionized construction.

The Strike Problem

Construction, as Professor Mills has noted, "is a strike-prone sector of the economy."[36] In many years, construction has accounted

---

the first edition which, of course, refutes some of these allegations of unionized construction superiority.

[35] William F. Maloney and James M. McFillen, *Unionized Construction Workers and Their Work Environment,* Technical Report No. 1 (Ann Arbor, Michigan: Center for Construction Engineering and Management, Department of Civil Engineering, University of Michigan, 1984). The quotations are from the "Summary of the Findings."

[36] D. Quinn Mills, "Construction," in Gerald G. Somers (ed.) *Collective Bargaining: Contemporary American Experience,* Industrial Relations Research Association Series (Madison, Wis.: The Association, 1980), p. 77.

for almost one-fourth of the all strikes and about 20 percent of all workdays lost during strikes, although construction accounts only for about 5 percent of the nation's total employment, and unionized construction accounts for only about one-third of that 5 percent.[37]

The reasons for this disproportion are twofold. First, the craft union structure means that each trade can negotiate separately in each jurisdiction and often each tries to outdo the others. This opens the way to a lot of strikes and considerable lost time.

Second, jurisdictional lines among the unions are often vague, variable from area to area, and altered over time by technology and new methods. Disputes arise as to which craft "owns the work" and strikes occur despite legal prohibitions and agreements. As contracting work is often subject to time constraints because of deadlines, interest costs, and other financial requirements, contractors are often unable to hold out against strikes. The buyer incurs an additional cost in the delay in the use of his purchase.

Of all complaints made by contractor groups to the unions, jurisdictional strikes are invariably near the top of the list. Open shop contractors do not have to contend with this difficulty, and are thus in a better position to assure buyers that a deadline can be met.

Advantages of Union Shop Contractors

Open shop contractors sometimes operate at a disadvantage in the labor market, particularly in the areas of recruitment and training. Because of the unions' long experience with referral services and apprenticeship programs, union contractors can better adjust to changing manpower needs. The open shop contractor will usually try to minimize these adjustments by maintaining as large a steady workforce as he can, sometimes even at an appreciable cost to him. Open shop contractors and their associations have worked out recruitment and training arrangements which in some respects emulate established union practices but in others are different and more appropriate. These arrangements are still being developed, and the cost advantages of the open shop could be somewhat tempered by these and other considerations if the development efforts of open shop contractors and associations fail.

METHODOLOGY

Most of the information presented in the following chapters was gathered through field interviews, telephone conversations, and

[37] Data from U.S. Bureau of Labor Statistics.

written correspondence with hundreds of contractors, association executives, and staff members of organizations connected with construction. For the original edition, about one-half of the interviews were conducted by the authors and the balance by staff members of the Wharton Industrial Research Unit. These interviews covered all parts of the country, except the Pacific states, during 1973 and 1974. The interviews for this edition were conducted almost entirely by David O. Northrup during 1981 and 1982, and included the entire continental United States. The interviews followed the interview guide set forth in Appendix A. For both editions, the interviewers usually began by making contact with key association executives in each area, particularly those affiliated with the ABC and the AGC, who then put them in touch with contractors active in local industry affairs. Because of its concentration of open shop activity, the South received the most coverage in 1973 and 1974, with visits to Texas, Oklahoma, Louisiana, Florida, South Carolina, North Carolina, Georgia, Virginia, the District of Columbia, and Maryland. Heavy coverage was also given to all northeastern and Great Lakes states, where unions are very strong. In 1982 and 1983, coverage was extensive in all areas of the country.

At the national level, we received cooperation from the headquarters of the major contractors' associations, again most notably the AGC and the ABC, but particularly in 1982 and 1983 from the specialty craft groups. Approximately two-fifths of AGC members operate open shops. In addition, we conducted wide-ranging interviews with officials from many of the largest contracting firms, both open shop and unionized, whose operations extend over broad geographic areas.

Wherever possible, information supplied in interviews was supplemented with documentation. In a number of instances we obtained private studies of open shop activity conducted by unions, by contractors' associations, and by government agencies or private researchers and/or research organizations. Such studies are much more accessible, complete, and methodologically sound today than they were a decade ago. In the area of labor market behavior, we assembled written descriptions of various plans for training programs, referral services, wage structures, and fringe benefit arrangements. We tried to determine the extent to which these plans were actually used. We noted cases where the activities seemed to have been more thoroughly developed on paper than in practice.

Our research included an analysis of all studies which even remotely dealt with open shop construction. An attempt was made to canvass as thoroughly as possible all available material, both within and without the public domain, including, in addition to the sources mentioned above: the published academic literature; the trade journals; the *Construction Labor Report,* a weekly information source issued by the Bureau of National Affairs, Inc.; McGraw-Hill's weekly magazine, *Engineering News-Record; Cockshaw's Construction Labor News + Opinion,* a monthly newsletter published by Peter Cockshaw; and the reports of the Construction Labor Research Council and the Business Roundtable.

Throughout this book, field work and literature search are supplemented by the results of two questionnaire surveys conducted by the Wharton Industrial Research Unit. A copy of each questionnaire is found in Appendix B. Responses were solicited from members of several contractors' associations, including the AGC, ABC, and several specialty contractor associations. In 1974, the questionnaires were sent to every other member, about 10,000 contractors in all. Usable responses were received from more than 1,500 firms, almost equally divided between union and nonunion. About 68 percent of the union respondents and 61 percent of the nonunion respondents were general contractors, with the rest scattered among various subcontracting specialties. About two-thirds of the subcontractors were in the electrical, heating, or plumbing fields.

In 1982, the questionnaires were sent to about 22,000 members of the ABC, AGC, and specialty contractor associations, but because of the size of the ABC (over 16,000 members), the questionnaire went only to every other member of this association. The response rate and the contractor/subcontractor mix were remarkably similar to those of the earlier questionnaire. It should also be noted that, as in all such questionnaires, the responses were heavily weighted by the larger companies. We have used the results of the questionnaires as indicators, rather than as the basis for precise statistical analyses.

Finally, two important sources of information were not tapped in this study. The first is the workers themselves. On certain subjects, such as skill acquisition, intersector mobility, and job search, worker input would have been invaluable. Unfortunately, a comprehensive canvass of any meaningful sample of workers is far beyond the resources of any nongovernmental body. The second untapped source is the building trades unions. In 1982 and 1983, we interviewed local union officials wherever possible. In some cases, we obtained internal letters and documents relating to local unions' concerns with open shop competition, and in other cases we have utilized speeches, position papers, convention proceedings, officers' reports, and the like from the public record. In addition, we

FIGURE II-2
Regions of the United States

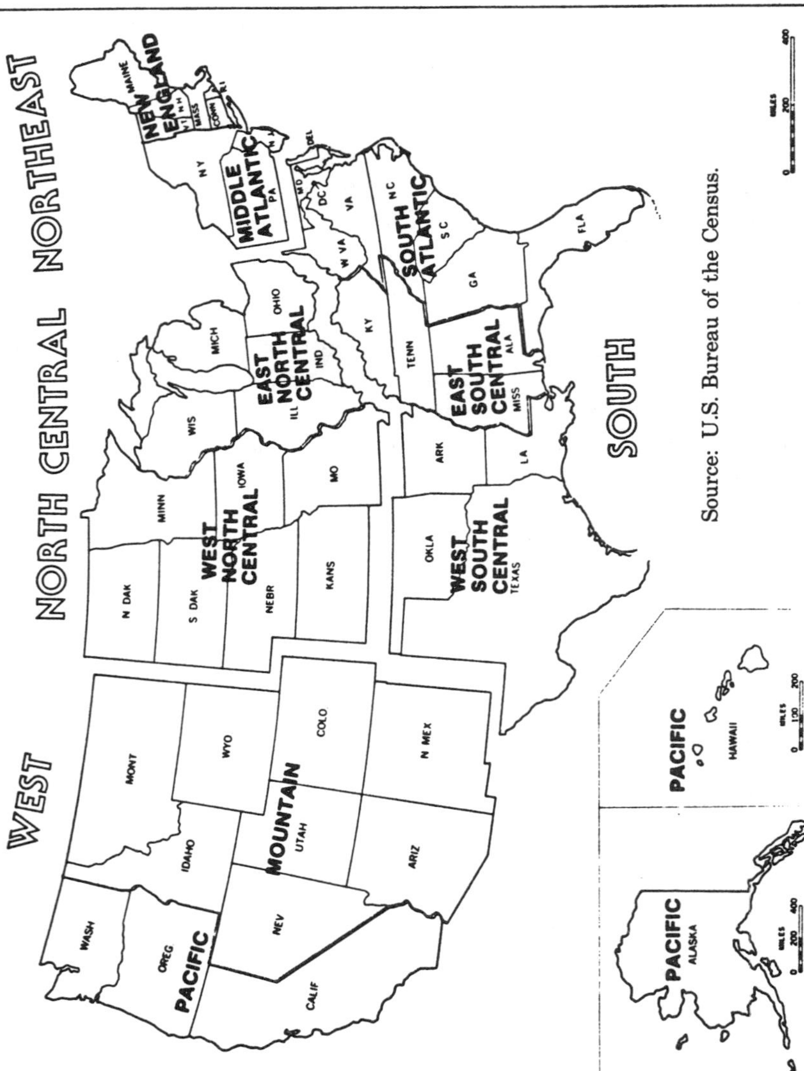

Source: U.S. Bureau of the Census.

had at our disposal the ample literature, both academic and governmental, dealing with unionized construction. In 1974, we also asked the national offices of twelve unions to answer questions on the extent and character of nonunion competition in their industry segments, and to indicate what steps they had taken in response. Only the Painters' union replied. In 1982, Robert Georgine, president of the Building and Construction Trades Department, AFL-CIO, was invited to present his views to the Wharton School's Labor Relations Council, and to speak to the students there. Unlike many other union officials, he not only declined this invitation; he refused to respond either to letters or to telephone calls.

The remainder of this volume is divided into three parts. Part Two describes and analyzes the extent and patterns of open shop activity in the subdivisions of the industry and the major geographical regions of the country. (Throughout the book, the regions of the country are defined in accordance with U.S. Bureau of Census practice, as set forth in Figure II-2, unless otherwise noted.) The section ends with an examination of the ways in which the building trades unions have sought to stem the growing penetration of open shops into other unionized areas. Part Three deals with various labor market issues and practices in the open shop sector. Part Four summarizes the findings and conclusions of this study.

PART TWO

Extent Of Open Shop Construction

CHAPTER III

Residential Construction

Residential construction has historically had a significant open shop component. Even in areas where unions remained firmly entrenched in other segments of the construction industry, the erection of homes and apartments has been performed almost exclusively with nonunion labor since 1950. There are, to be sure, important distinctions to be made within the residential sector. For example, in a few large cities and in some scattered small communities, the unions have maintained their strength throughout all of construction, including residential work. In many areas, moreover, union contractors have dominated the market for large, high-rise residential structures (i.e., apartments, hotels, dormitories), while being virtually shut out of the construction of single-family homes, small apartments and town houses, and repair and remodeling work generally.

As we shall document below, the open shop has steadily increased its share of multiple-unit construction since 1950 and took over probably more than 90 percent of the labor market for single-family dwellings. The reasons for the overwhelming success of the open shop in residential construction are both structural and economic. These will be discussed in this chapter, along with the extent of open shop growth in this sector.

EXTENT OF OPEN SHOP

Systematic attempts to estimate the extent of union representation in construction in general, and in residential construction in particular have until recently been few and far between. There is, nevertheless, some evidence that housing was at one time more heavily unionized than it is now. In 1936, the U.S. Bureau of Labor Statistics (BLS) surveyed the records of 5,450 contractors in 105 cities. These contractors employed 186,145 workers.[1] The bureau estimated that the sample included 25 to 30 percent of all building

[1] Edward P. Sanford, "Wage Rates and Hours of Labor in the Building Trades," *Monthly Labor Review,* XLV (August 1937), p. 283.

trades workers in the cities covered. Although the survey was concerned primarily with wages and hours, information was also gathered on union status. About one-third of the workers in the sample were engaged in residential construction. Of these, less than 57 percent were union members, compared to over 72 percent in nonresidential construction.[2] In larger cities (those with populations over 250,000), union workers constituted 62 percent of the residential workforce, but only 41 percent in smaller cities.

Other estimates of unionism in residential construction were made three decades ago by Richard Scheuch[3] and by William Haber and Harold M. Levinson.[4] Scheuch argued that the BLS survey overstated the incidence of unionism nationally, since the areas and employers in its sample were not representative of the industry as a whole. His conclusion was that "prior to World War II, single-family dwelling construction, with the exception of certain urban areas, was generally carried on in an essentially nonunion milieu. Since 1940, however, there has been a persistent increase in the unionization of the labor force."[5]

Relying on union membership reports to the AFL Building and Construction Trades Department (BCTD), Scheuch calculated that in 1947 the industry was 67.9 percent unionized and that the growth of organization since the 1930s was concentrated largely in the residential sector. In an interview with Scheuch in 1950, William Tobin, then industrial relations director for the National Association of Home Builders (NAHB), estimated that some 65 percent of workers employed by member firms were union members. Scheuch's own estimates of the extent of organization in twenty-seven areas throughout the country, based largely on personal interviews and congressional testimony, estimated that thirteen of the areas were almost entirely union and that six of these had changed from a predominantly nonunion character during or after the war. He noted further that even in many of those areas which were nonunion in the late 1940s, certain individual trades were heavily organized.

The Haber and Levinson study was based on a 1952 survey of sixteen cities, ranging in size from 49,000 (Battle Creek) to 3.6 million (Chicago). The authors, after conducting a total of 268 in-

---

[2] The reader is reminded that residential construction includes apartment building. The figure for homebuilding would doubtless be lower.

[3] Richard Scheuch, "The Labor Factor in Residential Construction," Ph.D. dissertation, Princeton University, 1951.

[4] William Habor and Harold M. Levinson, *Labor Relations and Productivity in the Building Trades* (Ann Arbor: Bureau of Industrial Relations, University of Michigan, 1956), esp. pp. 34–37.

[5] Scheuch, *"The Labor Factor in Residential Costruction,"* p. 2.

terviews primarily with local contractors and union representatives, concluded that union organization in both residential and nonresidential construction had grown considerably since 1936.

> In all sixteen cities surveyed in the summer of 1952, union strength in commercial, industrial, public, and semipublic work was close to 100 percent, with large apartment buildings only slightly weaker.[6]

The estimates for homebuilding were less uniform among the cities, although there was "substantial agreement" regarding the extent of unionization among those persons interviewed within each city. They estimated union strength to be close to 100 percent in such major cities as Chicago, St. Louis, and Cleveland and less than 25 percent in most of the smaller communities.

A major limitation of the Haber and Levinson estimates is that they are based on a relatively small and somewhat unrepresentative sample. The authors note that:

> As compared to what was probably a somewhat high estimate of 57 percent made by the Bureau of Labor Statistics in 1936, it is perhaps reasonable to estimate that approximately two-thirds of the homes being built in the early 1950's were under union control.[7]

Yet eleven of their sixteen cities are in regions identified by the BLS as relative union strongholds, and none of the larger cities in the South were included. Nevertheless, however much Haber and Levinson may have overestimated the extent of unionized residential building in 1952, it would appear highly unlikely that the actual nonunion component even approached the 80 percent figure commonly used twenty years later. The Haber and Levinson estimates for specific cities, as well as the earlier ones of Scheuch, are presented in Table III-1.

The period between World War II and the early 1950s probably represents the highwater mark of union representation in homebuilding for the past four decades. The years after World War II witnessed an explosive expansion in the demand for all kinds of construction, particularly residential, and the building trades unions for a sustained time had a congenial environment for their organizing efforts. Between 1954 and 1965, however, the average annual unemployment rate for construction workers never fell below 10 percent, and although the picture changed for the better in the late 1960s, it had already begun to darken again in 1970. Furthermore, residential building had already begun to move increas-

---

[6] Haber and Levinson, *Labor Relations,* p. 35.

[7] *Ibid.,* p. 36.

TABLE III-1
Estimates of Union Strength in Residential Construction by Scheuch and Haber/Levinson, by Selected Areas 1949 and 1952

Area	Scheuch 1949	Haber/Levinson Percent Range 1952
Atlanta, Ga.	mostly nonunion	—
Baltimore, Md.	open shop	—
Battle Creek, Mich.	——	0-25
Boston, Mass.	strongly unionized	50-75
Buffalo, N.Y.	unionized after war	75-90
Cape Cod, Mass.	nonunion	—
Charleston, S.C.	——	75-90
Charlotte, N.C.	——	0-25
Chicago, Ill.	strongly unionized	95-100
Cincinnati, Ohio	——	0-25
Cleveland, Ohio	——	95-100
Columbus, Ohio	open shop	0-25
Dallas, Tex.	mostly nonunion	—
Denver, Colo.	unionized after war	—
Detroit, Mich.	unionized after war	75-90
Grand Rapids, Mich.	——	0-25
Hartford, Conn.	strongly unionized	—
Indianapolis, Ind.	open shop	0-25
Kalamazoo, Mich.	——	0-25
Little Rock, Ark.	mostly nonunion	—
Long Island, N.Y.	unionized after war	—
Lynn, Mass.	mostly nonunion	—
Memphis, Tenn.	30 percent all union; rest union in some crafts	—
Milwaukee, Wis.	strongly unionized	—
New York, N.Y.	strongly unionized	—
Oklahoma City, Okla.	open shop	—
Philadelphia, Pa.	strongly unionized	—
Pittsburgh, Pa.	——	25-40
Portland, Oreg.	unionized after war	—
St. Louis, Mo.	strongly unionized	95-100
San Antonio, Tex.	open shop	—
San Francisco, Calif.	strongly unionized	—
Washington, D.C.	mostly open shop; some closed shop	25-40

Source: Richard Scheuch, "The Labor Factor in Residential Construction," Ph.D. dissertation, Princeton University, 1951, Chapter V; and William Haber and Harold M. Levinson, *Labor Relations and Productivity in the Building Trades* (Ann Arbor: University of Michigan, 1956), p. 252.

ingly away from the central cities, making it progressively more difficult for the unions to maintain control in some areas.

NAHB Surveys, 1969 and 1971

The only recent efforts, other than our own, to estimate the extent of union representation in housing have been undertaken by the NAHB. The association's periodic membership surveys asked respondents to identify themselves as either union, nonunion, or open shop. The 1969 canvass found that 18.6 percent were union employers, 32.7 percent were strictly nonunion, and the remainder were open shop, using union workers or subcontractors in some trades only.[8] Only 45 percent of the NAHB members responded to the survey, and it is probable that the larger and hence mostly union builders were somewhat more apt to respond. Thus, these figures on nonunion and open shop operations can be regarded as conservative estimates, but do demonstrate the downward trend of unionism in the residential sector during the 1960s.

The NAHB also conducts periodic surveys of its local executive officers to gather information on both union status and wage rates. Table III-2 summarizes the responses to such surveys in 1969 and 1971. Certain features of the table merit special note. First, the cities with extensive union coverage are almost entirely limited to major metropolitan areas, although even some large cities, such as Baltimore, Boston, and Washington, are all nonunion. The few small and medium-sized cities with high union representation are mostly on the West Cost. In the 1969 and the 1971 surveys, 89 of the 105 localities were reported to be at least 90 percent nonunion in single-family building.

The table also offers some clues as to how the extent of unionization in single-family building changed over time. Buffalo, for example, was estimated earlier by Scheuch as solidly union; by 1971 there was virtually no union representation in single-family building. The case of Boston is even more illuminating. Scheuch called the city "highly unionized"; Haber and Levinson put union strength there at 50-75 percent, noting that "while [Scheuch's characterization] is true for the city itself, a very large volume of homes is constructed in suburbs where union strength was reported to be much weaker."[9] In 1971, according to the NAHB, Boston was totally nonunion. But perhaps the more noteworthy changes are those

[8] Michael Sumichrast and Sara A. Frankel, *Profile of the Builder and His Industry* (Washington, D.C.: National Association of Home Builders, 1970), p. 202.

[9] Haber and Levinson, *Labor Relations,* p. 36, note 11.

TABLE III-2
Unionization in Single-Family Building as Estimated by NAHB Local Executive Officers 1969 and 1971

City	Percent of Single-Family Nonunion		City	Percent of Single-Family Nonunion	
	1969	1971		1969	1971
Birmingham, Ala.	100	100	Silver Spring, Md.	100	—
Gadsden, Ala.	—	98	Boston, Mass.	100	100
Montgomery, Ala.	100	—	Worcester, Mass.	100	—
Tucson, Ariz.	98	—	Ann Arbor, Mich.	0	—
Phoenix, Ariz.	5	5	Detroit, Mich.	0	—
Ft. Smith, Ariz.	100	100	Grand Rapids, Mich.	100	100
Mt. Home, Ariz.	—	100	Midland, Mich.	90	—
Harrison, Ariz.	100	—	Davison, Mich.	—	80
Hot Springs, Ariz.	100	—	Kalamazoo, Mich.	—	100
Pine Bluff, Ariz.	100	—	Duluth, Minn.	100	—
Los Angeles, Calif.	0	0	Minneapolis, Minn.	2	2
Sacramento, Calif.	0	—	Columbus, Miss.	90	—
San Diego, Calif.	2	5	Hattiesburg, Miss.	100	—
San Francisco, Calif.	0	0	Kansas City, Mo.	0	—
San Mateo, Calif.	0	—	St. Louis, Mo.	0	0
Modesto, Calif.	—	100	Billings, Mont.	50	80
Colorado Springs, Colo.	n.a.	90	Norfolk, Nebr.	100	—
Denver, Colo.	—	30	Dover, N.H.	—	95
Hartford, Conn.	100	—	Merrimack, N.H.	—	100
Mystic, Conn.	100	—	Lakewood, N.J.	0	—
Washington, D.C.	100	—	Basking Ridge, N.J.	—	100
Gainesville, Fla.	95	100	Binghamton, N.Y.	100	100
Tampa, Fla.	95	100	Spring Valley, N.Y.	50	—
W. Palm Beach, Fla.	50	75	Buffalo, N.Y.	—	95
Albany, Ga.	90	—	Syracuse, N.Y.	—	100
Valdosta, Ga.	100	100	Charlotte, N.C.	100	100
Atlanta, Ga.	—	98	Goldsboro, N.C.	100	—
Honolulu, Hawaii	n.a.	95	Greensboro, N.C.	100	100
Alton, Ill.	90	—	Rocky Mt., N.C.	100	—
Chicago, Ill.	1	0	Wilkesboro, N.C.	100	—
Quincy, Ill.	100	—	Fargo, N. Dak.	—	90
Indianapolis, Ind.	100	—	Canton, Ohio	97	95
Seymour, Ind.	90	—	Ashtabula, Ohio	100	—
Bloomington, Ind.	—	95	Cleveland, Ohio	5	10
Elkhart, Ind.	—	100	Norwalk, Ohio	100	—
Ames, Iowa	90	100	Toledo, Ohio	0	2
Des Moines, Iowa	100	—	Ardmore, Okla.	—	100
Iowa City, Iowa	95	100	Stillwater, Okla.	100	—
Sioux City, Iowa	90	100	Portland, Oreg.	10	—
Hutchinson, Kans.	100	—	Salem, Oreg.	100	—
Wichita, Kans.	—	100	Harrisburg, Pa.	100	—
Lexington, Ky.	100	100	Cresco, Pa.	—	95
Louisville, Ky.	100	100	Providence, R.I.	n.a.	100
Lafayette, La.	100	100	Anderson, S.C.	100	—
Shreveport, La.	75	—	Columbia, S.C.	100	100
Baltimore, Md.	100	100	Hartsville, S.C.	100	—

TABLE III-2 (continued)

City	Percent of Single-Family Nonunion 1969	1971	City	Percent of Single-Family Nonunion 1969	1971
Spartanburg, S.C.	100	100	Hampton, Va.	—	95
Sioux Falls, S. Dak.	75	100	Lynchburg, Va.	—	100
Chattanooga, Tenn.	—	90	Roanoke, Va.	—	100
Austin, Tex.	100	100	Kennewick, Wash.	25	50
Beaumont, Tex.	1	0	Seattle, Wash.	0	—
Dallas, Tex.	100	—	Spokane, Wash.	100	100
Victoria, Tex.	100	—	Tacoma, Wash.	10	20
Lubbock, Tex.	—	100	Morgantown, W. Va.	—	100
Wichita Falls, Tex.	—	100	Fond du Lac, Wis.	100	—
Salt Lake City, Utah	100	—	Madison, Wis.	100	—
Ogden, Utah	100	95	Milwaukee, Wis.	0	0
Petersburg, Va.	100	—	Wausau, Wis.	—	100

Source: National Association of Home Builders Labor Survey, mimeographed, 1969 and 1971.
n.a. = not available

covering only the short time period between 1969 and 1971 when some changes were reported in the extent of unionization in single-family building in sixteen cities. In all but three of these cases, involving small reductions in nonunion work in Chicago; Canton, Ohio; and Ogden, Utah, the changes were in the direction of more open shop building. Moreover, some of the increases were substantial; notice especially the figures for West Palm Beach, Florida; Billings, Montana; Sioux Falls, South Dakota; and Kennewick, Washington.

The NAHB surveys also verify that single-family housing was then, as it is now, more likely to be built on a nonunion basis than is high-rise apartment construction. Table III-3 summarizes the responses of executive officers in those cities where high-rise and single-family buildings were reported to differ in terms of their respective proportions of union representation. For comparative convenience, the table also includes the single-family estimates. In all but two of the twenty cases, high-rise building was reported to be significantly more unionized than single-family building. In the case of Denver, our field survey of 1974 indicated that residential work is both much more open shop than the NAHB reported and more open shop than high-rise apartment construction. But for the cases where there are numerical estimates for both single-family and high-rise construction, it seems sufficient to make the point that, in the aggregate, high-rise apartments tended more to be built by union workers than did single-family homes. As we shall see

TABLE III-3
Unionization in High-Rise Residential Building as Estimated by NAHB Local Executive Officers Selected Cities, 1969 and 1971

City	Percent High-Rise Nonunion	Percent Single-Family Nonunion	Year
Birmingham, Ala.	0	100	1969
Tucson, Ariz.	10	98	1969
Colorado Springs, Colo.	70	90	1971
Denver, Colo.	50	30	1971
Gainesville, Fla.	50	100	1971
W. Palm Beach, Fla.	25	75	1971
Honolulu, Hawaii	100	95	1971
Indianapolis, Ind.	25	100	1969
Seymour, Ind.	0	90	1969
Iowa City, Iowa	0	95	1969
Wichita, Kans.	90	100	1971
Silver Spring, Md.	95	100	1969
Boston, Mass.	0	100	1969
Worcester, Mass.	0	100	1969
Davison, Mich.	20	80	1971
Billings, Mont.	0	80	1971
Basking Ridge, N.J.	10	100	1971
Binghamton, N.Y.	0	100	1969
Canton, Ohio	0	97	1969
Columbia, S.C.	80	100	1971

Source: National Association of Home Builders, Labor Surveys, mimeographed, 1969 and 1971.

below, this is still the case, but unions have suffered further loss of market share in both multiple- and single-family dwellings.

There are a few further observations that need to be made about these NAHB estimates—or indeed about estimates of this kind generally—especially as they relate to the situation in very large cities. It can be observed from Table III-2 that most local NAHB officials estimated the extent of nonunion residential construction at either very close to zero or very close to 100 percent. Reports of total unionization of single-family housing, however, may be misleading for two reasons. First, in many of the large cities, NAHB chapters had contracts with the basic building trade unions. Thus, a homebuilder who opted to operate open shop may not have been a member of the association. In Detroit, for example, where unionization was reported by the NAHB to be virtually complete, a number of open shop builders belonged instead to the Associated Builders and Contractors (ABC). An official of the ABC in Detroit estimated in 1974 that 40 to 50 percent of residential work in the area was nonunion.

Furthermore, union contracts usually apply to a specific geographic jurisdiction in and around the city. As a suburban land becomes fully utilized, residential building may move into exurban areas outside the NAHB's territory. A good example is Cleveland, where the 1971 survey put the nonunion residential sector at 10 percent. Even if this estimate was accurate, it related only to the city and its immediate environs. Yet an increasing amount of residential building, mostly nonunion, was then taking place on the periphery of the metropolitan area.

On the other hand, NAHB estimates of zero union building did not always take into account the various subcontractors to whom the builder assigns work. In Houston, where virtually all homebuilding has long been open shop, there are nevertheless still one or two contractors who have national agreements with the United Brotherhood of Carpenters and Joiners, as will be discussed below. Also, although the builders themselves may not be signatory to any union contracts, some of them utilize union subcontractors in many areas for the specialty functions. These specialties will vary from locality to locality and even among builders in a single locality. Thus in Buffalo, builders may subcontract work to union firms in the electrical and mechanical fields; in New Hampshire, to masonry contractors; and in the Detroit area, to rough carpentry and drywall contractors. And, as will be detailed in subsequent chapters, this phenomenon is not uncommon in other branches of construction.

Industrial Research Unit Survey, 1974

The 1974 Industrial Research Unit questionnaire, which was sent to members of contractors' associations, excluding members of the NAHB, sought to obtain estimates from contractors as to the proportion of nonunion work performed in their localities for each sector of the industry. Not surprisingly, union respondents tended to estimate nonunion work at an appreciably lower level than did open shop respondents. In the low-rise residential sector (homes, town houses, and garden apartments), however, the responses from union and open shop firms were relatively consistent. Table III-4 summarizes the proportion of open shop low-rise residential work reported by the contractors surveyed. Since the figures in Table III-4 are somewhat distorted by the "no responses," Table III-5 is presented to indicate the proportion of those actually responding to the questionnaire who estimated open shop work in their areas at over 80 percent.

Since the union and open shop respondents do not necessarily represent the same local areas, the differences in their responses

TABLE III-4
Respondents' Estimates of the Proportion of Open Shop Work in Low-Rise Residential Construction by Region and Union Status, 1974

Region	Number	DK[a] NR	0-20 Percent	21-40 Percent	41-60 Percent	61-80 Percent	81-100 Percent
Union Respondents							
Not reported	89	41.6	10.1	4.5	4.5	5.6	33.7
New England	27	22.2	3.7	0.0	0.0	14.8	59.3
Middle Atlantic	125	33.6	1.6	0.8	4.8	5.6	53.6
E.N. Central	147	33.3	12.2	6.8	3.4	8.2	36.1
W.N. Central	60	33.3	15.0	3.3	3.3	11.7	33.3
S. Atlantic	68	17.6	0.0	0.0	8.8	8.8	64.7
E.S. Central	32	31.3	6.3	6.3	0.0	0.0	56.3
W.S. Central	45	13.3	4.4	2.2	2.2	8.9	68.9
Mountain	52	38.5	7.7	5.8	7.7	7.7	32.7
Pacific	133	44.4	15.8	6.0	10.5	9.0	14.3
Total	778	33.6	8.7	4.0	5.4	7.8	40.5
Open Shop Respondents							
Not reported	83	22.9	2.4	2.4	4.8	7.2	60.2
New England	63	17.5	1.6	0.0	1.6	7.9	71.4
Middle Atlantic	134	25.4	2.2	0.7	1.5	8.2	61.9
E.N. Central	75	16.0	2.7	2.7	1.3	9.3	68.0
W.N. Central	45	24.4	4.4	4.4	0.0	2.2	64.4
S. Atlantic	220	23.7	5.0	0.9	0.9	6.4	63.2
E.S. Central	45	8.9	0.0	0.0	0.0	8.9	82.2
W.S. Central	92	20.7	1.1	0.0	3.3	2.2	72.8
Mountain	21	19.0	4.8	4.8	9.5	14.3	47.6
Pacific	6	0.0	0.0	16.7	16.7	33.3	33.3
Total	784	21.1	2.9	1.4	2.0	7.0	65.4

Source: Industrial Research Unit Questionnaire, 1974.
[a] DK / NR = Do not know or not reported.

TABLE III-5
Proportion of Those Responding to Questionnaire Who Put Open Shop Work at Over 80 Percent by Region

Region	Union Respondents	Open Shop Respondents
New England	76.2	86.5
Middle Atlantic	80.7	83.0
E.N. Central	54.1	80.9
W.N. Central	50.0	85.2
S. Atlantic	78.5	82.8
E.S. Central	82.0	90.2
W.S. Central	79.5	91.8
Mountain	53.2	58.8
Pacific	25.7	50.0
Total	61.0	82.9

Source: Industrial Research Unit Questionnaire, 1974.

may have reflected actual differences in open shop housing in various areas within a region or simply conflicting perspectives between the two groups of respondents. In any event, the overall picture was clear, for even the responses of union firms showed low-rise residential building to be predominantly the province of the open shop by 1974. With the exception of the Pacific and Mountain states, housing construction appeared predominantly nonunion throughout the nation. With three-fifths of the responding union contractors estimating open shop building at over 80 percent, the common use of that figure is understandable for this period. Yet, in 1972, for example, the former labor counsel of the NAHB stated:

> . . . certainly some branches of the industry are more fully organized than others. Residential construction . . . is probably the least organized; perhaps 80 percent of the workers are nonunion.[10]

The now defunct Construction Employers Labor Relations Association in New York state, in an unpublished 1971 survey of open shop construction, also assumed residential work to be 80 percent nonunion.[11] As we shall see below, however, this figure is now an outdated underestimate and may have been an underestimate in the early 1970s as well.

The level of open shop work in high-rise housing construction was found by both union and nonunion respondents to the 1974 Industrial Research Unit survey to be appreciably lower than that in low-rise building. The estimates of the two groups, however, differed markedly. The median response among union contractors was about 29 percent, compared to about 68 percent for the open shop firms. About one-quarter of the difference was attributable to the varying distribution of respondents among regions. Thus, if the nonunion respondent totals by region were applied to the union respondent percentage distribution, the resulting median response would have been about 39 percent. This estimate is significant in light of the unions' traditional strength in high-rise housing.

The growth of open shops in residential building generally was also explored in the 1974 Industrial Research Unit questionnaire. Contractors were asked whether their current estimates of nonunion work were different from what they would have been a few years earlier. Since the responses indicated 20-percentage point ranges (i.e., 0–20 percent, 21–40 percent, etc.), the estimates of change were tabulated in terms of the movement over ranges for this period. Thus, a change to a different range represents, on average, a movement of some 20 percentage points. Table III-6 sum-

---

[10] Daniel Quinn Mills, *Industrial Relations and Manpower in Construction* (Cambridge, Mass.: MIT Press, 1972), quoting the former labor counsel of NAHB, p. 17.

[11] Copy in Industrial Research Unit library.

TABLE III-6
Respondents' Estimates of Change in Proportion of Open Shop Residential Construction Over Past Three Years by Region and Union Status

Region	Union								Nonunion							
	Total	Number Reporting Change	Change in Designated Range						Total	Number Reporting Change	Change in Designated Range					
			−3	−2	−1	+1	+2	+3			−3	−2	−1	+1	+2	+3
Low-rise building																
New England	27	7	0	0	1	1	4	1	63	5	0	0	0	3	1	1
Middle Atlantic	125	16	0	0	3	10	1	2	134	15	1	1	2	7	4	0
E.N. Central	147	21	0	2	1	12	5	1	75	7	0	0	1	4	1	1
W.N. Central	60	6	0	0	0	3	3	0	45	2	0	0	0	2	0	0
S. Atlantic	68	21	0	0	4	13	4	0	220	13	0	0	1	7	5	0
E.S. Central	32	2	0	0	0	1	0	1	45	6	0	0	2	2	2	0
W.S. Central	45	4	0	0	0	2	2	0	92	10	0	0	1	7	0	2
Mountain	52	8	0	0	2	5	1	0	21	3	0	0	0	1	2	0

Pacific	133	18	0	1	2	7	4	4	6	2	0	0	0	1	1	0
Not reported	89	10	0	0	0	7	1	2	83	9	0	0	0	7	2	0
Total	778	113	0	3	13	61	25	11	784	72	1	1	7	41	18	4
High-rise building																
New England	27	9	0	0	1	4	3	1	63	13	0	0	2	6	2	3
Middle Atlantic	125	27	1	0	2	18	5	1	134	13	0	0	3	7	3	0
E.N. Central	147	18	0	1	1	11	5	0	75	11	0	0	1	5	4	1
W.N. Central	60	9	2	1	2	3	1	0	45	4	0	0	0	4	0	0
S. Atlantic	68	19	0	1	4	8	6	0	220	28	0	0	5	16	4	3
E.S. Central	32	7	0	0	0	5	1	1	45	11	0	0	1	7	1	2
W.S. Central	45	8	0	0	2	2	4	0	92	12	0	0	1	9	1	1
Mountain	52	6	0	0	1	4	0	1	21	2	0	0	0	1	0	1
Pacific	133	7	0	0	1	3	3	0	6	0	0	0	0	0	0	0
Not reported	89	9	0	0	0	7	1	1	83	11	0	0	0	7	4	0
Total	778	119	3	3	14	65	29	5	784	105	0	0	13	62	19	11

Source: Industrial Research Unit Questionnaire, 1974.

marizes the responses for both low-rise and high-rise housing construction.

The responses in Table III-6 show a significant movement, considering the brief time span involved, in the direction of more open shop activity in both low-rise and high-rise residential building. Perhaps most noteworthy is the substantial agreement among union and open shop respondents regarding both the amount and direction of change. The relative propensity of open shop firms to respond "no change" in low-rise building is doubtless attributable to their judgment that low rise was already overwhelmingly open shop, and thus, the open shop share could not rise any further. It should be especially noted that about 13 percent of all union respondents reported a significant increase in open shop apartment building. Since these projects were long thought to be within the unions' domain, this development attests to the substantial gains made by the open shop in apartment construction in recent years.

NAHB Study, 1976–1977

The latest NAHB survey of its membership occurred in 1976–1977.[12] The results, set forth in Figure III-1, confirm both the continued decline of unionism in residential construction and the probability that the 1974 Industrial Research Unit survey underestimated the extent of open shop growth in this sector because it did not reach the small builders who responded to the association's questionnaire. The 1976–1977 NAHB survey found that only 8.1 percent of the responding homebuilders had "some employees unionized," that 92.1 percent of single-family builders were nonunion, and perhaps more astounding, that 82.9 percent of multifamily builders were nonunion. Moreover, only 18.8 percent of the 8.1 percent with "some employees unionized" used all unionized trades. Far more of the unionized builders employed union carpenters than any other trade—69 percent of the single-family builders and 67.3 percent of the multifamily ones. Union laborers were employed by 23.8 percent of the unionized single-family builders and by 41.2 percent of the multifamily ones. No other union trade was used by more than 20 percent of the unionized builders.[13]

---

[12] Michael Sumichrast, Gopal Ahluwalia, and Robert J. Sheehan, *Profile of the Builder* (Washington, D.C.: National Association of Home Builders, 1979). The questionnaire was mailed in 1977, but many questions related to activities in 1976. The union question asked "Are any of the firm's employees unionized?" This might have been answered as of 1977.

[13] *Ibid.*, p. 63.

FIGURE III-1

Homebuilders with Some Employees Unionized, 1976–1977

(percent of respondents to mail questionnaire)

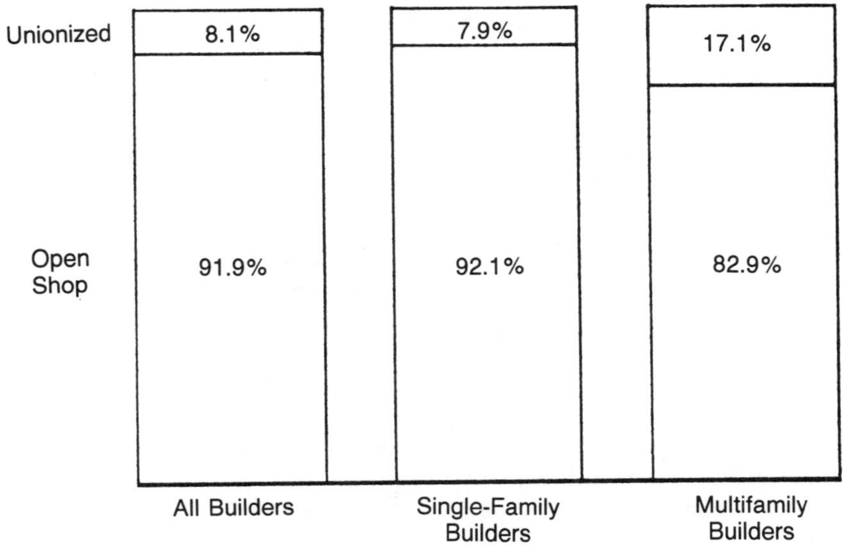

Source: Michael Sumichrast, Gopal Ahluwalia, and Robert J. Sheehan, *Profile of the Builder,* (Washington, D.C.: National Association of Home Builders, 1979), p. 63.

The dramatic sweep of the open shop in residential construction is further illustrated by the comparison of the 1969 and 1976–1977 NAHB surveys. Even granting that the 1969 survey overstated the extent of the union shop by a considerable amount, changes of the magnitude shown in Table III-7 demonstrate not only that unions are in danger of virtually disappearing from residential construction, but that open shop home and commercial builders have entered the multifamily dwelling subsector and are winning the bulk of the business.

A further indication of what occurred between 1969 and 1976–1977 is shown in Table III-8. There the two NAHB surveys are compared on a regional basis. The open shop gained in all regions over this period except in the Pacific, where the situation appeared to stabilize. In the already strongly open shop areas, such as New

TABLE III-7
Comparison of Responses, Open v. Union Shop National Association of Home Builders' Surveys, 1969 and 1976–1977 by Type of Residential Construction

Type of Operation	Percent All Respondents		Percent Single-Family Respondents		Percent Multi-Family Respondents	
	1969	1976–1977	1969	1976–1977	1969	1976–1977
Open shop	52.1	91.9	55.9	92.1	39.3	82.9
Union shop	47.9	8.1	44.1	7.9	60.7	17.1

Source: Michael Sumichrast and Sara A. Frankel, *Profile of the Builder and His Industry* (Washington, D.C.: National Association of Home Builders, 1970), p. 203; and Michael Sumichrast, Gopal Ahluwalia, and Robert J. Sheehan, *Profile of the Builder* (Washington, D.C.: National Association of Home Builders, 1979), p. 158.

England and the southern states, open shop builders expanded their market shares to well over 90 percent as they moved into the multifamily construction segments. More significantly, in the once union strongholds of the North Central regions, union control dissipated; the open shop rose from approximately 40 percent to 81 percent in the West North Central and to 87.1 percent in the East North Central. In the Mountain region, the open shop increased its share from 78.9 percent to a remarkable 99 percent; and in the Middle Atlantic region, the open shop control jumped from 62.2 percent to 97.9 percent. Thus, according to the data from the NAHB for 1976–1977, the open shop had virtually complete control of residential construction except in two regions—the Pacific area, where the union shop retained control of over 25 percent of the work, as compared with nearly 75 percent in 1969; and the West North Central area, where 19 percent of the work remained union, as compared with 60.5 percent in 1969.

REGIONAL STUDIES AND FIELD OBSERVATIONS

In 1982, when the Industrial Research Unit mailed its questionnaire for this study, the residential construction sector was at its lowest ebb since the 1930s. It was, therefore, determined that useful results would not likely be forthcoming if an attempt were made to include homebuilders in the survey. Accordingly, in our field work in this sector we concentrated on the areas which we had found in 1973–1975 to be the most strongly unionized. In addition,

TABLE III-8
Comparison of Responses, Open v. Union Shop National Association of Home Builders' Surveys by Region, 1969 and 1976–1977

Region[a]	Type of Employee Relations	Percent 1969	Percent 1976–1977
United States	Open Shop	52.1	91.9
	Union Shop	47.9	8.1
New England	Open Shop	85.5	94.3
	Union Shop	14.5	5.7
Middle Atlantic	Open Shop	62.2	97.9
	Union Shop	37.8	2.1
S. Atlantic	Open Shop	86.7	98.2
	Union Shop	13.3	1.8
E.S. Central	Open Shop	90.7	92.5
	Union Shop	9.3	7.5
W.S. Central	Open Shop	85.0	96.7
	Union Shop	15.0	3.3
E.N. Central	Open Shop	42.7	87.1
	Union Shop	57.3	12.9
W.N. Central	Open Shop	39.5	81.0
	Union Shop	60.5	19.0
Mountain	Open Shop	78.9	99.0
	Union Shop	21.1	1.0
Pacific	Open Shop	25.1	74.3
	Union Shop	74.9	25.7

Source: Michael Sumichrast and Sara A. Frankel, *Profile of the Builder and His Industry* (Washington, D.C.: National Association of Home Builders, 1970), p. 203; and Michael Sumichrast, Gopal Ahluwalia, and Robert J. Sheehan, *Profile of the Builder* (Washington, D.C.: National Association of Home Builders, 1979), p. 158.

[a] Regional definitions are consistent with those of the U.S. Bureau of Census. See Figure II-2.

we obtained a number of helpful regional studies which are summarized, together with our field observations, in this section.

Great Lakes Area

The industrial cities from Cleveland to Chicago comprised one of the few heavily unionized residential construction areas found in our 1973–1974 field surveys. Even there, however, homebuilders were increasingly moving toward the open shop, particularly in the suburbs of such cities as Cleveland and Toledo. Now, as the data in Table III-8 show and as our field interviews confirm, the open shop has won well over 80 percent of the work. The inner cities remain union for multiple dwellings; Chicago remains union for most construction, including residential, but little else is within the union purview. The Cleveland Chapter of the NAHB dealt with the Carpenters' union and other unions for many years, but ceased all collective bargaining activities in 1976 because, by that time, so few of its members were building union.

St. Louis and Kansas City remain union construction cities, but homebuilding is open shop except for the large multiple dwellings. The NAHB's Kansas City Chapter formerly handled labor negotiations, but ceased this activity in 1971 following a series of strikes. Most single-family dwelling builders are now open shop.

In Indianapolis in 1973–1974, the open shop built single-family residences and garden apartments up to three stories. High-rise dwellings were union constructed. Now the open shop builders are building some of the taller structures, particularly outside of the city.

Rocky Mountain Area

Arizona was a unionized homebuilding area in 1973–1974, but soon thereafter the local chapter of the NAHB ceased its collective bargaining activities and today the open shop is overwhelmingly dominant. In 1980, however, the Phoenix local of the International Brotherhood of Electrical Workers (IBEW) signed a special homebuilding concession agreement reducing wages $6.75 per hour to a rate of $8.00 per hour and granting other concessions. This permitted it to more than double its residential work over a seventeen-month period, thus reversing a long-term decline.[14] Homebuilding in New Mexico has been open shop for many years.

In 1973–1974, we found that in Colorado, single-family dwellings were built almost entirely nonunion, as were multiple-family dwellings outside of Denver. In that city, unionized contractors then built about one-half of the multiple dwellings. As described in Chapter VIII, several union officials confessed to, and were convicted of, setting fire to multiple-dwellings under construction by nonunion builders. Perhaps related to this is the fact that soon thereafter, the bulk of the multiple-dwelling contracts were won by the open shop. Table III-9 shows the percentage of awards and dollar volume listed in the F. W. Dodge *Daily Journal*[15] for the first six months of 1973, 1979, and 1982. This study was made by a University of

[14] "Phoenix IBEW Wins Back Residential Work from Open Shop under Special Agreement," *Construction Labor Report,* No. 1356 (December 2, 1981), pp. A-11-12.

[15] A fuller explanation of the Dodge Reports, their uses and limitations, is found in Chapter IV.

TABLE III-9
Residential Construction[a]
Union and Open Shop Percentage Share of Awards[b]
First Six Months, 1973, 1979, and 1982
Colorado

	Percent No. of Awards			Percent Dollar Volume		
	1973	1979	1982	1973	1979	1982
Union shop	11	7	2	11	11	1
Open shop	89	93	98	89	89	99

Source: David E. Moore, *Union and Open Shop Construction in the State of Colorado.* (Denver: Construction Industry Management Council and Colorado Building and Construction Trades Council, 1983).
[a] Includes both apartments and single-family residences.
[b] Awards as listed in the *Daily Journal,* published by F. W. Dodge, a subsidiary of McGraw Information Systems Company.

Colorado researcher for the Colorado Construction Research Project, a joint union-management undertaking. The data in Table III-9 show that by 1982, the open shop won 98 percent of the list awards and 99 percent of the dollar volume thereof. Since negotiated, unlisted awards are likely to be smaller ones going to open shop builders, the almost complete sweep of the open shop in Colorado is apparent.

California

The Pacific area was the high point of union residential building control, with 25.7 percent of the homebuilders unionized in 1976–1977 (Table VIII-8). In California, some chapters of the NAHB still negotiate contracts, but the number of homebuilders which they represent has steadily declined over the years. Moreover, there has been an increase in nonunion homebuilding since 1976, not only in the outlying suburban areas and in the more rural counties, but also in Los Angeles and its fast-growing counties. This is attested to by the study of the Construction Industry Research Board, which is summarized in Table III-10.

The data in this table show that for low-rise residential construction (single-family and multiple-family dwellings of three or four stories or less), the market share of union contractors in the eleven-county area covered declined from 58 percent in 1977 to 30 percent in 1981, a decrease of 48.3 percent in four years. In high-rise residential construction, the union sector declined from 70 to 58 percent,

TABLE III-10

Union-Open Shop Percent Market Share—Eleven County Southern California Area[a]
Low- and High-Rise Residential[b] *Construction*
1977 and 1981

County and No. of Responses	Percent Market Share[c]							
	Low Rise Residential				High Rise Residential			
	1977		1981		1977		1981	
	Union Shop	Open Shop	Union Shop	Open Shop	Union Shop	Open Shop	Union Shop	Open Shop
Regional Total[d] (284)	58	42	30	70	70	30	52	48
Los Angeles (159)	56	44	32	68	70	30	53	47
Orange (37)	68	32	38	62	77	23	53	47
Ventura (22)	72	28	44	56	84	16	59	41
San Bernardino (16)	61	39	31	69	74	26	57	43
Riverside (15)	49	51	19	81	63	37	48	42
Santa Barbara (15)	73	27	38	62	88	12	66	34
Kern (8)	55	45	22	78	61	39	39	61
San Luis Obispo (6)[e]	18	82	10	90	20	80	10	90
Mono (4)[e]	10	90	20	80	10	90	10	90
Imperial (2)[e]	50	50	30	70	50	50	30	70
Inyo (0)	n.r.	n.r.	n.r.	n.r.	n.r.	n.r.	n.r.	n.r.

Source: *A Survey of the Construction Industry Trends in the Eleven-County Southern-California Area* (Los Angeles: Construction Industry Research Board, January 31, 1983), Table 6.

[a] The authors of this study note that nonunion firms are underrepresented in their survey presumably because of a low response from small firms.

[b] Low rise is generally defined as single-family and multiple-family dwellings not in excess of three or four floors and not requiring elevator service.

[c] All responses are frequency-weighted averages of the midpoints of 20 percent intervals.

[d] The regional total for each category of construction is the weighted average for all responses. Responses were weighted by the actual county share of each type of construction for 1977 and 1981.

[e] Statistically unreliable because of small response.

n.r. = no response.

a fall of 25.7 percent. The decrease in the union construction share occurred in the major populated areas as well as in the outlying areas. The authors of this study state that nonunion firms were underrepresented in the survey because of a lower response rate by smaller firms.[16] Thus, if anything, the gains of the open shop could well be greater than the data in Table III-10 indicate.[17]

San Francisco remains strongly union, but suburban areas such as Alameda and Contra Costa counties are becoming steadily more open shop. According to a June 1981 study of the Greater Oakland Chapter of the Sheet Metal and Air Conditioning National Association (SMACNA), its members did 67 percent of new residential work in 1980, but could expect no more than 43.5 percent in 1981–1982. By 1983, its projections indicated that the unionized sector in these two counties would be left with only high-rise apartments and federal government housing subject to the Davis-Bacon Act.[18] This study was written prior to the changes in Davis-Bacon administrative rules, which, as noted in Chapter II, could further reduce union success.

Pacific Northwest

Like California, the Pacific Northwest was once a construction union stronghold, but this has changed dramatically for residential construction. For example, the union share of the value of electrical work permits issued in Seattle and surrounding areas for single-family dwellings in 1981 was only 35 percent of the total, and for multiunit housing only 37 percent.[19] Annual surveys made since 1976 (see table III-11) by the Inland Empire Chapter, AGC, Spokane, found that in Eastern Washington and Northern Idaho, 96 to 100 percent of the open shop builders held permits for single-family residences valued at $60,000 or more; 93 to 100 percent of those for duplex, triplex, and small apartment units; and, in all but one year,[20]

---

[16] *A Survey of the Construction Industry: Trends in the Eleven-County Southern California Region* (Los Angeles: Construction Industry Research Board, January 31, 1983), p. 6.

[17] For a graphic account of how and why an individual builder went open shop, see Robert W. Merry, "Struggle At the Site: Building-Trades Unions Battle to Recoup Losses with a Big Drive in Los Angeles," *Wall Street Journal,* April 11, 1979, p. 48. The data in Table III-10 indicate rather conclusively that this "big drive" failed.

[18] *State of the Industry: A Special Study Report and Assessment of the Impact of Open Shop Intrusion into the Heating, Ventilating, and Air Conditioning Construction Market in Alameda and Contra Costa Counties* (Oakland, Cal.: SMACNA Greater Oakland, June 1981), n. p.

[19] Data from the Puget Sound Chapter, National Electrical Contractors Association.

[20] The 1978 data in Table III-11 appear to reflect one or a few very large apartment complexes built union, especially one $10 million facility in Spokane.

TABLE III-11

Eastern Washington and Northern Idaho[a]

Union-Open Shop Residential and Apartment Construction

Building Permit Dollar Volumes, 1976–1983

	Residential[b]			Apartments[c]			Apartment Complex[d]		
Year	Value $ (million)	Percent Union	Percent Open Shop	Value $ (million)	Percent Union	Percent Open Shop	Value $ (million)	Percent Union	Percent Open Shop
1976	6.5	4	96	14.4	5	95	30.8	11	89
1977	38.0	2	98	49.1	7	93	51.1	45	55
1978	104.9	1	99	37.7	1	99	22.0	6	94
1979	97.8	2	98	23.3	1	99	58.2	6	94
1980	56.3	0[e]	100	8.3	0	100	28.1	16	84
1981	54.3	1	99	9.0	1	99	17.8	3	97
1982	40.7	1	99	3.7	0	100	8.8	42	58
1983	74.2	0	100	7.7	1	99	10.2	2	98

Source: Annual Surveys, Inland Empire Chapter, Associated General Contractors.

[a] Defined as cities and counties east of the 120th meridian in Washington and north of the 46th parallel in Idaho. Excludes some Idaho counties and Washington towns that lack a building permit or a comprehensive plan ordinance.

[b] Only building permits of $60,000 or more included.

[c] Includes permits for duplex, triplex, and small apartment units.

[d] Includes large units, usually in excess of $500,000 valuation.

[e] Actual union percentage was .001.

84 to 99 percent of those for apartment complexes valued at $500,000 or more. Finally, the Oregon-Columbia Chapter, AGC, Wilsonville, Oregon, reports that in Oregon and Southwest Washington almost all residential construction is nonunion.[21]

ANALYSIS OF RESIDENTIAL CONSTRUCTION TRENDS

The data presented in the preceding pages show a clear picture of the virtual elimination of union influence in the residential construction sector, except for large apartment construction in some areas and a few union enclaves in cities like New York, Chicago, and San Francisco and in other areas such as the Pacific states. Our data, however, demonstrate that today the Pacific area is between 80 and 85 percent open shop rather than to the approximate 75 percent found by the NAHB survey of 1976 (Table III-9). The decline of unions in single-family and low-rise multiple family housing has been a constant trend since the early 1950s. The erosion of union influence in high-rise multiple dwellings began in the 1960s, increased its pace in the 1970s, and continued in the 1980s at an even more accelerated rate. The ability of open shop builders to construct large apartment complexes is now thoroughly proved. It bodes ill for the construction unions because the firm that can put up a high-rise apartment or condominium complex is most likely to be equipped to construct large office buildings as well, and this has occurred in Baltimore, Washington, D.C., Florida, Texas, and many other places. The union loss in residential construction cumulates in other ways too, as builders move from residential into small commercial work.

Contributing Factors

Whatever period one studies, including the present one, residential construction has always been the least organized sector of the construction industry. Several factors contribute to the difficulties of unionism in this sector and to its exceptionally low level at the present time.

A key element is that residential contractors are relatively small employers with few employees. Moreover, the industry is very volatile, with firms entering and leaving at a relatively high rate. Thus in 1976, 40.2 percent of the single-family builders constructed less than ten units, another 26.6 percent built more than 500 units.[22]

---

[21] Associated General Contractors, Oregon-Columbia Chapter, *Construction Industry Profiles,* June 30, 1982, n. p.

[22] Sumichrast et al., *Profile of the Builder* p. 40.

Although the larger homebuilders have won an increased share of the market, small builders still have the bulk of the market.[23] Moreover, this situation has not changed materially from 1958. In 1976, the single-family builders averaged nineteen employees and thirteen construction workers,[24] and 35.1 percent were in the business less than five years and another 22.7 percent only five to nine years.[25] Multifamily unit builders tend to be larger and to have more employees, but their turnover in the business is also very high.

Actually, the NAHB study undoubtedly was biased toward the larger builder. Many establishments are one-man operations. It is not unusual for a builder to subcontract all phases of the work and to have no employees. This is typical in many cities. Small, inconspicuous firms are very difficult and costly for the unions to organize and to police.

A related factor alluded to earlier is the growing geographic dispersion of residential building. Housing construction has moved progressively away from the centers of large cities where the unions are strongest. Few local unions have the resources to pursue hundreds or even thousands of firms scattered throughout an area of hundreds of square miles.

The small local builder can enter the business, subcontract most of his work, and compete effectively with larger firms that may have union contracts. As a result, the region becomes increasingly nonunion as it both expands in area and increases in the number of individual enterpreneurs involved. This dispersion, moreover, must be kept in mind in evaluating estimates of union strength in certain cities. Both union and industry spokesmen may assert that their area is completely union, but the jurisdiction involved may encompass only a portion of the applicable market.

This move from the city is illustrated by what has occurred in Denver, where such movement involves many miles of travel for anyone in center city who wants work in homebuilding. There, single-family homebuilding is virtually all open shop, with an occasional union subcontractor or a national builder who has an agreement with the Carpenters' union. Until recently, Denver has been a boom building city almost continuously since World War II, thus reducing the interest of the unions in pursuing organization and jobs in homebuilding. Even if this interest had been high, however,

---

[23] "Homebuilding's New Look," *Engineering News-Record*, November 7, 1983, pp. 92–99.

[24] Sumichrast et al., *Profile of the Builder* p. 40.

[25] *Ibid.*, p. 26.

the sheer cost of attempting to unionize the myriad of building contractors and subcontractors in the area would have been enormous.

The increasing tendency of the homebuilding sector to go open shop since the early 1950s is also closely related both to the wage escalation in construction and to the difficulties of the unions in adapting their wage policies to fit the needs of this sector. In most areas, the industrial building scale has become much too high for the homebuilder to accept. Construction unions have recognized this in many instances and have established separate, lower rates for the homebuilding sector. Such rates in practice, however, do not work to the advantage of homebuilders. When the labor market is tight, union workers gravitate toward commercial and industrial work where the wages are higher and homebuilders have difficulty obtaining needed labor; when the opposite is true, homebuilders receive a surfeit of employees whose experience in homebuilding is limited and who are dissatisfied with the compensation received. Moreover, the lower union rate is usually much higher than the going rate in homebuilding.

Our interviews also revealed a strong feeling among homebuilders against the employment of personnel who traditionally were employed or trained on union commercial and industrial work. The homebuilders felt that such personnel, besides being unfamiliar with homebuilding construction techniques, were used to working at a more deliberate pace and were too insistent on observing narrow craft jurisdictional lines. Such reactions are perhaps in part merely a reflection of the very strong feeling against unions exhibited by small nonunion contractors. Nevertheless, the reactions indicate another reason why the homebuilding industry has tended to operate more and more on an open shop basis.

Another factor is that homebuilders use their labor across craft lines; for example, carpenters sometimes do cement work and various incidental duties in other trades in order to hasten the completion of their basic job. There is just not enough craft work in small home construction to make employment of specialists financially viable. A literal application of union jurisdictional demarcations is considered by most homebuilders to be an impossible burden for them to bear.

Subcontractors

Subcontracting firms that specialize in housing construction tend to be smaller than their counterparts in other branches of the industry. Many are family arrangements, small partnerships, or proprietorships. In southern and western areas as well as in nonurban areas, subcontractors often farm or do other work on the side. They are paid weekly by the homebuilding contractor with whom they have a close working arrangement, often following him from job to job in many different areas over long periods. They see little to be gained by unionization; indeed they often fear that unionization would jeopardize their working arrangement. The attempts of unions and unionized contractors to have such independent subcontractors legally determined to be employees subject to union jurisdiction have generally been rejected by the federal courts.[26]

The subcontractors that were formerly most likely to be unionized in homebuilding have traditionally been those in the electrical, plumbing, or mechanical trades. Craftsmen in these trades have always been more likely to be trained under union conditions, and nonunion firms have not been as available as in carpentry. For many years, however, the homebuilding sector has gained a surprisingly large number of small, often one- or two-man firms in these trades who specialize in homebuilding and repair and who make a very comfortable living operating on a nonunion or open shop basis. When interviewed, such subcontractors pointed to the continuity of employment and to the satisfaction of self-employment as more than compensating for their former union status.

In addition, the open shop has made enormous strides since the early 1970s in recruiting workers in highly skilled trades. The ABC, as discussed in Chapter IV, is now second only to the National Electrical Contractors Association (NECA) in membership of electrical contractors, and there is a separate open shop electrical contractors' association, the Independent Electrical Contractors, in addition to many open shop NECA and other highly skilled subcontractor association members. Our data in Chapter IV demonstrate that the open shop has made substantial gains in the electrical, mechanical, heating, ventilating, and air conditioning trades as well as in other trades, such as painting.[27] It is not likely, therefore, that homebuilders will find problems in filling their needs in these fields, except in a very few jurisdictions.

In the South, many of the open shop subcontractors specializing

---

[26] For a recent case, see Trustees of Sabine Area Carpenters Health and Welfare Fund v. Don Lightfoot Home Builder, Inc., ____ F. 2d ____ (CA5, May 12, 1983), *reprinted in Construction Labor Report,* Vol. 29 (January 1, 1983), pp. 465–66.

[27] See particularly Table IV-5 and Figures IV-6, IV-9, IV-10, and IV-11 in Chapter IV.

in homebuilding are black; in the Southwest, many are Mexican-American. Nearly all emphasized that discriminatory union policies had influenced their decision to move into homebuilding and to seek self-employment. In many areas individual subcontractor units, including the subcontractor as well as his employees or partners, are often members of a single racial or ethnic group. This enhances their cohesion as a working group, their interdependence, and their disinterest in seeking union membership or support. Some of these minority subcontractors, under the aegis of civil rights programs or contractor association affirmative action programs, have expanded from homebuilding into commercial building.

Homebuilders as a Union Threat

Finally, small-scale residential construction is characterized by a high degree of specialization by contractors.[28] A builder or a subcontractor in single-family housing is much less likely to do a significant amount of work outside his specialty than is a firm which concentrates in other kinds of construction. The result has been that unions have traditionally not looked upon the homebuilder as a direct competitive threat to union standards in other sectors of the industry. To the extent that unions were content to consolidate their strength in nonresidential construction and that housing contractors did not venture into commercial or other building, the unions could afford to acquiesce in a predominantly open shop residential sector. It may have offended their sensibilities, but it did not seem to imperil their existence or compromise their power elsewhere. Large apartment building, on the other hand, is more often done by contractors who can also erect commercial structures, as has already been noted. This work has traditionally been more apt to be located in or near large cities where union strength is concentrated, and it requires a scale of operation frequently beyond the resources and expertise of most homebuilding contractors. For these reasons, the relatively smaller penetration of open shops into apartment construction (which has been found consistently) was to be expected. The extent to which the large structures have been taken over by the open shop is, however, a new development.

The open shop has now moved conclusively into the larger residential structures. This began with garden-type apartments of three

---

[28] See Howard G. Foster, *Manpower in Homebuilding: A Preliminary Analysis*, Manpower and Human Resources Studies, No. 3 (Philadelphia: Industrial Research Unit, The Wharton School, University of Pennsylvania, 1974), Chapter 2, for a more detailed analysis of this point.

stories or less. Such buildings do not require elevators or heavy foundation work; in such work the subcontractors are often highly unionized. Moreover, such apartments spread early on to the suburbs and beyond.

The current interest in town house construction, again usually low-level multiple dwellings, has aided the movement of open shop contractors into apartment building. These homes are being built in cities, suburbs, and exurbia. In the Philadelphia area, for example, they are being constructed in the suburbs and outlying areas by both open shop and union contractors, but the former have made considerable gains. In New York, an open shop contractor from nearby Connecticut built a virtually new town of such residences in the previously union area of Westchester County, fifty miles from the center of Manhattan. In Denver, the open shop first captured at least one-half of all apartment construction, including high rises; then the open shop took over the entire field. In Indianapolis, the open shop had all the low rises (three stories and under); the union shop, the high rises. More recently, however, the open shop has moved into high-rise construction as well. In Texas—where space is not at a premium and where, consequently, high-rise apartments are found only in a few cities (and even there not in large numbers)—open shop contractors build most apartments.

Apartments today are thus being built both by predominantly commercial and industrial contractors and by homebuilders. Unions have seen homebuilders first move into apartment construction and then gravitate into nonresidential work. Since labor used in residential and nonresidential work is interchangeable to some degree, residential construction can, and probably does, provide a pool of manpower for open shop construction. Generally, low-rise multiple-family housing is now almost completely built open shop, as are over 80 percent of the high-rise structures.

In periods of extensive labor surplus, the housing sector is, at least for some unions, theoretically an inviting source of job opportunities for unemployed members. In launching the Coordinated Housing Organizing Program (Operation CHOP) in 1974, the Carpenters' union, which has lost the most from the erosion of union strength in homebuilding, acknowledged the "growing cancer" in the building trades:

> We must invade the housing construction industry's tremendous pool of nonunionism, and we must do the organizing job which is so necessary and vital to our survival. The greater the degree of organization which we accomplish in the construction industry overall, the greater the impact of our Brotherhood, its district councils and its local

> unions. The stronger we become, the better we will be able to secure job protection and benefits for our members.[29]

No discernible results came of this effort, but periodically the Carpenters' union has attempted, again without any noticeable success, to enroll the employees of homebuilders, or perhaps more accurately, to strike a bargain with the homebuilders themselves. In fact, as has already been pointed out, several key chapters of the NAHB that negotiated with the Carpenters in 1974, including those in northern Ohio and in Arizona, as well as some in California, no longer do so because most or all of their members have become open shop. When union members work for open shop companies, they usually do so under open shop conditions both in homebuilding and home improvement.

Perhaps because of this, the then president of the Carpenters' union called on his members in 1981 to improve the "scandalous conditions which exist in so many parts of the home improvement industry." [30] He further wrote:

> Occasionally, when work is slow, a commercial or residential carpenter will take on an odd home improvement job—add a room to a house, install a carpet, add closet space. . . . We expect such a member to abide by his long-standing obligation to the Brotherhood to use every honorable means to produce employment—no more, no less.[31]

Our field investigations revealed, however, few attempts of individual carpenters to promote their union in any aspect of residential construction.

In early 1982, the Carpenters and the International Brotherhood of Painters and Allied Trades signed an agreement covering on-site erection of prefabricated homes by the United Homebuilders of America, a Los Angeles firm which is 50 percent financed by a union-management pension fund. The fund also finances the homes to be built.[32] As we shall discuss in Chapter VIII, the use of pension funds to generate union jobs in homebuilding has a high union priority, but has yet to provide a significant number of jobs.

Finally, it must be borne in mind that residential construction is not merely an appendage to the industry; in recent years it has

[29] "Carpenters Launch Major Organizing Drive for Residential Construction." Construction Labor Report, No. 951 (January 2, 1974), quoting Carpenters' President William Sidell, p. 9.

[30] "Home Improvement Industry Scandals Must Be Exposed, Carpenters' President Says," Construction Labor Report, No. 1334 (June 24, 1981), quoting Carpenters' President William Sidell, p. 4-6.

[31] *Ibid.*, p. A-7.

[32] "Housing Breakthrough? National Labor Pact, Financing Set for Prefabricated Projects," *Engineering News-Record,* February 25, 1982, p. 67.

constituted approximately two-fifths of all building work. The preponderance of residential building, moreover, takes the form of single-family homes. The point here is that even though the open shop has been concentrated in small projects, and the average homebuilder constructs less than ten or twenty houses per year, these projects in the aggregate account for a substantial amount of expenditure and, concomitantly, a significant number of jobs. The open shop sector had over 90 percent of the residential work in 1983; it would, therefore, need less than one-quarter of non-residential work to capture the lion's share of all construction. As we shall demonstrate in the next three chapters, the open shop has taken over far more than this narrow margin.

NATURE OF UNION REPRESENTATION

Union strength in homebuilding today is concentrated in California and in isolated cities. It remains strong in the New York City metropolitan area, in the Chicago Standard Metropolitan Statistical Area and other areas of Illinois, in parts of California, and in several cities where unionization is strong. Its strength in California towns and cities, as formerly in those of Arizona, derived from the tremendous population explosion in those areas and from the advent of large builders who have mass-produced houses. Such builders found it advantageous to deal with unions; these unions, in turn, have organized other builders.

In some smaller cities, such as Peoria, Illinois, where factory workers are highly unionized, the construction unions remain strong and also have been able to maintain a grip on the homebuilding labor force. Nevertheless, in many other cities which have industrialized workforces—e.g., Louisville, Kentucky, Midland, Michigan, or Indianapolis—this is not the case. There seems to be no rationale for these variations. Likewise, New York, Chicago, and San Francisco unions have managed to maintain some hold on residential construction in their suburban areas, although this is definitely weakening. This is not the case in the Philadelphia or Boston areas, or indeed in most other metropolitan regions.

Variations in the Nature of Union Representation

Characterizing the residential construction industry in a locality as either "union" or "nonunion" tends to obscure several intermediate situations which may exist in different areas. The extent and impact of unionization in residential labor markets varies along a spectrum. At one end, both builders and subcontractors operate under contracts with the building trades unions which are based upon, or identical to, those negotiated in other sectors of the industry. At the other end, where most areas are situated, there is virtually no union representation among either builders or subcontractors. In between, some builders may be organized or some subcontractors may be organized, but in either case the residential work standards may differ from those in commercial building construction.

A few examples will illustrate the diverse forms of organization alluded to here. As already noted, in a very few areas, unions in the basic trades negotiate directly with the local chapter of the NAHB, and builders may be bound by the terms of the agreement by virtue of their membership in the association. In some cases the agreements provide for somewhat different conditions from those in nonresidential construction: wage rates may be lower and/or certain work rules relaxed. In other cases there are no differences; in Cleveland, for example, the contracts negotiated with the commercial employers' association were applied in their entirety to the homebuilders, but the homebuilders themselves were not directly represented in the bargaining process. The homebuilders found the results onerous and quit the arrangement as more and more went open shop. Today, a few homebuilders that operate in highly unionized northern Ohio areas remain union but there is no association bargaining.

In other areas, a builder may also adhere to a union contract to satisfy customers. In San Francisco and Northern California, the carpenters usually negotiate concurrently with the AGC for commercial work and with the Building Industry Association for residential construction. The negotiations, which cover a wide geographic area, have featured union tactics that frequently play off one association against the other and have forced commercial rates on the homebuilders, who, in 1981, employed 8,000 to 10,000 carpenters.[33] One result of this, of course, has been the already noted growth of the open shop in the suburban and outlying areas around the San Francisco Bay region.

In other areas where homebuilding is virtually all open shop, there still may be one or two national builders who maintain agreements with the Carpenters' union. Such agreements may permit the builders to utilize nonunion labor if the union cannot supply union employees, and also apparently do not require unionized sub-

[33] "Striking Carpenters, IBEW in N. Calif. Reach Tentative Agreements with Home Builders, NECA," *Construction Labor Report,* No. 1334 (June 24, 1981), pp. A-14-16.

contractors or laborers. Again, it is difficult to ascertain how common such arrangements are, but they cannot be widespread given the small coverage of unionism in the industry.

CONCLUSION

The homebuilding sector is the least unionized of any construction sector. The small size of homebuilding firms, the movement to suburbia and exurbia, and the inability of the homebuilding sector to operate under union wages and conditions have all contributed to a thirty-year trend of declining unionism in this sector. It is difficult to envision a reversal of this situation in the near future.

Multiple-family dwelling construction is somewhat more unionized, but a differentiation must be recognized between low- and high-rise structures. The former are analogous to family residential buildings and are being built overwhelmingly by open shop residential and commercial contractors.

High-rise apartments and condominiums were heavily union built until recently. The spread of such structures outside of the unionized cities; the increasing capability of open shop residential and commercial contractors to handle such large projects, as will be described in Chapter IV; and the sharp increases in union wage costs have all contributed in recent years to the ability of open shop contractors to capture the bulk of high-rise residential building. This in turn has aided open shop contractors in taking over an increasing share of commercial work, again as will be described in Chapter IV.

The unionized residential builders have had some of their best opportunities for obtaining work when the federal government has built subsidized housing under Davis-Bacon conditions. Now, however, the changes in Davis-Bacon administrative rules, as discussed in Chapter II, are likely to reduce substantially the advantages which the unions have had, thus contributing to a further deterioration of the union position in residential construction.

Union residential builders are thus confined to a few cities and areas as the open shop has solidified its hold on residential construction. By controlling well over 90 percent of all residential construction, which accounts for about 40 percent of all construction, the open shop is virtually assured of dominating more than a majority of all construction.

CHAPTER IV

Commercial Construction

Commercial construction is a heterogeneous mix, with projects varying greatly in size, type of building, and value. The next chapter will discuss the union status in the construction of some of the largest industrial structures, such as oil refineries, chemical plants, and power generating facilities. This chapter explores union and open shop involvement in the remaining components of the commercial sector, which cover a broad range, including such diverse buildings as retail stores, warehouses, factories, office buildings, nursing homes, hospitals, shopping centers, schools, and post offices. These projects may each represent expenditures from a few thousand to tens of millions of dollars.

There is little doubt that commercial construction is still less hospitable to the incursions of the open shop contractor than residential building, as the studies described in the previous chapter have shown. The relative persistence of union strength in the commercial sector is attributable, in part, to factors of size and location. Since many of these projects represent large concentrations of employees, unions have the incentive to expend effort and resources to keep them organized; and since they are conspicuous because of their size, it is less likely that they will escape the unions' attention. Moreover, commercial, more than residential buildings, traditionally have been in or near metropolitan centers where the unions' economic and political strength is greatest.

It should not be inferred, however, that union control over commercial building was once complete. Smaller structures like service stations, individual stores, small warehouses, or modest professional offices are especially prone to open shop competition even in the most tightly organized locality. Nonetheless, it is not with the small job that the significant challenge to the union lies, but rather with the spread of open shop competition into the market for buildings in the range of several million dollars and up. As far back as 1972, union leaders were expressing concern over this problem. One local leader of the Sheet Metal Workers' International Association (SMW), in an unusually frank communication to the membership entitled "The Honeymoon is Over," described it this way:

> In Syracuse they started with a 4½ million dollar Shopping Center. The Building Trades took them on and made a full scale effort to get the job switched to Union. The Trades put pickets on the project for 9 months. When injunctions were issued another Trade would proceed to picket. Unfortunately the Building Trades did not remain strong and eventually some unions even went in there to work. Needless to say, the project was completed without going too far beyond the anticipated completion date. As a matter of fact the job went so well A.B.C. went to the other side of Syracuse and built a 6½ million dollar Shopping Center and this job went that much easier. They use each completed project as advertisement and leverage to gain the next. They are presently involved in a Students Housing project on one of the Universities in the Syracuse area. The final estimate of this project will exceed 30 million dollars. You can see where it would be relatively easy for them to convince an owner to have his construction done by them especially when they have built up such an impressive track record.[1]

In a similar vein, the Iron Workers Employers Association of Washington, D.C., commented in mid-1974 in an "Open Letter to All Members of Local #5" of the International Association of Bridge, Structural and Ornamental Iron Workers:

> The total construction volume in the United States last year was 109 billion dollars. Of that total, members of the Associated Builders and Contractors, commonly called the ABC, did 45 billion. Most of you know that ABC is a strong, growing open-shop group. When you add to this the large volume of housing and other residential work done by non-union contractors not affiliated with any organization, you can quickly see that the building business in the United States is fast becoming predominantly non-union.
>
> In the Washington, D.C. area the situation is worse. Statistics prove that union construction represents a smaller percentage every year. Only the Davis-Bacon Act with its artificial price supports on government work keeps union construction alive.
>
> There are many reasons for the growth of non-union companies. Practically all of these relate to the higher cost and the lack of productivity union contractors are experiencing. Much has been said about the cost of living. Figures available from your International will show that the rate of pay for Local #5 members over the last ten years has increased twice as fast as the cost of living, even taking into account the inflation of the last year.
>
> Some costs are not tied directly into wages. The recent increases in Workmens Compensation for the District of Columbia extend to Maryland and Virginia for union contractors only. The recent rate increases initiated by the insurance companies because of the increased benefits have cost companies erecting structural steel the equivalent

[1] Communication from Henry Landau, president-business manager, Local #83, Sheet Metal Workers' International Association, Albany, New York, to membership, October 5, 1972, in author's possession.

> of a raise of $1.40 per hour for every employee. This one item alone is turning more and more of your work in the suburbs to the nonunion companies because we cannot compete.[2]

This description of the union situation in Syracuse, New York, and Washington, D.C., could be applied, changing only names and details, to many areas. To be sure, union control remains tight in many metropolitan areas, but outside of the city limits the proportion of open shop commercial construction tends to rise. Since a considerable number of major construction projects (such as office buildings, warehouses, shopping centers, and industrial plants) have moved outside the cities, the opportunities for the open shop contractor have grown substantially.

Nearly everyone interviewed in 1974 and 1982, including both union and open shop contractors, dated the big jump in open shop construction in the commercial sector to the period of wage inflation in the construction industry which began in the late 1960s. Prior to this time, the dispersion of people and jobs to suburbia and exurbia, as well as the growing industrialization of the South and West, aided open shop contractors in establishing themselves. It was the increased building costs of the late 1960s and early 1970s, however, that gave the open shop its big economic advantage and consequent opportunity. Moreover, as "The Honeymoon is Over" clearly showed, each opportunity created another, especially as union construction wages continued to inflate. A successful $200,000 job puts a contractor in a position to bid on a $500,000 one; when he completes that, he can raise his sights to $750,000 and then perhaps double that, etc. The larger the job, the larger becomes the capability and the better the track record to sell. With users looking for cost relief and with a growing cost gap between union and open shop, both the size of projects and the number of projects for the open shop contractor in the commercial sector have grown substantially since 1968, and this growth has accelerated in more recent years.

EARLY STUDIES

Prior to the publication of our first edition of *Open Shop Construction* in 1975, there were only two comprehensive efforts to

[2] Letter in author's possession, not dated, but obviously written in mid-1974. The $45 billion reference is clearly hyperbole. For an analysis of Davis-Bacon impact, see Armand J. Thieblot, Jr., *The Davis-Bacon Act,* Labor Relations and Public Policy Series, Report No. 10 (Philadelphia: Industrial Research Unit, The Wharton School, University of Pennsylvania, 1975). Professor Thieblot is currently revising and updating this study for 1984 publication.

gauge the magnitude of open shop activity in commercial and industrial construction. The 1936 U.S. Bureau of Labor Statistics (BLS) survey put the proportion of union workers in nonresidential building construction at 72.2 percent.[3] The later study, by Haber and Levinson, offered no precise estimate, but nevertheless concluded: "There is no doubt that the strength of union organization in all branches of the industry was considerably greater in the postwar period than it had been in 1936. In 1950, union membership was approximately three times its 1936–1937 level, whereas the number of persons employed in contract construction had somewhat more than doubled."[4] These same authors surveyed sixteen cities in 1952 and put commercial work at "close to 100 percent" union. Although the overall focus of their study was on the residential sector, they do not specify whether there were open shop contractors among their sources of information. When considering the extent of union commercial work, the Haber and Levinson study ignored commercial construction in small communities, as well as in the South, since many of these areas did not appear to be unionized to any degree at the time of their limited survey. Certainly this is the recollection of many persons whom we interviewed in 1974 and 1982.

RECENT NATIONAL STUDIES

Since the publication of our original edition of *Open Shop Construction,* there have been attempts by the U.S. Bureau of Labor Statistics (BLS), the U.S. Department of Housing and Urban Development in collaboration with the Massachusetts Institute of Technology (Bourdon-Levitt), and the Construction Labor Research Council (CLRC) to estimate the relative position and size of open and union shop construction. We shall examine these studies in this section. In addition, a number of studies have been made, especially by contractor groups, or contractor-sponsored research organizations, either on a regional or a national craft basis. We shall examine these surveys in the following sections. The final section will discuss our own questionnaire surveys of 1974 and 1982. Throughout the discussions we shall add information based upon our nationwide field interviews of 1974 and 1982.

---

[3] Edward P. Sanford, "Wage Rates and Hours of Labor in the Building Trades," *Monthly Labor Review,* Vol. 45 (August 1937), p. 297.

[4] William Haber and Harold M. Levinson, *Labor Relations and Productivity in the Building Trades* (Ann Arbor, Mich.: Bureau of Industrial Relations, University of Michigan, 1956), p. 35.

U.S. Bureau of Labor Statistics and the Construction Labor Research Council

In 1972, 1973, and 1977, the BLS surveyed the construction industry in seventeen major cities. Establishments included in its survey were limited to those with eight employees or more. Table IV-1 shows the percentage of union construction workers in city areas surveyed in 1973 and 1977 and the percentage change during the four years between 1973 and 1977. These data undoubtedly include industrial construction, and perhaps highway and heavy as well, but are not likely to include many residential contractors because of their small employee complement. It is believed, however, that most of those companies surveyed fit within our broad category of commercial builders.

The data do contain some interesting information. New York, Cleveland, Chicago, and San Francisco remained predominantly union construction cities while Washington, Baltimore, Miami, Atlanta, and Dallas-Fort Worth continued to move toward open shop. A union decline of 15 percent in the Philadelphia area reflects the expansion of the open shop in the suburbs, since the inner city

TABLE IV-1
Percentage of Union Construction Workers
Selected Cities

	March 1973	March 1977	Change
17 Area Average	75–79	65–69	−10
Boston	75–79	70–74	− 5
Long Island	80–84	85–89	+ 5
Newark	80–84	70–74	−10
New York	95+	85–89	−10
Philadelphia	80–84	65–69	−15
Atlanta	45–49	35–39	−10
Baltimore	35–39	30–34	− 5
Dallas-Ft. Worth	30–34	20–24	−10
Miami	55–59	45–49	−10
Washington	40–44	25–29	−15
Chicago	95+	95+	—
Cleveland	95+	90–94	− 5
Detroit	90–94	95+	+ 5
Denver	70–74	70–74	—
Los Angeles	90–94	90–94	—
San Francisco	95+	95+	—
Seattle	90–94	90–94	—

Sources: *Industry Wage Survey: Contract Construction, September 1973,* Bulletin 1911 (Washington, D.C.: Government Printing Office, 1976); *Compensation in the Construction Industries in Large Metropolitan Areas,* Report 616 (Springfield, Va.: National Technical Information Service, 1980).

FIGURE IV-1
Construction Industry
Union Employment As a Percentage of Total

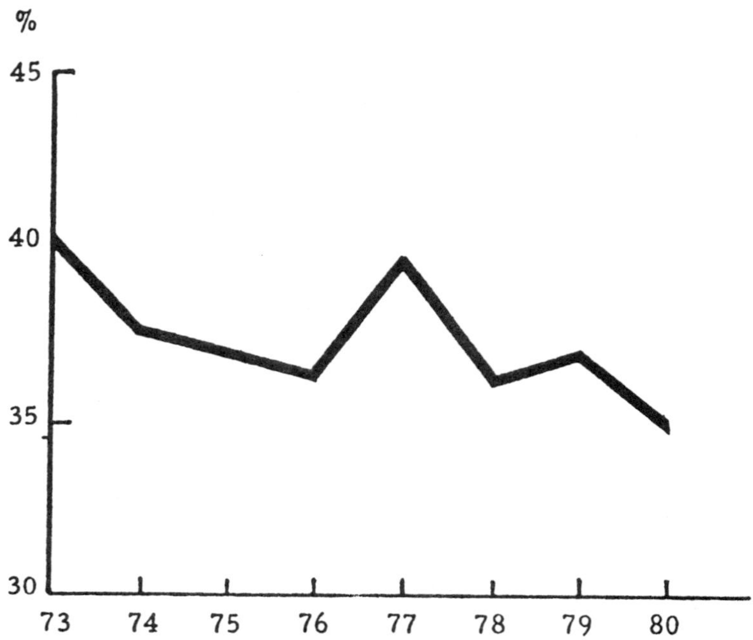

Source: U.S. Bureau of Census, Current Population Survey, questions supplied by U.S. Bureau of Labor Statistics, May 1973–1980; compiled and charted by Construction Labor Research Council.

remains heavily union. Overall, the average decline in the percentages of union construction workers for the seventeen cities was approximately 10 percent between 1973 and 1977.

A second survey has been conducted between 1973 and 1980 by the U.S. Bureau of Census monthly Current Population Survey (CPS) using supplemental questions supplied by the BLS.[5] The CLRC, a research organization sponsored as a data source by construction associations, contractors, and users, has charted the trends in union employment using CPS data. These trends are set forth in Figures IV-1 and IV-2.

The CPS data contain a few shortcomings which detract somewhat from their accuracy. For example, they include such nonconstruc-

[5] The purpose of the CPS is to determine employment and unemployment on a monthly basis. The May survey each year carries additional questions including, since 1973, union representation.

FIGURE IV-2
Construction Employment, Union & Nonunion

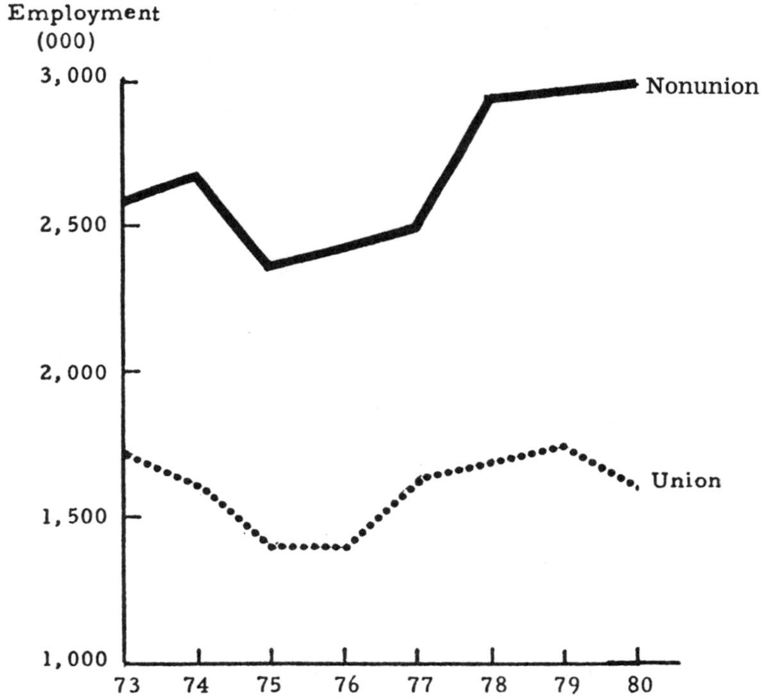

Source: See Figure IV-1.

tion, largely nonunion employees as administrators and clerical workers. This overstates the percentage of nonunion employees at any given time. It would not, however, be likely to affect the trend of union and nonunion employment. More important is the fact that:

> the data from 1977 on are not strictly comparable to those of earlier years because of a change in the wording of the section of the questionnaire concerning union membership. The effect of the change was to increase the number of workers whose response was entered on the union side of the tally. While the magnitude of this difference has not been measured by BLS, the jump in union employees in 1977 and the figures for subsequent years are, therefore, an overstatement relative to the 1973 through 1976 data. The decline in union employment from 40 to 35 percent may then be conservative.[6]

[6] "Union/Non-union Employment Trends in Construction," *Construction Labor Research Council,* April 1981, p. 3.

Figure IV-3 presents the CPS data for 1977 and 1980 on a regional basis. The West North Central states, which according to the data were less than 40 percent union in 1980, showed a slight union gain between 1977 and 1980; the South Atlantic states were found to be about 20 percent union in both years; all other regions showed a clearly discernible decline in union construction between the two years. The Pacific and East North Central were the only regions over 50 percent union in 1977, but both these percentages fell appreciably by 1980.

A conservative reading of Table IV-1 and Figures IV-1, IV-2, and IV-3 shows that union construction lost almost 250,000 jobs between 1973 and 1980. As the CLRC notes:

> While total construction industry employment has increased from about 4.3 million in 1973 to under 4.6 million in 1980, union employment declined 125,000. If union members had retained 1973's 40 percent market share of employment in 1980, rather than dropping to 35 percent, there would have been 237,000 more union jobs . . . [in 1980].[7]

A third series of studies by the BLS, *Selected Earnings and Demographic Characteristics of Union Members, 1970* and *Earnings and Other Characteristics of Organized Workers, May 1977* and *May 1980,* also documents the decline of unionism in construction during the 1970s.[8] Data in the reports divided skilled workers in construction into two categories, "carpenters" and "construction craft workers except carpenters." Figure IV-4 shows that the percentage of unionized carpenters declined over the ten-year period from 44.5 percent to 31.3 percent. It also shows that the other skilled construction crafts, the group with the higher degree of unionization, declined from 55.5 percent to 46.4 percent.[9]

The 1970 and the later BLS studies define union representation differently. The 1970 study used figures for union membership; the later two for union membership *plus* representation.[10] This makes a difference particularly in the twenty states which outlaw com-

[7] *Ibid.,* p. 1.

[8] *Selected Earnings and Demographic Characteristics of Union Members, 1970,* Report No. 417 (Washington, D.C.: Government Printing Office, 1972); *Earnings and Other Characteristics of Organized Workers, May 1977,* Report No. 556 (Washington, D.C.: Government Printing Office, 1979); *Earnings and Other Characteristics of Organized Workers, May 1980,* Bulletin No. 2105 (Washington, D.C.: Government Printing Office, 1981).

[9] "Construction Industry Worker Characteristics," *Construction Labor Research Council,* July 1979, p. 1.

[10] *Ibid.*

FIGURE IV-3
Percentage of Union Representation in Construction
U.S. and Regions[a]

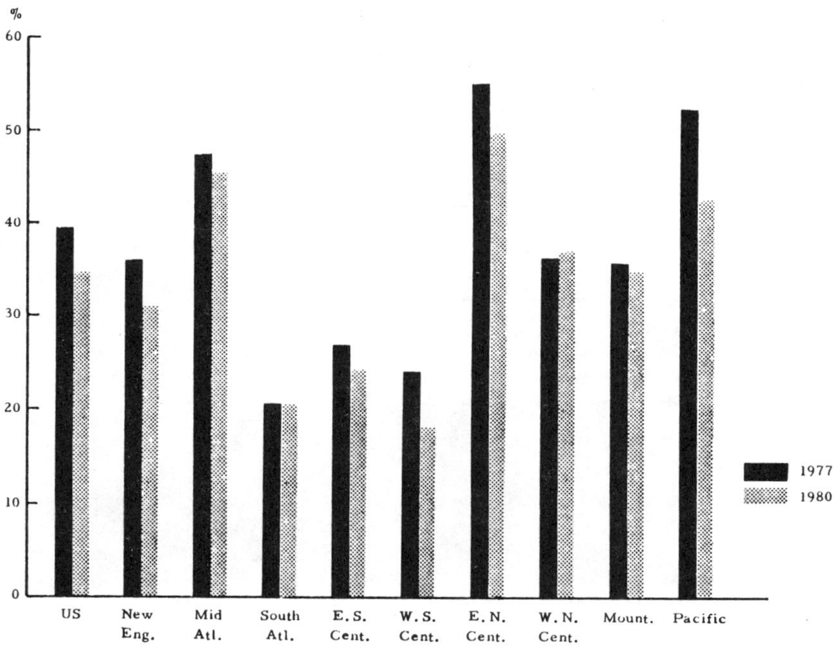

Source: U.S. Bureau of Census, Current Population Survey; U.S. Bureau of Labor Statistics listings from the May supplemental questions; charted by the Construction Labor Research Council.

[a] The nine regions are composed of the following states:
New England = Maine, New Hampshire, Vermont, Massachusetts, Rhode Island, and Connecticut; Middle Atlantic = New York, New Jersey, and Pennsylvania; E.N. Central = Ohio, Indiana, Illinois, Michigan, and Wisconsin; W.N. Central = Minnesota, Iowa, Missouri, North Dakota, South Dakota, Nebraska, and Kansas; South Atlantic = Delaware, Maryland, District of Columbia, Virginia, West Virginia, North Carolina, South Carolina, Georgia, and Florida; E.S. Central = Kentucky, Tennessee, Alabama, and Mississippi; W.S. Central = Arkansas, Louisiana, Oklahoma, and Texas; Mountain = Montana, Idaho, Wyoming, Colorado, New Mexico, Arizona, Utah, and Nevada; Pacific = Washington, Oregon, California, Alaska, and Hawaii.

FIGURE IV-4

Percentage of Union Represented Construction Workers[a]
Carpenters and Other Skilled Craftsmen
1970, 1977, and 1980

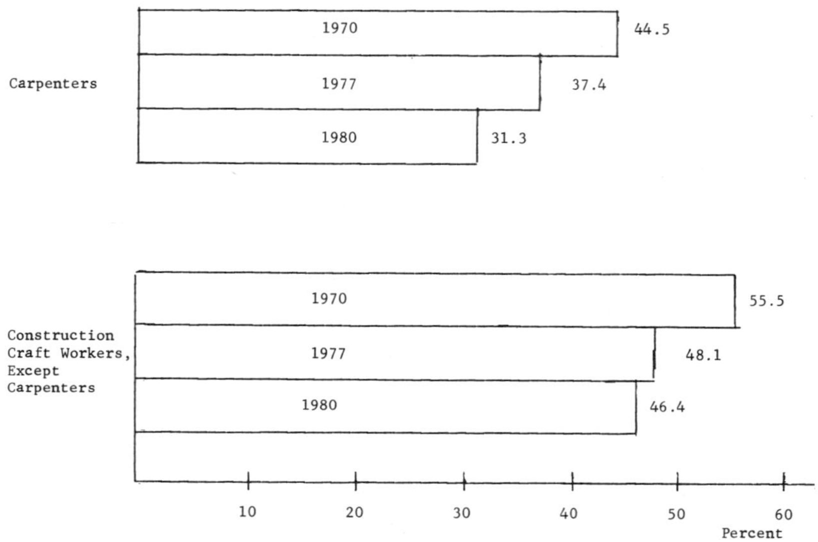

Source: *U.S. Bureau of Labor Statistics, Selected Earnings and Demographic Characteristics of Union Members, 1970,* Report 417 (Washington, D.C.: Government Printing Office, 1972) Table 3, p. 9; *Earnings and other Characteristics of Organized Workers, May 1977,* Report 556 (Washington, D.C.: Government Printing Office, 1979), Table 7, p. 25; and *Earnings and Other Characteristics of Organized Workers, May 1980,* Bulletin 2105 (Washington, D.C.: Government Printing Office, 1981), Table 7, p. 27.

[a] The 1970 data include union members only; the 1977 and 1980 data, union members and union represented employees. Hence, the decline in unionism is somewhat understated.

pulsory unionism,[11] where representation of nonmembers does occur. Therefore, the data in Figure IV-4 probably understate union losses.

U. S. Department of Housing and Urban Development-MIT

In 1976, the U.S. Department of Housing and Urban Development contracted with the Department of Civil Engineering, Massachusetts Institute of Technology (MIT), to make a study of union and

[11] Alabama, Arizona, Arkansas, Florida, Georgia, Iowa, Kansas, Louisiana, Mississippi, Nebraska, Nevada, North Carolina, North Dakota, South Carolina, South Dakota, Tennessee, Texas, Utah, Virginia, and Wyoming.

open shop construction.[12] A few years later the authors of that research report, Clinton C. Bourdon and Raymond E. Levitt, wrote a book summarizing their findings which are, at least in part, quite different from all the other evidence that we have uncovered.[13]

Bourdon and Levitt acknowledge that open shop contractors have gained considerably, pointing out that in 1979 they were "well distributed" among the 400 largest firms on the *Engineering News-Record* list.[14] They emphasize, however, that the largest open shop firms are headquartered and principally work in the South and that the average open shop firm is likely to be smaller than its union shop counterpart.[15]

Bourdon and Levitt also correctly note that residential construction is heavily nonunion and that many homebuilders have moved into commercial construction.[16] Beyond that, they find that unions still dominate commercial construction in the cities which they investigated. Table IV-2 summarizes their data for eight metropolitan areas. In none of these areas is the percentage of open shop firms or employment (in residential and highway and heavy construction, as well as commercial construction) in excess of 50 percent.

The Bourdon-Levitt data in Table IV-2, which are for 1976, are not supported by the data from BLS studies of 1973 and 1977 which appear in Table IV-1. Table IV-3 compares these studies for the four cities which are included on both. The BLS data include the entire standard Metropolitan Statistical Area (i.e., the cities and their contiguous areas) and Bourdon-Levitt claimed to cover these same limits. It does appear, however, that for three of the four cities, the open shop is quite underestimated by Bourdon-Levitt. According to all other evidence, Baltimore was open shop-dominated by the mid-1970s, and the situation in Atlanta is believed to be properly reflected by the BLS data. The Atlanta situation is discussed at length below, with data showing the open shop trend. A 1978 study of the commercial construction scene in Baltimore noted:

> All interviewees acknowledged that in the Baltimore area within the present decade a dramatic increase in the size of the open-shop sector

---

[12] Clinton C. Bourdon and Raymond E. Levitt, *A Comparison of Wages and Labor Management Practices in Union and Nonunion Construction,* HUD Contract No. H. 2327-R (Cambridge, Mass: MIT Department of Civil Engineering, June 1, 1978).

[13] Clinton C. Bourdon and Raymond E. Levitt, *Union and Open-Shop Construction* (Lexington, Mass.: D.C. Health and Company, Lexington Books, 1980).

[14] *The Engineering News-Record* publishes its "400" list every year. We shall analyze those data in Chapter V.

[15] Bourdon and Levitt, *Union and Open-Shop Construction,* pp. 12-19.

[16] *Ibid.*

TABLE IV-2
Union and Open Shop Employment Distribution by Metropolitan Area and Product Market 1976 Survey

	Product Market (percent)					Percent Nonunion By Area	
	Total Firms	Residential	Commercial Building	Heavy and Highway	Total Employment	Firms	Employment
Boston						28	15
Union							
General contractor	257						
Subcontractor	276	11	72	17	12,980		
Nonunion							
General contractor	107						
Subcontractor	116	55	42	3	2,291		
Baltimore						43	47
Union							
General contractor	179						
Subcontractor	216	2	63	35	4,246		
Nonunion							
General Contractor	151						
Subcontractor	144	49	46	5	3,685		
Grand Rapids						37	20
Union							
General contractor	80						
Subcontractor	77	1	48	51	3,136		
Nonunion							
General contractor	48						
Subcontractor	44	48	50	2	796		

Kansas City						12	2
Union							
General contractor	68						
Subcontractor	146	7	61	32	6,624		
Nonunion							
General contractor	11						
Subcontractor	17	68	32	0	92		
Atlanta						38	49
Union							
General contractor	130						
Subcontractor	177	1	91	8	5,884		
Nonunion							
General contractor	101						
Subcontractor	88	73	24	3	5,761		
New Orleans						41	36
Union							
General contractor	82						
Subcontractor	87	1	92	7	7,162		
Nonunion							
General contractor	69						
Subcontractor	50	16	82	2	4,018		
Denver/Boulder						36	35
Union							
General contractor	157						
Subcontractor	215	6	73	21	6,411		
Nonunion							
General contractor	110						
Subcontractor	98	67	30	3	3,464		

TABLE IV-2—Continued

	Total Firms	Product Market (percent)			Total Employment	Percent Nonunion By Area	
		Residential	Commercial Building	Heavy and Highway		Firms	Employment
Portland						25	10
Union							
General contractor	167						
Subcontractor	200	11	72	17	7,224		
Nonunion							
General contractor	69						
Subcontractor	55	66	24	10	784		

Source: Clinton C. Bourdon and Raymond E. Levitt, *Union and Open-Shop Construction*, Lexington Books (Lexington, Mass.: D.C. Heath and Company, 1980), pp. 20–21.

TABLE IV-3
Percentage of Union Construction Workers, Four Cities Comparison of U.S. Bureau of Labor Statistics and Bourdon-Levitt Findings

City	U.S. Bureau of Labor Statistics 1973	1977	Bourdon-Levitt 1976
Boston	75–79	70–74	85
Baltimore	35–39	30–34	53
Atlanta	45–49	35–39	51
Denver	70–74	70–74	65

Source: Tables IV-1 and IV-2.

> had taken place at the expense of the closed-shop sector. Even the trade unions have recently recognized the magnitude of this expansion. . . . The vice-president of the Maryland chapter of the AGC came closest to citing numbers. As opposed to fifteen years ago when there were only small open-shoppers, he claims that today over eighty percent of the work under his chapter is open- or mixed-shop.[17]

Boston was already seeing its union strength decline in the mid-1970s. As for Denver, we believe that varying union construction employment at large office centers during the two BLS and the Bourdon-Levitt studies may be responsible for the lower figure found in 1976, or this figure may simply reflect selective sampling in the Bourdon-Levitt study. Denver office construction has been the last big union stronghold in Colorado, as will be noted below.

The Bourdon-Levitt study claimed that open shop gains have been great but cannot be measured and that unionized builders still dominated commercial construction in the mid-1970s. The authors clearly imply that open shop construction does not have the economic advantages which we have found and which are attested to by the growth of the open shop since the late 1960s. They also attempt to prove that differences in the employment, deployment, and handling of the workforce do not vary substantially between open and union shop contractors. We shall examine this claim in Chapter IX.

It is, of course, true in a narrowly literal sense that the gains of the open shop cannot be precisely measured at any one time. Nevertheless, the swing to the open shop has been of such magnitude over the years, as has been supported by so much evidence, that it is difficult to believe that union shop employment was as great a

[17] Robert D. Cochrane, "Open- vs. Closed-Shop: The Construction Industry in Baltimore," *Harvard College Economists,* Vol. 2, No. 2 (1978), p. 35.

percentage of the total in 1976, even for most of the cities listed in Table IV-2, as Bourdon and Levitt claim. The BLS data shown in Table IV-1 and Figures IV-1, IV-2, and IV-3 make the Bourdon-Levitt findings of questionable validity. It is, moreover, noteworthy that the Bourdon-Levitt study ignores the local and regional studies which we cited in the first edition of *Open Shop Construction*. Whatever the merit of these studies individually—and, as discussed in a later section of this chapter, many have obvious limitations—the fact remains that all point in the same direction to the growth and increasing dominance of the open shop in commercial construction. We must conclude, therefore, that the Bourdon-Levitt data are not as representative for the mid-1970s period as are the BLS data and other documentation.

GENERAL CONTRACTOR ASSOCIATION NATIONAL STUDIES

Other than the studies just noted, the only national studies of the open shop-union shop relative market share have been either done by, or sponsored by, contractor associations. In addition, contractor and union organizations have made or sponsored a number of regional and city studies, as have two academics. These will all be analyzed in this and the following sections of the chapter since, although these studies do not always distinguish commercial construction from other sectors of the industry, they are largely directed to that branch.

Associated General Contractors of America

For several years, questionnaires sent by the national office of the Associated General Contractors of America (AGC) to its membership have included questions relating to the union, open shop, or doublebreasted status of the contractor members. Through the courtesy of the AGC national staff, the Industrial Research Unit obtained data from these surveys for the years 1969 and 1973. It was felt that a comparison of these years would be particularly useful since it was in this period that our field interviews indicated that the open shop sector experienced considerable growth.

The 1969 and 1973 AGC Surveys. The results of the AGC surveys must be evaluated with some caution for several reasons. First, the AGC does not include all general contractors within its membership, although at the time of these surveys it was the largest of all contractor groups. It is also true that smaller firms, which are more likely to be open shop than larger outfits, are more likely to belong to the Associated Builders and Contractors (ABC) or to no association than to the AGC. In addition, some AGC chapters make no provision for nonunion members, and until recently even required that all members give the association power of attorney to negotiate for them in labor relations matters. Where such AGC chapters exist, a contractor who wished to operate open shop would of necessity eschew membership in the AGC. (In recent years, however, a separate AGC district or subchapter has been established in some states [for example, Louisiana] for nonunion builders, or those who decline to permit the association to bargain for them, where the original AGC chapter requires bargaining authority.) These factors may well cause an understatement of the percentage of open shop contractors as surveyed by the AGC.

Two other caveats suggest the possibility of a response bias. The 1969 survey elicited a total of 3,044 responses; the 1973 survey, 5,938 responses. These numbers represent approximately 38 and 72 percent, respectively, of the total AGC membership in the two years covered. It is generally believed that smaller firms, with meager managerial and administrative staff, are less likely to respond to surveys and that these firms are more apt to be nonunion. If this reasoning is accurate, the effect will be to understate open shop activity. In addition, the much larger proportional response in 1973 could mean that open shop members of the AGC, by then accorded much greater recognition in the association, were less reluctant to state their status; or that being on the defensive, contractors operating union were anxious to assure a good response for their views.

The data, however, do offer some clues as to the relative strength of the unions in commercial building among the various regions of the country in the years 1969 and 1973, as well as to trends pertaining to the growth of nonunion competition. A contractor is defined as a commercial builder if he reported any volume in that category during 1969 or 1973. Data on total volume, however, are limited specifically to that work which falls within the commercial building classification. Table IV-4 summarizes the union / nonunion breakdown by number of contractors as well as by dollar volume. (Dollar volume data were obtained only for the 1969 survey.)

There are two features of the findings in the table that merit mention. First, the breakdowns show the familiar pattern of union strength in the northern and western states with substantial open shop activity in the South. Second, and perhaps most significantly, the change from 1969 to 1973 in the direction of a substantially greater open shop presence is dramatic. (Unfortunately, the dollar volume figures for 1973 which would enable us to assess the change

TABLE IV-4

Union Status of AGC Members Engaged in Commercial Building Construction, by Region, 1969 and 1973

Region	1969 Number of Contractors by Dollar Volume Percent Union	1969 Number of Contractors			1973 Number of Contractors		
		Union	Nonunion	Percent Union	Union	Nonunion	Percent Union
New England	58.7	153	34	81.8	158	74	68.1
Middle Atlantic	98.4	141	3	97.9	342	45	88.4
E.N. Central	99.4	367	11	97.1	536	47	91.9
W.N. Central	82.7	165	84	66.3	275	183	60.0
S. Atlantic	57.9	107	178	37.5	159	496	24.3
E.S. Central	20.1	78	49	61.4	80	136	37.0
W.S. Central	50.8	159	49	76.4	199	216	48.0
Mountain	98.6	108	21	83.7	169	67	71.6
Pacific	95.3	219	13	94.4	409	49	89.3
Total United States	72.2	1,497	442	77.2	2,327	1,313	63.9

Source: Associated General Contractors of America.

in greater depth are not available.) The reduction in the percentage of contractors who operated exclusively union is across-the-board, although especially marked in the South. It must also be noted, however, that in 1973 some chapters of the AGC were just beginning to admit open shop members and that the open shop membership percentages are undoubtedly affected somewhat by such organizational policies. Nevertheless, the bottom line is that in 1973 over one-third of the association's members in commercial building were open shop contractors. A statistical bias or error cannot account for this very strong trend, especially since it conforms not only to the other data presented in this chapter, but also to the findings of our field interviews.

Recent AGC Information. We were unable to obtain a more recent poll by the AGC of its members, but we have received information to the effect that by 1983 nearly 40 percent of this association's 8,400 contractor members were open shop contractors,[18] and many union shop AGC members had doublebreasted operations. The AGC now has a large open shop department as well as a major open shop committee and staff, offers seminars specially designed either for open shop contractors or those considering becoming open shop, and provides a variety of other services for its nonunion members.[19] The Carolinas' Branch, the AGC's largest chapter covering both North and South Carolina, is avowedly open shop.

The Industrial Research Unit's 1982 questionnaire, described in Chapter I, was sent to all 8,400 AGC members and received a response of 1,373, or 16.3 percent. Of these, 844, or 61.5 percent, reported that they were union, whereas 38.5 percent declared that they operated nonunion. The respondents from the AGC membership included 1,011 who classified themselves as commercial contractors, including some who indicated that they operated in more than one sector of the industry. The AGC commercial respondents were slightly more unionized; 639, or 63.2 percent, were union, and 372, or 36.8 percent, nonunion. Again, given the propensity of larger contractors to respond more readily to questionnaires than for smaller ones to do so and the tendency of the AGC's union contractors to be larger than their open shop counterparts, it is not unlikely that this questionnaire could understate the AGC open

---

[18] This statement is based upon discussions with officials at AGC headquarters. See also "AGC Membership Rolls Continue to Show Growth of Open Shop," *Construction Labor Report,* Vol. 28 (October 23, 1982), p. 1089.

[19] For examples of reports of the AGC Open Shop Committee, see *Construction Labor Report,* Vol. 28 (March 24, 1982), pp. 119-20; (October 13, 1982), pp. 1113-14; and Vol. 29 (January 12, 1983), pp. 1558-59.

shop representation by a few percentage points. Nevertheless, it does appear that the AGC members in the commercial sector are a higher percent union than is the membership as a whole (which the AGC officials approximate at 60 percent union and 40 percent open shop).

ABC Membership Trend and Analysis

The ABC was founded in Baltimore as an organization of open, or as the ABC terms it "Merit Shop," builders. Its growth has mirrored the expansion of the open shop. From a small local group in 1952, the ABC has grown into the country's second largest contractor organization after the National Association of Home Builders. Figure IV-5 charts the ABC's growth from 1952 to 1983. Its spurt in membership since 1970 clearly reflects the growth of the open shop. The ABC suffered membership declines in only 1970 and 1982, both sharp recession years in the construction industry, but each time the setback was slight and the rebound fast.

Unlike the AGC, which is an organization of general contractors only, the ABC includes both subcontractors and contractors, as well as others in the building business. Three samplings (or "waves") of the ABC by the Opinion Research Corporation showed that the subcontractors actually comprise a majority of the ABC's membership, as charted in Figure IV-6. The admission of both contractors and subcontractors is one reason for the rapid increase in the ABC's membership. Nevertheless, the growth must be considered primarily a reflection of the expansion of the open shop, and it is all the more remarkable because the ABC has no chapter in North or South Carolina, probably the two states most completely open shop. This is because, as already noted, the Carolinas' AGC chapter is strongly open shop and services contractors throughout these states.

Although the ABC includes one of the country's largest contractors, Brown & Root, and a fair number of homebuilders, its membership is overwhelmingly composed of commercial contractors and subcontractors. In 1980, 85 percent of ABC member firms were engaged in commercial work and about 40 percent of those surveyed did more than approximately one-half of all their work in this sector.[20] Thus, the growth of the ABC is also a reflection of the increased open shop share of commercial building work.

---

[20] "Survey Shows Steady Growth of Merit Shop Contractors," ABC News Release containing results of Opinion Research Corporation Survey of 1980, September 1981. See also "Surveys Show Open Shop Surge," *Engineering News-Record,* September 17, 1981, p. 232.

FIGURE IV-5
Overall ABC Membership Growth from 1952–1983

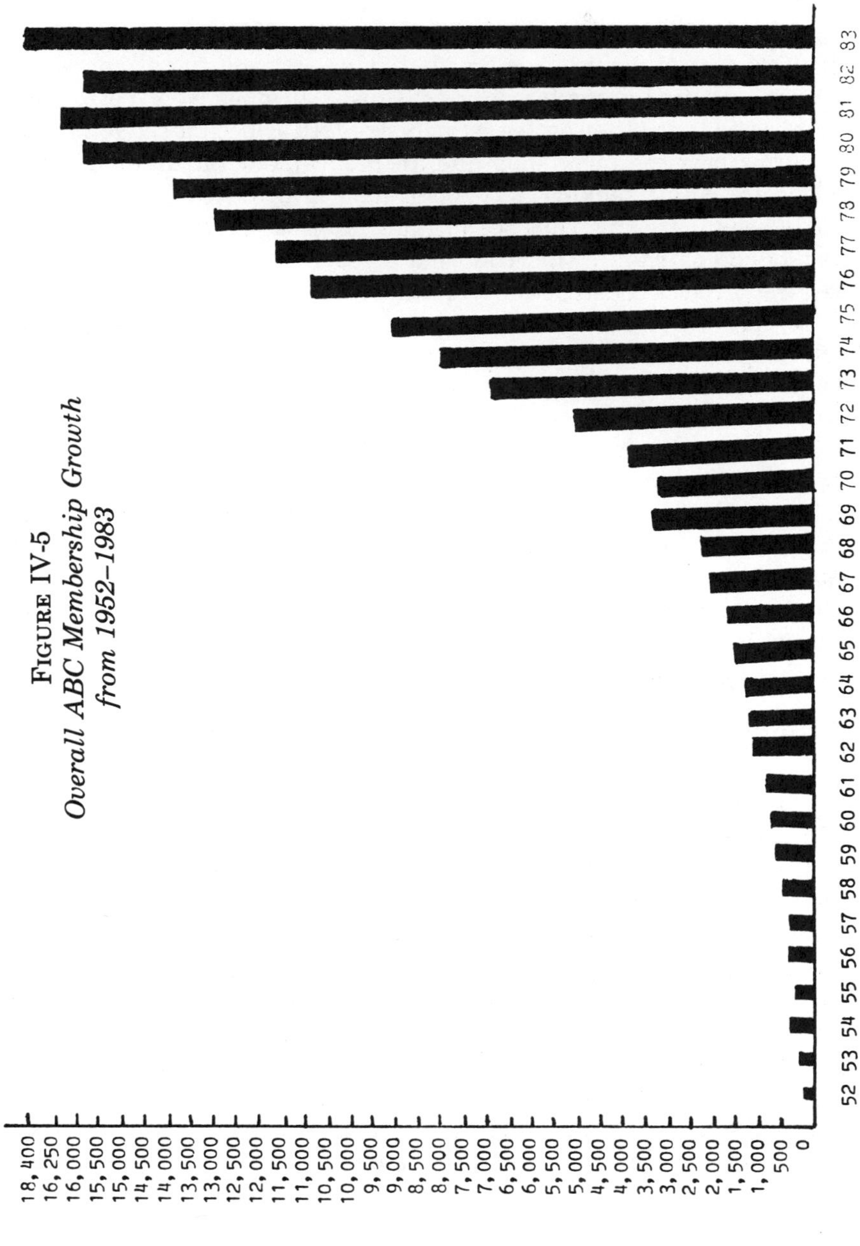

Source: Associated Builders and Contractors, Inc.

Another such indication is the geographic expansion of the ABC, first along the Eastern Seaboard, then to the Midwest, and since the mid-1970s, all across the country. Today the ABC has its largest chapters in Colorado, Southern California, and Florida. It has five chapters in California, four in Michigan, and additional chapters in such one-time union bastions as Kansas, Illinois, New York, and Ohio, as well as other states.[21] Such strength is both a reflection of the growth of the open shop and an inducement for builders who either are, or are seeking to be, open shop to affiliate with the ABC.

ABC members have also been taking on progressively larger jobs. In 1980, 65 percent of the ABC's surveyed general contractor members reported a construction volume of more than $1 million and 24 percent reported a volume of more than $5 million, as compared with 59 percent and 17 percent, respectively, in 1977. In addition, 25 percent of its subcontractor members reported an annual volume between $1 and $2.5 million and another 23 percent between $500,000 and $1 million.[22] This steady growth in volume indicates that ABC commercial sector members have an increasing capability to handle larger and larger jobs, although in many areas they would likely rank below the average size of unionized contractors. The number of open shop contractors able to handle the big industrial jobs is growing, but still limited. Nevertheless, the capability indicated by this survey demonstrates that the ABC's member firms are in a position to compete for a large share of commercial work.

SPECIALTY CONTRACTOR ASSOCIATION NATIONAL STUDIES

In the first edition of *Open Shop Construction* we noted that the open shop contractors were relatively weak in the highly skilled electrical, mechanical, and plumbing trades, and that this posed problems for the expansion of the open shop, particularly in bidding for the larger commercial contracts. In our field work for the present book, we were surprised to find that the open shop contractors were having little difficulty in utilizing qualified nonunion subcontractors in these skilled trades. The membership of the ABC, shown in Figure IV-6, with its surprisingly large percentage of electrical, heating, ventilating, air conditioning and mechanical contractors, is certainly an indication of this change. Studies of key subcon-

---

[21] List of ABC chapters, founding dates, and locations thereof provided by the Association.

[22] "Survey Shows Steady Growth of Merit Shop Contractors," p. 2.

FIGURE IV-6

Survey of Contractor and Subcontractor Members Associated Builders and Contractors

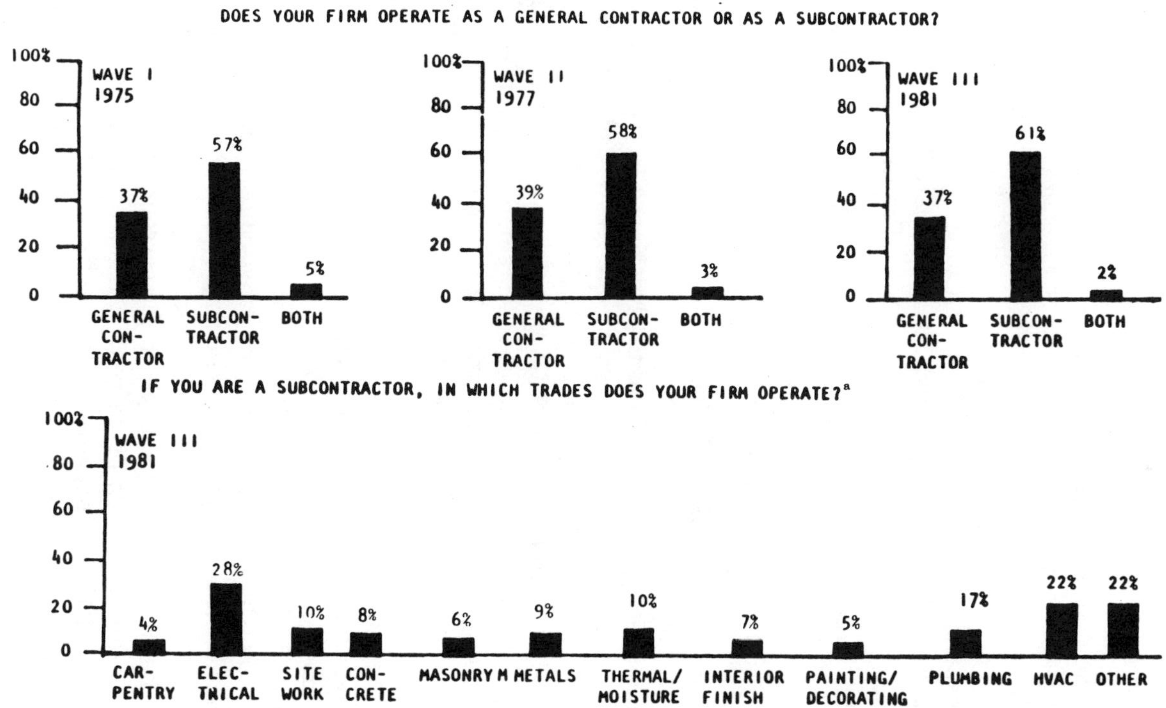

Source: Opinion Research Corporation, *Associated Builders and Contractors Trend Study III,* May 1981 (Princeton, N.J., 1981).

[a] Waves I and II were not reported in a manner directly comparable to Wave III.

tractors' organizations, which are presented in this section, provide further proof of open shop gains in key sectors.

National Electrical Contractors Association

The National Electrical Contractors Association (NECA) is the largest of the specialty contractors' associations and it has maintained a bargaining relationship with the International Brotherhood of Electrical Workers (IBEW), both nationally and through its local chapters, for many years.[23] Historically, NECA members and chapters have been unionized, although there have been exceptions (e.g. in the Carolinas). Now NECA's studies show a clear open shop trend.

The NECA annually develops data on the number of electrical contractor employers and employees in inside electrical work (i.e., construction excluding highway, heavy, or other outside work) by IBEW contract coverage or "other" and by NECA membership or "other." (See Table IV-5.) These data have been carefully compiled by the NECA using Bureau of the Census unemployment compensation data and reports from the National Electrical Benefit Fund (which is jointly administered by the IBEW and the NECA). The category "other" in relation to IBEW membership is, of course, largely nonunion, but it could include coverage with contracts of other unions. The likelihood of this is small, however, and for trend purposes, it is clearly appropriate to assume that "other" and nonunion are virtually synonymous insofar as the data in Table IV-5 are concerned.

The data in Table IV-5 reveal a steady erosion of IBEW strength except for a few year-to-year variations. Despite an increase in the number of electrical employers of over 11,000 from 1972 to 1981, those dealing with the IBEW declined by about 3,000 and the employers who operated nonunion increased by approximately 12,500. Thus, whereas 37 percent of the employers were IBEW represented in 1972, only 22 percent dealt with this union in 1981.[24]

The IBEW also lost sharply in the percentage of employees which it represented, from 55 percent of the total in 1972 to 39 percent in 1981. This decline, however, leaves the IBEW with a substantially higher percentage of employees represented than employers signed

---

[23] For background, see Michael A. Mulcaire, *The International Brotherhood of Electrical Workers,* Studies in the Social Sciences, Vol. 5 (Washington, D.C.: Catholic University of America, 1973); and Daniel Quinn Mills, *Industrial Relations and Manpower in Construction* (Cambridge, Mass.: MIT Press, 1972), p. 135.

[24] National Electrical Contractors Association, *Inside Electrical Construction Survey,* 1972-1981.

TABLE IV-5

IBEW And Other Inside Electrical Employers and Employees United States, 1972–1981

Year	Employers					Employees				
	Total	IBEW	Percent IBEW	Other	Percent Other	Total	IBEW	Percent IBEW	Other	Percent Other
1972	33,189	12,421	37	20,768	63	327,411	178,735	55	148,676	45
1973	33,129	12,187	37	20,942	63	346,593	181,884	52	164,709	48
1974	35,491	11,775	33	23,716	67	346,844	176,071	51	170,773	49
1975	35,989	11,400	32	24,589	68	317,818	159,499	50	158,319	50
1976	37,172	11,425	31	25,519	69	315,920	162,598	52	153,322	48
1977	38,760	9,900	26	28,860	74	332,888	158,523	48	174,365	52
1978	40,563	9,048	22	31,515	78	364,018	164,188	45	199,830	55
1979	42,902	10,853	25	32,049	75	397,605	188,138	47	209,467	53
1980	43,249	10,308	24	32,941	76	407,295	173,450	43	233,845	57
1981	42,722	9,489	22	33,233	78	417,158	160,782	39	251,376	61

Source: National Electrical Contractors Association, *Inside Electrical Construction Survey,* March 1982.

to agreements. This indicates that the IBEW has maintained its relations with the larger electrical firms while losing the smaller ones. From our field work, we have ascertained that the IBEW has been unable to unionize a large percentage of new electrical contractors, many of whom, paradoxically, are former union members but are anxious to avoid restrictive work rules and other impediments to productivity.

Interestingly, the NECA membership share of employers with IBEW contracts rose from 39 percent in 1972 to 48 percent in 1980, but the total number of such employers declined by almost 2,000 while NECA's actual employer members in this category remained almost constant.[25] Meanwhile, the ABC greatly expanded its membership among electrical contractors and had about 3,200 such members in mid-1983. The NECA reported a membership of 6,100 electrical contractors in November 1982.[26]

The Independent Electrical Contractors (IEC) (formerly the Associated Independent Electrical Contractors—AIEC), an open shop group organized in 1966, has expanded to twenty-eight state and local chapters with the greatest growth being in the South. In 1983, the IEC had about 2,000 members and was growing rapidly. About 60 percent of its members were in commercial work, 10 percent in residential, and the balance in industrial and miscellaneous. The IEC, like the electrical contractor members of the ABC, contends that although the NECA has opened its doors to nonunion members, the NECA's programs are directed primarily to the problems of collective bargaining with the IBEW.[27] This is in accord with our field findings.

The gains of the open shop in the electrical trade have altered what was once a key construction union citadel. As we shall note in the following chapter, the businesses of open shop electrical contractors are growing; several are among the largest, although most are still smaller on average than those of their union counterparts. Their continued growth seems assured—particularly because of the policies of the IBEW relating to the hours question as described below.

---

[25] *Ibid.*

[26] ABC membership information from ABC national headquarters. NECA membership figures from "NECA's Focus: Economy, Suit," *Engineering News-Record*, November 18, 1982, p. 78.

[27] Interviews, IEC headquarters, Washington, D.C., September 1982 and August 1983; letter to the author from Mark Levin, executive vice-president, IEC, September 19, 1983.

The IBEW and the Ten-Hour Day

A key issue in construction today is the right of contractors to work a ten-hour day, four-day week without paying overtime for the ninth and tenth hour on any day as long as the workweek is not greater than forty hours. Some employers in the construction industry prefer this schedule because work curtailed by inclement weather can be made up on the fifth day. Many employees prefer this schedule also because they do not lose pay if work is not scheduled on one day, and because a three-day weekend without loss of pay is possible. The four-day, ten-hour schedule is used rather widely by open shop builders and is found on some union jobs. It cannot be utilized economically on federal government projects because the Public Contracts (Walsh-Healey) Act of 1936 requires that time and one-half be paid for all work in excess of eight hours in one day if government funding of the job exceeds $10,000. Efforts to modify this provision have been thwarted by the opposition of the AFL-CIO, especially its construction union affiliates.[28]

The IBEW, behind President Charles H. Pillard, has been the leader among construction unions in the opposition to the ten-hour day, four-day week. Local officials who signed agreements accepting this schedule have been suspended from the union and the agreements have been repudiated by the national union.[29] Pillard was reelected to office in "a tightly controlled convention" which also voted down local IBEW negotiation of ten-hour days without overtime pay.[30]

The net effect of the IBEW's stance on this issue has been the acceleration of union decline. As we shall discuss in Chapter VIII, the IBEW's refusal to compromise on this issue has negated project agreements in major construction and cost the IBEW a considerable number of jobs.[31] In one case, an IBEW local in Kentucky was forced to use an open shop contractor to build its own office.[32] In southern

---

[28] In July 1982, the U.S. Senate tabled an amendment to the Walsh-Healey Act which would have allowed federal contractors to schedule workweeks without having to pay overtime premiums until forty hours have been worked in one week. The bill was sponsored by building contractors and opposed by the AFL-CIO. See "Senate Tables Amendment That Would Authorize Four Tens Work Schedule," *Construction Labor Report,* Vol. 28 (July 14, 1982), p. 682.

[29] "IBEW Chief Reelected," *Engineering News-Record,* September 23, 1982, pp. 176-77; and "Union's Stand on Hours is Costly," *Engineering News-Record,* July 1, 1982, pp. 10-11.

[30] "IBEW Chief Reelected," p. 176.

[31] "Unions Stand on Hours is Costly," p. 10.

[32] "Disarray and Panic in the Organized Electrical Sector," *Cockshaw's Labor News + Opinion,* Vol. 13, No. 6 (June 1983), p. 2.

Illinois, despite a decline in the employment of unionized inside electricians from 428 in June 1980 to 180 in 1983, wage and productivity concessions were blocked by the National NECA/IBEW Council on Industrial Relations.[33] In southern Indiana unionized electrical contractors lost 76 percent of the market during the four years prior to 1983.[34] Attempts to negotiate a special agreement with a helper classification and a flexible-hours schedule were vetoed by Pillard who has insisted on maintaining "standards." In Sacramento, California, the NECA chapter signed an agreement with the National Association of Independent Unions, following a long strike by the IBEW, which called for wages, hours, and conditions considerably below the IBEW's demands.[35] These incidents are not isolated, but reflect the general downward trend of unionization in the electrical trade.

Sheet Metal and Air Conditioning Contractors' National Association

In June 1982, the board of directors of the Sheet Metal and Air Conditioning Contractors' National Association (SMACNA) engaged the consulting firm, Arthur D. Little, Inc., to make a study which was designed in part to determine "the important national and regional trends in the use of non-union labor in the sheet metal industry and in the construction industry in general" and to discern "what are the comparative strengths and weaknesses of SMACNA members, versus their industry competitors."[36] The study was presented to the SMACNA in June 1982.

The SMACNA is composed of approximately 2,400 contracting firms with 96 percent of its membership having labor contracts with the SMW, a key building construction union.[37] The Arthur D. Little study found that the membership of SMACNA was about evenly divided as to the advisability of soliciting membership among non-union firms. The report also found that approximately 125,000 firms operate businesses within SMACNA's jurisdiction, of which about

[33] "Despite Alarming Loss of Market Share in Illinois, Labor Rejects NECA's Pleas for Contract Concessions," *Cockshaw's Construction Capsules,* Vol. 13, No. 9 (September 1983), p. 1.

[34] "Disarray and Panic in the Organized Electrical Sector," p. 2.

[35] "Open Shop Firms in Sacramento Join NECA to Work under NAIU Agreement," *Construction Labor Report,* No. 1355 (November 25, 1981), pp. A-13-14.

[36] Arthur D. Little, Inc., *A Study of the Competitive Positions of Union vs. Nonunion Contractors in the Sheet Metal Industry,* Report to SMACNA, June 1982, p. 1.

[37] *Ibid.,* p. 4.

21,000 have more than ten employees, and about 4,000 are estimated to have annual revenues in excess of $500,000.[38] Table IV-6 shows the union status of these large firms by SMACNA and non-SMACNA membership, as estimated from an Opinion Research Corporation telephone survey performed for the Arthur D. Little study.

The data in Table IV-6 confirm that the nonunion sector is much stronger outside of the SMACNA than within its ranks—82 percent vs. 3 percent—and that in all cases, unionized contractors control a much larger percentage of the market in firms with larger revenues than in those with smaller incomes. Nevertheless, the larger nonunion firms are capable of growth and pose a direct threat to SMACNA's members. Figures IV-7 and IV-8 support this finding.

Figure IV-7 shows the union-nonunion division of work in sheet metal contracting from 1970 to 1982 in current and in constant 1982 dollars. In current dollars, nonunion contractors have increased in volume much more rapidly than union contractors. In constant dollars, however, which removes the effect of inflation, the incomes of unionized contractors are shown to have declined from $9 billion in 1973 to approximately $6.5 billion in 1982 while nonunion contractors increased their incomes from about $1 billion to about $5 billion in the same period. Figure IV-8 shows that a loss of market share for unionized contractors occurred in all seven states surveyed for the study. Nationwide, the decline for unionized contractors was estimated at 35 percent, with large union contractors accounting for 25 percent of this loss.[39]

The respondents to the Arthur D. Little survey found the economic advantages of the nonunion sector to be enormous. These advantages include wages that are 30 percent lower and staffs which are 50 percent journeymen (in contrast to 75 percent journeymen for union contractors), both of which make it exceedingly difficult for unionized firms to compete and contribute to the belief that the unionized market share will continue to decline in the future.[40]

The Arthur D. Little study estimated that for 1972 nonunion sheet metal contractors accounted for only 10 percent of the work, but by 1982, this had risen to 40 percent. If this trend continues, the nonunion share could be expected to exceed 50 percent within five years.[41] As Figure IV-9 demonstrates, it is predicted that the

[38] *Ibid.*, pp. 5-6.
[39] *Ibid.*, p. 21.
[40] *Ibid.*, p. 2.
[41] *Ibid.*, p. 21.

TABLE IV-6
Large Sheet Metal Contractors[a]
Open vs. Union Shop
by SMACNA Membership and Revenue Size, 1982

Total Firm Revenue ($ millions)	SMACNA Members[b]			Non-SMACNA Members			Total		
	Union	Nonunion	Percent Nonunion	Union	Nonunion	Percent Nonunion	Union	Nonunion	Percent Nonunion
Over 10	215	5	2	55	35	38	270	40	12
5-10	275	10	3	35	80	69	310	90	22
2-5	670	20	2	80	420	84	750	440	36
Under 2	960	50	4	160	980	85	1,120	1,030	47
Total	2,120	85	3	330	1,515	82	2,450	1,600	39

Source: Projections from an Opinion Research Corporation telephone survey of the sheet metal contracting industry made for Arther D. Little, Inc., *A Study of the Competitive Positions of Union vs. Non-union Contractors in the Sheet Metal Industry,* Report to SMACNA, June 1982. p. 6.

[a] Sheet Metal revenue in excess of $500,000 and more than ten employees.

[b] About 8 percent, or 185 of SMACNA members have revenues of less than $500,000 and / or less than ten employees.

FIGURE IV-7

Sheet Metal Contracting Revenues, by Type of Firm

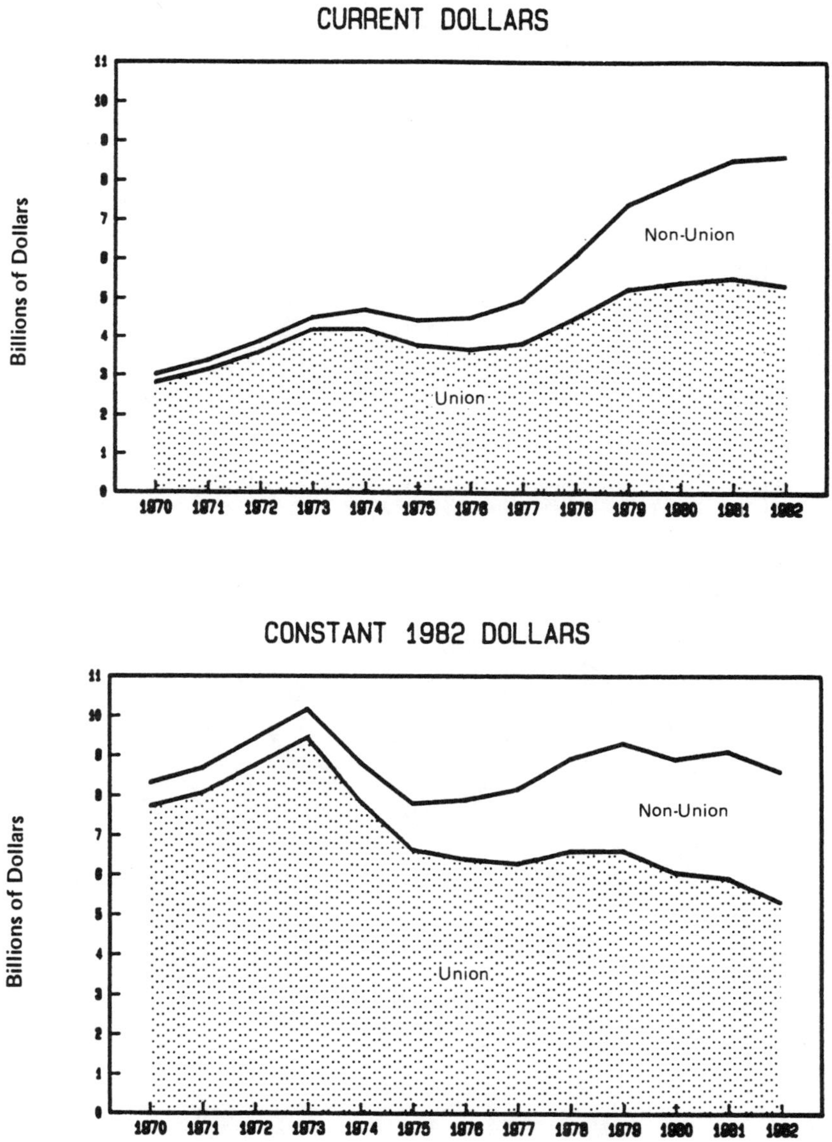

Source: Arthur D. Little, Inc., *A Study of the Competitive Positions of Union vs. Non-Union Contractors in the Sheet Metal Industry,* Report to SMACNA, June 1982, p. 22.

FIGURE IV-8
Sheet Metal Contracting
Change in Union Market Share
Selected States, between 1977 and 1982

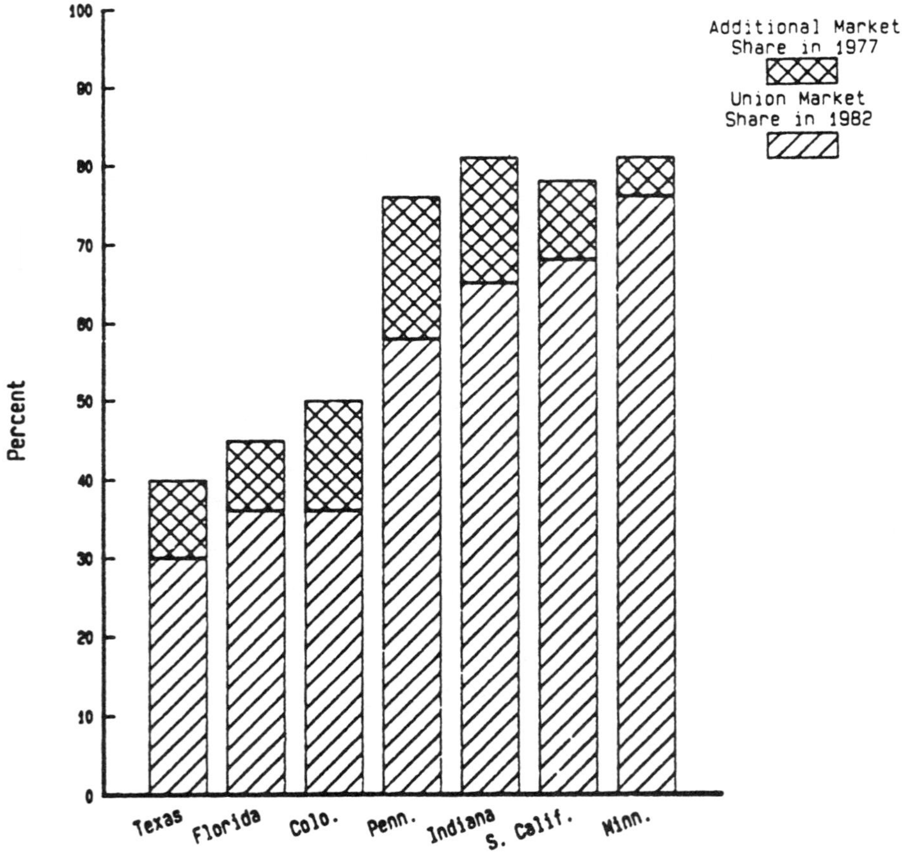

Source: Telephone Survey by Opinion Research Corporation for Arthur D. Little, Inc., *A Study of the Competitive Positions of Union vs. Non-Union Contractors in the Sheet Metal Industry,* Report to SMACNA, June 1982, p. 24.

unionized contractors will still dominate the largest jobs, but this could also change. As the authors of the study contend:

> the union contractors dominate the "speciality" jobs where the skill level of the firm is critical and the very large jobs where capitalization is a key factor. We expect that the nonunion contractors will gradually

FIGURE IV-9
Distribution of Sheet Metal Contracting Revenues by Job Size and Type of Firm
(thousands of dollars)

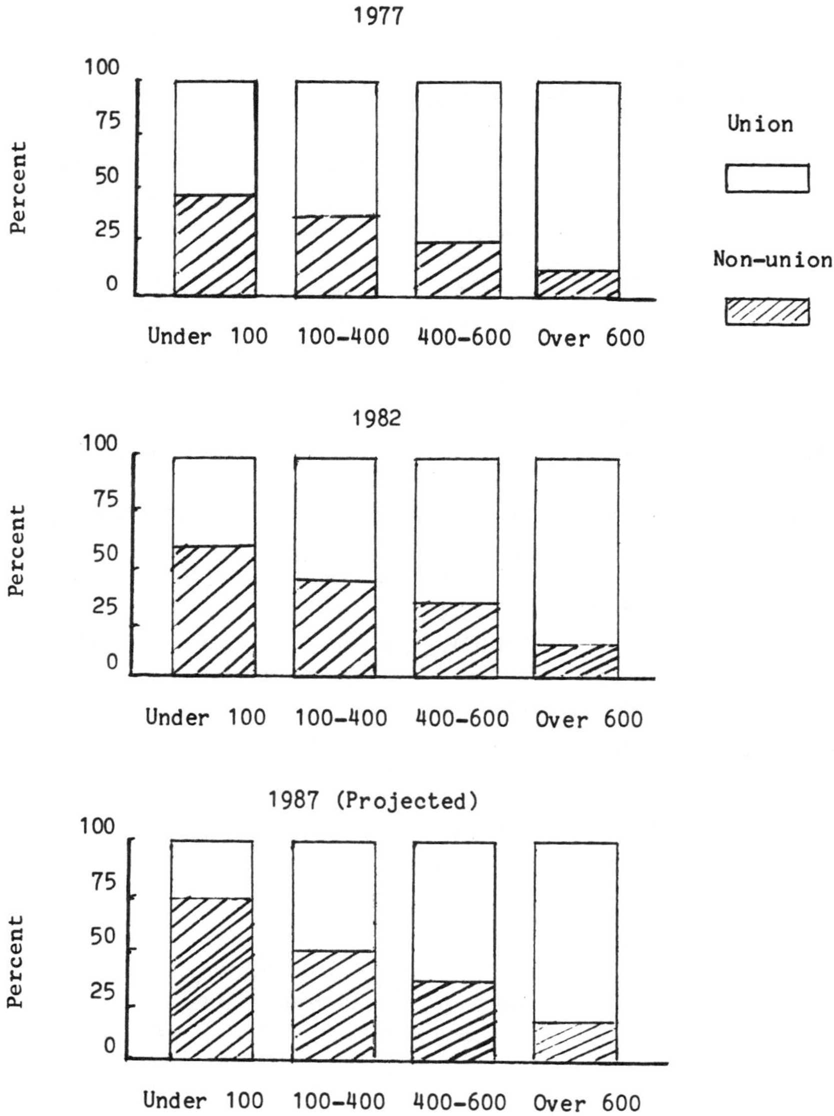

Source: Arthur D. Little, Inc., *A Study of the Competitive Positions of Union vs. Nonunion Contractors in the Sheet Metal Industry,* Report to SMACNA, June 1982, p. 27.

> become more skilled and financially powerful allowing for their continued growth.[42]

For residential and industrial construction, the large open shop sheet metal contractors already have a larger share of the work than do their union shop counterparts, as seen in Figure IV-10. The clear trend indicates that commercial work in this field could well be dominated by the open shop in the future unless the union shop contractors and the SMW take stronger measures in order to stem the tide.

At its 1982 convention, the SMW passed a resolution which gave its local officers " 'wide discretion' to 'take whatever steps necessary' to 'recoup work for our members by making union contractors more competitive.' "[43] This was followed in April 1983 by a renegotiation of the SMACNA-SMW standard agreement permitting contractors to employ one "preapprentice" for every three apprentices, and allowing small contractors to retain a preapprentice as long as at least one apprentice and three journeymen are working. Preapprentice wages were set at 30 percent of the journeyman rate plus some form of health and welfare benefits. In addition, the SMW leadership has indicated a willingness to approve more competitive deployment standards in order to offset open shop cost advantages. For example, a crew composed of one journeyman, one regular apprentice, and three preapprentices was accepted on low-rise residential construction in Atlanta.[44]

The national SMACNA-SMW agreement is a guideline for local or area agreements. It is, therefore, up to the local unions and contractor associations to determine how much they will alter traditional practices to meet open shop competition. Given the significant wage, benefit, and worker-deployment advantages of the open shop, as confirmed by the Little study, it may well require more drastic measures than the local unions are willing to support if the trend toward the open shop is to be reversed.

The growth of the open shop sector and the reluctance of the SMACNA's membership to open its doors to the nonunion contractors have had another impact. The ABC now has more heating, ventilating, and air conditioning (HVAC) contractor members than does the SMACNA—about 3,000 to the SMACNA's 2,400—al-

---

[42] *Ibid.*, p. 25.

[43] "Sheet Metal Workers, SMACNA to Discuss Revisions in Their National Agreement," *Construction Labor Report,* Vol. 28 (November 10, 1982), p. 1228.

[44] "Sheet Metal Union Agrees to Preapprentice Category which 'Opens the Door' to Much Needed Cooperative Efforts," *Cockshaw's Construction Capsules,* Vol. 13, No. 4 (April 1983), p. 1.

FIGURE IV-10
Sheet Metal Contracting Revenues by Type of Work and Type of Firm

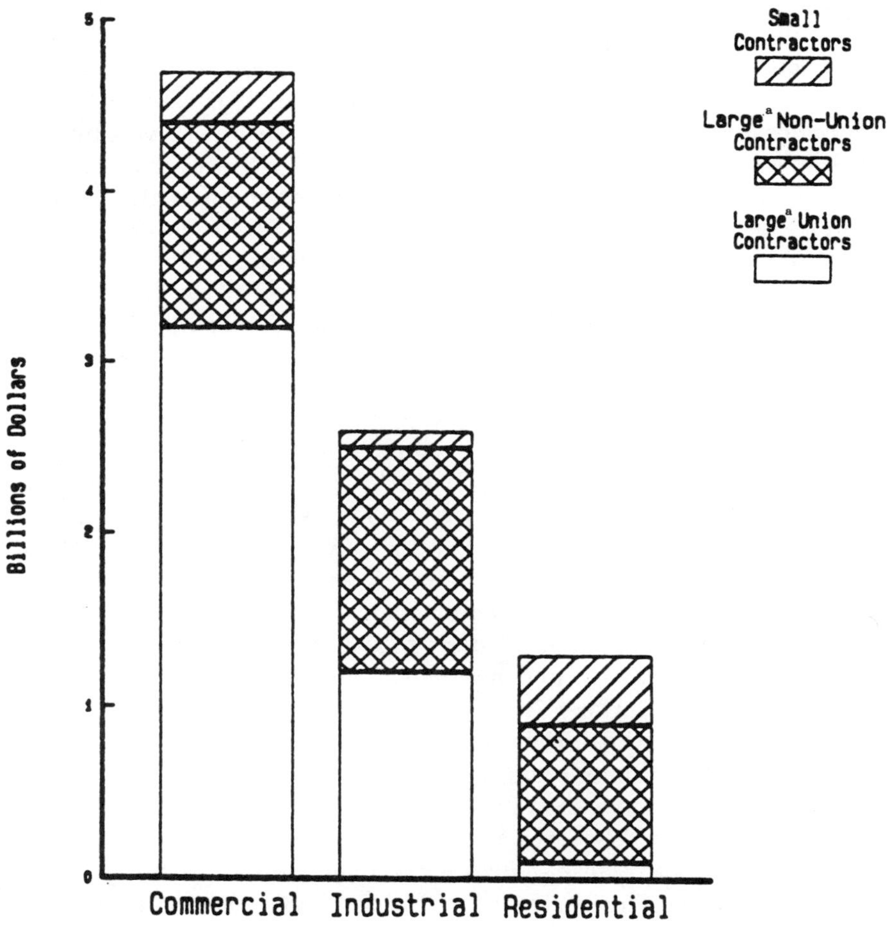

Source: Telephone Survey by Opinion Research Corporation for Arthur D. Little, Inc., *A Study of the Competitive Positions of Union vs. Non-Union Contractors in the Sheet Metal Industry,* Report to SMACNA, June 1982, p. 23.

[a] More than 10 employees and more than half a million dollars in sheet metal revenues.

though it is likely that the SMACNA's members do a larger volume of business than their ABC competitors.[45]

The Mechanical Contractors Association of America

The Mechanical Contractors Association of America (MCAA) has a membership of 1,600 unionized plumbing, pipe fitting, and related mechanical contractors. Almost 60 percent of the MCAA's members do an annual volume of work below $5 million, 17.5 percent have volumes in the $5 to $10 million range, and 22.5 percent have annual volumes which exceed that amount.[46] The MCAA surveyed its members in September 1982 on the subject of nonunion competition. The report that followed, entitled *The Unionized Mechanical Construction Industry in Crisis,*[47] was pessimistic both in its tone and its conclusions: ". . . . if we are to remain competitive, MCAA and the United Association of Journeymen and Apprentices of the Plumbing and Pipe Fitting Industry [UA] and their affiliates must act decisively, and now, to avoid . . . the continual decline of work for MCAA members as UA employers."[48]

The study estimated, both on a regional and a national basis, the extent of open shop mechanical work in 1977 and 1982, and then projected the data for 1985. Figure IV-11 shows that, by 1982, nonunion mechanical construction had already passed the 50 percent mark and it is estimated that as much as 70 percent of the mechanical work will be performed by open shop contractors by 1985. The MCAA completed another study in 1983 which found that these trends had accelerated, and predicted that the 70 percent figure would be achieved prior to 1985.[49] As in the case of all construction work thus far analyzed, the open shop is strongest in low-rise residential work, but commercial work and high-rise apartments were already majority open shop in 1982, and every sector is estimated to be dominated by the open shop by 1985.

Table IV-7 shows these figures broken down on a broad regional basis for each of the three years. Interestingly, the most highly unionized areas were found to be the Plains and Rocky Mountain regions. (This conflicts somewhat with our interviews, which indi-

---

[45] Membership information from the ABC headquarters staff, August 1983.

[46] "Diversity of Size and Work Reported by Mechanical Contractors," *MCAA News Release,* October 7, 1981.

[47] Mechanical Contractors Association of America, *The Unionized Mechanical Construction Industry in Crisis,* November 1982.

[48] *Ibid.,* p. 1.

[49] Mechanical Contractors Association of America, *The Unionized Mechanical Construction Industry Continues in Crisis,* December 1983.

FIGURE IV-11
Nonunion Mechanical Construction
Percentage by Construction Sector
1977, 1982, and 1985 (Estimated)

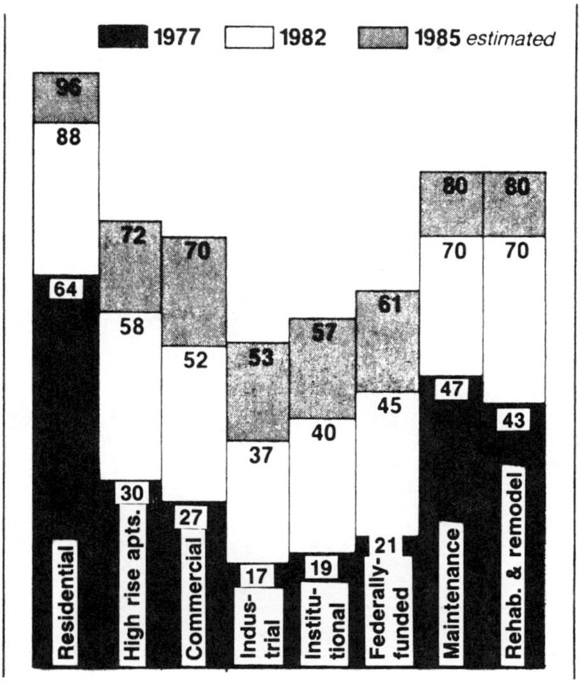

Source: Reprinted from *Engineering News-Record,* December 9, 1982, copyright McGraw-Hill, Inc.

cated that the Pacific states were the most highly unionized, and may be the result of sampling errors.) The regional divisions are broad, probably because of the small survey sample (approximately 320 contractors). This grouping does not allow for the separation of highly organized areas (for example, New Jersey) from highly open shop areas (for example, Virginia). What is significant in each region, however, is the dramatic growth of nonunion mechanical construction, both on a regional and on a nationwide basis, which occurred between 1977 and 1982, and which is projected to continue through 1985.

TABLE IV-7
Nonunion Mechanical Construction
Percentage by Region[a] *and Construction Sector*
1977, 1982, and 1985 (Estimated)

Description	FW	P&R	SW	MW	SE	NE	Average
1977							
Residential & Garden Apts.	39%	51%	84%	62%	79%	70%	64%
High Rise Apts.	21	20	39	28	43	28	30
Commercial	17	18	35	24	35	32	27
Industrial	19	7	26	14	21	17	17
Institutional	18	12	21	14	25	22	19
Fed. Asst./Davis-Bacon	21	11	26	14	31	24	21
Serv. & Maintenance	34	38	67	43	51	51	47
Rehab. & Remodel	32	34	58	45	44	47	43
1982							
Residential & Garden Apts.	85%	80%	97%	83%	93%	89%	88%
High Rise Apts.	68	45	68	50	60	55	58
Commercial	47	42	68	44	59	55	52
Industrial	34	25	60	24	43	38	37
Institutional	41	32	47	26	51	46	40
Fed. Asst./Davis-Bacon	45	34	56	30	53	51	45
Serv. & Maintenance	76	64	81	59	69	69	70
Rehab. & Remodel	69	69	79	63	70	69	70
1985							
Residential & Garden Apts.	94%	93%	100%	96%	98%	95%	96%
High Rise Apts.	77	51	74	81	83	68	72
Commercial	67	57	83	66	76	71	70
Industrial	46	37	71	44	64	55	53
Institutional	57	43	68	46	71	59	57
Fed. Asst./Davis-Bacon	66	52	66	48	72	67	61
Serv. & Maintenance	85	72	90	75	75	83	80
Rehab. & Remodel	79	79	82	78	80	82	80

Source: Mechanical Contractors Association of America, *The Unionized Mechanical Construction Industry in Crisis,* November 1982, p. 3.

[a] Regions are defined as follows:

FW=Far West: Washington, Oregon, and California

P&R=Plains and Rockies: Idaho, Nevada, Utah, Wyoming, Montana, North Dakota, South Dakota, Nebraska, Kansas, Iowa, Missouri, and Colorado

SW=Southwest: Arizona, New Mexico, Oklahoma, and Texas

MW=Midwest: Minnesota, Wisconsin, Michigan, Illinois, Indiana, and Ohio

SE=Southeast: Arkansas, Louisiana, Kentucky, Tennessee, Alabama, Georgia, North Carolina, South Carolina, Florida, and Mississippi

NE=Northeast: Maine, Vermont, New Hampshire, Massachusetts, Connecticut, Rhode Island, New York, Pennsylvania, New Jersey, Delaware, Maryland, Virginia, West Virginia, and District of Columbia

The problems of the unionized mechanical contractors are similar to those of their sheet metal counterparts: large union-open shop wage differentials, restrictions on the use of helpers, the requirement that journeymen do semiskilled work, local union sabotage of the right to use subjourneymen even when permitted by contract concessions, and strikes, all of which, according to the perceptions of the unionized contractors responding to the MCAA surveys, give the open shop firm an overwhelming advantage.[50]

Although the United Association does not seem to have adopted a considered national approach like the SMW, wage concessions in the craft have been widespread. According to the MCAA, 63 of the 170 agreements negotiated during the first half of 1983 call for wage freezes and 10 for wage rollbacks. The MCAA found that the smallest wage increases were in the South and Midwest, and the largest in western states. The MCAA attributed "these responsible settlements . . . to the slow economy and a growing sense of awareness of increased competition from nonunion contractors."[51] Moreover, the MCAA has found that 45 of its 78 affiliated state and local associations have one helper classification in their contracts. This, however, represents only 10 percent of the approximate number of existing MCAA contracts.[52] The 1983 study found that although most respondents believed that a subjourneyman clause in their contracts would be helpful, those who did have such a clause did not find it so in competition with nonunion firms. As a result of this, 236 of the 270 respondents were considering nonunion operations and a majority of union journeymen who had established businesses were said to be operating nonunion.[53]

The ABC has enrolled 2,000 plumbing and other mechanical contractors, giving it a larger membership in this craft than the MCAA, although the ABC's mechanical contractor members are likely to have smaller firms than those in the MCAA and employ fewer tradesmen.[54] Nonunion mechanical contractors, however, are more likely to join the National Association of Plumbing-Heating-Cooling Contractors (NAPHCC), the largest and oldest association in the craft.

---

[50] *The Unionized Mechanical Construction Industry in Crisis,* p. 6.

[51] *Engineering News-Record,* September 15, 1983, p. 73.

[52] "Wage Moderation Holding," *Engineering News-Record,* September 15, 1983, p. 73; and "MCAA Survey Finds Use of Helpers on the Rise," *Engineering News-Record,* April 22, 1982, p. 198.

[53] *The Unionized Mechanical Construction Industry Continues in Crisis,* pp. 5-6.

[54] Membership information from the ABC headquarters staff, August 1983.

National Association of Plumbing-Heating-Cooling Contractors

The NAPHCC has about 5,900 mechanical contractor members. This is more than the MCAA and the ABC combined. The NAPHCC members are concentrated in commercial and light industrial work whereas those belonging to the MCAA are more likely to be in the heavy industrial sector; however, there is considerable overlap in membership with many contractors belonging both to the NAPHCC and the MCAA.

The NAPHCC includes both union and open shop contractors, and the organization provides services for both. For example, in 1983, it had 200 open shop apprentice training programs in addition to those for union shop members. The NAPHCC has surveyed its membership and found a steady increase in the open shop group from 50 percent of the organization in 1978 to about 60 percent in 1983.[55] Some of the NAPHCC open shop members are believed to have doublebreasted concerns associated with union contractors who are members of the MCAA.

Specialty Contractor Surveys—Final Comment

Other specialty contractor associations contacted confirmed the open shop trend. For example, the Painting and Decorating Contractors of America (PDCA) saw its membership decline from a peak of 10,000 in the late 1950s to less than 3,500 in mid-1984. Although the PDCA has long had some open shop members, it is generally perceived as a unionized contractors' association. This has caused many members to resign after going open shop. PDCA officials report that many areas have few, if any, unionized contractors. In June 1984, its bylaws were changed so that it could solicit unionized contractors "openly."[56]

The growth of the open shop in the highly skilled electrical, HVAC, and mechanical trades has been truly significant, and has gone a long way toward overcoming the serious shortage of open shop subcontractors in these key areas of construction which we perceived in the first edition of *Open Shop Construction*. All signs point to the continued erosion of union influence in these trades and the increased capacity of the open shop.

It is likely that many of the new open shop contractors in these skilled trades, as well as their employees, were trained under union-

---

[55] Interview with NAPHCC officials, Washington, D.C., August 17, 1983; letter from NAPHCC technical director, October 7, 1983.

[56] Interview with PDCA officials, Washington, D.C., August 17, 1982; "PDCA to Woo Open Shop," *Engineering News-Record,* May 10, 1984, p. 59.

management apprentice systems. To ensure a continued supply of craftsmen for their firms and for the industry, the open shop companies must expand their training. In Chapter X, we shall assess the extent to which this is being done.

AREA STUDIES

In addition to the national studies discussed above, significant information is provided by a number of area studies. These studies attempt to quantify the status of open shop vs. union shop commercial construction within a state, a portion thereof, or a city and its suburban counties (Standard Metropolitan Statistical Area—SMSA). Most of these studies have been made or underwritten by contractor agencies, although in some cases a union, a university, or a research agency have been involved. These studies are scattered in coverage pattern, apply to different time periods, and some of their results are fragmentary; nevertheless, together with our field work and questionnaires, they can be put into perspective. Some of the studies amount to little more than a listing of jobs won by open shop contractors, but even these lists can serve to dispel the notion that open shop commercial construction is limited to small jobs in which unions have little interest.

A few cautionary remarks are in order before examining the results of contractor and union surveys of open shop work. To the extent that such a survey purports to go beyond a simple listing of nonunion projects, it must attempt to gauge open shop activity in relation to some measure of the total volume of construction in an area. Obtaining comparable and meaningful measures of nonunion construction and total construction is not an easy task. A common approach is to ascertain the union status of contractors who, according to Dodge Reports for an area, are awarded contracts.[57] The problem is that coverage of construction work in these reports is not exhaustive. Smaller projects, which are more likely to be built open shop, are underrepresented. Projects which are negotiated, rather than bid, are not represented at all. There is, unfortunately, no way of knowing whether negotiated building contracts are more unionized than those awarded to the lowest bidder. Further, the Dodge Reports identify only prime contractors—those who are responsible directly to the purchaser—and thus, the task of measuring

---

[57] The F. W. Dodge Division of McGraw Information Systems Company issues reports on the status of construction projects out for bid. These "Dodge Reports" are made for individual localities and are a widely used source of information about the construction industry.

TABLE IV-8
Nonresidential Building Construction by Union Status
Philadelphia Area, 1966–1980[a]

Year	Union				Nonunion			
	Dollar Volume[b]	Percent	Number of Jobs	Percent	Dollar Volume[b]	Percent	Number of Jobs	Percent
Philadelphia								
1966	238.8	95.9	194	89.8	10.1	4.1	22	10.2
1967	257.5	99.0	265	94.6	2.5	1.0	15	5.4
1968	601.4	99.3	239	92.6	3.9	0.7	19	7.4
1969	329.7	93.5	236	93.3	22.9	6.5	17	6.7
1970	529.3	99.0	213	93.8	5.6	1.0	14	6.2
1971	560.4	99.6	194	94.6	2.4	0.4	11	5.4
1972	341.5	99.0	158	90.8	3.6	1.0	16	9.2
1973	590.0	98.8	151	85.3	7.0	1.2	26	14.7
1974	392.2	98.9	161	92.0	4.5	1.1	14	8.0
1975	502.8	97.7	194	88.6	11.7	2.3	25	11.4
1976	275.4	96.6	130	86.1	9.6	3.4	21	13.9
1977	274.3	98.6	157	91.8	4.0	1.4	14	8.2
1979	436.7	98.0	151	89.9	8.7	2.0	17	10.1
1980	287.4	96.2	92	82.9	11.5	3.8	19	17.1

Surrounding Counties[c]							
1966	237.1	84.7	306	65.1	42.9	15.3	164	34.9
1967	188.3	88.9	232	66.3	23.4	11.1	118	33.7
1968	246.4	87.4	218	63.2	35.6	12.6	127	36.8
1969	283.2	83.9	493	76.1	54.3	16.1	155	23.9
1970	308.5	85.9	150	52.6	50.6	14.1	135	47.4
1971	335.3	81.7	126	45.8	74.9	18.3	149	54.2
1972	269.7	77.4	146	51.8	78.8	22.6	136	48.2
1973	231.7	74.9	134	46.2	77.6	25.1	156	53.8
1974	159.8	71.1	119	44.4	64.9	28.9	149	55.6
1975	182.0	72.5	91	40.4	69.1	27.5	134	59.6
1976	96.5	67.0	76	40.0	47.5	33.0	114	60.0
1977	96.0	64.2	74	38.7	53.6	35.8	117	61.3
1979	267.8	81.5	140	52.8	60.6	18.5	125	47.2
1980	124.2	73.1	130	56.8	45.7	26.9	99	43.2

Source: General Building Contractors Association of Philadelphia, Inc.
[a] 1978 data not available.
[b] In millions of dollars.
[c] Bucks, Chester, Delaware, and Montgomery.

the union status of the entire job is complicated. Since the situation of a union general contractor using open shop subcontractors is less common than its converse, the net effect of this factor is probably to understate the aggregate magnitude of union representation in an area.

Another factor which must be considered is that a single large project awarded in any one year in an area can distort the area trend for that period. Consequently, if the volume of union shop projects in an area increased in one year, for example, from $300 million to $325 million, this could be the impact of just one award and not indicate a trend.

Nevertheless, Dodge Reports are widely used to develop measures of total construction activity because of their convenience. The alternative—to trace the value and the identity of the contractor for every project—is usually beyond the capacity of the individual researcher or analyst. Thus, the findings of some of the studies described below should be taken only as approximations, with due regard for the methodological and practical obstacles which stand in the way of precise estimates.

Philadelphia

The General Building Contractors Association of Philadelphia, Inc., a chapter of the AGC, has compiled information on the union status of contracts in the Philadelphia area for various years since 1966. The projects covered are those reported in the Dodge Reports and valued at $50,000 or more. Table IV-8 summarizes the figures separately for the City of Philadelphia and four surrounding counties (Bucks, Chester, Delaware, and Montgomery) for the years 1966 to 1980. There appear to be three major conclusions to be drawn from Table IV-8, all reflecting situations or phenomena noted earlier: (1) open shop work represents a larger proportion of projects than of volume, demonstrating again that open shops operate more freely in the small job market; (2) open shop work is much greater in suburban Philadelphia than in the city itself (although it is far from dominant in the suburbs); and (3) the open shop share of all building appears to have grown in recent years. The percentage of the volume dropped off in 1979 and 1980, as the volume of open shop construction first gained less, then declined faster than did its union shop counterpart, but since 1980, our impression is that the open shop has rebounded faster from the recession.

The Philadelphia statistics emphasize the differences between city and suburban areas. Since 1966, the percentage of jobs and volume for the open shop in the suburban areas increased over fivefold.

Union contractors have a decided edge in the dollar volume of work, but split the volume of jobs on average. Because many smaller jobs are not reported in the Dodge Reports, these figures probably understate the open shop share, but they also indicate that even in the suburbs, the unions in the Philadelphia area are still dominant on the larger jobs.

In Philadelphia, the unions remain powerful, holding on to 96.2 percent of the volume in 1980, about the same as in 1966. Philadelphia construction unions are politically powerful and have resorted to violence to maintain their position, as described in Chapter VIII. It is unlikely that any major project will be built nonunion in Philadelphia in the immediate future, although the AGC affiliate is in the process of changing its bylaws to accept nonunion firms for the first time (particularly in order to accommodate those who build in the suburban counties).

Washington, D.C.

In contrast to Philadelphia, Washington, D.C., has been an enormous success for the open shop. Open shop contractors first captured work in the Maryland and Virginia suburbs of the city. These open shop contractors built first low-rise, then high-rise apartments, and from this experience, began to build multistory commercial buildings and offices. By the early 1970s, these open shop builders were equipped to bid on larger office and apartment buildings within the city itself. The fact that office buildings within the District of Columbia are limited in height by zoning restrictions aids those open shop contractors who had previously had limited high-rise construction experience.

The open shop segment, bolstered by doublebreasted concerns of unionized companies, began eroding the union share of business in Washington in 1972. During this period, union members were fully employed on the construction of the Washington subway and on other large projects. At the same time, the unions won substantial wage increases and other benefits from the employers which spurred demands from users for open shop bidders on jobs. Poor union worker productivity and lack of adequate mechanics was also a factor.

Figure IV-12 shows that the unionized contractors' share of work in Washington eroded rapidly in 1975 and 1976, although some of this was recovered during the next two years. After 1978, however, the share of work done by the unionized contractors once again deteriorated, and a major push initiated by the unions in 1980 did not change the situation. A study based upon Dodge Reports, made

FIGURE IV-12
Union and Open Shop Construction Volume
Percent Share
Washington, D.C., Area
1972–1983

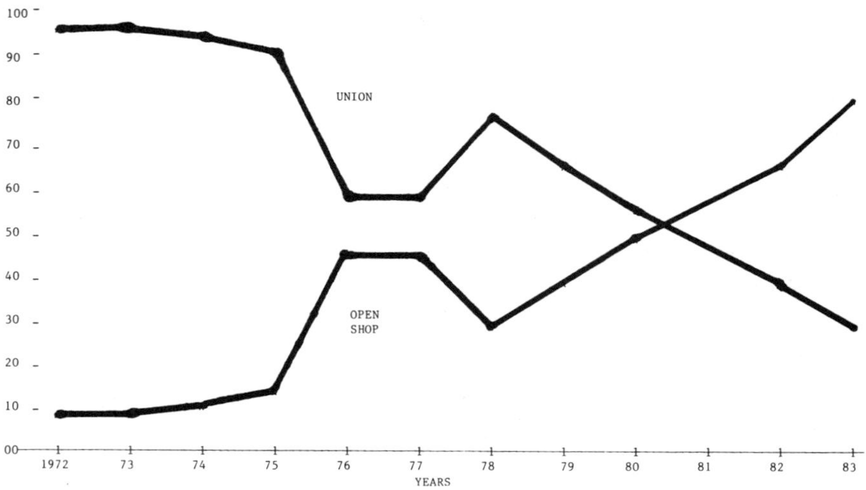

Source: Adapted from *The Business Review of Washington,* September 5, 1980, p. 19; and *Washington Post,* November 2, 1983, pp. A1, A10.

by the Washington Building and Construction Trades Council, showed that in the first ten months of 1983, in Washington and the contiguous Virginia and Maryland areas, only 239 projects valued at $686 million, of a total of 2,650 projects valued at $2.9 billion, went to unionized contractors. Within the city itself, the unionized group won 25 percent of the jobs and 50 percent of the volume in this period. In the neighbouring Maryland and Virginia counties, the union share was only 13 percent of the jobs and 4 percent of the volume.[58]

Columbus, Ohio

Since 1972, the Central Ohio Division of the AGC has compiled data on the inroads of open shop contractors into the commercial construction market. It begins with the projects reported in the

[58] A summary of these developments, which corroborates our field interviews, is found in Esther Smith, "Unions Want to Close Open Shops," *The Business Review of Washington,* September 5, 1980, pp. 19, 21; and in Peter Perl, "Unions Lose Jobs, Clout in Construction," *Washington Post,* November 2, 1983, pp. A-1, 10.

Dodge Reports for the Columbus area, and on a monthly basis identifies the union or nonunion status of each general contractor. The projects surveyed are limited to commercial and industrial building. The findings for the years 1972 through 1982 are given in Table IV-9.

The figures in Table IV-9 show a steady nonunion share of about 40 percent of dollar volume over the period covered (which may well have succeeded the years of the most dramatic open shop gains). According to the former executive of the Central Ohio Division of the AGC, if negotiated building awards were included in this survey, the open shop would have at least 50 percent of the dollar volume.[59] The 50 percent estimate was confirmed by the secretary-treasurer of the Columbus Building Trades.[60]

The union sector in Columbus has managed to retain a strong position for two reasons. First, as the capital of the state of Ohio and the home of Ohio State University, Columbus enjoys considerable state and federal government work which comes under the purview of the Davis-Bacon Act and a similar state law. With unions still in control of a majority of the large commercial and industrial work, union wage rates are required under such construction which blunts open shop competition.

Second, the construction unions in the Columbus area have endeavored through cooperative efforts with the unionized contractors to control costs and to improve productivity and the quality of work performed through a joint organization, Management and Organized Labor Striving Together, known as "Operation Most." This organization is devoted to promoting the use of union construction labor by eliminating disputes and improving productivity. This approach, which eschews the direct action and violence featured in the Philadelphia area, has been copied in a number of other areas. Nevertheless, serious strikes have occurred in Columbus as recently as June, 1982, because some crafts have demanded higher than pattern settlements.[61] This, of course, provides the open shop with a key advantage.

Detroit and Michigan Electrical Work

We have discussed at length the emerging strength of the open shop in the highly skilled trades as it pertains to the national scene.

---

[59] Interview, Columbus, Ohio, August 20, 1981.

[60] "Iron Workers, Plumbers and Fitters Resist Columbus Pattern Settlements," *Construction Labor Report*, Vol. 28 (June 23, 1982), p. 615.

[61] *Ibid.*, pp. 615-16.

TABLE IV-9
Permit Awards, Union and Nonunion
Columbus, Ohio Area, 1972–1982

	Union				Nonunion			
Year	Dollar Volume[a]	Percent	Jobs	Percent	Dollar Volume[a]	Percent	Jobs	Percent
1972	190.4	59.3	276	39.4	130.5	40.7	425	60.6
1973	220.5	62.8	266	38.7	130.4	37.2	422	66.3
1974	228.7	62.7	258	48.7	136.2	37.3	272	51.3
1975	133.1	61.6	120	35.3	84.9	38.4	220	64.7
1976	273.3	68.7	174	38.5	124.7	31.3	278	61.5
1977	174.0	52.6	214	44.8	157.3	47.4	264	55.2
1978	264.5	62.7	177	37.3	157.9	37.3	297	62.7
1979	303.0	62.8	190	40.4	178.9	37.2	280	59.6
1980	312.9	68.4	105	37.9	144.4	31.6	205	62.1
1981	317.4	66.9	115	29.5	156.8	33.1	275	70.5
1982[b]	130.2	52.2	55	29.2	119.2	47.8	133	70.8

Source: Associated General Contractors of Ohio, Central Ohio Division.
[a] Volume in millions of dollars.
[b] First six months only.

TABLE IV-10
Building Permits[a] *by Union Status and Type of Construction Southeastern Michigan, May 1973–May 1974*

Number and Percent	Number of Permits Classified	Residential (Excluding homeowner)	Commercial	Industrial	Institutional
Number of permits	74,163	63,407	8,684	1,125	947
Percent union	25.3	20.2	55.1	57.7	58.2
Percent nonunion	74.7	79.8	44.9	42.3	41.8

Source: *Construction Labor Report,* No. 1000 (December 11, 1974), p. A-3.
[a] Of the 88,000 permits, 74,163 disclosed name of company. These did, however, provide information on type of work.

The situation in the electrical craft in Detroit and Michigan affords an opportunity to examine this development at the local and state levels.

In mid-1974, the International Brotherhood of Electrical Workers, Local No. 58, which has jurisdiction over Detroit and southeastern Michigan, and the Southeastern Michigan Chapter of the National Electrical Contractors Association, made a study to determine the amount of work being done nonunion within Local No. 58's jurisdiction.

> Essentially, the survey consisted of examining every electrical permit issued by some 123 municipalities in our wage area during the period of May, 1973, through May, 1974. Approximately 88,000 permits were examined and classified as residential, commercial, institutional and industrial. Also it was determined whether they were drawn by union or nonunion contractors.[62]

In this period, the total electrical contracting was $249,623,500 for the area, of which $135,619,400, or 54.3 percent, was classified as union, and $114,004,100, or 45.7 percent, as nonunion (see Table IV-10).

In a letter to the membership, Jack Jones, business manager of Local No. 58, stated that considerable business had been lost to ABC contractors who "can guarantee a prospective customer that there will be no jurisdictional work stoppages, no strikes . . . and can generally provide his services at a substantially cheaper price." He also felt that Local No. 58 had "lost much small work to nonunion firms that belong to no organization and we obviously lost much

[62] "Detroit Electricians Local Warns Members of Inroads Being Made By Open Shop Contractors," *Construction Labor Report,* No. 1000 (December 11, 1974), p. A-3.

change-over and renovation types of work to inplant forces, some of whom are organized and some of whom are not."[63]

That the union share of electrical work has continued to erode in Michigan and, presumably, in Detroit, which is the strongest union construction area in Michigan, is demonstrated by the data in Table IV-11. It shows the number and percentage of Michigan inside electrical employers and employees who are under contract with the International Brotherhood of Electrical Workers (IBEW). These data are from the NECA surveys discussed above in connection with Table IV-5, and cover the period 1973 to 1980. As discussed there, "other" and nonunion are largely, if not entirely, synonymous.

Table IV-11 generally follows the trends shown in Table IV-5. The percentage of employers under contract to the IBEW declined from 50 to less than 40 percent. Nevertheless, the IBEW held on to the biggest jobs and, therefore, represented a majority of the employees in most years with the percentage varying as affected by large job awards. It is very significant that in the most highly unionized state in the country, and a state in which the Big Three automobile companies always build union, this highly skilled craft continues to see its position decline in terms of employers unionized to the point where in 1980 the IBEW represented only about 50 percent of the employees under its jurisdiction.

Atlanta: Penick Study

A careful study of open shop activity within the city of Atlanta was done at the Georgia Institute of Technology for the period from 1961 to 1972 by Joel H. Penick.[64] This study examined building permits for projects valued at $80,000 or more. After presenting these data, more recent developments will be examined.

By excluding small projects, negotiated awards, and suburban construction, the Penick study tends to understate the overall extent of nonunion construction. The union status of firms identified in building permits was ascertained through consultations with a nonunion contractor and an official of the local building trades council. In a number of cases, however, the union-open shop status of the job could not be specified, either because the work was performed entirely by subcontractors and/or by individuals unattached to the

---

[63] *Ibid.*, p. A-4.

[64] Joel H. Penick, "Building Construction Awards in Atlanta, 1961-1972," mimeographed report prepared by Georgia Institute of Technology, Atlanta, Georgia, 1973.

TABLE IV-11

IBEW and Other Inside Electrical Employers and Employees

Michigan, 1973–1980

	Employers					Employees				
Year	Total	IBEW	Percent IBEW	Other	Percent Other	Total	IBEW	Percent IBEW	Other	Percent Other
1973	1,303	652	50	651	50	13,822	6,502	47	7,320	53
1974	1,325	656	50	669	50	13,400	6,432	48	6,968	52
1975	1,290	629	49	661	51	10,222	6,056	59	4,166	41
1976	1,332	634	48	698	52	10,500	5,788	55	4,712	45
1977	1,405	502	36	903	64	11,952	7,413	62	4,539	38
1978	1,471	515	35	956	65	13,916	8,851	64	5,065	36
1979	1,483	573	39	910	61	14,050	7,695	55	6,355	45
1980	1,440	555	39	885	61	12,616	6,553	52	6,063	48

Source: National Electrical Contractors Association, *Inside Electrical Construction Survey for the Years 1973-1979 and 1980*, March 1982.

general contractor (day labor) or because the general contractor was unknown to either of the author's consultants. For the years covered, the proportion of unknown contractors ranged between 14 and 40 percent of all jobs, accounting for 7 to 31 percent of total volume. Since the building trades council spokesman believed that most day labor work was nonunion and that contractors unknown to him were most likely open shop, Penick calculated the union-nonunion distribution with the assumption that 90 percent of the unknown category was open shop.

Table IV-12 summarizes the Penick study's findings for 1961 to 1972. With the exceptions of 1963 and 1972, the data show a steady increase in open shop activity. Penick explains the large 1963 jump in the open shop share of building in terms of a sharp increase in apartment construction. The 1972 increase in the union construction share was attributed to a few exceptionally large projects, including a $36 million office, hotel, and apartment complex. The study showed that, among projects of $50,000 or more, the nonunion share of jobs rose from 25.2 percent in 1965 to 53.6 percent in 1971. Even among the very largest jobs there is a nonunion presence. Thirty percent of the $2 million and over projects (which include large apartment complexes) during all years of the study were nonunion, including three of the twenty-four which were valued at over $5 million.[65]

Inexplicably, Penick concludes that "although union contractors have been obtaining fewer jobs, they have generally maintained their dollar share of the market."[66] His own data, partially reflected in Table IV-12, appear to contradict such a conclusion. Except for what Penick himself notes to be the unusual year of 1972, his findings reveal a fairly steady decline in the union share of commercial building, even among projects valued at $500,000 or more. For smaller jobs ($80,000 to $500,000), the union share was virtually cut in half (to 26 percent in 1972), and these data do not include a predominantly open shop market below $80,000. These small jobs, moreover, become significant in the aggregate; in the years covered by the survey they constituted from about one-third to over one-half of all building construction. And finally, as observed earlier, the limiting of the data to Atlanta itself almost certainly served to understate the penetration of open shop construction in the metropolitan area for the period studied.

---

[65] *Ibid.*, p. 24.
[66] *Ibid.*, p. 63.

TABLE IV-12
Percent Distribution of Union and Open Shop[a] Construction in Atlanta
Selected Years, 1961–1972

	Volume		Number of Projects	
Year	Percent Union	Percent Open Shop	Percent Union	Percent Open Shop
1961	74.7	25.3	60.4	39.6
1963	54.2	45.8	41.6	58.4
1965	73.9	26.1	55.7	44.3
1967	69.7	30.3	49.2	50.8
1969	60.3	39.7	43.3	56.7
1971	53.4	46.6	38.6	61.4
1972	72.7	27.3	39.9	60.1

Source: Joel H. Penick, "Building Construction Awards in Atlanta: 1961-1972," mimeographed report prepared by Georgia Institute of Technology, Atlanta, Georgia, 1973.
[a] Assumes unknown to be 90 percent nonunion.

Atlanta Developments, 1972–1983

Confirmation that the union contractors' share of work in the Atlanta area is declining has come from the unions themselves. In June 1974, unions affiliated with the North Georgia Building and Construction Trades Council signed a new productivity agreement calling for "the elimination of featherbedding, illegal strikes and nonworking stewards," and dealing also with "employer hiring prerogatives, overtime, safety, jurisdictional disputes and grievance procedures."[67] Previous agreements along these lines had been negotiated one year earlier covering the electrical, mechanical, plumbing, and steamfitters trades. According to the article reporting these agreements, open shop contractors were doing 10 to 30 percent of the work in Atlanta in 1973, but as high as 51 percent in mid-1974.[68] Loss of work is, of course, the key factor which induces unions to grant concessions.

By 1971, the open shop was strong enough in Atlanta to support a chapter of the ABC. At that time, all major jobs inside the perim-

[67] "Atlanta Building Trades Offer Productivity Agreement," *Engineering News-Record,* June 13, 1974, p. 43.
[68] *Ibid.*

eter highway (Interstate 285), which extends well out into the suburbs, were built union. The open shop began with just two and three story buildings outside of the perimeter highway. During the next several years, however, there was a tremendous amount of federally funded prevailing wage (Davis-Bacon) work, including the building of roads, highways, a mass transit system, and an airport expansion. Many companies moved to Atlanta and the population burgeoned. Because the unions were occupied with the prevailing wage work, the open shop grew in capability and built numerous office parks just outside of the city. In 1981, the ABC contractors built three projects within the city with a total square footage of 432,000. The first large open shop project in the city of Atlanta was finished in 1982. From then on, open shop construction in the inner perimeter grew rapidly. In October 1983, ABC contractors had forty-three projects under construction in Atlanta totalling nearly 10 million square feet. This work included an extensive highway reconstruction project, much of the area's new transit system, plus hotels, office buildings, shopping centers, department stores, and other commercial facilities.[69]

The percentage of work in Atlanta which was open shop in 1982 cannot be estimated with the accuracy of a study but both union and nonunion contractor and association officials interviewed put the figure at close to 50 percent within the city, with the open shop in the clear majority on awards and 40 to 50 percent by volume. If the suburbs are included, the open shop was estimated to have 65 percent of the jobs, and 55 percent of the volume. By late 1983, the open shop claimed 90 percent of the Atlanta SMSA volume.[70] The unions demonstrated their serious problem again in 1983 by agreeing to substantial concessions, including 16 percent average pay reductions and such rule changes as working Saturdays straight time on make-up days, permitting four-day, ten-hour schedules without daily overtime rates, allowing special lower wages on residential and smaller commercial jobs, initiating a new "starter" classification at sub-apprentice wages, eliminating the use of unnecessary personnel, and other productivity improvements.[71] Prior to this

---

[69] "Merit Shop Blooms as Atlanta Booms," *Builder & Contractor,* October 1983, pp. 23-28.

[70] *Ibid.,* and Atlanta field interviews, July 1982; and "Atlanta Building Trades' Project Phoenix Is Response to ABC's 'Open Shop War'," *Construction Labor Report,* Vol. 29 (November 30, 1983), pp. 1326-28.

[71] "Atlanta Carpenters Agree to Wage and Work-rule Cuts to Regain Jobs," *Construction Labor Report,* Vol. 28 (January 19, 1983), pp. 1597-98; and "Building Trades in Atlanta Sign Pacts with Pay Cuts Averaging 16%," *Construction Labor Report,* Vol. 29 (July 20, 1983), pp. 638-85.

agreement, the North Georgia Building Trades Council had begun a program to "revitalize organized labor in the [Atlanta] construction industry," but the trend continues toward the open shop, as the 1983 wage concessions and a second revitalizing program indicate.[72]

New Orleans

On October 16, 1974, the chairman of the Labor Committee of the Associated General Contractors of Louisiana, Inc., New Orleans District, composed exclusively of union contractors, wrote the building trades unions in the New Orleans area as follows:

> From time to time we have advised you of the growing influence of open shop contractors in the New Orleans area. We would like now to furnish you with a list of jobs which have been awarded to open shop contractors in our area during the months of January through September 1974, which totals $105,097,541. The list encompasses those types of jobs or projects which have heretofore been performed by union contractors.
>
> Our contractors feel that open shop contractors have a distinct competitive advantage over them in that they, the open shop contractors, are not required to abide by the restrictive and costly conditions contained in our collective bargaining agreements. This information is being sent to you in the hope that you will recognize this situation and agree to modify the respective agreements in order to make our contractors more competitive. If the present conditions are allowed to continue, open shop contractors will eventually dominate the construction market as we have known it and union contractors will, for the most part, cease to exist.
>
> We will, of course, be delighted to discuss this subject further with you, including any of the projects contained in the enclosed list. We would like to point out that the information contained in the list was obtained by us through the reporting services which we subscribe to and, consequently, there may be other projects which have been awarded to open shop contractors of which we are not aware. We would also like to point out that the contractors included in the list were classified by your office as being primarily open shop contractors.[73]

Included in the list, which ran seventeen pages, were contracts as low as $3,000 and several over one million. The largest was a $7 million apartment development, and several were large highway jobs. The bulk, however, consisted of every imaginable commercial construction project: offices, stores, shopping centers, service sta-

---

[72] "North Georgia Building Trades Revitalize Unionized Construction Industry," *The Journal of Labor (Atlanta),* December 18, 1981.

[73] Copy of letter and survey in author's possession.

TABLE IV-13
New Orleans Area
Open and Union Shop Construction Contract Awards
1977–1983
(in millions of dollars)

		Union Shop		Open Shop	
Year	Total Volume	Dollar Volume	Percent	Dollar Volume	Percent
1977	353.5	166.4	47.1	187.2	52.9
1978	354.2	199.0	56.2	155.2	43.8
1979	479.3	274.3	57.2	205.0	42.8
1980	546.6	317.9	58.2	228.7	41.8
1981	755.8	462.2	61.2	293.6	38.8
1982	728.8	480.7	66.0	248.1	34.0
1983[a]	270.4	115.2	42.6	155.2	57.4

Source: Associated General Contractors of Louisiana, Inc., At-Large District.
[a] First six months.

tions, warehouses, banks, factories, schools, college buildings, churches, etc. By this time the ABC had formed a fast-growing chapter in New Orleans and had already established several training programs.

Since then, the AGC has formed an "at large" district, a pseudonym for an open shop branch which in 1984 merged with the New Orleans district after the latter changed its bylaws to admit open shop concerns. This district has tracked the construction awards by volume since 1977 and their results are presented in Table IV-13. The number of awards are not available, but our interviews indicate that they would show a majority for the open shop. Moreover, if negotiated contracts were included in Table IV-13, the open shop share may well have been higher.[74] Table IV-13 shows, however, that the open shop share declined steadily from 1977 through 1982 and then rose dramatically during the first six months of 1983. Whether this is indicative of a new trend or the accident of a few awards is not clear. The committee for the forthcoming New Orleans World Fair has asked foreign participants to build union which could revive union construction in the area.

As in many other areas, the capability of the open shop continues to grow in the New Orleans region. Between 1981 and 1983, an open shop contractor built Heritage Plaza, a $20 million, twenty

[74] This was the comment both of union and open shop association personnel in May 1982 interviews.

story office complex in Metairie, a New Orleans suburb.[75] Downtown high rise office buildings remain a union province, but our 1982 field interviews and 1983 telephone contacts indicate a further erosion of union control outside of the city and in other parts of Louisiana.

Houston

Houston is basically an open shop area, but building construction in the city itself has remained union for major projects, particularly large office buildings which have been built there in great numbers since 1970. While the unions were concentrating on work in the city, the open shop grew dramatically in outlying Harris county (where some high rise office buildings have also been built), in the process industries' plant construction, and in maintenance work on the Texas gulf coast, particularly south and west of the city. Table IV-14 shows the percentage of union and open shop jobs for the city and Harris county beyond the city limits for 1977, 1979, 1980 and 1982, and the volume for the last two years, as developed by the Houston Chapter of the AGC. We shall examine the Gulf Coast process industry data in the following chapter.

The data in Table IV-14 show that the large innercity projects are clearly union-dominated with unionized contractors taking 64 percent of the volume in 1982, but only 36 percent of the jobs, whereas the open shop won 80.5 percent of the much smaller volume and 70.8 percent of the jobs in the surrounding county. Unionized construction had 52.1 percent of the total volume but only 36 percent of the jobs for 1982 projects in the combined city and county. The data for earlier years follow the same general pattern although varying from year to year.

The union-open shop commercial construction picture in Houston is actually more complicated than these figures indicate. A large number of jobs are being built by mixed union and nonunion forces. The mixture has become so prevalent in the area that the general contractors often do not bother to set up separate gates for the two groups. There were, moreover, recurrent rumors tht open shop builders would invade the downtown office construction market, but the reluctance of owners and the collapse of the office building boom in 1982 thus far has precluded this. It is clear, however, that open shop builders have the capability to build large office complexes in Houston because of their work in Harris county, and it

[75] "Merit Shop Stands Tall in New Orleans," *Builder & Contractor,* May 1983, p. 30.

TABLE IV-14
Houston Area Permit Survey, Commercial Construction
Number of Jobs and Dollar Volume, by Union and Open Shop
1977, 1979, 1980 and 1982

(Houston)

	Jobs					Volume (millions)				
		Union		Open Shop			Union		Open Shop	
Year	Total	No.	Percent	No.	Percent	Total	Amount	Percent	Amount	Percent
						$	$		$	
1977	3,314	858	25.9	2,456	74.1	n.a.	n.a.	54.0	n.a.	46.0
1979	3,552	866	24.4	2,686	75.6	n.a.	n.a.	61.0	n.a.	39.0
1980	3,546	966	23.2	2,580	76.8	1,350.2	850.6	63.0	499.7	37.0
1982	3,293	1,184	36.0	2,109	64.0	1,865.8	1,206.7	64.7	659.1	35.8

(Harris County outside of Houston)

Year	Jobs					Volume (millions)				
		Union		Open Shop			Union		Open Shop	
	Total	No.	Percent	No.	Percent	Total	Amount	Percent	Amount	Percent
						$	$		$	
1977	1,025	96	9.0	929	91.0	n.a.	n.a.	39.0	n.a.	61.0
1979	987	56	6.0	931	94.0	n.a.	n.a.	38.5	n.a.	61.5
1980	801	43	30.0	758	69.8	204.8	61.9	30.0	142.9	69.8
1982	873	34	3.9	839	96.1	410.0	80.0	19.5	330.0	80.5

TABLE IV-14 (continued)

(Total Harris County and Houston)

	Jobs					Volume (millions)				
		Union		Open Shop			Union		Open Shop	
Year	Total	No.	Percent	No.	Percent	Total	Amount	Percent	Amount	Percent
						$	$		$	
1977	4,339	954	22.0	3,385	78.0	n.a.	n.a.	49.5	n.a.	50.5
1979	4,539	922	20.0	3,617	80.0	n.a.	n.a.	57.0	n.a.	43.0
1980	4,347	1,009	23.2	3,338	76.8	1,555.1	912.5	58.7	642.6	41.3
1982	4,166	1,218	29.2	2,948	70.8	2,275.8	1,286.6	52.1	1,089.1	47.9

Source: Houston chapter, Associated General Contractors of America.
n.a. = not available.

should be noted that they have already completed large projects in Dallas.

The decline of energy prices which began in 1982, together with the recession, led to cutbacks by numerous companies and the collapse of the office building boom in Houston, Denver, and other cities. By October 1982, Houston had a vacancy rate of 11.4 percent and a planned space increase of 34.4 percent (that is, office construction under way as a percentage of existing space). Among large cities, only Denver had a larger overcapacity.[76] By July 1983, the situation in Houston had worsened because of a huge amount of space available for sublease.[77] Just what effect this will have in the long run to the union hold on large inner city commercial construction remains to be determined. The open shop, as we shall point out in Chapter V, has also been hard hit by the decline in industrial plant construction, but is showing signs of renewed growth. The Houston unionized trades have accepted area-wide project agreements which first froze wages and then in 1984 provided for a 15 percent wage reduction.[78] The unionized contractors and unions have also agreed on a productivity plan which, among other things, eliminates pay for some time not worked and the employment of unnecessary personnel. This was deemed necessary because "nonunion contractors have been securing large shares of construction work once considered union-controlled. Labor lost its edge [according to officers of the Houston AGC and construction unions] . . . because 'archaic' contracts kept union companies from maintaining a competitive edge."[79]

The building trades also participated in the much publicized Houston organizing plan of the AFL-CIO. After claiming initial success, however, the trades have apparently transferred their effort to Dallas which is perhaps more open shop in large commercial construction than Houston.[80] But unless the unions can hold their edge

---

[76] "End of the Office Boom. Developers Face Financial Ruin. Bankers Are Left Holding the Bag," *Business Week,* October 4, 1982, p. 97.

[77] Randall Smith, "Glutted Office Space Market may not Rebound Until 1984," *Wall Street Journal,* July 13, 1983, p. 33.

[78] Copy of the Agreement, dated January 19, 1983, in author's possession; and telephone interview, AGC Houston executive, May 17, 1984.

[79] Mike Cinelli, "Labor, Contractors Design Productivity Plan," *Houston Chronicle,* May 12, 1983, sec. 1, p. 1.

[80] Susan Dunlop, "Houston Project Lays Solid Base for Organizing," *AFL-CIO News,* January 8, 1983, pp. 1, 13. The article claims that the drive had "been recruiting new members among skilled journeymen, in addition to their already extensive apprenticeship programs." Since then, no claims have been made about construction recruiting; and Laurie Lewis, "Union Shifting Emphasis from Houston to Dallas," *Carrollton [Texas] Chronicle,* June 29, 1983, pp. 1-2. The claim in this article is that

in Houston's office building construction, they will have no center of work control there. As we shall see in Chapter V, the unions are, however, making a determined effort to recapture Gulf Coast industrial work and have made substantial concessions to do so.

Colorado

As discussed in Chapter III, the Colorado Construction Industry Management Council and the Colorado Building and Construction Trades Council created the Construction Research Project and utilized the University of Colorado to study the union-open shop division of jobs in the state. For the study, Dodge Report awards were examined for the first six months of 1973, 1979, and 1982.[81] As presented in Table IV-15, this study divided commercial construction into four subdivisions—building construction less than $7.5 million, building construction greater than $7.5 million, school construction, and tenant finishing construction[82]—and presented the data separately for the Denver metropolitan area and for Colorado outside of this area. The study found that commercial construction comprised 62 percent of the jobs and 69 percent of the dollar volume of all construction reported in the first six months of 1982 in the Dodge Report for Colorado. Within the Denver metropolitan area, commercial construction accounted for 65 percent of the jobs and 81 percent of the dollar volume, whereas in other Colorado areas the percentages were 58 percent of the jobs and only 46 percent of the dollar volume.[83] Large structures in cities and heavy, highway, and industrial work outside of cities can account for major differences in types of construction between cities and outlying areas.

The data in Table IV-15 show that throughout the period studied, unionized builders controlled the projects over $7.5 million in the Denver metropolitan area, accounting for 55 percent of the awards and 66 percent of the dollar volume. It is interesting, however, that in this category the open shop provided strong competition in all years, winning 38 to 50 percent of the awards and 17 to 47 percent of the dollar volume. This means that the capability of the open

---

there was less unemployment in Dallas than Houston, and the unions felt that this gave them a better opportunity in the former city where they had lost substantially in commercial construction.

[81] David E. Moore, *Union and Open Shop Construction in the State of Colorado* (Denver: Construction Industry Management Council and Colorado Building and Construction Trades Council, 1983).

[82] Tenant finishing construction is the interior finishing of a building by the owner or lessor.

[83] Moore, *Union and Open Shop Construction*, p. 5.

TABLE IV-15
Percentage Union and Open Shop Commercial Construction by Category and Area, Colorado 1973, 1979, and 1982

Area		Construction under $7.5 million Percentage						Construction over $7.5 million Percentage						School Construction Percentage						Tenant Finishing Percentage	
		No. of Awards			Dollar Volume			No. of Awards			Dollar Volume			No. of Awards			Dollar Volume			No. of Awards	Dollar Volume
		1973	1979	1982	1973	1979	1982	1973	1979	1982	1973	1979	1982	1973	1979	1982	1973	1979	1982	1982[b]	1982
Denver Metro	Union	44	35	26	40	39	42	66	55	57	69	73	68	75	75	13	83	94	12	31	46
	Open Shop	56	65	74	60	61	58	33	45	43	31	27	32	25	25	87	17	6	88	69	54
State except Denver Metro	Union	37	19	8	48	26	23	—[a]	—[a]	50	—[a]	—[a]	52	62	25	0	54	27	0	0	0
	Open Shop	67	81	92	52	74	77	—[a]	—[a]	50	—[a]	—[a]	48	38	75	100	46	73	100	100[c]	100
Total Colorado	Union	42	28	18	44	34	34	50	62	55	53	83	66	70	44	7	71	63	7	22	30
	Open Shop	58	72	82	56	64	64	50	38	45	47	17	34	30	56	93	29	37	93	78	70

Source: David E. Moore, *Union and Open Shop Construction in the State of Colorado* (Denver: Construction Industry Management Council and Colorado Building and Construction Trades Council, 1983), Tables 2, 3, and 4, and Figure 5.

[a] Insufficient data for analysis.
[b] Earlier data not available.
[c] Just five awards.

shop is increasing and poses a threat to the union builders in their last stronghold.

In all other categories of commercial construction, Table IV-15 shows the open shop to be dominant and increasing its share. In 1982, it held 82 percent of the awards and 64 percent of the dollar volume of commercial construction under $7.5 million for all of Colorado. From awards of 30 percent and dollar volume of 29 percent in 1973, the open shop literally took over school construction in Colorado, obtaining 93 percent of the awards and an equal percentage of the dollar volume in 1982. Finally, in 1982, the open shop also had about 75 percent of the tenant finishing work in the state.

This same study also attempted to determine subcontractor awards for the first six months of 1982 by a mail and telephone survey of general contractors. The data thus collected showed that unionized mechanical contractors received 35 percent of the awards and 56 percent of the dollar volume, unionized electrical subcontractors 37 percent and 68 percent, respectively, and unionized sheet metal subcontractors 37 percent and 46 percent, respectively.[84] Using these data, and assuming that subcontractors perform 70 percent of the dollar volume, the study estimated that unionized contractors controlled 60 percent of all commercial construction dollar volume in the state for 1982.[85]

Again, it should be emphasized that the data exclude negotiated awards which could alter the results. In addition, the methodology employed to obtain subcontractor data may be somewhat biased since unionized general contractors who utilize open shop subcontractors are not anxious to broadcast that information. Moreover, open shop contractors sublet less than their unionized counterparts and therefore the assumption of 70 percent of volume for subcontractors may not apply to the open shop general contractors. Given the highly unionized subcontractor dollar volume reported, the methodology used could well overstate the total unionized dollar volume.

Other data would also support a lower unionized dollar volume of commercial construction in Colorado. In 1980, the NECA annual survey found that only 24 percent of the inside electrical employers in Colorado had IBEW contracts and 38 percent of the inside electrical employees were IBEW represented, as compared with 39 percent and 43 percent respectively in 1973.[86] In seeming contrast, a

---

[84] *Ibid.,* Tables 8a, 9, and 10.

[85] *Ibid.,* Table 13.

[86] National Electrical Contractors Association, *Inside Electrical Construction Survey For the Years 1973-1979 and 1980,* March 1982, Colorado data.

study by NECA's Rocky Mountain chapter—covering twenty-one of the most populous counties in Colorado, including Denver—found that union electrical contractors had 65 percent of the commercial work in 1980.[87] This latter study, however, is based on permits and is heavily weighted toward Denver, the state's union stronghold.

In the sheet metal trade, the information is also mixed. The Arthur D. Little study for SMACNA showed that unionized sheet metal contractors had about 50 percent of the Colorado construction business in 1977, but only 37.5 percent in 1982.[88] Within Denver, however, the unionized sheet metal contractors filed 38.8 percent of the work permits, but retained over 60 percent of the dollar volume in the period from 1981 through 1982. These data are drawn from a study by SMACNA's Denver, Pueblo and Colorado Springs chapter (see Table IV-16). This study again covers the traditional union stronghold, but it also shows a diminution of the unionized share since 1975. Perhaps more significantly, this SMACNA chapter reports that from 1976 to 1982, twenty-one union sheet metal contractors went out of business, the same number went open shop, and thirteen became union contractors, for a net loss of twenty-nine unionized contractors.[89]

On the basis of all these data, we can conclude that commercial construction is about evenly divided between union and open shop forces in Colorado. Since mid-1982, however, the office building boom, in which union construction has been concentrated, has come to a halt. Denver, with a planned space increase of 40.4 percent, the highest of any large city, had a vacancy rate of 15.8 percent in October 1982, again the highest in the country.[90] All new construction of office buildings and complexes in Denver not already under way has been halted and virtually no new office construction is expected to be commissioned till at least 1985, especially since by mid-1983 the Denver vacancy rate had climbed to 19.3 percent.[91] The only possibility of earlier revival would be a turnaround in the energy industries which had originally sparked the boom in Denver office construction.

The net effect of the decline of office building construction in Denver has been a sharp decline in the extent and percentage of

---

[87] National Electrical Contractors Association, *1980 Market Survey* (Denver: Rocky Mountain Chapter, National Electrical Contractors Association, 1980), p. 81.

[88] See Figure IV-8, above.

[89] Data from Denver, Pueblo, and Colorado Springs Chapter, Sheet Metal and Air Conditioning Contractors' National Association.

[90] "End of the Office Boom," p. 97.

[91] Smith, "Glutted Office Space Market," p. 33.

TABLE IV-16
Building Permits
City and County of Denver
Heating and Other Sheet Metal Contractors

Year	Total		Union				Open Shop			
	No. of Permits	Dollar Volume	No. of Jobs	Percent	Dollar Volume	Percent	No. of Jobs	Percent	Dollar Volume	Percent
1975	1424	17,398,180	657	46.1	14,844,880	85.3	767	53.9	2,553,300	14.7
1976	1577	17,688,972	726	46.0	14,585,622	82.5	851	54.0	3,103,350	17.5
1977	1701	18,505,575	767	45.1	13,304,099	71.9	934	54.9	5,201,476	28.1
1978	2070	21,740,478	912	44.1	16,132,029	74.2	1158	55.9	5,608,449	25.8
1979	2722	30,640,387	1051	38.6	22,203,175	72.5	1671	61.4	8,437,212	27.5
1980	2426	21,788,226	966	39.8	15,144,000	69.5	1460	60.2	6,644,226	30.5
1981	2569	38,490,173	828	32.2	28,618,441	74.4	1741	67.8	9,871,732	25.5
1982, (1st Qtr.)	456	9,979,645	177	38.8	7,850,767	78.7	279	61.2	2,128,878	21.3

Source: Denver, Pueblo, and Colorado Springs Chapter, Sheet Metal and Air Conditioning Contractors' National Association, Inc.

union construction in the state. Since, as we shall document in Chapters V and VI, the open shop has one-half of the work in Colorado's industrial sector and is dominant in the highway and heavy sector, and, as we have set forth in Chapter III, in residential construction as well, the narrow base of union construction in the depressed large office market means that unionized builders and union members are in dire straits, with government work their sole large employment source. They have agreed to wage reductions and rule changes and are working with unionized contractors to improve productivity. The secretary of the Colorado Building Trades Council blames the open shop for the necessity of the givebacks, declaring that unions "will bleed for a while but they will come back stronger."[92] With open shop contractors continuing to increase their capability and union members, as the union officials admit, seeking work with such contractors, it remains to be seen whether the unions can regain their Denver share and, as their officials promise, "end up keeping their best people."[93]

California

The labor situations in Northern and Southern California are quite different from one another and therefore will be discussed separately. In principal cities of the state, however, the long and thorough dominance of the unionized contractors and unions has been shaken by gains of the open shop which began in the 1970s. A primary reason for the open shop gains has been the extraordinary wage increases won by unions. Table IV-17 shows wage and benefit rates for four key crafts in four California cities in or around June 1983. It is not remarkable given rates of this character, which had been occurring throughout the 1970s, that users and contractors commenced looking to the open shop as an alternative. Both Northern and Southern California are highly unionized, with recent open shop gains.

Southern California. The study prepared by the Construction Industry Research Board (CIRB), Los Angeles, of industry trends in eleven Southern California counties is, as noted in the previous chapter, a careful analysis of the changes that have occurred in this once union-dominated area.[94] Table IV-18 shows the percentage

---

[92] "Colorado Crafts Agree on Rollbacks; Carpenter Concession Talks Continue," *Construction Labor Report,* Vol. 29 (May 4, 1983), p. 324.

[93] *Ibid.*

[94] Construction Industry Research Board, *A Survey of the Construction Industry Trends in the Eleven-County Southern-California Region* (Los Angeles: Construction Industry Research Board, January 31, 1983).

TABLE IV-17
Wage and Benefit Rates, Four Crafts
Four California Cities, June 1983[a]

Craft	Los Angeles		San Francisco		Sacramento		Fresno	
	Wage & Benefit Rate $	Wage Rate $	Wage & Benefit Rate $	Wage Rate $	Wage & Benefit Rate $	Wage Rate $	Wage & Benefit Rate $	Wage Rate $
Carpenter	22.82	16.29	25.16	18.68	25.16	18.68	25.16	18.68
Electrician, inside	28.28	22.97	31.20	25.59	21.07	14.15	23.01	18.40
Plumber	26.12	18.69	36.29	24.64	26.43	19.72	26.33	20.90
Sheet metal worker	25.75	21.09	27.95	22.45	24.26	17.54	22.48	16.71

Source: *Handbook of Wages and Benefits for Construction Unions,* 1983 (Washington, D.C.: Construction Data Services, 1983), pp. 445–447, 452–454.

[a] Time is as close to June 30, 1983, as available; all rates are within sixty days of this time.

TABLE IV-18
Union Contractors' Market Share[a]
Commercial Construction
Eleven-County Southern California Region
1977 and 1981

Responses	Percentage Share 1977	Percentage Share 1981
Regional Total[b] (284)	74	54
Los Angeles (159)	74	56
Orange (37)	77	58
Ventura (22)	85	58
San Bernardino (16)	72	49
Riverside (15)	63	36
Santa Barbara (15)	79	54
Kern (8)	75	50
San Luis Obispo (6)[c]	46	18
Mono (4)[c]	15	15
Imperial (2)[c]	70	50
Inyo (0)	n.r.	n.r.

Source: *A Survey of the Construction Industry Trends in the Eleven-County Southern-California Region* (Los Angeles: Construction Industry Research Board, January 31, 1983), Table 6.
[a] All percentages are frequency weighted averages of the midpoints of 20 percent intervals.
[b] The regional total for each category of construction is the weighted average of all responses. Responses were weighted by the actual county share of commercial construction.
[c] Results for small-response counties are statistically unreliable. Variations as large as 20 percentage points could result with very small changes in the number of responses.
n.r. = no response.

share of the unionized contractors for 1977 and 1981 for each of the eleven counties and for the region as a whole.

In all counties, there was a decided decline in the unionized contractors' share. Overall, their share fell from 74 percent in 1977 to 54 percent in 1981, a decline of 20 percentage points. The decline was most pronounced in Riverside County, east of Los Angeles, but it was widespread throughout each region.

According to the CIRB report, a key factor precipitating the decline was the failure of the union sector to enroll new firms and to offset the number of former union firms which went nonunion.[95] Moreover, the CIRB believes that its report underestimates the extent of the open shop because "nonunion firms are underrepre-

[95] *Ibid.*, p. 3.

sented in the responses to the survey."[96] It also estimates that the "24 firms that left the union ranks since 1976 are representative of a much larger number of firms in the region." The CIRB estimates that number to be 443 with a combined volume of $323.4 million in 1976. Of these firms, 27.7 percent were in the commercial sector.[97]

The CIRB report compares union and nonunion firms in low-rise and high-rise commercial building as well as in the institutional subsectors in the eleven-county area. Again, it should be noted that nonunion firms are underrepresented in the survey. In the low-rise building industry, the distribution of work by dollar volume was fairly equal between union and nonunion firms, as were participation in public construction and county of work. The open shop firms, however, tended to be newer than their union shop counterparts.[98]

In the high-rise commercial building industry, however, the situation is very different. Fewer nonunion firms are represented; they do business in smaller volume categories, and do little in public works.[99] To what extent the underrepresentation of open shop firms in this grouping skewed the results is unknown. We do know that major office buildings in the area are a union preserve.

The industrial building data are also limited because of the small open shop response. Here again the open shop respondents handle smaller volume business, perform a smaller percentage of their volume on publicly-supported construction, and are newer firms.[100]

As Table IV-19 shows, the perception that the unionized firms have lost a considerable share of work is prevalent among both open shop and unionized respondents to the CIRB questionnaire. Although a greater proportion of nonunion firms believed that the unionized firms had lost ground, the unionized firms did not materially disagree, with very close to one-third of the latter finding that unionized firms had had their share of the market "substantially" decreased between January 1977 and December 1981.

Data from the Arthur D. Little study for the SMACNA seem to support the CIRB's general finding that unions in Southern California construction are still strong but have lost position. Thus, Figure IV-8 shows that unionized sheet metal contractors had about 78 percent of the Southern California business in 1977, but slipped to 66 percent in 1982.

---

[96] *Ibid.*, p. 6.
[97] *Ibid.*, pp. 16-17, 28.
[98] *Ibid.*, Table 17.
[99] *Ibid.*, Table 18.
[100] *Ibid.*, Table 19.

TABLE IV-19
Union and Open Shop Contractors' Responses Regarding Local Competition Eleven Southern California Counties

Commercial Construction Subsector and No. of Firms	How Share of Work Going to Union Contractors Has Changed Between January 1977 and December 1981					Average Value (−2 to +2)
	Percent Decreased		Percent Stayed the Same	Percent Increased		
	Substantially (−2)	Somewhat (−1)	(−0−)	Somewhat (+1)	Substantially (+2)	
Commercial low-rise:						
Union firms (79)	44.3	43.0	5.1	3.8	3.8	−1.20
Nonunion firms (33)	45.5	33.3	18.2	0	3.0	−1.18
All firms (projected)[a]	44.7	39.1	10.4	2.3	3.5	−1.19
Commercial high-rise:						
Union firms (22)	31.8	45.4	13.6	4.6	4.6	−0.96
Nonunion firms (4)	50.0	25.0	25.0	0	0	−1.25
All firms (projected)[a]	35.6	41.1	15.9	3.7	3.7	−1.02
Institutional:						
Union firms (20)	34.6	50.0	11.5	3.9	0	−1.15
Nonunion firms (6)	66.7	16.7	16.6	0	0	−1.50
All firms (projected)[a]	43.6	40.7	12.9	2.8	0	−1.25

Source: *A Survey of the Construction Industry Trends in the Eleven-County Southern-California Region* (Los Angeles: Construction Industry Research Board, January 31, 1983), p. 41.

[a] Values for all firms were calculated based on the number of responses from union and nonunion firms, adjusted for the underrepresentation and smaller size of nonunion firms. The number of nonunion firms responding was increased by a factor of 1.6 for calculation purposes.

The open shop has made other gains in Southern California. It now has one of the largest, if not the largest, chapters of the ABC, indicating the interest and success of open shop builders. An area overwhelmingly union in the early 1970s has moved substantially away from that category. Nevertheless, unionized firms still control a clear majority of the commercial work and nearly all of the high-rise commercial building. The fact, however, that there are open shop builders even in commercial high-rise construction certainly threatens further erosion of the unionized builders' position.

The unions have recognized this and reacted in diverse ways. The IBEW settled for a "modest increase" in 1983 after striking for six weeks in 1981 to obtain a $7.35 per hour increase over two years. The plumbers also granted concessions.[101] These and other unions, however, have struck to enforce a rule forbidding contractors to subcontract any work to nonunion contractors, or as in the case of the IBEW, won that provision in exchange for productivity improvements. We shall discuss the impact and legality of such restrictive clauses in Chapter VIII. It is sufficient to note here that it is debatable whether the unions gain more than they lose from restrictive clauses, since such clauses provide a strong inducement for contractors to go open shop or doublebreasted while probably increasing the cost differential between union and open shop work. This is true because such clauses prohibit acceptance of the lowest bidder and, more directly, rule out bids by the lowest potential bidders.

Northern California. Traditionally, Northern California is more unionized than Southern California. San Francisco is one of the strongest union cities in the country. Nevertheless, the open shop has made gains in commercial construction, particularly in the counties north, south, and east of San Francisco. This change, as in other parts of the country, became especially noticeable in the latter half of the 1970s. A study made by an industry group in Northern California in 1977 reported that 35 percent of the HVAC permits in Alameda and Contra Costa counties and 61 percent of the plumbing permits in the San Francisco Bay area, in a several months' sample in 1976, were held by open shop contractors. The report further found that 30 percent of all active licensed construction firms in 1966 were open shop, but that most worked on small

[101] "L.A. IBEW Trades Modest Pay Increases for Restrictive Subcontracting Clause," *Construction Labor Report,* Vol. 29 (June 8, 1983), pp. 516-17; and "S. Calif. Plumbing Contractors Seek Concessions in Contract Negotiations," *Construction Labor Report,* Vol. 29 (June 8, 1983), pp. 509-10.

"add-on" and renovation projects, presumably residential and small commercial. Finally, the report found that 58.2 percent of all building permits in excess of $30,000 in the Santa Clara Valley, south of San Francisco, went to nonunion contractors during a three month period in 1976.[102] About this time also, the ABC established its Northern California Chapter, and later the Northern California AGC Chapter ceased requiring adherence to its master agreement as a condition of membership.

In 1981, the Greater Oakland Chapter of SMACNA made its study covering Alameda and Contra Costa counties. For commercial work within the HVAC crafts, it found that union sheet metal subcontractors did 78.9 percent of the work in 1980, and 73 percent was projected for 1981-82. If, however, the three largest union firms were not counted, the small and medium-sized commercial work division would have been 65 percent union in 1980 and a projected 56.9 percent in 1981-82. This study further found that seventeen HVAC contractors, 23.2 percent of the total, would not sign new union contracts, a fact which threatened to send many of the small and medium-sized jobs to the open shop.[103]

The recession and the open shop incursion into their territory have induced the Northern California unions to cut back their very high wages, including plumber rates of over $30 per hour, and to reduce their counterproductive conditions, including pay for "Black Fridays"—days on which no work is done.[104] These unions still control the major work in the city, as well as much of the home-building and small commercial work, but they are losing out in the outlying areas. As noted, the ABC now has a chapter in San Francisco which has over 400 members, mostly small commercial builders. Thus there are cracks in one of the most unionized construction areas in the country, and the cracks are slowly widening.

In Sacramento, the state capital located in the central valley, the open shop has made greater strides. The NECA branch broke with

---

[102] As reported in "Open Shop Grows in the West," *Engineering News-Record,* June 23, 1977, p. 203.

[103] *State of the Industry: A Special Study Report and Assessment of the Impact of Open Shop Intrusion into the Heating, Ventilating, and Air Conditioning Construction Market in Alameda and Contra Costa Counties* (Oakland, Cal.: SMACNA Greater Oakland, June 1981, n.p.).

[104] See, e.g., "San Francisco Bay Area Crafts Adapt to Uncertain Times, Open Shop Specter," *Construction Labor Report,* Vol. 28 (May 12, 1982), pp. 363-64; "N. Calif. Laborers Agree Early on Wage Freeze / Rollback Plan with AGC," *Construction Labor Report,* Vol. 28 (January 19, 1983), pp. 1604-05; "AGC, North Calif. Carpenters Agree on Innovative Four-Year Contract," *Construction Labor Report,* Vol. 28 (September 15, 1982), pp. 1012-13; and "North Calif. Carpenters Agree on Three-Year Extension of Agreement," *Construction Labor Report,* Vol. 29 (May 4, 1983), pp. 330-32.

the IBEW and signed with an independent union at a lower rate.[105] The ABC chapter there estimates that commercial construction is only 20 percent open shop in its areas and that building in central Sacramento remains union. In 1982, however, a ten-story downtown building was being built by an ABC chapter member using some union subcontractors.[106]

Summary. California is probably the most unionized construction state but in the past decade there have been definite gains made by the open shop in this state. These gains have been smaller than those in many other areas, but they are important considering the relatively late start of the open shop development in California. The importance of the open shop in California is highlighted by one additional statewide statistic. In 1973, 79 percent of the inside electrical contractors and 63 percent of the inside electrical employees were covered by IBEW contracts; by 1980 these figures had fallen to 47 percent of the employers and 45 percent of the employees.[107] This loss, combined with the open shop gains among plumbers and sheet metal workers, surely indicates a problem for union construction in the state where it is the strongest overall but where the unionized wage structure and stringent working conditions are a constant incentive to seek open shop construction.

Pacific Northwest

The Pacific Northwest was also long one of the strongest unionized construction areas in the country. As in the case of California, we assumed for our original work that it was highly unionized and did not visit the area for field interviews. Our 1974 questionnaire (see Tables 23-25, below) confirmed our assumption. Thus, Table VI-23 shows that, in the Pacific region, 133 replies were received from union contractors, but only six from nonunion contractors. Moreover, only eight union contractors and none of the open shop ones believed that the open shop was increasing its share of work in 1974.

As we shall document, this situation has changed enough that in 1981 the ABC "targeted the Northwest as fertile ground for what

[105] "Striking a Blow for the Open Shop," *Engineering News-Record,* October 8, 1981, p. 78; "Electricians' War Continues," *Engineering News-Record,* October 14, 1982, p. 101; and "Divided NLRB Orders Hearing on Sacramento NECA-IBEW Situation," *Construction Labor Report,* Vol. 28 (September 22, 1982), pp. 1023-24. The matter is still in litigation.

[106] Interview with ABC official, Sacramento, July 26, 1982.

[107] National Electrical Contractors Association, *Inside Electrical Construction Survey,* California data.

has clearly become a national open shop movement."[108] As in California, exorbitantly high wages provided a strong incentive for change. The upward pull on wages in the Northwest was generated in the 1970s by the Alaska pipeline construction. By mid-1983, wages in the principal cities of the Northwest were close to those in the major cities of California.[109] By then, however, open shop contractors had won a significant portion of the work.

Oregon and Southern Washington. The Oregon-Columbia Chapter, AGC, located at Wilsonville, near Portland, Oregon, has studied contract awards for Oregon and southwest Washington since 1978. Table IV-20 shows these data through the first five months of 1982. As has generally occurred, the open shop made its first gains in the smaller commercial work, winning a majority of the job awards in 1980 while increasing its percentage of the volume from 9 percent in 1978 to 32 percent in 1980. Additional gains were made the following years, but large union jobs prevented the open shop from gaining a bigger percentage of volume even though they won about two-thirds of the awards during 1981 and the first five months of 1982. If negotiated awards had been included in these studies, the open shop's share might have been larger.

Additional confirmation of the decline of unionized construction in Oregon and southwestern Washington is provided in Table IV-21 which shows a decline in the number of contractors who were signatory to the master agreements negotiated by the Oregon-Columbia Chapter, AGC, between 1975 and 1982. The percentage decline for the various crafts ranged from 36 to 63 percent. Of course, some contractors went out of business and a few others bargained on an individual basis. The major reason for the decline, according to the staff of the Oregon-Columbia Chapter, is that so many companies went open shop or lost business to open shop contractors. By early 1984, the construction unions in Oregon agreed that open shop contractors had won a majority of work in that state and in southwestern Washington.[110] As a result, the chapter now admits open shop contractors and provides them with a hiring mechanism, group employee benefits, training, and other services.[111]

---

[108] "Taking Aim at Spokane's Unions," excerpted from *Spokane Magazine* in *Builder & Contractor,* April 1981, p. 16.

[109] *Handbook of Wages and Benefits for Construction Unions, 1983* (Washington, D.C.: Construction Data Services, 1983), pp. 469-83.

[110] "New Construction Labor Pact in Oregon Offers Several Innovative Features," *Construction Labor Report,* Vol. 29 (December 14, 1983), pp. 1371-72.

[111] Interviews, Oregon-Columbia Chapter, AGC, Wilsonville, Oregon, July 1982; and information supplied by chapter staff, August 1983.

TABLE IV-20
Union and Open Shop Contract Awards
Oregon and Southern Washington
January 1978–May 1982[a]
(in millions of dollars)

Year	Total		Union				Open Shop			
	No. of Awards	Dollar Volume	No. of Awards	Percent	Dollar Volume	Percent	No. of Awards	Percent	Dollar Volume	Percent
1978	—[a]	956.3	—[a]	—[a]	872.0	91	—[a]	—[a]	84.2	9
1980	1,808	770.5	813	45	521.9	68	995	55	248.6	32
1981	1,587	950.7	569	36	577.1[b]	61	1,018	64	363.6	39
1982[d]	485	382.3	182	38	251.9[c]	66	303	62	130.5	34

Source: *Construction Industry Profiles* (Wilsonville, Ore.: Oregon-Columbia Chapter, Associated General Contractors of America, June 30, 1982).

[a] 1979 data and number of awards for 1978 unavailable.
[b] Includes one $40 million job; union share would drop to 53 percent without this job.
[c] Includes one $60 million job; union share would drop to 59 percent without it.
[d] First five months only.

TABLE IV-21
Number of Contractors Signatory to Master Agreement, Associated General Contractors, Oregon and Southwestern Washington by Contractor Craft, 1975, 1980 and 1982

Craft	Number of Signatories 1975	1980	March 1982	Percent Decline 1975-1982
Carpenters	229	142	134	41
Cement Masons	154	104	96	38
Laborers				
Oregon	225	169	144	36
Washington	225	137	121	46
Operating Engineers	230	144	126	45
Teamsters	170	105	92	46
Ironworkers	169	72	63	63

Source: Oregon-Columbia Chapter, Associated General Contractors of America.

Seattle and Northern Washington. The open shop has also made gains in northern Washington. The Seattle Chapter, AGC, claimed that open shop contractors, "private and public constituted 41 percent of the markets."[112] In the electrical craft, however, a study by the Seattle Chapter of the NECA showed that, whereas union electrical contractors in that area were maintaining about the same minority share of work permits in 1981 that they had in 1977 (35 to 37 percent), they increased their proportion of the overall dollar volume of the work from 47 percent in 1977 to 64 percent in 1981. Moreover, in commercial construction alone in the Seattle area, the union contractors increased their share of work permits from 68 percent in 1977 to 75 percent in 1981 and increased their share of dollar volume from 83 to 89 percent in the same period.[113]

Despite this seemingly excellent position, the electrical contractors in Seattle saw their position deteriorate substantially after 1981. One reason was that substantial work on nuclear power facilities disappeared first with the cancellation of several facilities and then with the bankruptcy of the Washington Public Power Supply System.[114] In part because of potential work on this and

[112] *Construction Industry Profiles* (Wilsonville, Ore.: Oregon-Columbia Chapter, Associated General Contractors of America, June 30, 1982), Introduction, n.p.

[113] National Electrical Contractors Association, *Union Versus Nonunion Permit Report* (Seattle: Puget Sound Chapter, National Electrical Contractors Association, 1982), n.p.

[114] The public utility situation will be discussed in Chapter V. Five plants were begun, three were cancelled, and work on the balance ended when the system went bankrupt in mid-1983.

other major projects, and in part because of wages of $22.12 per hour, expensive fringe benefits, and low productivity, smaller jobs were taken over by open shop contractors. Even by 1980, the situation in the surrounding area, and to some extent in Seattle itself, was deteriorating for the IBEW and unionized contractors. In 1973, the IBEW dealt with 53 percent of the inside electrical employers and represented 66 percent of the inside electrical employees in Washington State. By 1980, these percentages had declined to 33 and 54 percent, respectively, a decisive drop-off in both categories, but there was, again, a clear indication that the IBEW still dealt with the largest contractors. Inexplicably, in Oregon the IBEW was somewhat stronger, with 77 percent of the inside electrical contractors a party to its agreements in 1973 and 52 percent in 1980, and with 78 percent of the inside electrical employees enrolled as members in 1973 and 63 percent in 1980.[115]

When negotiations began in 1983 between the Puget Sound Chapter of NECA and IBEW Local No. 46, headquartered in Seattle, forty-nine of the NECA chapter's seventy-three union firms cancelled their assignment of bargaining requests. Of those forty-nine firms, forty-five announced that they were planning to terminate their relations with the IBEW. The IBEW was able to reduce these defections by about 50 percent by agreeing, as of June 1, 1983, to a $1.50 per hour wage reduction, to the use of "stockmen" (delivery men and helpers) on jobs, and to other wage and productivity improvements for contractors. At the time that the agreement was signed, it covered about 1,400 electricians of whom 600 were unemployed.[116]

The Mechanical Contractors Association of Washington looked at their own situation by studying permits in the state's sixteen largest cities and by examining manhours for plumbing and mechanical contractors reported to the Washington State Department of Labor and Industries and the Washington State Plumbing and Pipefitting Pension Trust. Since the State Department includes both union and nonunion data and the Pension Trust only union data, it was possible to estimate quite accurately the relative position of the two groups. Both sources also excluded administrative and clerical personnel from their data.[117]

---

[115] National Electrical Contractors Association, *Inside Electrical Construction Survey*, Oregon data.

[116] "Seattle IBEW Takes $1.50 Wage Cut, Approves Modified Helper Category," *Construction Labor Report*, Vol. 29 (May 25, 1983), pp. 436-37.

[117] Mechanical Contractors Association of Washington, *A Survey of the Mechanical Contracting Market in Washington State Union Versus Non-Union Comparison* (Seattle: Mechanical Contractors Association of Washington, 1982).

FIGURE IV-13
Market Share of
the Union Mechanical Contractor
in Washington State
by Manhour

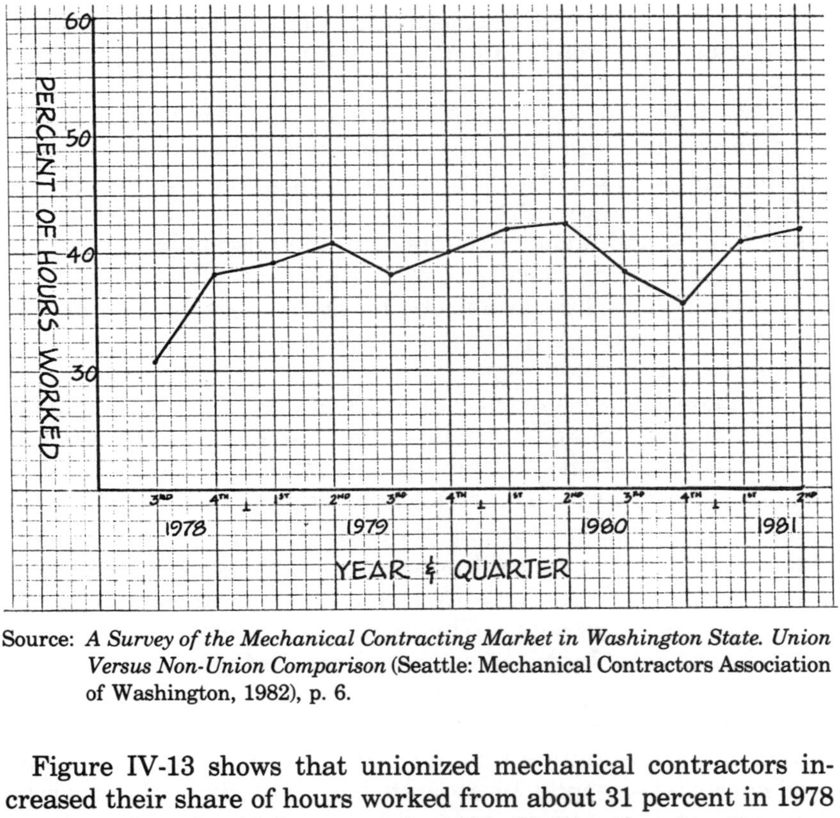

Source: *A Survey of the Mechanical Contracting Market in Washington State. Union Versus Non-Union Comparison* (Seattle: Mechanical Contractors Association of Washington, 1982), p. 6.

Figure IV-13 shows that unionized mechanical contractors increased their share of hours worked from about 31 percent in 1978 to approximately 42.5 percent in 1981. Unlike the situations we have found in so many areas, the unionized sector in Washington State did not hold a majority of the larger jobs. In the examination of permits it was found that for the 1979 to 1981 period the unionized mechanical contractors had 35 to 36.9 percent of the permits and 39.7 to 41.5 percent of the dollar volume of the jobs. Moreover, the open shop's share of the dollar volume grew slighlty in the same two-year period.[118]

The Washington mechanical contractors also sent a questionnaire to users as part of their study. Twenty-two major companies and

[118] *Ibid.*, p. 7.

institutions replied, 75 percent of which had utilized nonunion mechanical and plumbing contractors. Of these, 90 percent were satisfied with open shop performance, and 83 percent indicated a high probability of using open shop forces in the future. High wages, low productivity, and inflexible work schedules were cited as reasons for not employing unionized workers.[119] This indicates that while the share of the unionized mechanical contractors has been relatively stable in recent years, it could very well decline in the near future.

Eastern Washington and Northern Idaho. These areas of the Northwest have also seen major gains by the open shop, as shown by a careful tracking of developments by the Inland Empire Chapter, AGC, since January 1976.[120] Table IV-22 covers the period from 1976 to 1983 and includes commercial and miscellaneous (actually small commercial) construction. Work below $30,000 in permit value for 1976 and below $40,000 for other years, as well as work in areas not requiring permits are excluded from the table. These projects are likely to be open shop.

The data follow a typical pattern. The open shop was dominant in the miscellaneous category in all years covered except 1980 when one large job gave the union group the edge. Unionized construction had a majority of the commercial work prior to 1981, but it appears that 1981 marked the turning point. Several craft unions in other parts of Idaho have agreed that considerable work has been lost to the open shop and they have been forced to make substantial concessions in order to compete.[121] The 1980 labor agreements in that area provided for very large increases and resulted in contractors from other areas who had jobs in eastern Washington or northern Idaho either going open shop or demanding major concessions. The once all-union Inland Empire Chapter, AGC, was 15 percent open shop and nearly 50 percent doublebreasted by mid-1982.[122] Moreover, the ABC has established a chapter in this area. The region thus appears to be moving toward a strong open shop position.

---

[119] *Ibid.*, pp. 15-16.

[120] Associated General Contractors of America, *Construction Industry Survey* (Spokane: Inland Empire Chapter, Associated General Contractors of America, 1976-1981).

[121] "Electricians and Contractors in Idaho Agree on One-Year Wage and Benefit Freeze," *Construction Labor Report*, Vol. 28 (April 7, 1982), p. 181; and "SMW Business Manager Lands More Jobs for Idaho Members," *Construction Labor Report*, Vol. 28 (January 12, 1983), pp. 1551-52.

[122] Interview with AGC official, Spokane, July 27, 1982.

TABLE IV-22

Eastern Washington and Northern Idaho[a]

Union-Open Shop Commercial[b] *and Miscellaneous*[c] *Construction*

Building Permit Dollar Volumes[d]

1976–1983

	Total Dollar Volume	Union		Open Shop		Total Dollar Volume	Union		Open Shop		Total Dollar Volume	Union		Open Shop	
		Dollar Volume	Percent	Dollar Volume	Percent		Dollar Volume	Percent	Dollar Volume	Percent		Dollar Volume	Percent	Dollar Volume	Percent
			1976					1977					1978		
Commercial	69.7	48.6	70	21.1	30	124.1	78.1	63	46.0	37	135.5	94.1	70	41.4	30
Miscellaneous	37.2	10.8	29	26.4	71	42.7	14.1	33	28.6	67	41.2	12.0	29	29.2	71
Total commercial and miscellaneous	106.9	59.4	56	47.5	44	166.8	92.2	55	74.6	45	176.7	106.1	60	70.6	40
			1979					1980					1981		
Commercial	249.6	176.1	71	73.5	29	216.8	144.2	67	72.6	33	157.4	74.6	47	82.8	53
Miscellaneous	11.8	1.3	11	10.5	89	15.8	9.2	58	6.6	42	9.5	2.5	26	7.0	74
Total commercial and miscellaneous	261.4	177.4	68	84.0	32	232.6	153.4	66	79.2	34	166.9	77.1	46	89.8	54
			1982					1983							
Commercial	163.4	85.1	52	78.3	48		173.0	86.6	50	86.5	50				
Miscellaneous	4.3	.5	11	3.9	89		4.8	.1	3	4.6	97				
Total Commercial and Miscellaneous	167.7	85.6	51	82.2	49		177.8	86.7	49	91.1	51				

Source: Inland Empire Chapter, Associated General Contractors, Spokane, Washington.

[a]Defined as cities and counties east of the 120th meridian in Washington and north of the 46th parallel in Idaho. Excludes some Idaho counties and Washington towns that lack a buildng permit or comprehensive plan ordinance.

[b]Commercial construction is defined as stores, offices, buildings, schools, hospitals, banks, restaurants, hotels, and taverns.

[c]Miscellaneous construction includes churches, recreation facilities, warehouses, grain and potato storage, storage sheds, service stations, and other buildings which are included in commercial construction in this book.

[d]In millions of dollars of permits, $40,000 or more for all years except 1976, for which the limit is $30,000.

Area Studies—Final Comment

Despite their limitations, the area studies all point in the direction of the growth of the open shop during the 1970s and early 1980s. Areas once considered almost completely unionized, such as California and the Pacific Northwest, have seen a significant rise in the nonunion sector. Thus, despite continuing union strength in large downtown office building, the unionized segment is obviously in deep difficulty. After examining the results of our 1974 and 1982 questionnaires, the overall position of the open shop in commercial construction will be estimated.

INDUSTRIAL RESEARCH UNIT QUESTIONNAIRES

The 1974 and 1982 Industrial Research Unit questionnaires, although not identical, are sufficiently similar to give corroborative testimony to the relative positions of the union and open shop in those years. Table IV-23 shows the 1974 respondents' estimates of the proportion of open shop work in the commercial sector by region and union status. Since the percentages in Table IV-23 are somewhat distorted by the inclusion of nonrespondents, Table IV-24 more accurately illustrates the disparity between union and nonunion responses by showing the proportion of those actually responding who put open shop commercial work at over 60 percent.

The 1974 Questionnaire Data

There are several possible explanations for the differing responses offered by union and open shop contractors in 1974. First, and perhaps most obvious, is the different perspective brought to the question by members of the two groups. They were likely to have belonged to different associations, have had different business acquaintances, and generally have been traveling in different circles. In short, they were more likely to be aware of work being done by contractors like themselves. Second, the respondents surely did not proportionately represent the same market areas within the broad regions. Open shop respondents from New England were more representative of the northern New England states than of the metropolitan Boston area. Open shop respondents in the Midwest were more likely to be referring to Columbus, Indianapolis, and Grand Rapids, than to Chicago, Milwaukee, or Cleveland. Third, the distribution of the two groups of 1974 respondents among regions was very different. If the regional distribution of union respondents were applied to the percentage estimates of nonunion activity given by

TABLE IV-23

Percent Distribution of Respondents' Estimates of the Proportion of Open Shop Work in Commercial Construction by Region and Union Status, 1974

	Union							Nonunion						
Region	Number	NR/DK[a]	0-20	21-40	41-60	61-80	81-100	Number	NR/DK[a]	0-20	21-40	41-60	61-80	81-100
Not reported	89	36.0	29.2	7.9	15.7	6.7	4.5	83	20.5	4.8	12.0	22.9	18.1	21.7
New England	27	29.6	11.1	11.1	25.9	7.4	14.8	63	11.1	1.6	7.9	31.7	30.2	17.5
Middle Atlantic	125	28.8	20.0	18.4	19.2	9.6	4.0	134	26.1	6.0	8.2	22.4	17.9	19.4
E.N. Central	147	27.9	37.4	14.3	12.9	4.1	3.4	75	13.3	4.0	17.3	25.3	22.7	17.3
W.N. Central	60	28.3	36.7	11.7	16.7	1.7	5.0	45	22.2	6.7	6.7	15.6	13.3	35.6
S. Atlantic	68	17.6	10.3	19.1	36.8	11.8	4.4	220	20.4	2.3	4.1	15.9	24.1	33.2
E.S. Central	32	15.7	18.8	12.5	37.5	9.4	6.3	45	8.9	2.2	8.9	6.7	33.3	40.0
W.S. Central	45	6.6	24.4	22.2	28.9	13.3	4.4	92	16.3	8.7	13.0	21.7	23.9	16.3
Mountain	52	34.7	34.6	9.6	9.6	11.5	0.0	21	23.8	19.0	9.5	19.0	9.5	19.0
Pacific	133	40.6	42.9	11.3	3.8	1.5	0.0	6	50.0	33.3	16.7	0.0	0.0	0.0
Total	778	29.0	29.6	13.9	17.2	6.7	3.6	784	19.2	5.0	8.9	20.0	22.1	24.7

Source: Industrial Research Unit Questionnaire, 1974.

[a]NR/DK—Do not know or not reported.

Note: Percents may not add to 100.0 because of rounding.

TABLE IV-24
Proportion of Those Responding to Questionnaire Who Put Open Shop Commercial Work at Over 60 Percent by Region, 1974

Region	Union	Nonunion
New England	31.6	53.6
Middle Atlantic	19.1	50.5
E.N. Central	10.4	46.1
W.N. Central	9.3	62.8
S. Atlantic	19.7	72.0
E.S. Central	18.6	80.5
W.S. Central	19.0	48.0
Mountain	17.6	32.2
Pacific	1.5	0.0
Total	14.5	57.9

Source: Industrial Research Unit Questionnaire, 1974.

nonunion respondents, the median response would fall from 66 percent to 56 percent. Finally, union and open shop contractors may have interpreted the question differently. In our 1974 interviews with industry spokesmen, those representing open shop firms would frequently estimate open shop activity in terms of the proportion of jobs, whereas those representing union contractors would speak in terms of total volume. Since open shop construction is relatively more prevalent among smaller projects, these respective approaches resulted in markedly differing estimates, as many of the tables in this chapter demonstrate.

These factors preclude any precise conclusions as to the overall proportion of open shop commercial construction in the nation in 1974. The figures do suggest, however, that this proportion was far from negligible. In the southern states, over 59 percent of the union firms responding to this question in 1974 put open shop activity at over 40 percent of all commercial building. Only in the West did commercial work appear to have resisted substantial open shop incursions into the strong union market, but as noted above, interest in the open shop appeared to be growing in 1974 even in this region, and continued to grow throughout the decade. Both union and nonunion respondents, moreover, agreed that any changes then occurring were in the direction of more open shop competition for commercial building work.

The survey also solicited respondents' views on whether the amount of open shop activity in their areas had changed over the few years preceding 1974 (see Table IV-25). Eighteen percent of the union respondents believed that the open shop share had increased substantially prior to 1974, and less than 4 percent felt that it had actually fallen. The comparable figures for nonunion respondents were 23 percent and 2.5 percent, respectively. The remaining respondents estimated that there had been no change—or at least not enough of a change to move the proportion to a different percentage range.

The responses reflected in Table IV-25 are perhaps most significant in terms of the relative agreement between union and open shop contractors. Of all the types of construction specified on the questionnaire, the commercial sector was estimated to be the one having undergone the most change. Although the two groups dif-

TABLE IV-25
Respondents' Estimates of Change in Open Shop Share of Commercial Construction over Past Few Years by Region and Union Status, 1974

Region	Total	Total Indicating Change[a]	Lower by 3	Lower by 2	Lower by 1	Higher by 1	Higher by 2	Higher by 3
Union								
New England	27	11	0	0	1	3	7	0
Middle Atlantic	125	36	0	1	5	25	5	0
E.N. Central	147	30	1	2	2	16	9	0
W.N. Central	60	10	0	1	3	5	1	0
S. Atlantic	68	23	1	0	4	12	6	0
E.S. Central	32	12	0	1	1	8	1	1
W.S. Central	45	15	1	0	2	7	4	1
Mountain	52	9	2	0	0	4	3	0
Pacific	133	8	0	0	1	5	2	0
Not reported	89	15	0	0	0	11	2	2
Total	778	169	5	5	19	96	40	4
Nonunion								
New England	63	23	0	2	1	12	7	1
Middle Atlantic	134	31	0	0	5	20	6	0
E.N. Central	75	29	0	1	3	18	6	1
W.N. Central	45	7	0	0	0	5	2	0
S. Atlantic	220	37	0	0	4	24	8	1
E.S. Central	45	20	0	0	1	15	2	2
W.S. Central	92	30	0	0	2	22	4	2
Mountain	21	5	0	0	0	3	1	1
Pacific	6	0	0	0	0	0	0	0
Not reported	83	19	0	0	1	14	3	1
Total	784	201	0	3	17	133	39	9

Source: Industrial Research Unit Questionnaire, 1974.
[a]Change expressed in terms of movement between ranges of 20 percentage points (i.e., 0-20, 21-40, etc.).

fered markedly in their estimates of the current share of open shop commercial activity, many among them concurred in the view that there had been an appreciable increase in the penetration of open shop firms into the commercial building market, a view consistent with the findings of the studies described above. It appeared, both from these studies and from our interviews with spokesmen for both union and open shop contractors during 1973 and 1974, that open shops in many areas had not only consolidated and augmented their hold on the small job market, but that they were also moving increasingly into the market for larger commercial projects. When asked about the extent to which nonunion contractors have captured commercial work, respondents often answered in terms of a monetary cutoff point: that open shops were active among jobs costing less than $1 million, but were not likely to compete for bigger projects. As already clearly documented, however, this cutoff point not only was steadily rising then, but has continued to do so.

The responses in this earlier period also suggested that open shop commercial construction was more prevalent among certain types of projects. Such structures as apartments, nursing homes, and shopping centers seemed especially prone to open shop incursions, more so than office buildings. Two major reasons for this were postulated. First, the role of the purchaser was found to be significant. A governmental agency or large industrial corporation was likely to be more sensitive to economic and political pressures exerted by organized labor. Second, there was undoubtedly a connection between the location of a project and its union status. Whereas major shopping centers tend to follow residential populations toward the peripheries of metropolitan areas, large office buildings are more likely to favor central locations. As noted, in virtually every branch of construction, union strength wanes with distance from the central city.

Some of these factors have altered since 1974, but not all. Government construction and city center work are still union strongholds, but the walls have been pierced. The 1982 Industrial Research Unit questionnaire provides an additional barometer of the changes which have occurred.

The 1982 Questionnaire Data

As was explained in Chapter I, the 1982 Industrial Research Unit questionnaire was modeled closely on that of 1974, but with some changes made as dictated by the changing times. The memberships of the AGC and the ABC were polled in both years. In 1974 every other member of each organization, which then had approximately the same number of members, was sent a questionnaire. In 1982, the ABC had about twice the membership of the AGC. In order to prevent an open shop imbalance, only one-half the ABC members were sent questionnaires, while all AGC members received them. In addition, approximately 10,000 questionnaires were mailed to members of specialty contractor associations, a majority of whose membership is unionized. A total of 3,972 completed questionnaires were returned, a 15 percent response.

The 1982 questionnaire results for commercial construction are summarized in Tables IV-26, IV-27, and IV-28. Because of the different samples used in the two years, some variation in answers would be expected. Therefore, our interest must be concentrated on the trends which the questionnaire results reveal.

Comparing Tables IV-23 and IV-24 with Tables IV-26 and IV-27, we find a wide variety of responses by region but a virtually identical result in total. According to Tables IV-24 and IV-27, which eliminate the distortions caused by the inclusion of nonrespondents, 16.7 percent of the union respondents in 1982, as compared with 14.5 percent in 1974, put open shop commercial work at over 60 percent, while 55.6 percent of the open shop respondents in 1982, but 57.9 percent in 1974 did likewise. The similarities between the two years are remarkable.

Between regions, however, the responses to both of the questionnaires are remarkable for their differences. In New England those in the union sector who felt that open shop commercial work was over 60 percent, declined from 31.6 percent in 1974 to 18.6 percent in 1982, and in the Middle Atlantic region, from 19.1 percent to 13.8 percent. For other regions, however, the perceived percentage increased, with variations from one percentage point in the East North Central to 32.6 percent in the South Atlantic region. The nonunion respondents tended to be somewhat more consistent over a regional basis. They were slightly more conservative in their estimates in 1982 than in 1974, with the exception of the Pacific region where the response rose from zero to an optimistic 38.9 percent.

The data in Tables IV-26 and IV-27 indicate considerable regional gains by the open shop in former union strongholds, as perceived by union shop contractors, particularly in the West North Central and Pacific regions. This is in accordance with the findings set forth earlier in this chapter. On the other hand, the overall questionnaire results indicate stability in the open shop-union shop national ratio and this is at variance with most, but certainly not all, of the other

TABLE IV-26

Percent Distribution of Respondents' Estimates of the Proportion of Open Shop Work in Commercial Construction by Region and Union Status, 1982

Region	Union							Nonunion						
	Number	NR/DK[a]	0-20	21-40	41-60	61-80	81-100	Number	NR/DK[a]	0-20	21-40	41-60	61-80	81-100
Not Reported	44	9.1	13.6	31.8	15.9	25.0	4.5	49	16.3	0.0	8.2	18.4	40.8	16.3
New England	62	4.8	8.0	22.6	46.8	16.1	1.6	111	9.9	9.9	5.4	29.7	27.0	18.0
Mid Atlantic	275	12.7	22.2	21.8	31.3	9.8	2.2	228	13.2	7.5	14.0	35.1	18.4	11.8
E.N. Central	474	8.6	34.8	22.8	24.3	7.4	2.1	205	8.3	10.7	12.2	33.2	24.9	10.7
W.N. Central	220	9.5	31.4	20.0	24.1	10.0	5.0	157	9.6	7.6	14.0	24.2	20.4	24.2
S. Atlantic	109	11.0	12.8	28.4	35.8	27.5	10.1	550	10.9	4.7	4.2	14.5	21.8	43.8
E.S. Central	69	13.0	4.3	15.9	29.0	29.0	7.2	111	7.2	4.5	3.6	18.0	52.4	34.2
W.S. Central	105	2.9	10.5	22.9	40.0	21.0	2.9	284	7.4	4.6	7.4	25.0	5.9	19.7
Mountain	108	10.2	12.0	22.2	34.3	17.6	3.7	184	7.0	4.3	10.3	34.2	28.3	15.8
Pacific	362	9.1	26.2	30.4	27.1	5.0	2.2	143	11.9	9.1	18.9	25.9	23.1	11.2
Total	1,828	9.7	24.2	22.5	28.8	11.7	3.3	2,022	9.9	6.3	9.1	24.7	25.6	24.5

Source: Industrial Research Unit Questionnaire, 1982.
[a] NR/DK Do not know or not reported.
Note: Percents may not add to 100.0 because of rounding.

TABLE IV-27
Proportion of Those Responding to Questionnaire Who Put Open Shop Work in Commercial Construction at Over 60 Percent by Region, 1982

Region	Union	Open Shop
New England	18.6	50.0
Mid Atlantic	13.8	34.8
E.N. Central	10.4	38.8
W.N. Central	16.6	49.3
S. Atlantic	42.3	73.7
E.S. Central	39.1	71.8
W.S. Central	24.5	60.1
Mountain	23.7	47.4
Pacific	7.9	38.9
Total	16.7	55.6

Source: Industrial Research Unit Questionnaire, 1982.

evidence presented. To what extent these results are a function of sampling and respondent biases is difficult, if not impossible, to determine.

Table IV-28 gives the estimates made by the questionnaire repondents of the change in the open shop share of work between 1978 and 1982. The attempt to elicit information was largely unsuccessful because 33.3 percent of the union and 56.6 percent of the open shop respondents either said that they did not know or declined to answer. Of the unionized contractors, 28 percent felt that the open shop had gained up to 10 percent during this five year period and another 15.6 percent believed that the gains were from 11 to 20 percent. These gains were found to be spread over all regions except the South Atlantic, where the open shop was already dominant by 1978. Only 22.1 percent of the unionized respondents believed that the gains of the open shop exceeded 20 percent during this period.

A majority of the open shop respondents did not reply to this question, and those that did were evenly spread in their totaled responses (excluding the 76 to 100 percent category). Regionally, the responding open shop group answers reflected open shop gains in the East North Central, West North Central, and Pacific regions. Thus 12.2 percent believed that the open shop had gained up to 10 percent in the East North Central region, and 9.3 percent believed that the increases there were between 11 and 20 percent. The reported percentages for these categories in the West North Central

TABLE IV-28
Respondents' Estimates of Changes in Open Shop Share of Commercial Construction[a] *by Region and Union Status, 1978-1982*

Region	Union								
	Number	NR/DK[a]	0-10	11-20	21-30	31-40	41-50	51-75	76-100
Not reported	44	18.2	27.3	6.8	9.1	29.5	6.8	2.3	0.0
New England	62	17.7	21.0	22.6	16.1	17.7	1.6	1.6	1.6
Mid Atlantic	275	39.6	20.7	18.5	8.4	4.0	3.6	4.0	1.0
E.N. Central	474	37.3	32.9	13.5	6.7	4.4	2.1	1.2	1.2
W.N. Central	220	36.8	27.7	14.1	7.3	5.5	1.2	2.3	1.8
S. Atlantic	109	32.1	5.5	16.5	9.2	9.2	11.0	11.0	5.5
E.S. Central	69	23.2	23.2	11.6	4.3	13.0	11.6	10.1	2.9
W.S. Central	105	25.7	24.8	20.0	5.7	8.6	7.6	4.8	2.8
Mountain	108	23.1	27.8	22.2	12.0	9.2	2.7	1.8	0.9
Pacific	362	32.9	37.3	13.8	6.9	2.8	1.3	1.9	3.0
Total	1,828	33.3	28.0	15.6	7.7	5.7	3.4	3.2	2.1

Region	Nonunion								
	Number	NR / DK[a]	0-10	11-20	21-30	31-40	41-50	51-75	76-100
Not reported	49	71.4	2.0	2.0	4.1	4.1	8.2	6.1	2.0
New England	111	64.9	2.7	7.2	8.1	6.3	4.5	4.5	1.8
Mid Atlantic	228	52.6	7.0	11.4	7.5	6.6	7.0	5.3	2.6
E.N. Central	205	47.8	12.2	9.3	8.3	9.3	5.9	5.9	1.5
W.N. Central	157	52.2	12.7	10.2	8.3	4.5	3.2	5.1	3.8
S. Atlantic	550	72.4	2.2	2.9	4.4	5.1	4.9	5.8	2.4
E.S. Central	111	55.0	2.7	9.9	7.2	6.3	9.0	7.2	2.7
W.S. Central	284	54.2	3.9	8.5	8.5	8.5	7.4	6.7	2.5
Mountain	184	39.7	12.0	7.1	15.8	8.7	8.7	4.3	3.8
Pacific	143	35.7	18.2	15.4	9.8	4.2	5.6	7.0	4.2
Total	2,022	56.6	6.9	7.7	7.8	6.5	6.1	5.9	2.7

Source: Industrial Research Unit Questionnaire, 1982.
NR / DK Do not know or not reported.
[a] Change expressed in terms of movement between averages of 10 percentage points (i.e., 0-10, 11-20, etc.).

were 12.7 and 10.2 percent, respectively; and for the Pacific, 18.2 and 15.4 percent, respectively.

We can conclude that the data in Table IV-28 reflect the gains of the open shop in those regions where the union shop has been historically dominant, and may support a modest open shop expansion overall in the five-year period according to the views of the union shop respondents. Overall, however, the Industrial Research Unit 1982 questionnaire gives less support to strong open shop growth in commercial construction during the 1970s than do other data presented in this chapter. Rather this questionnaire indicates a basic stability in the open shop—union shop share of commercial construction although there seems to have been a slight increase in the open shop percentage over the years studied.

CONCLUSION

From this mass of information, certain trends do stand out:

1. The open shop share of work in commercial construction has continued to increase since 1975 although at a slower rate in more recent years.
2. The most significant open shop advances have come in two areas: a) in the highly skilled electrical, mechanical, plumbing, and pipefitting trades, where in the early 1970s the open shop was very weak; and b) in the former regional union bastions: the East North Central, West North Central, and Pacific.
3. In terms of number of jobs, the open shop undoubtedly has a clear majority of the commercial work. There are thousands of small jobs which are done nonunion, and over the years unions have been receiving less of such work as their wage rates and other costs have risen. In addition, the open shop captures a clear majority of the smaller awards that are open for bid, and therefore of the total bid awards, as well as an increasing number of the larger ones.
4. The union strength in commercial construction is concentrated in center city office construction, particularly in high-rise buildings in major cities. The tremendous dollar volume of such structures put unions in a position to claim a large share of the dollar volume of commercial construction.
5. Buoyed by the Davis-Bacon Act and similar state laws, unions have traditionally held a major share of governmental commercial construction. In recent years, the open shop has won an increasing share of such work and is now dominant in many subsectors, such as school construction. The new rules govern-

ing the Davis-Bacon Act administration, and particularly the elimination of the 30 percent rule and the use of government-financed construction in surveys, as explained in Chapter II, seem certain to have an adverse impact on the union share of government work.

On the basis of these facts, it is estimated that the open shop has more than 65 percent of the number of jobs in commercial construction. The division on the basis of dollar volume is heavily a function of the extent of high-rise office building. When that volume is high, as in the period from 1978 to 1981, it gives the unionized sector a share of the market varying from an estimated 35 to 50 percent of the total. When, as in 1983, high-rise office construction is depressed, the union share falls significantly. Thus the high-rise office market is crucial to unions. In view of the increasing ability of open shop contractors to perform this work, as they have in various cities and suburbs, control of such jobs will be a key element for the open shop and union shop contractors in determining the future of each in commercial construction.

CHAPTER V

Industrial Construction

Large-scale industrial construction of major industrial plants, dams, refineries, and power plants has, at least since World War II, been the special domain of the biggest unionized construction companies, such as Bechtel, Ebasco, Lummus, Morrison, Kaiser Engineers & Constructors, and several others. The largest of these companies operate throughout the country, and often, all over the free world as well. They usually have design and engineering divisions or affiliated companies and can bid and execute a major job from drawing board to the last construction detail. About forty of such companies are members or associate members of the National Constructors Association (NCA), have contracts that gross from $100 million to several billion dollars, and are called "signatory" firms because they have national agreements with all or several of the building trades unions.

Today, major industrial construction remains the stronghold of unionism, but beginning in the 1960s, the hegemony of the unions has been challenged. The challenge has come from five sources: (1) the increasing capacity of the two largest open shop contractors, Brown & Root and Daniel International (now a division of Fluor), to win major industrial jobs, not only in the South where they originally rose to prominence, but elsewhere as well; (2) the growth of other open shop contractors, including some doublebreasted operations of NCA contractors, and their capability to take on the larger jobs; (3) the industrialization of the South where large open shop contractoring firms were first concentrated; (4) the strength of open shop contractors in such specialties as offshore oil drilling rigs, petroleum refineries, and chemical plants, and their subsequent expansion into other industrial work; and (5) the increasing willingness of large industrial users to permit open shop contractors to bid on jobs. We shall explore these causal factors as we examine the data relevant to the open shop penetration of industrial construction.

THE ENGINEERING NEWS-RECORD "400"

The *Engineering News-Record* annually compiles a list of the 400 largest construction firms in terms of new work booked for the previous year. The list covers all sectors of construction except residential construction. Identification of the union versus open shop status of the companies has become increasingly difficult for a number of reasons: 1) there is no central source for such information; 2) many companies operate on a doublebreasted basis; 3) some companies, for fear of union reprisals, do not want information about their open shop status publicized and will not answer or reply correctly to such questions; 4) some companies list their doublebreasted operations separately in the *Engineering News-Record* (ENR) 400, but others combine results without revealing the volumes of the separate union and open shop units; and 5) over the years many large industrial constructors have purchased open shop concerns in order to be able to operate doublebreasted. Thus Daniel has been acquired by Fluor, Yeargin by Perini, and such firms as Nello Teer and Watkins Engineering & Construction are among those that have been taken over by other major signatory firms.

In addition to these limitations, the firms listed are not necessarily in industrial construction because the list also includes the volume leaders in the commercial, highway, and heavy sectors. Nevertheless, the industrial constructors are the volume leaders, and therefore, the ENR data is one reasonable gauge, among many, to use in examining open shop construction status in the industrial sector.

The 1970-1973 ENR Data

In the first edition of *Open Shop Construction,* we compared the ENR data for 1970-1973, identifying the open shop contractors in each year. Table V-1 shows that in 1973, thirty-one primarily open shop firms were listed in the 400. Of these, twenty-five were headquartered in the South, three in Maryland (the birthplace of the Associated Builders and Contractors—ABC), one in West Virginia, and two in New England. These figures, of course, do not include some doublebreasted operations of union firms.

The 1973 data also illustrated another phenomenon that still holds true, but to a lesser extent: the concentration of open shop firms at the top and bottom of the ENR 400. Two of the open shop firms, Brown & Root and Daniel International, were among the largest in the industry; two others, J. Ray McDermott and H. B. Zachry, were among the top fifty; and most of the balance were concentrated among the smaller members of the 400. Moreover, as Table V-2 shows, open shop contractors on the 400 list during this period grew faster than their union shop counterparts, but the greatest growth was produced by the two largest followed by the

TABLE V-1
Engineering News-Record *"400" List*
Open Shop Contractors, 1971–1973

Company Headquarters	1973		1972		1971	
	Rank	Contracts in $ Millions	Rank	Contracts in $ Millions	Rank	Contracts in $ Millions
Brown & Root, Houston, Texas	1	4,740.4	4	1,466.8	3	1,491.5
Daniel International, Greenville, S.C.	4	1,882.8	3	1,603.8	5	888.0
J. Ray McDermott & Co., Inc., New Orleans, La.	24	452.0	9	498.7	20	346.0
H. B. Zachry Co., San Antonio, Texas	43	214.6	77	100.1	40	164.3
BE&K, Inc., Birmingham, Ala.	64	131.0	285	33.1		
Yeargin Construction Co., Greenville, S.C.	100	86.3	117	69.5	75	83.5
Ruscon Corp., Charleston, S.C.	123	73.0	126	64.9	94	72.2
Nello L. Teer Co., Durham, N.C.	124	72.8	217	41.4	140	51.4
Joe M. Rodgers & Assoc., Inc., Nashville, Tenn.	137	66.6	102	80.2	262	32.0
Blythe Brothers Co., Charlotte, N.C.	149	62.2	144	56.1	151	48.9
Ballenger Corp., Greenville, S.C.	164	55.1	230	39.8	108	65.0
Dickerson, Inc., Monroe, N.C.	187	49.9	182	46.7	141	51.1
McCrory-Sumwalt Construction Co., Columbia, S.C.	198	46.7	207	42.1	226	35.7
Vecellio & Grogan, Inc., Beckley, W. Va.	216	43.9	268	35.5	270	30.5
M. B. Kahn Construction Co., Columbus, S.C.	217	43.8	114	71.0	179	42.2

TABLE V-1 (continued)

Company Headquarters	1973 Rank	1973 Contracts in $ Millions	1972 Rank	1972 Contracts in $ Millions	1971 Rank	1971 Contracts in $ Millions
Delta Engineering Corp., Houston, Texas	223	43.2	228	39.9	384	22.1
Dargan Construction Co., Inc., Myrtle Beach, S.C.	227	42.5				
Hardaway Construction Co., Inc., Nashville, Tenn.	239	41.7	250	37.1	301	27.8
George W. Kane, Inc., Durham, N.C.	249	40.6	224	40.8	276	30.3
Allen M. Campbell Co., Tyler, Texas	303	34.5				
Thomas P. Harkins, Inc., Silver Spring, Md.	315	33.0				
Gulf Contracting Inc., Sarasota, Fla.	319	32.7				
Cianbro Corp. Pittsfield, Me.	323	32.1				
Capeletti Bros., Inc., Hialeah, Fla.	329	31.4			342	25.1
Paul N. Howard Co., Greensboro, N.C.	331	31.3	375	24.7	344	25.0
Charles E. Brohawn & Bros., Inc., Cambridge, Md.	339	30.3				
Seaward Const. Co., Inc., Portsmouth, N.H.	352	28.3	314	30.2		
The Arundel Corp., Baltimore, Md.	363	27.0	238	38.6	138	51.9
D. R. Allen & Son, Inc., Fayetteville, N.C.	375	26.2				
Demetree Builders, Orlando, Fla.	389	25.0				
N. C. Monroe Construction Co., Greensboro, N.C.	396	24.5	319	29.9		

Source: *Engineering News-Record,* (April 11, 1974), pp. 49–58; (April 12, 1973), pp. 51–67; (April 6, 1972), pp. 49–63.

TABLE V-2
Engineering News-Record *"400" List*
Amounts, Growth, and Percent of the Total Contracts for Open Shop Contractors, 1971–1973

Group	1973		1972		1971		Growth, 1971–1973	
	Contracts in $ Millions	Percent of Total	Contracts in $ Millions	Percent of Total	Contracts in $ Millions	Percent of Total	Amount	Percent
Total "400"	$55,600.7	100.0	40,678.9	100.0	36,600.7	100.0	19,000.0	51.9
Open shop contractors on list for year	8,545.4	15.4	4,490.9	11.0	3,584.5	9.8	4,960.9	138.4
Open shop contractors on lists all three years	8,073.9	14.5	4,397.7	10.8	3,559.4	9.7	4,514.5	126.8
Largest two open shop contractors[a]	6,623.2	11.9	3,070.6	7.5	2,379.5	6.5	4,243.7	178.3
Largest four open shop contractors[b]	7,289.8	13.1	3,669.4	9.0	2,889.8	7.9	4,400.0	152.3
Open shop contractors except largest two	1,922.2	3.5	1,420.3	3.5	1,205.0	3.3	712.2	59.5
Open shop contractors except largest four	1,255.6	2.3	821.5	2.0	694.7	1.9	560.9	80.7
Open shop contractors on list all three years except largest two	1,450.7	2.6	1,327.1	3.3	1,179.9	3.2	270.8	23.0
Open shop contractors on list all three years except largest four	784.1	1.4	728.3	1.8	669.6	1.8	114.5	17.1

Source: See Table V-1
[a] Brown & Root and Daniel International.
[b] Brown & Root, Daniel International, J. Ray McDermott, and H. B. Zachry.

TABLE V-3
Engineering News-Record *"400" Survey*
1975 Analysis by Region[a]

	Total	New England	Middle Atlantic	South	Middle West	Miss. River to Rockies	Far West
Total respondents	731	40	156	142	128	148	117
1975 U.S. contract in $ millions	$35,909	1,418	6,869	8,323	5,647	5,962	7,690
No. of union contrs	449	22	103	31	115	78	100
Contract volume	$24,099	698	5,486	1,730	5,356	4,029	6,800
No. of open shop contrs	225	13	42	93	9	57	11
Contract volume	$6,519	135	740	3,931	209	1,127	378
No. of doublebreasted contractors	57	5	11	18	4	13	6
Contract volume	$5,291	586	643	2,662	83	806	511
Average volume of work performed open shop	30%	38%	28%	51%	10%	31%	21%
Ratio open shop vs. all-union contractors	33%	37%	29%	75%	7%	42%	10%

Ratio open shop volume vs. all-union volume	21%	16%	12%	69%	4%	22%	5%
Ratio open shop volume plus average percent of doublebreasted volume performed open shop vs. all-union volume	25%	34%	14%	77%	4%	25%	7%
Dollar volume of 1975 awards to be performed with nonunion labor	$7,230	359	480	4,855	64	1,201	271

Source: *Engineering News-Record,* May 27, 1976, p. 8.

[a] Regional definitions:

New England = Maine, New Hampshire, Vermont, Massachusetts, Rhode Island, and Connecticut

Middle Atlantic = New York, New Jersey, Pennsylvania, Maryland, Delaware, and District of Columbia

South = Virginia, West Virginia, North Carolina, South Carolina, Georgia, Florida, Alabama, Mississippi, Louisiana, Kentucky, and Tennessee

Middle West = Ohio, Indiana, Illinois, Michigan, and Wisconsin

Mississippi River to Rockies = Minnesota, Missouri, Iowa, Kansas, Arkansas, Nebraska, North Dakota, South Dakota, Oklahoma, Texas, Montana, Wyoming, Colorado, and New Mexico

Far West = Idaho, Utah, Arizona, Nevada, California, Oregon, Washington, Alaska, and Hawaii

four largest open shop firms. If the top four were eliminated from the calculation, the growth was less than that of the total. The conclusion at that time was that the open shop contractors, other than the top four, had a long way to go before they would be able to compete on the largest jobs. As we shall note below, part of that route has been traveled.

The ENR 1975-1976 Analysis

In 1975, the editors of the *Engineering News-Record* made the first of several analyses of their 400 list in an effort to track the progress of the open shop. This study compared their 1970 and 1975 lists, and reported as follows:

> The number of open shop contractors on the ENR 400 has increased from 19% of 300 contractors responding to the question in 1970 to 31% of 264 respondents in 1974. Of the total 57 open shop contractors in 1970, 22 or 39% were only open shop and 61% reported they did some portion of their work on an open shop basis. Last year [1973], the 82 contractors doing some or all of their work open shop was split into 35% working solely open shop and 65% or 53 companies that do only some part of their work open shop.
>
> More than three quarters of the exclusively open shop contractors are in the South, and among the ENR 400 there are none that say they work only open shop in the far western states. Of the 53 open shop companies that do some of their work using union labor, 22 do so through a separate operating unit, and 11 of the 53 do at least 50% of their work on an open shop basis.
>
> Geographically, the 82 contractors on the 400 that work open shop are divided as follows: New England, 5; Middle Atlantic, 7; Midwest, 9; South, 37; Mississippi to the Rockies, 18; and Far West. 6.[1]

In 1976, the editors of the *Engineering News-Record* made one of their most detailed surveys of open shop progress, studying the penetration of the open shop for 1975 in conjunction with the 400 survey of that year, the results of which are summarized in Table V-3.[2] This survey found that "one third of the top contractors in the U.S. have turned to the open shop as a means of securing new business in today's increasingly competitive marketplace . . . and . . . these open shop contractors claimed 25 percent of all non-home-building contracts awarded in 1975."[3] On a regional basis, the survey revealed that the most completely open shop area was the South;

---

[1] "The ENR 400," Reprinted from *Engineering News-Record,* April 10, 1975, pp. 52–53, copyright, McGraw-Hill, Inc. All rights reserved.

[2] "Open Shop Survey Shows Penetration at the Top," *Engineering News-Record,* May 27, 1976, pp. 8–9.

[3] *Ibid.,* p. 8.

TABLE V-4
Engineering News-Record *"400" Survey*
Comparison of Open Shop Representation
by Decile[a], *1969 and 1979*

Decile	1969	1979
First	3	9
Second	2	3
Third	0	2
Fourth	1	6
Fifth	0	2
Sixth	3	4
Seventh	1	5
Eighth	1	8
Ninth	0	7
Tenth	2	5
Total	13	51

Source: *Engineering News-Record* as compiled by Clinton C. Bourdon and Raymond E. Levitt, *Union and Open-Shop Construction* (Lexington, Mass.: D. C. Heath and Company, Lexington Books, 1980), p. 19.
[a] Each decile, 40 companies by size, largest in first.

the most unionized areas, the Middle West and the Far West. Because diverse states were combined into the "Mississippi River to Rockies" area, several different patterns were probably obscured by this part of the 1976 report.

In terms of size, the survey found that "two-thirds of the country's biggest construction companies, those with contracting volumes have $23 million in 1975, work with union labor only."[4] Since then most of these firms have established or acquired open shop subsidiaries. Nevertheless, open shop firms still account for a disproportionately small volume of the total industrial business—but a volume that is growing.

The 1969–1979 Comparison

Bourdon and Levitt compared the 1969 and 1979 ENR 400 data. They identified fifty-one open shop firms on the 1979 list (exclusive of doublebreasted companies), as compared with only thirteen in 1969. Moreover, as set forth in Table V-4, they found that the open shop firms in 1979 were well represented among the forty largest firms and well distributed throughout the list. In both years, the

[4] *Ibid.*

firms which they identified as open shop were heavily concentrated in the South.[5]

1975–1980 Analysis

A similar study covering the years 1975 to 1980 was made by a large signatory firm. This study compared domestic sales trends for twelve large open shop firms with those of thirteen large union shop firms. International sales were eliminated from the data since many of the firms do not compete abroad and estimates were made for Daniel after its acquisition by Fluor. These data are set forth in Table V-5 which shows that the growth rate of the open shop contractors was much faster than that of the unionized contractors. Thus between 1975–1977 and 1977–1980 the open shop group increased their sales by 56 percent while the union constructors increased theirs by 36 percent.

The ENR Data for the 1980s

In a November 1981 report the *Engineering News-Record* noted the following in summarizing its findings for the 1980 top 400:

> Five of the country's largest contractors are either nonunion or part of companies that have separate union and nonunion operations [doublebreasted]. In all, 77 of ENR's top 400 contractors say they are open shop and 73 indicate they are part of doublebreasted firms. These numbers are surprisingly large to industry observers who note that nonunion contractors historically have had their strongest hold on smaller types of projects.[6]

The 1983 *Engineering News-Record* survey, covering projects awarded in 1982, revealed continued gains for the open shop. Twelve new open shop companies joined the 400. The *Engineering News-Record* calculated that 95 of the 400 companies, or 23.8 percent, were open shop and that these contractors accounted for 16 percent of the $13.5 billion in new work which was awarded to the members of the 400.[7] Thus, the open shop firms have still not gained a share of the volume equal to their numbers.

Nevertheless, the data summarized in Table V-6 demonstrate a major change in the open shop situation since 1973. Table V-6 compares union, open shop, and doublebreasted firms, who an-

---

[5] Clinton C. Bourdon and Raymond E. Levitt, *Union and Open-Shop Construction* (Lexington, Mass.: D.C. Heath & Co., Lexington Books, 1980), pp. 18–19.

[6] "Union Construction in Trouble," Reprinted from *Engineering News-Record,* November 5, 1981, p. 26, copyright, McGraw-Hill, Inc. All rights reserved.

[7] "The Top 400 Contractors," *Engineering News-Record,* April 28, 1983, p. 73.

TABLE V-5
Comparison of Large Open Shop and Union Contractor Domestic Sales Trends, 1975–1980

	Domestic Sales Reported to ENR of Large Open Shop Firms (millions)						Sales Trends 1975–1977 vs. 1977–1980		
	1975	1976	1977	1978	1979	1980	1975–1977	1977–1980	Percent Change
Brown & Root	4,627	3,735	3,723	4,790	5,551	9,175	12,085	19,516	+62
Daniel / Fluor[a]	3,086	788	1,868	1,333	1,100	4,153	5,742	6,586	+15
McDermott	900	266	300	600	819	931	1,466	2,350	+60
BE&K	190	234	251	383	630	659	675	1,672	+148
Austin	469	384	502	502	755	562	1,355	1,819	+34
Fish	249	110	383	185	40	445	742	670	−10
Zachry	106	256	295	293	276	377	657	946	+44
Yeargin	65[b]	98	212	316	290	321	375	927	+147
Blount	274	188	410	804	597	1,322	872	2,723	+212
Rogers	70	76	83	109	107	158	229	374	+63
Cianbro	10[b]	15[b]	20[b]	36	41	111	45	188	+317
Pizzagalli	20[b]	39	38	39	93	87	97	2,191	+126
Total	10,066	6,189	8,085	9,390	10,299	18,301	24,340	37,990	+56

TABLE V-5 (continued)

	Domestic Sales Reported to ENR of Large Union Firms (millions)						Sales Trends 1975–1977 vs. 1977–1980		
	1975	1976	1977	1978	1979	1980	Sales 1975–1977	Sales 1977–1980	Percent Change
United Engineers & Contractors	800	141	3,681	1,479	1,615	821	4,622	3,915	−15
Stone & Webster	2,100	1,800	3,200	3,200	1,500	390	5,000	5,090	+2
Bechtel	1,907	1,076	1,966	703	262	2,122	4,949	3,087	−38
Ebasco	577	624	2,541	1,750	2,330	3,496	3,742	7,576	+102
Morrison-Knudson	655	628	832	715	919	1,931	2,115	3,565	+69
Jacobs	233	499	238	616	624	2,465	970	3,715	+283
Badger	311	350	248	202	330	330	909	862	−5
Stearns-Roger	1,818	978	931	2,416	1,127	1,009	3,727	4,552	+22
Rust	691	895	1,125	1,697	1,255	1,967	2,711	4,919	+81
Dravo	146	196	189	324	243	1,502	531	2,069	+290
C-E Lummus	501	600	1,025	417	832	1,600	2,126	2,849	+34
Parsons	1,106	1,092	650	365	1,373	1,140	2,848	2,878	+1
Catalytic	103	160	159	201	175	580	422	956	+127
Total	10,948	9,039	16,785	14,085	12,595	19,353	34,672	47,033	+36

Source: *Engineering News-Record* "400" data, compiled by major signatory firm.
[a] Daniel acquired by Fluor in 1977.
[b] Estimated.

TABLE V-6
Engineering News-Record *1982 Data*

ENR's Top 400

Union • • • versus • • • Open shop

	Union: No. of firms		Union: Av. awards[a] $ million				Open shop: No. of firms		Open shop: Av. awards[a] $ million		
	Total	Double breasted	1982	1981	% chg. 1982–1981		Total	Double breasted	1982	1981	% chg. 1982–1981
Total group	272	43	115.1	132.6	−13.2	Total group	95	20	101.7	111.3	−8.6
1.	26	6	—	—	—	1.	5	2	—	—	—
2.	25	9	347.2	353.5	−1.8	2.	9	2	259.0	309.4	−16.3
3.	25	5	161.6	167.8	−3.7	3.	9	2	123.0	139.3	−11.7
4.	29	6	115.1	133.5	−13.8	4.	10	1	114.6	106.7	+7.4
5.	26	5	93.0	115.9	−19.7	5.	12	3	94.0	93.4	+0.6
6.	28	4	78.4	100.4	−21.9	6.	9	2	77.0	69.9	+10.2
7.	33	3	72.4	79.9	−9.4	7.	6	0	59.8	67.1	−10.9
8.	26	2	54.8	77.0	−28.8	8.	11	1	55.5	64.4	−13.8
9.	26	1	47.3	72.8	−35.1	9.	13	4	47.1	58.9	−20.0
10.	28	2	39.9	64.9	−38.6	10.	11	3	38.1	44.8	−14.9

Source: Reprinted from *Engineering News-Record,* May 5, 1983, copyright, McGraw-Hill, Inc. All rights reserved.

[a] Open shop awards exclude those for double breasted firms and firms not ranked in 1981. Union awards include only those for firms reporting 100% of their workers under collective bargaining agreements. Excludes firms not ranked in 1981. Group 1 is top 40 firms; group 2 second 40 firms, etc.

swered pertinent questions on the *Engineering News-Record* questionnaire. The companies are divided by volume into ten groups of forty each. As has historically been true, union shop contractors dominated the "megaprojects" awarded in 1982, and therefore the top 10 percent of the list in Table V-6. In all other categories, however, open shop contractors are evenly distributed at all levels of the 400 and account for 23 percent of the award volume, which is their appropriate number among the 400. If the awards to the sixty-three doublebreasted contractors could be sorted out and properly divided among union and nonunion firms, the percentage of open shop companies and their volumes would be higher.

As in previous surveys, a large majority of open shop firms were headquartered in the South in 1982–1983 and did much of their work there. By this time, however, open shop firms had expanded in the West, particularly in the Rocky Mountain states, and both southern and especially southwestern contractors held important contracts there.

Further Analysis of 1982 ENR *Data*

The widespread takeover of open shop firms by signatory companies, and / or the creation of open shop entities by such concerns, has made it impossible to develop meaningful data for 1982 to compare with those of a decade earlier. We have, however, identified seven open shop companies in the largest fifty of the 1982 ENR 400: Brown & Root, McDermott, Austin Industries, Blount International, BE&K, Becon, and H. B. Zachry. In addition, about twenty of the largest unionized companies in the 1982 top fifty have open shop subsidiaries of which we are aware, and some have more than one such affiliate. These subsidiaries include Daniel, owned by Fluor, and Becon, the open shop arm of Bechtel. Most large open shop companies have union shop affiliates. Table V-7 shows the top companies in the 1982 ENR 400 together with the doublebreasted subsidiaries of those that we have been able to identify.

The open shop lost some ground at the top in 1983 as Brown & Root was hurt by the sharp drop in the construction of petroleum-related and other process industries, by the cancellation of Exxon's synthetic fuel project on Colorado's west Rocky Mountain slope, and by its withdrawal as contractor for a Texas utility coalition which was building a nuclear power plant after the users dismissed Brown & Root as the design contractor.[8] As we shall note below, however,

[8] The unions spread considerable propaganda about Brown & Root's problems with this nuclear project which was subsequently awarded to a union firm, Ebasco. As

Brown & Root did win the largest contract offered for bid in process construction in 1983; Exxon's Baytown, Texas, refinery upgrading project.[9]

If Daniel International were listed separately, we believe that it would show awards of $2 to $2.5 billion for 1982, thus ranking it close to Brown & Root. Daniel has the great bulk of Fluor's domestic work, in addition to some international work. It too has been hurt by the construction depression, but will undoubtedly come back strongly as plant investment rises.

The 1983 data, published in 1984, shows some small open shop gains. Five new open shop contractors joined the list, and more doublebreasted companies were found by the journal (most of which were previously uncovered by our research and are listed for the largest contractors in Table V-7). The *Engineering News-Record* found that the open shop contractors in the 400's middle range had the most success in gaining new business. On the other hand, more open shop than union contractors finished the year with operating deficits, undoubtedly a function of the slump in the process industries. The continued glut in office building space will undoubtedly hurt union contractors in 1984. Membership in the Houston Chapter of the AGC, for example, was down 30 percent in May 1984 from 1983. Overall, open shop contractors did better than their union shop competitors in the 400 in raising their backlogs.

ENR 400 Data—Final Comment

For reasons already noted, the 400 data are subject to some error but they do show the steadily increasing strength of the open shop not only in achieving 400 status, but also in the growing capability of open shop contractors to handle larger projects. In the early 1970s, only Brown & Root and Daniel International could compete with the signatory companies in bidding for huge industrial complexes. Two other companies, J. Ray McDermott and H. B. Zachry, also had great capability by the early 1970s and could bid for very large jobs. By 1983, several other companies had joined this group, including BE&K, Austin, Fish Engineering, Blount International, Payne and Kellor, and Becon. Moreover, very few of the signatory companies lack a doublebreasted subsidiary. As these companies lose jobs to the open shop, they become more inclined to shift re-

discussed below, however, nearly every other ongoing nuclear project has had serious cost overruns since the Three Mile Island accident. The Brown & Root job is the only one in this category being built by an open shop contractor.

[9] See the discussion below concerning the Houston area for an analysis of why Brown & Root was awarded this contract.

TABLE V-7
The Fifty Largest Contractors by Union Status, 1983

Rank	Holding Company and Principal Firm If Known	Union (U) Open Shop (OS)	Doublebreasted Affiliate If Known[a]	1982 Contract (in $ millions) Total	Foreign
1.	Kellogg Rust, Inc., Houston, Tex.—Kellogg Rust Constructors	U	KRI	8,200.0	5,000.0
2.	Fluor Corp., Irvine, Calif.—Fluor Constructors	U	Daniel Int'l	6,507.1	4,097.3
3.	The C. T. Main Corp. Boston, Mass.	U	Main Erbauer	5,704.0	5,201.0
4.	Bechtel Group, Inc., San Francisco, Calif.—Various Bechtel companies	U	Becon	5,490.0	3,866.0
5.	The Parsons Corp., Pasadena, Calif.	U		4,319.7	3,074.4
6.	Raymond International, Inc., Houston, Tex.	U	Ind. Co. of Steamboat Springs; Sayler Marine Corp.	4,217.3	2,687.9
7.	Stearns-Catalytic World Corp., Allentown, Pa.	U	PCC	4,045.0	384.0
8.	Santa Fe Int'l—C. F. Braun Constructors, Alhambra, Calif.	U	San Jacinto Div. of Santa Fe Int'l	3,315.0	3,300.0
9.	Lummus Group, Inc., Bloomfield, N.J.	U		3,252.0	2,392.0
10.	Foster Wheeler Corp., Livingston, N.J.	U	Energy Plant	3,200.0	2,500.0
11.	Morrison-Knudsen Co., Inc., Boise, Idaho	U		3,104.4	1,325.8
12.	Enserch Corp., Dallas, Tex.—Ebasco Serv.	U	Enserch Eng. & Constructors	2,880.6	508.4
13.	Guy F. Atkinson Co., S. San Francisco, Calif.	U	Tyger	2,654.3	1,993.9
14.	Brown & Root, Inc., Houston Tex.	OS	Mid Valley[b]	2,472.9	1,250.1
15.	McDermott, Inc., New Orleans, La.	OS		2,367.2	1,454.1
16.	J. A. Jones Const. Co., Charlotte, N.C.	U	Metric; Tiber	1,927.4	90.0
17.	Stone & Webster Engrg. Corp., Boston, Mass.	U		1,889.4	978.4
18.	Burns & Roe, Inc. Oradell, N.J.	U		1,876.4	0
19.	Turner Corp., New York, N.Y.—Turner Const.	U	Belco; TransCon; BFW; Universal	1,590.9	240.2
20.	The Austin Co., Cleveland, Ohio	U		1,316.9	336.0
21.	Dravo Corp., Pittsburgh, Pa.	U	Weyher/Livsey	1,150.0	307.3
22.	Perini Corp., Framingham, Mass.	U	Yeargin	1,040.0	50.0
23.	Austin Corp., Inc., Dallas, Tex.—Five Austin companies: Bridge, Commercial, Paving, Power, and Road	OS	Coastal Const.; Coastal Ind. Const.; Nat'l Valve & Mfg.	1,039.4	0

24.	Blount Int'l, Ltd., Montgomery, Ala.—Blount Constructors Div.	OS	Blount Bros. Corp.[b]	1,012.4	48.1
25.	Gilbane Building Co., Providence, R.I.	U		962.3	0
26.	Tishman Realty & Const. Co., Inc., N.Y.C.	U		952.5	0
27.	Peter Kiewit Sons, Inc., Omaha, Neb.	U	Gilbert Corp.; South Prairie	952.0	123.7
28.	CBI Industries, Inc., Oak Brook, Ill.	U		910.0	519.0
29.	Jacobs Engrg. Group, Inc., Pasadena, Calif.	U	UMC of Louisiana, Inc.	905.0	8.5
30.	Morse/Diesel, Inc., New York, N.Y.	U		840.7	0
31.	Dillingham Corp., San Francisco, Calif.—Hawaiian Dredging & Const.	U	Watkins Eng. & Const.; Odin Const. Co.	788.5	239.5
32.	Barton-Malow Co., Detroit, Mich.	U		738.2	0
33.	McCarthy Brothers Co., St. Louis, Mo.	U		732.0	0
34.	Huber, Hunt & Nichols, Inc., Indianapolis, Ind.	U		719.0	0
35.	Davy Inc., Cleveland, Ohio	U	Olsen; DMI Int'l	704.8	0.6
36.	Centex-Bateson-Rooney-Golden, Dallas, Tex.—Frank J. Rooney, Inc.[c]	U	Great Southwest Corp.	704.0	0
37.	George A. Fuller Co., New York, N.Y.	U		667.5	261.0
38.	Lehrer/McGovern, Inc., New York, N.Y.	U		640.0	140.0
39.	Ford, Bacon & Davis, Inc., Dallas, Tex.	U	MeriCon Int'l	636.0	162.0
40.	BE&K, Inc., Birmingham, Ala.	OS		634.5	0
41.	Mellon-Stuart Co., Pittsburgh, Pa.	U		629.2	0
42.	HBE Corp., St. Louis, Mo.	U		552.8	0
43.	Procon Int'l, Inc., Des Plaines, Ill.	U		535.3	448.2
44.	Becon Const. Co., Inc., Houston, Tex.	OS	Bechtel[b]	515.0	0
45.	Pullman Power Products Corp., Williamsport, Pa.	U		515.0	5.0
46.	The Badger Holding Co., Inc., Cambridge, Mass.—Badger-Am., Inc.	U	Process Plants, Inc.	512.0	425.0
47.	The Heyward-Robinson Co., Inc., N.Y.C.	U		491.0	95.0
48.	H. B. Zachry Co., San Antonio, Tex.	OS		485.0	64.0
49.	CEI Const., Inc., Bethesda, Md.	U	Omni	481.1	0
50.	Dick Corp., Pittsburgh, Pa.	U		442.0	0

Source: *Engineering News-Record,* April 28, 1983, pp. 69, 72; and for doublebreasted affiliates, author's research.

[a] Other doublebreasted affiliates probably exist but are not admitted or known.

[b] Union affiliates of open shop companies.

[c] Other affiliated union and open shop companies exist.

Table V-8
Engineering News-Record *1983 Data*

ENR's Top 400

Union contractors • • • versus • • • Open shop contractors

	Union: No. of firms		Union: Av. awards[a] $ millions					Open shop: No. of firms		Open shop: Av. awards[a] $ millions			
	Total	Double breasted	1983	1982	% chg. '83–'82	Loss[b]		Total	Double breasted	1983	1982	% chg. '83–'82	Loss[b]
Total group[b]	267	54	126.5	124.2	+ 1.9	10	Total group[b]	110	25	125.7	123.1	+ 2.2	21
1.	29	16	—	—	—	8	1.	4	1	—	—	—	—
2.	23	9	324.1	376.7	−14.0	11	2.	14	3	323.3	368.1	−12.2	15
3.	29	4	176.6	146.2	+20.8	7	3.	9	3	165.5	99.8	+65.8	25
4.	27	4	132.9	112.0	+18.6	8	4.	12	3	138.3	115.1	+13.6	36
5.	24	5	111.4	80.2	+38.8	4	5.	13	4	108.9	86.2	+26.3	9
6.	27	3	91.4	94.8	− 3.6	7	6.	10	1	75.6	66.8	+13.2	20
7.	28	3	71.8	76.0	− 5.5	0	7.	11	1	62.1	62.7	− 0.9	18
8.	25	2	59.0	62.5	− 5.6	13	8.	13	6	58.6	62.9	− 6.8	15
9.	29	6	50.3	52.1	− 3.5	17	9.	11	1	41.9	49.7	−15.8	25
10.	26	2	43.9	54.2	−19.0	23	10.	13	2	44.5	45.0	− 1.7	33

Source: Reprinted from *Engineering News-Record,* May 3, 1984, copyright, McGraw-Hill, Inc. All rights reserved.

[a] Average awards excludes group one, doublebreasted firms and firms not ranked in 1982.

[b] Losses are % of firms responding to survey. Group one is top 40 firms; group two, second 40 firms, etc.

sources to the open shop subsidiary. Some of the results are found in data developed by the NCA.

NATIONAL CONSTRUCTORS ASSOCIATION DATA

Since 1961, the NCA has examined the extent to which open shop contractors have won contracts which this association believes would otherwise have gone to NCA members. The NCA data for 1963–1981 are charted in Figures V-1 and V-2.

The NCA found that in 1963, the signatory companies lost about 75 contracts and by 1971 that number had risen to 180. The open shop share then dropped sharply until 1973 when it again took off arriving at a total of 540 jobs in 1981 (see Figure V-1). In terms of

FIGURE V-1
Number of Major Construction Jobs Lost by NCA Members to Open Shop Contractors 1963–1981

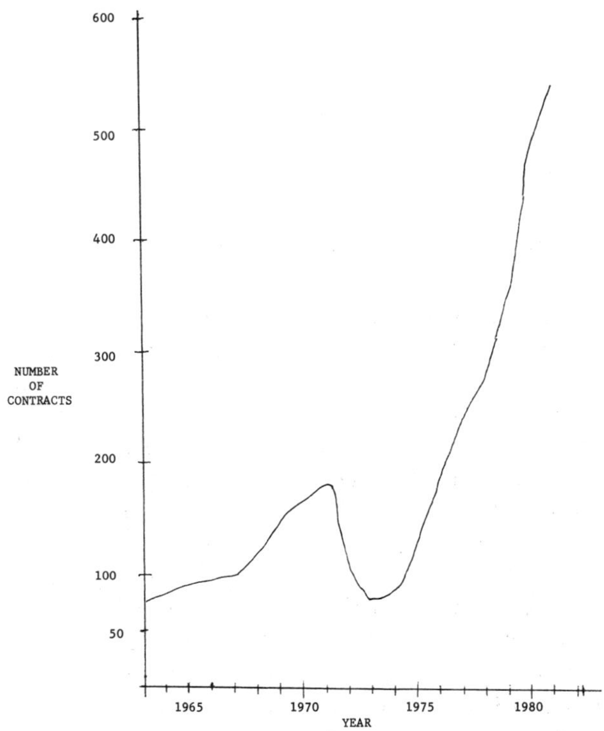

Source: National Constructors Association.

FIGURE V-2
Dollar Volume of Major Construction Jobs Lost by NCA Members to Open Shop Contractors 1963–1981

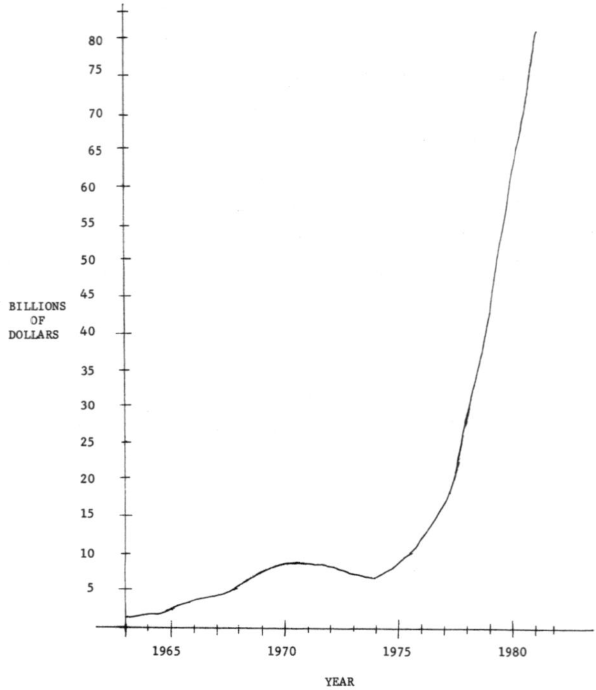

Source: National Constructors Association.

dollar volume, Figure V-2 shows that the open shop share has gone up steadily since 1963 except for a slight dip in 1973–1974. In eighteen years, the dollar volume of the open shop jobs lost by the NCA members rose from less than $2 billion to nearly $80 billion.

Table V-9 presents these data for the years 1961 to 1981, broken down by three types of jobs—petrochemical plants, power plants, and others—as well as by region. These data are worth careful scrutiny on a number of counts. They show the great strength of open shop contractors in the construction of petrochemical plants and complexes, and their continued and increasing expansion in construction of such plants in the Southeast and Gulf Coast regions. What is also very significant, however, is the steady rise in open shop construction of power plants, once an exclusive unionized contractor preserve, and of other industrial jobs. We shall discuss these developments in more detail below, but it is obvious from the data

TABLE V-9
Major Construction Contracts Won by Open-Shop Contractors by Project Type and Region[a]
1961–1981

	Number of Jobs	Dollar Volume[b]	Petro Chem	Power Plant	Other Jobs	East Coast	Great Lakes	Mid-Continent	Gulf Coast	South East	West Coast
1961–1963	75	$1.4	40	3–5	32	1	3	1	33	36	1
1963–1965	90	2.5	54	5	31	2	3	2	40	40	3
1965–1967	100	4.2	63	12	25	1	2	2	48	43	4
1967–1969	150	7.6	72	37	41	1	3	2	92	47	5
1969–1971	181	8.5	97	37	47	4	3	4	98	65	7
1971–1973	78	7.2	22	25	31	2	6	4	37	24	3
1973–1977	n.a.	n.a.	n.a.	n.a.	n.a.	n.a.	n.a.	n.a.	n.a.	n.a.	n.a.
1978	270	28.1	185	40	45	7	5	20	163	55	20
1979	343	42.7	211	45	70	6	8	37	203	62	27
1980	482	59.8	289	53	140	7	15	65	238	110	47
1981	549	77.3	346	59	134	5	16	80	250	128	59

Source: National Constructors Association.

[a] Regions are defined as follows:

East Coast = New England states, New York (east of Rochester), Pennsylvania (Altoona and East), Delaware, Maryland, and the District of Columbia

Great Lakes = Balance of New York and Pennsylvania, Ohio, Kentucky, Indiana, Illinois, Wisconsin, Minnesota, Michigan, and West Virginia

Mid-Continent = Montana, Wyoming, North Dakota, South Dakota, Colorado, Iowa, Missouri, Kansas, and Nebraska

Gulf Coast = Texas, Oklahoma, Arkansas, Louisiana, and Mississippi

Southeast = Florida, Georgia, South Carolina, North Carolina, Tennessee, Alabama, and Virginia

West Coast = Washington, Oregon, California, Idaho, Nevada, Utah, Arizona, New Mexico, Hawaii, and Alaska.

[b] In billions of dollars.

n.a. = not available.

in Table V-9 that the open shop capability has expanded considerably, especially since 1978.

Equally of interest has been the steady increase in the open shop share of industrial jobs in all regions except the East Coast. From 1978 to 1981, the number of contracts won by open shop contractors went from 5 to 16 in the Great Lakes region, 20 to 80 in the Mid-Continent region, and 20 to 59 in the West Coast region. Even greater strides were made in the Gulf Coast and Southeast regions. Clearly, the signatory firms have significant competition, especially in the South, but increasingly in other parts of the country as well.

In 1983, the NCA reviewed the situation again and found, as set forth in Table V-10, that large open shop contractors had lost about $20 billion in contract volume since 1981. No data were given for losses of union shop contractors, although the data released by the NCA intimated that the open shop losses might be greater than those of the signatory contractors. The data in Table V-10 show a decline in these large jobs for open shop contractors in all states except Colorado and Washington, and again emphasize the declining fortunes of the process industries. Absent much better evidence, including data for union shop concerns, it is difficult to assess this information other than as a result of recession in one year.

INDUSTRIAL DATA

The NCA data also provide information about the open shop penetration in particular industries. In this section, details are set forth concerning developments in the public utility, the petrochemical, and the paper and other industries.

Public Utilities

The data in Table V-9 show that during the period from 1978 to 1981 open shop contractors won 197 power plant construction jobs. Power plant construction, and particularly the building of nuclear power plants, represents the largest and most sophisticated work of the industrial contractor. In the late 1960s, the top two open shop contractors, Brown & Root and Daniel International, became major forces in this work. H. B. Zachry soon entered the field winning several contracts, and more recently BE&K, Austin, and Becon have taken on power plant construction work. A few companies, in particular Duke Power, developed their own workforces for such jobs.

The Edison Electric Institute, the trade association of investor-owned electric utilities, periodically makes surveys of the "labor

TABLE V-10
States with Major Changes in Nonunion Industrial Work[a]

State	Number of projects		Estimated cost ($ millions)	
	1981	1983	1981	1983
Alabama	18	10	1,360	742.5
Colorado	21	25	6,889.4	7,524.4
Georgia	18	8	2,222.7	854.7
Louisiana	54	56	6,405	5,218.5
Mississippi	17	11	3,780	2,621
Nebraska	6	2	1,214	440
New Mexico	17	12	2,680.5	2,149.5
North Carolina	31	12	7,932.9	5,945
Ohio	2	1	740	5
South Carolina	18	9	6,030.5	3,831
Texas	149	135	18,508.9	14,791.1
Utah	7	3	2,504.5	790
Virginia	18	5	2,113	527
Washington	3	7	44	580

Source: Reprinted from *Engineering News-Record,* May 3, 1983, copyright, McGraw-Hill, Inc. All rights reserved.
[a] Changes totalling more than $500 million.

relationship" in utility construction. Table V-11 shows the results of a 1973 survey by the Institute Construction Problems Subcommittee, Industrial Relations Division, and Table V-12 contains the data from a 1980 study covering construction starts and projects starts from 1975 to 1985. The data in these tables serve as a supplement to the NCA information.

A number of cautionary points must be made in evaluating the data in Table V-11. The questionnaire responses were not always clear on the degree of open shop; therefore, we placed respondents in the category where most of the work was done. Also, in the transmission and distribution category, many in-house forces are unionized, some are nonunion, and often they are both. Moreover, some of the work could be overhaul and maintenance rather than new construction. On the other hand, the construction of power plants by in-house forces on the part of a southern utility undoubtedly refers to Duke Power, which is unique in that its forces have the capacity to build nuclear power plants.

According to the Edison Electric survey, by the first quarter of 1973 open shop concerns were being used to build only six facilities, five in the South and one unidentified (not including the Duke Power in-house arrangement). In addition, nine utilities were utilizing open

TABLE V-11
Edison Electric Institute Power Facility Construction Survey by Facility Type, Union Status, and Region,[a] *1973*

Region	Power Plants				Transmission and Distribution			
		Type of Labor				Type of Labor		
	Number of Responses	Union	Open Shop	In-house Forces[b]	Number of Responses	Union	Open Shop	In-house Forces[b]
United States	87	79	6	2	89	26	9	54
New England	2	2	—	—	4	1	1	2
Middle Atlantic	15	15	—	—	15	3	1	11
Midwest	28	27	—	1	27	11	—	16
South	28	22	5	1	26	5	6	15
West	12	12	—	—	14	4	—	10
Unidentified	2	1	1	—	3	2	1	—

Source: Edison Electric Institute, Industrial Relations Division, Construction Problems Subcommittee.

[a] Regions are defined as follows:

New England = Maine, New Hampshire, Massachusetts, Connecticut, and Rhode Island
Middle Atlantic = New York, New Jersey, and Pennsylvania
Midwest = Ohio, Indiana, Michigan, Illinois, Iowa, Wisconsin, Minnesota, Kansas, Missouri, and West Virginia
South = Virginia, North Carolina, South Carolina, Georgia, Florida, Arkansas, Kentucky, Mississippi, Texas, and Oklahoma
West = California, Washington, Oregon, Montana, Idaho, Utah, Nevada, Colorado, and New Mexico.

[b] In-house forces are either union, nonunion, or some of each.

shop contractors on transmission and distribution construction, and others, like Duke Power, were utilizing nonunion in-house forces for this work.

The Midwest utility reported in Table V-11 as using in-house personnel for construction facilities is probably Dayton Power and Light. The company established a special subsidiary, UCON, Inc., to handle its construction "more efficiently, with better productivity and obviously at a lower cost . . . to better meet our construction schedules, which formerly were not met by most contractors except at extreme cost and the utmost pressure by us."[10]

Duke Power Company, operating in the Piedmont of North and South Carolina, has had its in-house construction division since 1925. In the construction of fossil power plants it utilizes its own forces plus those of contractors. The latter are generally specialty types, operating particularly in the fields of boiler installation, high pressure piping, heat insulation, etc., which are usually purchased from major suppliers on an erected basis. During the 1925 to 1980 period, Duke Power used these contractors in an attempt to maintain stable employment. This contractor labor was usually union.[11]

When Duke Power went into nuclear power it decided to eliminate contract labor because of the rising costs and the problems beginning to develop over jurisdiction and work rules. Since Duke Power had its own design and engineering departments, it was felt that company labor would be able to provide better integration with engineering. Two nuclear plants were in operation in early 1974 and eleven more were in progress. The construction forces then numbered 5,000 in addition to 550 engineers and designers.

Several other utility companies have in-house construction forces but they do not take on the major projects. For example, Pennsylvania Power and Light does its own construction, excluding power plants and large office buildings, with an in-house company which it purchased soon after its formation. The employees of this company are represented by an independent union. The Arizona Public Service Company's in-house force is similar, but it is unionized by the AFL-CIO construction trades.

Overall, Table V-11 shows that open shop construction had made some gains by 1973, but power plant building remained overwhelmingly the province of major union shop contractors, most of whom were members of the NCA. There is, however, reason to believe

[10] The information on UCON, Inc., is based on a talk by L. J. Renas, Vice President, Dayton Power and Light Company, St. Louis, November 14, 1974.

[11] Interviews, Duke Power Company personnel, Charlotte, North Carolina, February 7, 1974, and July 15, 1982.

Table V-12
Electric Utility Fossil Unit Power Plant Survey[a]
Construction Starts, 1975–1980 and
Projected Starts, 1981–1985

Year Construction Starts	Total No. of Units	Union				Open Shop	
		Project Agreement		Local Agreement			
		No.	Percent	No.	Percent	No.	Percent
1975	28	16	57.1	8	28.6	4	14.3
1976	23	12	52.2	9	39.1	2	8.7
1977	7	6	85.7	0	0	1	14.3
1978	14	4	28.6	6	42.9	4	28.6
1979	9	2	22.2	4	44.4	3	33.3
1980	8	4	50.0	4	50.0	0	0
Projected starts as of 1980:							
1981[b]	15	8	53.3	4	26.7	3	20.0
1982	11	2	18.2	2	18.2	7	63.6
1983	13	5	38.5	2	15.4	6	46.2
1984	7	2	28.6	0	0	5	71.4
1985[c]	10	4	40.0	2	20.0	4	40.0
Totals							
1975–1980	89	44	49.5	31	34.8	14	15.7
1981–1985	56	21	37.5	10	17.9	25	44.6
1975–1981	145	65	44.8	41	28.3	39	26.9

Source: Edison Electric Institute, Construction Committee Survey, 1980.
[a] Ninety-four companies surveyed. Seventy-eight responses. Seventeen had no fossil starts. Sixty-one had fossil starts.
[b] Undecided - 2.
[c] Undecided - 6.

that this survey understated open shop penetration. Table V-5, based on NCA data, lists 120 power plant jobs lost to NCA contractors in twelve years including 25 for the years 1971 to 1973 when only partial data were available. It does not appear likely that all of these were in the South. In addition, we know that power plants were being constructed on an open shop basis in the Rocky Mountain and Midwest regions during this period. These facilities may, however, have been begun after the Edison Electric survey was made.

The Edison Electric 1980 survey deals only with fossil-fired units (principally coal-fired units) and found only thirty-nine started or projected open shop operations in this area. Yet the NCA counted over 100 in 1980 and 1981 alone. The reason for this apparent contradiction is that the NCA data include major modifications and nuclear and hydroelectric facilities as well as fossil-fired units. The data in Table V-12 are further complicated by the fact that many of the projected units have been cancelled, and some of the started ones abandoned or delayed. Moreover, since the percentage of open shop for the projected group is almost three times that of those that were actually started, it raises questions of accuracy.

The construction of nuclear power facilities requires even more capability and expertise than does that of fossil units, and the number of contractors who can handle these jobs include only the largest. Table V-13 shows that the open shop group was not a factor in nuclear power plant construction until 1972, but that between 1972 and 1981 open shop contractors, including Duke Power's in-house crew, captured about 45 percent of the nuclear work in the Southeast and the Southwest plus two jobs in the West.

Since the accident at Three Mile Island in 1979, construction at about fifty nuclear plants in the United States has been canceled, no new facilities have been begun, and sixty remain under construction. Some of the latter may also be canceled or construction postponed.[12] Public fears whipped up by the press since the Three Mile Island accident, increased labor costs, inconsistent regulation by the Nuclear Regulatory Commission, and cost overruns have combined with energy conservation to put nuclear power at a distinct cost disadvantage to fossil-fired units.[13] The latter can now also incorporate techniques to conform with environmental regulations and are thus more acceptable.

Actually, the future for all power plant construction has been considerably dampened by economic, regulatory, and environmental problems so that construction in this industry offers considerably less job potential than was once the case. Nevertheless, electric power systems continue to be required to spend large sums—about $40 billion both in 1982 and 1983—on construction, replacement,

---

[12] Thomas J. Lueck, "Nuclear Dilemma for Ohio Utility," *New York Times,* October 8, 1983, p. 37.

[13] These cost overruns have affected both union and open shop contractors and have been enormous. See, e.g., Lueck, *ibid.;* Mathew L. Wald, "New Delays Seen for Power Plant," *New York Times,* October 9, 1983, p. 53; "Indiana Reactor May End Era," *New York Times,* September 14, 1983, p. D-1; "WPPSS Cutbacks at Hanford Have Severe Impact for Crafts," *Construction Labor Report,* Vol. 28 (October 27, 1982), pp. 1159–61; Peter W. Bernstein, "A Nuclear Fiasco Shakes the Bond Market," *Fortune,* Vol. 105 (February 22, 1982), pp. 100–115; and "Nuclear Plant Owners Sue Brown & Root," *Engineering News-Record,* January 7, 1982, p. 18.

TABLE V-13
Nuclear Power Plants Built and Under Construction as of July 1981

Region[a]	Total	Prior to 1972		1972–1981		Percent
		Union	Open Shop	Union	Open Shop	Open Shop
Northeast	35	2	0	33	0	0
Midwest	43	9	1	33	0	0
Southeast	56	1	0	32	23	41.8
Southwest	10	0	0	4	6	60.0
West	19	1	0	16	2	11.1
Total U.S.	163	13	1	118	31	20.8

Source: *Nuclear News,* Vol. 24 (August 1981), pp. 85–92, 97–103; and Industrial Research Unit union-open shop information.

[a] Regions are defined as follows:

Northeast = New England, New York, New Jersey, Maryland, Delaware, and District of Columbia

Midwest = Ohio, Indiana, Illinois, Wisconsin, Minnesota, Iowa, Missouri, Kansas, Nebraska, North Dakota, and South Dakota

Southeast = Virginia, West Virginia, Kentucky, Tennessee, North Carolina, South Carolina, Georgia, Florida, Alabama, Mississippi, and Louisiana

Southwest = Texas, Oklahoma, Arkansas, and New Mexico

West = California, Oregon, Washington, Nevada, Utah, Idaho, Montana, Colorado, and Arizona.

and improvement.[14] It is likely, however, that open shop growth in this field will not be rapid. Utilities are often under intense pressure by the political supporters of unions to favor unionized contractors.[15] Moreover, because the rates charged by public utilities are controlled by state, and sometimes local governmental commissions, and because of pressures by consumer and other interest groups to keep these rates low, utility managers are extremely anxious to avoid antagonizing organized labor lest unions retaliate by opposing rate adjustments.[16] In addition, the labor forces of most utilities are represented by the International Brotherhood of Electrical Workers

[14] See John S. Alberici, "The Next Decade: Outlook for Construction in the Electric Utility Industry," *Constructor,* September 1983, p. 39. This article is a reprint from *Electrical World,* March 1983.

[15] A case in point is the award of the large facility being built in Utah by a consortium of Los Angeles and Utah utility companies. The contract went to Bechtel over Daniel, despite an alleged lower bid by Daniel, supposedly because of the insistence of the Los Angeles group.

[16] An illustration of this point is the Edison Electric Institute's nationwide, full-page advertisements designed to spur utility construction and rate adjustments, which featured Robert A. Georgine, President, Building and Construction Trades Department, AFL-CIO.

(IBEW) which is also a key construction union. This adds to the pressure to build union.

These political factors weigh against the heightened economic pressures to build open shop. Considering all these aspects, plus the existence of a large number of qualified unionized concerns and the inability to build open shop in a number of highly unionized states, it seems likely that open shop involvement in power plant construction work will remain heavily concentrated in the less-unionized southern states, will continue to expand in the central Midwest and Rocky Mountain states, and will probably amount to about 34 to 38 percent of the work by 1990, up from approximately 25 to 30 percent in 1980. This is less than that projected by the Edison Electric Institute survey (Table V-12), but seems more realistic.[17]

Chemical, Petrochemical, and Refinery Plants

The greatest gains of the open shop in industrial construction have been made in the construction of processing facilities—chemical, petrochemical, petroleum refining and related operations. There are three basic reasons why this has occurred: 1) the major expansion of those industries has occurred in the Sun Belt, where unions have traditionally been the weakest; 2) this is the area where the largest open shop companies are headquartered and these companies have grown with, and specialized in, this work; and 3) large users in these industries have turned to open shop contractors because the costs of union construction became inflated and an alternative was sought, and because some user companies have sought to keep their industrial facilities nonunion and so have avoided having union construction workers involved in plant construction or maintenance work.

A major gain for the open shop occurred when E. I. DuPont de Nemours & Company, the largest chemical concern in the United States, changed its policies. DuPont for many years operated its own construction company. During World War II it began operating union in construction, continuing to do so until the early 1970s. DuPont did not sign union contracts but entered into verbal agreements with some unions and operated through subcontractors with others. The jobs, no matter how they were handled, were strictly union. Engineering and design was done by DuPont; and the labor, except that of subcontractors, was put directly on DuPont payrolls.

---

[17] These estimates are our own but are in accord with those of the National Constructors Association.

Construction labor received all union benefits plus company benefits.[18]

By 1970, DuPont felt that its costs and productivity were so out of line that significant changes were required. It gradually withdrew from direct employment of construction labor and began awarding contracts in some areas to three open shop contractors: Brown & Root, Daniel International, and H. B. Zachry. By late 1974, about 50 percent of DuPont's huge construction jobs—employing approximately 10,000 employees—were being done on an open shop basis. This, of course, meant a major gain for the industrial sector of the open shop and it encouraged other chemical companies to move in the same direction. The nearly 350 petrochemical plans won by open shop contractors between 1961 and 1973, as shown in Table V-9, are indicative of shifts by DuPont and other concerns.

Although continuing to deal with unions in several areas, DuPont has become about 60 percent open shop in its construction work. Union construction for DuPont received a further blow in 1982–1983 when union pickets caused the company to terminate union contractors on a new life-science experimental station project near Wilmington, Delaware. DuPont had awarded the concrete work to a union contractor joint venture and other contracts both to union and open shop contractors. The concrete portion began in August 1981 and continued until early November 1982 while open shop crews worked elsewhere on the site. In November 1982, DuPont awarded a $5.5 million contract to an open shop mechanical contractor, whereupon pickets from Pipe Fitters Local No. 80, United Association of Journeymen and Apprentices of the Plumbers and Pipe Fitting Industry (UA) appeared. The unionized concrete workers then stopped working, despite a clear understanding with the company that there would be no strikes or other stoppages on the project.[19]

DuPont warned the contractors and union that they would be terminated if they did not go back to work. After this warning went unheeded past two DuPont extension deadlines, the unionized contractors were terminated and the remaining concrete work was

[18] This information is based upon talks by Trumbull Blake, Director, Construction Division, E. I. du Pont de Nemours & Company, before the Associated Builders and Contractors convention, Freeport, Grand Bahamas, November 9, 1973, and before the Business Roundtable Conference, St. Louis, November 14, 1974, and upon personal interviews in 1974.

[19] "Termination Threat Carried Out by Du Pont Against Union Contractor," *Construction Labor Report,* Vol. 28 (January 5, 1983), pp. 1525–1526.

awarded to an open shop firm, Wohlsen Company.[20] In January 1983, a union mechanical firm was terminated and replaced by a nonunion firm because Local No. 80 refused to man the job or to guarantee that the union workers would cross their own picket line, or even use a special reserved gate where there was no picketing.[21] The net impact of this action has been a major loss for union construction in Delaware.

Most chemical and petroleum companies, especially in the South but also in other regions where it is feasible to utilize open shop contractors, now permit both open and union shop companies to bid on the job and give the contract to the lowest bidder. A few, not wanting unionized personnel on the premises, limit their bids to open shop contractors as a means of encouraging the plant employees to remain nonunion. Although this latter group of companies is a distinct minority, these policies have assisted the open shop companies to grow. Overall, however, it has been the lower costs of open shop contractors which have led to their gains and to their clear hegemony over construction in the Gulf Coast processing industry.

Paper and Other Industries

Four areas are significant to paper manufacturing in the United States: the South, the West Coast, northern Minnesota-Wisconsin-Michigan, and Maine. Of these, the South and the West Coast are by far the most important, with the South taking the lead from the West Coast where the wood supply is declining. Although the South is the most nonunion area in the country, paper manufacturing is an exception. The industry is highly unionized in this region with only a few newly built manufacturing facilities, and perhaps one or two other plants, operating nonunion.

In part because of a desire not to antagonize in-plant forces, and because of the capability of unionized contractors, paper manufacturing plants were traditionally built union in the South. In the late 1960s and early 1970s, Brown & Root and Daniel entered the field. In the 1970s, BE&K was formed by former executives of Rust Engineering (now part of Kellog-Rust), which has been a leading contractor in the field. BE&K added another capable open shop

[20] In Chapter VIII, we note that later the head of this firm was brutally attacked and critically hurt by union pickets.

[21] "Pipe Fitters Local 80 Refuses to Man Union Firm's Work at Du Pont Project," *Construction Labor Report,* Vol. 28 (January 26, 1983), pp. 1629–1630.

contractor to those available at a time when paper companies were looking for ways to reduce construction costs. Today, a majority of paper companies in the South surveyed in our 1982 field investigation either are building, or plan to build, open shop, or at the very least will permit open shop builders to bid on all projects. In Maine, the situation is similar. Daniel built a facility under a union project agreement in the 1970s, but in recent years nearly all the construction work at paper mills has gone open shop.

In other plant construction in the South, open shop contractors now predominate. Yeargin built a large Firestone Tire plant in North Carolina; Daniel built the Nissan plant in Tennessee; and many other examples could be given. In New England, and in parts of the Midwest and Rocky Mountain areas, open shop contractors are also building factories and other industrial structures, as the regional and local data presented below will demonstrate. These gains are significant and growing. The unionized contractors are still well ahead in such construction on a national basis even though they have lost ground, but the open shop group is gaining. This is in part because of its strength in those areas of the country, particularly the Sun Belt, where industry is growing the fastest, but primarily because of cost factors.

SPECIALTY CONTRACTOR STUDIES

In Chapter IV, data from specialty contractor studies were presented which showed that the open shop has gained dramatically in the highly skilled electrical, sheet metal, mechanical, and plumbing trades. Those studies covered industrial as well as commercial sectors, and are cited here because they demonstrate that the last gains of the open shop have been made in industrial construction, even though these gains are serious enough to pose significant problems for the unionized sector. For electrical work, the data in Table IV-5, based upon studies by the National Electrical Contractors Association (NECA), showed that by 1981 only 22 percent of the inside electrical employers had IBEW contracts, but that this group employed 39 percent of the employees. This indicates that the largest contractors, who are most likely to be engaged in industrial construction, work under union agreements.

The situation is somewhat similar in the sheet metal, heating, and air conditioning trades. Figure IV-10, from the Arthur D. Little study made for the Sheet Metal and Air Conditioning Contractors' National Association (SMACNA), shows that the large nonunion contractors had a significant share of the industrial work amounting to almost 50 percent. Moreover, Figure IV-8 showed that the non-union share of such work was growing in all states examined for this study.

For the mechanical and plumbing contractors, Figure IV-11 showed steady open shop gains in industrial construction. In 1977, a study by the Mechanical Contractors Association of America (MCAA) found that the open shop had 17 percent of the industrial work in 1977 and 37 percent in 1982, but projected a figure of 53 percent in 1985.

REGIONAL AND LOCAL STUDIES

Several of the regional and local studies presented in Chapter IV also included separate analyses of industrial construction, and these are presented in this section. Also included are two special studies of Gulf Coast areas, Mobile, and Houston.

The Mobile Study

Mobile, Alabama, is a Gulf port city surrounded by a large industrial area. The Mobile construction industry covers not only the city, in which few large buildings exist, but also most of the southern one-third of Alabama, the panhandle of Northwest Florida, and the Gulf Coast of Mississippi. In this area is a heavy concentration of chemical, petrochemical, and paper manufacturing plants.[22]

This area was once one of the most thoroughly unionized areas in the Southeast. Although residential and highway construction have long been open shop, commercial, industrial, and heavy construction were considered 90 percent union as late as 1975. Then the situation changed dramatically. Data from the monthly reports of the Mobile AFL-AGC Pension and Welfare Plans show that total hours worked by the area construction unions was 5.7 million in 1975, and declined each year thereafter to less than 2 million in 1980.[23] A survey of the thirteen largest area users by the Mid-Gulf Business Roundtable in early 1981 found that 86 percent of the 3,545 area construction workers on new construction and contract

---

[22] *Mobile, Alabama. A Microcosm of the Problems Facing the Unionized Construction Industry.* (This study carries no publisher or date. It has been ascertained, however, that it was compiled by contractors of the Mobile Chapter, AGC in 1980–1981. Data found therein have been checked and found accurate).

[23] *Ibid.*, p. 3.

maintenance[24] were nonunion.[25] Overall, the Mobile AGC estimated that the union contractors' share from 1975 to 1980 has declined from 90 percent of the area market volume to 20 to 25 percent at most.[26] Of the ten major construction projects of $10 million or more in 1981, it was found that nine were being built open shop and the tenth by Daniel under a union project agreement.[27] The study blamed this situation on high wage and benefit costs, loss of confidence in union construction, refusal of unions to compromise in order to regain their market, and a reduction of available skilled union manpower and apprentices as work declined.[28]

Houston Gulf Coast Area Business Roundtable Reports

Perhaps the most accurate information in regard to the relative open and union shop shares of work is compiled on a regular basis by the Houston Gulf Coast Area Business Roundtable. It has been tracking new construction and contract maintenance employment by union and open shop for each quarter of the year since the early 1970s. Table V-14 gives these statistics for Houston and the neighboring Gulf Coast areas from 1975 to 1983.

Table V-14 shows that in 1975 the number of union employees exceeded those employed by open shop contractors in new construction in the Houston area, but that this lead dropped the following year and continued to drop in subsequent years. What occurred was that the Houston construction unions negotiated an area project agreement which was designed to improve productivity and to lower costs. The IBEW, however, refused to ratify it and the agreement was never put into effect. Meanwhile, there was ample work in center city large office construction for the unionized trades, and they lost interest in competing for the process industry jobs.

More recently, the oil glut and the recession severely cut back on the expansion of process industry facilities. This reduced the number of open shop jobs in new construction by about 60 percent, but the union shop share was considerably less affected in total.

---

[24] Contract maintenance involves the use of outside contractors rather than company employees to perform regular maintenance operations. It is very common in the Gulf Coast processing industries.

[25] *Mobile, Alabama,* p. 4.

[26] *Ibid.,* pp. 4–6.

[27] *Ibid.,* pp. 6–7. This report also put forward a program of action for the unions and unionized contractors.

[28] See "Union Recovery Suffers a Setback," *Engineering News-Record,* October 6, 1983, pp. 12–13; and "Building Trades Rebuff Exxon," *ibid.,* June 21, 1984, pp. 27–28. Also telephone interviews, Houston, April–June, 1984.

TABLE V-14
Houston–Gulf Coast Area
Industrial Construction Union and Open Shop Employment[a]
New Construction and Contract Maintenance, 1975–1983

Year	New Construction Total	Union	Percent Union	Open Shop	Percent Open Shop	Contract Maintenance Total	Union	Percent Union	Open Shop	Percent Open Shop	All Construction Reported Total	Union	Percent Union	Open Shop	Percent Open Shop
1975	16,148	9,142	56.6	7,006	43.4	3,975	1,086	27.3	2,889	72.7	20,123	10,228	50.8	9,895	49.2
1976	26,064	12,737	48.9	13,327	51.1	7,827	2,025	25.9	5,802	74.1	33,891	14,762	43.6	19,129	56.4
1977	27,013	8,361	31.0	18,652	69.0	6,157	748	12.1	5,409	87.9	33,170	9,109	27.5	24,061	72.5
1978	25,884	5,222	20.2	20,662	79.8	7,450	929	12.5	6,521	87.5	33,334	6,151	18.5	27,183	81.5
1979	23,202	4,617	19.9	18,585	80.1	8,555	1,232	14.4	7,323	85.6	31,757	5,849	18.4	25,908	81.6
1980	19,924	3,750	18.8	16,174	81.2	9,844	1,441	14.6	8,403	85.4	29,768	5,191	17.4	24,577	82.6
1981	14,204	3,891	27.4	10,313	72.6	11,707	1,488	12.7	10,219	87.3	25,911	5,379	20.8	20,532	79.2
1982	11,064	3,675	33.2	7,389	66.8	7,100	887	12.5	6,213	87.5	18,164	4,562	25.1	13,602	74.9
1983	12,147	4,490	37.0	7,657	63.0	7,326	626	8.5	6,700	91.5	19,473	5,116	26.3	14,357	73.7

Source: Contractor and Manpower Status Report, Houston Gulf Coast Area Business Roundtable, August 1983.
[a] Data are for second quarter of each year.

Nevertheless, as of the second quarter of 1983, the open shop still accounted for 63 percent of the new construction work. At the height of employment, from 1978 to 1980, the open shop had 79.8 to 91.2 percent of the jobs.

In contract maintenance, the open shop has been dominant throughout the years covered and had over 90 percent of this work by the second quarter of 1983. As a result, open shop employment in industrial construction was 73.7 percent of the total reported to the Houston Gulf Coast Area Roundtable in the second quarter of 1983.

The decline in Houston office building construction described in Chapter IV and the decline in process and other industrial construction noted here resulted in tremendous competition for any new work available. Under the leadership of Robert A. Georgine, president of the AFL-CIO Building & Construction Trades Department, the Houston Ship Channel Industrial Construction Recovery Agreement was signed, which reduced wages to about 80 percent of their former levels and brought them close to Brown & Root's wage schedule (which is the open shop industrial construction norm in the area); permitted use of nonjourneymen (helpers) up to 40 percent of the number of journeymen; and made other productivity improvements. It did not, however, reduce fringe benefits or alter jurisdictional lines, and it still required overtime after eight hours in one day. The unions and unionized contractors hoped that this would gain increased industrial construction work for union contractors commencing with the largest job put out for bid in 1983: Exxon's refinery upgrading project, which is designed to improve the efficiency of its Houston area Baytown refinery and to permit it to process lower grade crude oil.

Exxon asked three unionized contractors—Foster & Wheeler, Kellog-Rust, and Bechtel—and three open shop contractors—Brown & Root, Daniel, and Zachry—to bid. All three open shop contractors came in with bids lower than any of those made by the union shop contractors. The differences were found in three key areas: fringe benefits, overtime, and management fees.[29] Union fringe benefits were high because union contractors have been declining in number and fear excessive liabilities unless the pension funds are kept fully funded or nearly so.[30] Overtime for the union contractors was higher because of the IBEW's refusal to permit a

[29] *Ibid.*

[30] In Chapter VIII, the impact of multiemployer pension legislation on open and union shop contracting is discussed.

ten-hour, four-day schedule and make up days at straight time, as discussed in Chapter IV. Why the union contractors' management fees were higher is not clear. One reason could be that they feared the local unions would balk at the use of subjourneymen to the extent the contract permitted or would otherwise not adhere to the contract, thus raising labor costs. In such an emergency, higher management fees would give them a cushion against higher labor costs.

Exxon awarded its initial contract to Brown & Root for this job, much to the chagrin of the construction trades. Basically, as long as jurisdictional lines are maintained, overtime is required after eight hours per day, and fringe benefit costs are high, unionized contractors will have difficulty competing with their major open shop competitors, who are also never subject to jurisdictional disputes, have no noneconomic limitations on their labor deployment, and can train employees without rigid rules.

Although the union construction trades did win one award with the Ship Channel Agreement—a co-generation plant being built by Atlantic-Richfield Corporation which was awarded to Bechtel on a negotiated, nonbid basis—the unions have refused to offer it to union contractors except on one other occasion. This involved a power plant for a public power authority on the Colorado River of Texas. Under the aegis of the Texas council of construction unions, prospective labor costs were lowered even more by the unions to offset their higher fringe benefit costs. This contract, however, was won by Zachry, which was able to underbid the unionized competitors because its fringe benefit costs were lower than the average of the three nonunion firms that bid on the Exxon job, which was the rate that the unions attempted to meet.

When Exxon was ready to request bids for the second stage of its refinery improvement, it again requested bids both from unionized and open shop firms. It seemed that Exxon, after it received such harsh criticism from the unions for letting the original contract to Brown & Root, might very well give the second contract to a union contractor if the cost was reasonably close to that of the lowest open shop contractor. The unions, however, hurt themselves by refusing to agree to the Ship Channel Agreement use for Parsons, one of the largest signatory contractors, and Zachry won this award also. Similarly, the open shop arm of Stearns-Catalytic, another large signatory firm, won an Exxon job for a gasification plant in Wyoming when the unionized firms were refused any concessions or a satisfactory project agreement.

TABLE V-15
Percentage of Union and Open Shop Industrial Construction by Area, Colorado, 1971, 1979, and 1982

Area		Percentage No. of Awards			Percentage Dollar Volume		
		1973	1979	1982	1973	1979	1982
Denver Metro	Union	48	38	28	52	61	59
	Open Shop	52	62	72	48	39	41
State except Denver Metro	Union	39	20	9	45	65[a]	30
	Open Shop	61	80	91	55	35	70
Total Colorado	Union	44	30	19	48	62	52
	Open Shop	56	69	81	52	38[a]	48

Source: David E. Moore, *Union and Open Shop Construction in the State of Colorado* (Denver: Construction Industry Management Council and Colorado Building and Construction Trades Council, 1983), Table 5.
[a] The "increase in dollar awarded to union contractors in 1979 . . . is solely attributable to one very large job. . . ." (Moore, p. 17).

The Colorado Study

The study sponsored by the Colorado Construction Industry Management Council and the Colorado Building and Construction Trades Council, which was utilized in the two previous chapters, provided information on the open shop share of industrial construction. The date given in Table V-15 are divided into those for Denver, for Colorado excluding Denver, and for the state as a whole.

The data in Table V-15 show that the open shop has a majority of the awards even in the Denver metropolitan area. In volume, the division statewide has been fairly even over the decade studied, except for 1979 when one large contract gave the union shop a decided advantage. The fact that open shop contractors secured over 40 percent of the industrial sector's dollar volume in the Denver metropolitan area, the state's most heavily unionized region, as well as 70 percent in the rest of the state, is indicative of the strength of the open shop in the state.

This study found that the most significant open shop growth in industrial construction in Colorado was in the building of water treatment plants. The number of awards to open shop firms for such work grew from 61 percent in 1973 to 79 percent in 1982, and the dollar volume went from 29 percent to 94 percent in the same period.[31]

[31] David E. Moore, *Union and Open Shop Construction in the State of Colorado* (Denver: Construction Industry Management Council and Colorado Building and Construction Trades Council, 1983), Figure 7a, p. 18.

The growth of the open shop in Colorado has been spurred on by synthetic fuel development, and would have been substantially enhanced if the decline in fuel prices had not curtailed the program. The first facility to produce synthetic oil in the state has been that of Union Oil, for which Daniel was the contractor. Exxon had a much bigger project, also in Colorado, for which Brown & Root was awarded the contract. The Exxon project was, however, cancelled before the major portion of the construction began. It had had the potential of increasing the open shop's share of the state's industrial construction by a significant amount for several years. Nevertheless, as shown in Table V-10, open shop contractors gained in Colorado in 1983.

California

The industrial construction data for the eleven county Southern California area covered by the survey of the Construction Industry Research Board are summarized in Table V-16. It shows that union

TABLE V-16
Union Contractors' Market Share[a]
Industrial Construction
Eleven County Southern California Region
1977 and 1981

Area and No. of Responses	Percentage Share 1977	Percentage Share 1981
Regional Total[b] (284)	79	63
Los Angeles (159)	79	66
Orange (37)	75	59
Ventura (22)	90	69
San Bernardino (16)	79	63
Riverside (15)	74	48
Santa Barbara (15)	84	72
Kern (8)	84	56
San Luis Obispo (6)[c]	70	35
Mono (4)[c]	n.r.	n.r.
Imperial (2)[c]	50	70
Inyo (0)	n.r.	n.r.

Source: *A Survey of the Construction Industry Trends in the Eleven-County Southern California Region* (Los Angeles: Construction Industry Research Board, January 31, 1983), Table 6.

[a] All percentages are frequency weighted averages of the midpoints of 20 percent intervals.

[b] The regional total for each category of construction is the weighted average of all responses. Responses are weighted by the actual county share of industrial construction.

[c] Results for small-response counties are statistically unreliable. Variations as large as 20 percentage points could result with very small changes in the number of responses.

n.r. = no response.

market share in this once very strongly union area declined from 79 percent in 1977 to 63 percent in 1981. (Again, it should be noted that "nonunion firms are underrepresented in the responses to the survey.")[32] The open shop share of the market increased in all the counties for which statistically reliable data were secured, including Los Angeles county. The greatest open shop gains occurred in the fast-growing Riverside area, but perhaps the most surprising finding is that in 1981 industrial construction in Los Angeles was one-third open shop. The high union rates are undoubtedly a factor as is the continued propensity of the unions there to hurt the industry by strikes.[33]

The perception that unionized industrial construction has lost market share since 1977 is shared by union and open shop firms in Southern California. Their responses, set forth in Table V-17, show that 43.6 percent of the union firms and 60 percent of the nonunion firms believe that the union share of work has decreased substantially. In addition, 43.6 percent of the union firms and only 26.7 percent of the nonunion firms believe that the union share has decreased somewhat. Very few in either category perceived the union share to have increased or to have remained stable.

Industrial construction thus remains heavily union in Southern California, but the open shop has made significant gains, capturing one-third of the business volume by the end of 1981. Continued strikes, exorbitant wages, and increased capability of open shop contractors are likely to increase the open shop share of the sector in the 1980s.

In Northern California, our interviews indicate a similar situation, but with smaller open shop gains than in Southern California. San Francisco remains a union preserve for the industrial sector, but construction of smaller plants and factories in outlying areas is increasingly done by open shop contractors. Overall, therefore, it appears that the gains of the open shop in California have progressed at a significant rate, and are likely to continue to do so in the years ahead.

The Pacific Northwest

The Pacific Northwest has seen increased open shop activity in the industrial sector as it has in other construction segments. Un-

---

[32] *A Survey of the Construction Industry Trends in the Eleven-County Southern California Region* (Los Angeles: Construction Industry Research Board, January 31, 1983), p. 6.

[33] A four-month strike by the Carpenters' union in late 1983 is a case in point.

TABLE V-17
Union and Open Shop Contractors' Responses Regarding Local Competition, Eleven Southern California Counties: How Has the Percentage of Work Going to Union Industrial Contractors Changed January 1977–December 1981

	Share of Work Going to Union Contractors Percent of Responses in Each Category					
	Decreased			Increased		
No. of Firms	Substanially (−2)	Somewhat (−1)	Stayed the same (−0−)	Somewhat (+1)	Substanially (+2)	Average Value (−2 to +2)
Union firms (39)	43.6	43.6	5.1	5.1	2.6	−1.20
Nonunion firms (15)	60.0	26.7	13.3	0	0	−1.47
All firms (projected)[a]	49.8	37.1	8.3	3.2	1.6	−1.30

Source: *A Survey of the Construction Industry Trends in the Eleven-County Southern California Region* (Los Angeles: Construction Industry Research Board, January 31, 1983), Table 12.

[a] Values for all firms were calculated based on the number of responses from union and nonunion firms, adjusted for the underrepresentation and smaller size of nonunion firms. The number of nonunion firms responding was increased by a factor of 1.6 for calculation purposes.

fortunately, the data of the Oregon-Columbia Chapter, AGC, do not distinguish commercial and industrial construction. Our field interviews indicated, however, that the open shop has made considerable gains in Oregon and southwest Washington, particularly in small plant construction and in industrial construction work outside of the principal cities. This is verified by the data in Table V-10.

Data for eastern Washington and northern Idaho are available, however, and are presented in Table V-18 for the years 1976 to 1981. These data show tremendous fluctuation in the union-open shop ratio. The union share of industrial work jumped from 20 percent in 1976 to 90 percent in 1978, declined to 46 percent in 1979, rose to 73 percent in 1980, and fell to 19 percent by 1981. The explanation for this is that unionized contractors obtained very large jobs in 1978 and 1980 which greatly increased their share of the dollar volume. The open shop builds the great bulk of smaller plants and sawmills in this area, and these jobs show up as a clear majority of the volume of work when large union jobs are not under construction.

TABLE V-18
Eastern Washington and Northern Idaho[a]
Union-Open Shop Industrial Construction
Building Permit Volumes[b]
1976–1983

Year	Total	Union		Open Shop	
	Dollar Volume	Dollar Volume	Percent Union	Dollar Volume	Percent Open Shop
1976	7.2	1.5	20	5.7	80
1977	14.0	7.2	52	6.8	48
1978	40.8	36.7	90	4.1	10
1979	17.4	7.9	46	9.4	54
1980	43.1	31.6	73	11.5	27
1981	23.3	4.4	19	18.9	81
1982	14.6	4.0	27	10.7	73
1983	32.9	9.7	29	23.2	71

Source: Annual Studies, Inland Empire Chapter, Associated General Contractors, Spokane, Washington.
[a] Defined as cities and counties east of the 120th meridian in Washington and north of the 46th parallel in Idaho. Excludes some Idaho counties and Washington towns that lack a building permit or comprehensive plan ordinance.
[b] In millions of dollars of permits, $30,000 or more for 1976; $40,000 or more for other years.

The biggest industrial jobs in the Pacific Northwest in recent years involved the construction of five nuclear plants by the Washington Public Power Supply System. As already noted, the system collapsed and defaulted on its bonds before completing any of the facilities.[34] The rather poorly managed project was manned by union labor, and its failure has greatly reduced union industrial construction in this region. The data in Table V-18 do not include employment in these plants.

INDUSTRIAL RESEARCH UNIT QUESTIONNAIRES

Responses to the 1974 Industrial Research Unit questionnaire support the findings that open shop contractors had made gains in the industrial sector by that time, and also confirmed the strength of the union shop in this sector. (See Table V-19.) Once more, the open shop contractors believed, for reasons already set forth, that their own constituency was larger than that of the union shop contractors, while the union contractors believed the opposite. Within each region, there were both agreements and major dis-

[34] See note 12 above, particularly Bernstein, "A Nuclear Fiasco."

TABLE V-19
Percent Distribution of Respondents' Estimates of the Proportion of Open Shop Work in Industrial Construction by Region and Union Status, 1974

Region	Number	NR/DK[a]	0–20	21–40	41–60	61–80	81–100
Union							
Not reported	89	42.7	36.0	9.0	6.7	1.1	4.5
New England	27	29.6	37.0	7.4	7.4	11.1	7.4
Middle Atlantic	125	31.2	36.8	14.4	10.4	3.2	4.0
E.N. Central	147	26.5	53.7	8.2	4.8	2.7	4.1
W.N. Central	60	30.0	41.7	11.7	6.7	5.0	5.0
S. Atlantic	68	25.0	33.8	17.6	14.7	2.9	5.9
E.S. Central	32	18.8	28.1	25.0	9.4	9.4	9.4
W.S. Central	45	22.3	46.7	13.3	11.1	2.2	4.4
Mountain	52	36.6	44.2	5.8	1.9	5.8	5.8
Pacific	133	43.6	50.4	3.0	1.5	0.0	1.5
Total	778	32.4	43.1	10.3	6.8	3.1	4.4
Nonunion							
Not reported	83	33.7	16.9	8.4	16.9	18.1	6.0
New England	63	20.6	12.7	15.9	23.8	19.0	7.9
Middle Atlantic	134	32.8	16.4	14.2	11.9	12.7	11.9
E.N. Central	75	17.3	24.0	21.3	10.7	17.3	9.3
W.N. Central	45	35.5	13.3	11.1	8.9	11.1	20.0
S. Atlantic	220	33.6	9.1	8.6	14.5	13.2	20.9
E.S. Central	45	8.9	11.1	13.3	15.6	24.4	26.7
W.S. Central	92	23.9	21.7	17.4	17.4	12.0	7.6
Mountain	21	28.6	28.6	4.8	28.6	0.0	9.5
Pacific	6	50.0	33.3	16.7	0.0	0.0	0.0
Total	784	28.4	15.4	12.8	15.1	14.4	13.9

Source: Industrial Research Unit Questionnaire, 1974.
[a] NR/DK = Do not know or not reported.

crepancies in their answers. Nevertheless, the responses demonstrate the surprisingly strong showing of open shop contractors in this sector at this early date, even though the nonunion contractors acknowledge less impact in the industrial sector than in the commercial sector.

As in the case of commercial construction, we asked respondents to the questionnaire whether they believed that the open shop share of work had increased or decreased; and if so, to estimate the direction of that change between ranges of 20 percentage points. Table V-20 summarizes the results of this inquiry for union and nonunion respondents.

More open shop than union contractors in 1974 believed that the open shop was gaining, and they also tended to believe that the

TABLE V-20

Respondents' Estimates of Change in Open Shop Share of Industrial Construction Over Past Few Years by Region and Union Status, 1974

Region	Total	Total Indicating Change[a]	Lower by 3	Lower by 2	Lower by 1	Higher by 1	Higher by 2	Higher by 3
Union								
New England	27	5	0	0	0	3	2	0
Middle Atlantic	125	23	0	0	4	17	2	0
E. N. Central	147	11	1	0	1	9	0	0
W. N. Central	60	8	1	1	1	3	1	1
S. Atlantic	68	13	1	1	2	6	3	0
E. S. Central	32	8	0	1	0	4	2	1
W. S. Central	45	5	1	0	1	2	0	1
Mountain	52	4	0	0	0	4	0	0
Pacific	133	7	0	1	0	3	3	0
Not reported	89	8	0	0	1	5	1	1
Total	778	92	4	4	10	56	14	4

Nonunion								
New England	63	20	0	1	1	10	5	3
Middle Atlantic	134	24	0	0	7	15	2	0
E. N. Central	75	21	0	0	2	15	2	2
W. N. Central	45	3	0	0	0	3	0	0
S. Atlantic	220	30	2	1	5	15	6	1
E. S. Central	45	17	0	0	1	11	2	3
W. S. Central	92	26	1	1	1	17	5	1
Mountain	21	4	1	0	0	2	0	1
Pacific	6	0	0	0	0	0	0	0
Not reported	83	13	0	0	2	8	3	0
Total	784	158	4	3	19	96	25	11

Source: Industrial Research Unit Questionnaire, 1974.

[a] Change expressed in terms of movement between ranges of 20 percentage points (i.e., 0–20, 21–40, etc.).

TABLE V-21

Percent Distribution of Respondents' Estimates of the Proportion of Open Shop Work in Industrial Construction by Region and Union Status, 1982

Region	Union							Nonunion						
	Number	NR/DK[a]	0–20	21–40	41–60	61–80	81–100	Number	NR/DK[a]	0–20	21–40	41–60	61–80	81–100
Not Reported	44	22.7	25.0	31.8	13.6	4.5	2.3	49	28.6	12.2	12.2	14.3	20.4	12.2
New England	62	4.8	14.5	33.9	30.6	14.5	1.6	111	15.3	14.4	11.7	26.1	20.7	11.7
Middle Atlantic	275	14.2	38.5	28.4	14.2	2.5	2.2	228	18.4	18.0	20.6	21.9	12.3	8.8
E.N. Central	474	8.9	58.2	19.4	9.1	2.7	1.7	205	12.7	22.9	22.9	19.5	15.1	6.8
W.N. Central	220	13.6	50.0	19.1	10.0	4.1	3.9	157	17.8	21.6	15.3	14.6	17.8	12.7
S. Atlantic	109	13.8	18.3	21.1	30.3	9.2	7.3	550	17.1	9.6	7.6	16.4	21.6	27.8
E.S. Central	69	7.2	18.8	27.5	18.8	21.7	5.8	111	11.7	8.1	11.7	27.0	19.8	21.6
W.S. Central	105	4.8	20.0	20.0	29.5	21.0	4.8	284	19.0	9.2	14.4	21.8	22.9	12.7
Mountain	108	11.1	35.2	27.8	13.9	7.4	4.6	184	14.1	21.2	21.7	20.1	14.1	8.2
Pacific	362	12.4	45.9	24.3	13.6	1.9	2.2	143	16.8	32.2	22.4	15.4	9.8	3.5
Total	1,828	11.3	42.1	23.4	14.7	5.6	2.9	2,022	16.7	15.6	15.1	19.3	18.1	15.1

Source: Industrial Research Unit Questionnaire, 1982.

[a] NR/DK = Do not know or not reported.

gain was somewhat greater. Nevertheless, among those respondents who believed that there had been any change in the share of open shop contractors in industrial construction we again found agreement that the open shop was gaining, although not as greatly in this sector. Significantly, only 18 union and 26 nonunion respondents of a total of 1,562 respondents believed that open shop industrial construction was losing ground.

The 1982 Industrial Research Unit questionnaire had a much larger response than that of 1974. As shown in Table V-21, 42.1 percent of the union respondents estimated the open shop share of industrial construction at between 0 and 20 percent and another 23.4 percent believed it to be between 21 and 40 percent. The open shop respondents were much more widely dispersed in their answers with from 15 to 20 percent of the respondents showing up in each of the percentage categories.

On a regional basis, the questionnaire is in accord with our general findings. The strength of the open shop in the southern regions, its growth in the eastern and midwestern areas, and its emergence as a factor in the Pacific region are all apparent from the data.

A comparison of Tables V-19 and V-21 shows that the union respondents in 1982 consistently estimated a higher open shop penetration than did those in 1974. If, for example, one examines the total of the 21–40 and the 41–60 percentage categories in both tables

TABLE V-22

Comparison of Questionnaire Responses in 1974 and 1982 Which Put Open Shop Share of Industrial Construction at 21–40 and 41–60 Percent, by Region and Union Status

	Total Percentage in 21–40 and 41–60 Percent Categories			
	Union		Open Shop	
Region	1974	1982	1974	1982
New England	14.8	64.5	39.7	37.8
Middle Atlantic	24.8	42.6	26.1	42.5
East North Central	13.0	28.5	32.0	42.4
West North Central	18.4	29.1	20.0	29.9
South Atlantic	32.3	51.4	23.1	24.0
East South Central	34.4	46.3	28.9	38.7
West South Central	24.4	49.5	34.8	26.2
Mountain	7.7	41.7	33.4	41.8
Pacific	4.5	37.9	16.7	37.8
Total	17.1	38.1	27.9	34.4

Source: Tables V-19 and V-21.

TABLE V-23
Respondents' Estimates of Change in Open Shop Share of Industrial Construction[a] *by Region and Status, 1978–1982*

Region	Union									Nonunion								
	Number	NR/DK[b]	0–10	11–20	21–30	31–40	41–50	51–75	76–100	Number	NR/DK[b]	0–10	11–20	21–30	31–40	41–50	51–75	76–100
Not reported	44	38.6	34.1	9.1	2.3	4.5	6.8	0.0	4.5	49	73.5	6.1	2.0	4.1	4.1	6.1	2.0	2.0
New England	62	19.4	30.6	22.9	9.7	14.5	1.6	1.6	0.0	111	66.7	5.4	8.1	8.1	4.5	2.7	1.8	2.7
Middle Atlantic	275	40.4	37.5	12.3	3.6	1.1	0.4	3.3	2.5	228	53.9	15.8	8.3	6.1	3.5	4.4	3.1	4.8
E.N. Central	474	39.1	46.0	6.1	3.4	1.1	1.5	1.3	1.7	205	49.8	20.0	11.7	6.3	2.0	3.9	3.4	2.9
W.N. Central	220	40.0	43.6	7.3	2.7	2.7	0.0	2.7	0.9	157	54.8	21.0	6.4	3.8	4.5	1.3	3.8	4.5
S. Atlantic	109	35.8	18.3	15.6	2.8	9.2	7.3	5.5	5.5	550	74.4	4.0	4.4	3.8	2.7	4.0	4.0	2.7
E.S. Central	69	21.7	31.9	11.6	7.2	5.8	8.7	7.2	5.8	111	55.9	11.7	9.0	5.4	3.6	2.7	6.3	5.4
W.S. Central	105	26.7	32.4	12.4	10.5	2.9	5.7	4.8	4.8	284	59.2	7.4	7.7	8.1	4.6	5.3	4.2	3.5
Mountain	108	25.0	50.9	11.1	5.6	2.8	1.9	0.0	2.8	184	42.9	22.8	12.0	8.7	2.2	4.3	1.6	5.4
Pacific	362	34.8	48.3	7.7	1.7	0.6	0.8	1.7	4.4	143	38.5	35.7	7.0	4.9	2.8	1.4	6.3	3.5
Total	1,828	35.4	41.4	9.4	3.8	2.6	2.0	2.4	2.9	2,022	59.1	13.3	7.5	5.8	3.3	3.8	3.8	3.7

Source: Industrial Research Unit Questionnaire, 1982.

[a] Change expressed in terms of movement between ranges of 10 percentage points (i.e. 0–10, 11–20, etc.).

[b] NR/DK Do not know or not reported.

(see Table V-22), one finds a remarkable increase in the percentage totals from the union respondents, more than doubling for the country as a whole, and increasing many times for the New England, Middle Atlantic, Mountain and Pacific regions. The responses from open shop contractors were more consistent for the two questionnaires, showing only a small increase in the percentage of the two categories for 1982. (The lower percentages for New England and West South Central for 1982 are attributable to the fact that a large number of open shop contractors responded in the categories of 61–80 and 81–100 percent for these regions.)

Table V-23 gives the respondents' estimates of change in the open shop share of industrial construction from 1978 to 1982. The union respondents' estimates were concentrated at the low end—41.4 percent and 9.4 percent believed that the open shop had increased its share from 0–10 percent and 11–20 percent, respectively, while only 13.7 percent felt that the change had been greater. Almost 60 percent of the open shop respondents did not reply to this question, but the largest percentage that did reply put the change in the 20 percent or lower category. This adds weight to the view that, overall, open shop contractors have made some small gains in industrial construction, but that the unionized contractors continue to maintain a hold on the majority of such work.

DAVIS-BACON IMPACT

The foregoing analysis was made prior to the already discussed changes in the administrative rules governing the Davis-Bacon Act. Where the federal government participates in, or subsidizes, industrial construction, the Davis-Bacon Act governs wages and conditions. This has, of course, aided the unionized contractors. Assuming that the new rules are in effect for several years, one can predict gains for open shop contractors that will give them added experience to solidify their position. This should further erode the union position in industrial construction.

CONCLUSION

The construction of large industrial facilities was once a union stronghold. Substantial open shop gains have now been made particularly in the South where industry in general and the process industry, paper manufacturing, and related industries are especially concentrated. The decade of the 1980s might see greater gains elsewhere by open shop contractors who increasingly have added to their experience by winning work throughout the country. It is believed that industrial plant construction in particular will be built more and more frequently by open shop contractors unless the construction unions can better control the factors which increase costs and decrease productivity. Given the manner in which craft organization adds to construction costs, such a turnabout seems unlikely.

CHAPTER VI

Highway and Heavy Construction

Highway and heavy construction activities are generally considered together because of their similarities of equipment, labor, and methods of work. Highway construction involves the construction of roads, which includes earth moving and site preparation, and concrete, asphalt, or other road material laying; it often includes the construction of bridges, tunnels, or other accesses through, under, over, and around natural and man-made obstacles. Heavy construction includes site preparation for large projects: the building of dams, airports, underground facilities, missile sites, etc. We have also included within this chapter the so-called utility work, which involves laying water, gas, and underground electrical lines and other work needed in order to bring these necessities to consumers.

Both highway and heavy work require large numbers of heavy equipment operators, truck drivers, and laborers, plus some carpenters, iron workers, electricians, and other craftsmen for special jobs. Where the work is unionized, the largest craft union is the International Union of Operating Engineers (IUOE), representing the equipment operators, oilers, and helpers. The Laborers' Union and the Teamsters are also well represented. Carpenters are involved in form and related work, and iron workers may be used for rod-setting. Special contractors are likely to perform bridge, tunnel, electrical, and sewer work, and have agreements with the appropriate specialty crafts.

Open shop contractors organize their labor quite differently. For example, oilers are not employed; equipment operators are expected to lubricate their own machines, assisted when necessary by laborers. Laborers may be upgraded to helpers, and then to operators of small machines or trucks. If qualified, they continue up the ladder to the larger equipment. Laborers or helpers do much of the carpentry work under supervisors or lead carpenters, and they also do the rod-setting work and assist journeymen in the other crafts. As in commercial construction, the labor force is compensated over a wide range within a smaller number of classifications, and journeymen are not employed to do such unskilled or semiskilled work as, for example, rod-setting, very rough carpentry, or the unloading of materials.

FACTORS AFFECTING THE EXTENT OF OPEN SHOP

Before analyzing open shop competition in highway and heavy construction, it is necessary to recognize certain salient features of that sector. Some features work toward greater open shop penetration, while others serve to neutralize what might place the union contractor at a disadvantage. Unlike most residential and commercial work, much highway construction is performed away from major centers of population. As we have noted previously, union strength generally tends to wane with distance from the metropolitan areas. Thus, the construction of highways, especially new highways connecting metropolitan areas rather than serving any one of them, might be expected to represent a particularly attractive market for the nonunion contractor.

The very remoteness of these projects, however, has traditionally raised questions about the ability of open shop contractors to find crews for them. As we have noted earlier, the union contractor has historically had the advantage of ready access to the large pools of manpower that can be assembled on relatively short notice from many distant areas. Without this extensive network of contacts with different labor markets, the open shop contractor initially relied on less systematic means of recruitment or sought to maintain a steady workforce that was willing to travel. Most often, he used some combination of the two approaches. Moreover, the task of maintaining a steady workforce is difficult for highway contractors because of the seasonal nature of work in colder climates. Recently, as the open shop firms have grown, many have developed access to the human resources needed to compete with unionized contractors. With significant exceptions, however, open shop firms in this sector tend to be smaller than their unionized rivals.

In highway construction, the lack of a constant workforce is abated to some degree by the relatively heavy reliance on unskilled and semiskilled labor. Compared to the construction industry as a whole, highway projects typically employ a disproportionate number of laborers, helpers, truck drivers, and small-equipment operators. Even within the skilled ranks there are many jobs, such as cement masonry and concrete form building, that require less skill and training than others, such as electrical and mechanical work. The latter are more predominant in building construction. Wages for unskilled labor in highway work are substantially higher than those for unskilled labor in other industries (owing in part to the effects of the Davis-Bacon Act and similar state prevailing wage laws), and it is presumably easier to recruit unskilled workers than trained craftsmen. Because of these wage and recruitment factors, the open shop highway contractor usually has little difficulty in recruiting the required labor force.

Davis-Bacon Impact

Another important feature of highway construction is that it is almost totally financed by public funds, and much of it is supported by the Federal Highway Administration. As a consequence, contractors are subject to the provisions of the Davis-Bacon Act or other prevailing wage statutes which often require that employers pay the equivalent of union wages. The exceptions are those states, such as Texas or the Carolinas, where open shop contractors have been represented by vigilant associations that regularly provide irrefutable wage surveys to the administrators of these laws.[1] Indeed, in no segment of the industry is the Davis-Bacon Act more significant in maintaining union power. This was recognized early on by the construction unions. The Operating Engineers in particular utilized the act to its fullest extent, even employing a former U.S. Department of Labor staff member who was previously involved in administration of the act to coordinate its organizing activities.[2] According to Professor Garth L. Mangum, historian of the Operating Engineers, in *The Operating Engineers: The Economic History of a Trade Union:*

> The major organizing tools were federal prevailing wage laws. The procedure was to approach a contractor with whom one or more of the unions already had contractual relationships, usually either an interstate contractor or one who was also engaged in building work.

---

[1] See, in general, Armand J. Thieblot, Jr., *The Davis-Bacon Act,* Labor Relations and Public Policy Series, Report No. 10 (Philadelphia: Industrial Research Unit, The Wharton School, University of Pennsylvania, 1975). Professor Thieblot is currently revising and updating this study. See also U.S. Comptroller General, *The Davis-Bacon Act Should Be Repealed* Report HRD-79-18 (Washington, D.C.: General Accounting Office, 1979); Robert S. Goldfarb and John F. Morrall III, "The Davis-Bacon Act: An Appraisal of Recent Studies," *Industrial & Labor Relations Review,* Vol. 34 (January 1981), pp. 191–206; Morgan O. Reynolds, "Understanding Political Pricing of Labor Services: The Davis-Bacon Act," *Journal of Labor Research,* Vol. III (Summer 1982), pp. 295–310; and Michael R. Metzger and Robert S. Goldfarb, "Do Davis-Bacon Minimum Wages Raise Product Quality?," *Journal of Labor Research,* Vol. IV (Summer 1983), pp. 265–72.

[2] Garth L. Mangum, *The Operating Engineers: The Economic History of a Trade Union,* Wertheim Publications in Industrial Relations (Cambridge: Harvard University Press, 1964), p. 257.

> Here the engineers took the lead because the scarcity of skilled operators often brought them into contractual relationships before the other crafts. Laborers and truck drivers could be obtained where operating engineers could not. The state committee would agree to whatever concessions were necessary to allow the union contractor to become successful bidder on a heavy or highway project. These agreements also promised no stoppages.
>
> The Davis-Bacon Act provides that a wage rate common to the majority of the employees in a particular classification in an area (or 30 percent if there is not a majority paid at the same rate) shall be determined as the prevailing rate which all in that classification must be paid. This being the case, once the Joint Heavy and Highway Construction committee was successful in signing the employers of a majority of the members of their crafts in an area, they could appeal to the Department of Labor and have those wage agreements become binding upon all other employers upon construction projects in which the federal government was involved. The unions then suggested that the contractors form a statewide association, if they had not already done so, and bargain as a group. Whatever wages were negotiated became the prevailing rate, and, since the contractors were more interested in relative than absolute costs as long as they were established in advance of the bidding, rates could be increased to a level more acceptable to the union member.
>
> Once the union scale became prevailing, there were advantages to using the union as a labor supply source which usually brought in the nonunion contractor. If not, interrelations between union and nonunion contractors could be used as a lever for organization. The familiar pattern of persuading a union contractor to break off relations with a nonunion subcontractor is well known. An example of a more direct technique was one case where union engineers refused to hoist some quick-drying concrete which a nonunion subcontractor was attempting to pour.[3]

Professor Mangum gives the Davis-Bacon Act substantial credit for the unionization of much of the nation's highway work, which, prior to the establishment of the federal program in the early 1950s, had been substantially open shop.[4] Thus, he states, "Heavy and highway construction was reasonably well organized in twenty states in 1954. By 1958, four more which had been partially organized and four more which had little or no organization in 1954 were well organized. Seven had progressed from little or no organization to partial organization. Little progress was made in the other thirteen states."[5]

---

[3] *Ibid.*, pp. 257–58.

[4] There are apparently no early studies of the degree of unionism in highway work such as those cited in the chapters on homebuilding, commercial, and industrial construction, except for those of the Associated General Contractors, cited below.

[5] Mangum, *The Operating Engineers*, p. 258.

In *The Operating Engineers,*[6] Mangum relates the story of the Operating Engineers from its origin to 1958. For the next fifteen years, as Armand J. Thieblot, Jr., in *The Davis-Bacon Act,* clearly demonstrates,[7] the Act continued to aid unions in maintaining a strong position in highway construction. Since 1975, our field work has indicated heavy losses for unionized firms in the private sector and reduced strength for such companies in publicly supported highway projects. In heavy construction the situation is similar, since much of the work is either federally funded and thereby covered by the act,[8] or supported by other public funds and hence often covered by state legislation that, as noted below, is similar in content and administration to the Davis-Bacon Act. Sometimes, however, such legislation is even more overtly union-oriented.

The Davis-Bacon Act does have a few supporters among the open shop highway contractors. Two large contractors, one in North Carolina and the other in New York State, told us that they can win jobs under the act's provisions because they know their competitors' labor costs and can deploy manpower more cheaply. This is a limited asset, however, because the act's regulations have historically required that the open shop contractor follow union job designations unless, as in Texas or the Carolinas, the open shop job titles have been found prevailing. Thus, potential labor-saving manpower utilization by open shop contractors has been limited.

Yet even in such cases, open shop employers have been able to maintain a real advantage by deploying labor more productively. For example, the same employee can be utilized in several jobs (such as oiler, small-equipment driver, and laborer) and paid the rate for the job for which he was hired. The open shop employer saves on waiting time, call-in pay, recruiting, training, and, especially, on fringe benefit costs, which are based largely on the individual employee rather than on the number of hours that he works. Moreover, by using the same employees in different jobs, the open shop employer reduces turnover, training costs, and delays. He also avoids lost time from strikes and has unfettered control of first-line supervisors.

Nevertheless, despite some advantages to the open shop under traditional Davis-Bacon rules, it has restricted nonunion competition and raised their costs toward the union level. The fact that the

---

[6] *Ibid.*

[7] Thieblot, *The Davis-Bacon Act,* pp. 109–20, 126–29. See particularly his descriptions of the "Metro Segment C-7" and "Delta Electric" cases.

[8] Dams, bridges, airports, etc., are almost exclusively federally funded today, although other government bodies may contribute some funds.

open shop Associated Builders and Contractors (ABC) has long been in the forefront of the movement to repeal the Davis-Bacon Act is an obvious indication that most open shop contractors dislike the law.[9] It is believed, however, that the new regulations instituted by the Reagan administration, which were discussed in Chapter II, will reduce unions' advantages under the act.[10]

Other Characteristics of Public Highway Construction

The public funding of highway construction may have other consequences with respect to the prospect of open shop contractors obtaining the work. Since the laws of many states require that contracts be awarded to the lowest qualified bidders, the influence of unions on governmental agencies is, theoretically, somewhat neutralized. Furthermore, bidding procedures will sometimes provide that various segments of a project (e.g., grading, paving, bridge building) be contracted separately rather than through a single general contractor; thus, the awarding of the main portion of a project to a union contractor does not necessarily preclude open shop bidders from winning work. In fact, however, union pressures (including the refusal to work with nonunion concerns, the threat of and actual acts of violence against highway contractors' equipment and men, and the political power of the crafts) all tend to reduce open shop work on predominantly union jobs.[11]

Some public agencies, however, perform their own highway work, especially repair and maintenance. Since these jobs tend to be smaller projects for which open shop contractors are usually most competitive, such activity probably reduces job opportunities for

---

[9] For the generally negative reactions of contractors to the Davis-Bacon Act in the early 1970s, see Thieblot, *The Davis-Bacon Act,* Chapter IX. Our 1981–1983 field surveys found, if anything, more emphatic opposition to the act by contractors, both open and union shop.

[10] As noted in Chapter II, the regulations governing the Davis-Bacon Act were substantially revised by Secretary of Labor Raymond J. Donovan in 1983 to lessen both their inflationary and prounion bias. As approved by the courts, these regulations substitute a weighted average of wages paid in a given area instead of a wage received by 30 percent as prevailing; eliminate use of eleven counties in determining prevailing wages for rural areas; exclude federal projects from prevailing wage determinations in other than highway and heavy projects, where there is little nonfederal construction, in order to reduce the upward skewing of results; expand the right of contractors to use helpers; and simplify record keeping for contractors. See Building and Construction Trades' Department, AFL-CIO v. Raymond J. Donovan, Secretary of Labor, 712 F.2d 611 (D.C. Cir. 1983), cert. denied 1045. S. Ct. 975 (1984). See also *Daily Labor Report,* No. 130 (July 6, 1983), pp. D-1-14.

[11] See Mangum, *The Operating Engineers,* pp. 256–59 for examples of certain union pressure tactics, and Chapter VIII below for examples of violence and additional union pressure tactics.

open shop contractors more than for union contractors. In recent years, municipal employees have become increasingly unionized, especially by the American Federation of State, County, and Municipal Employees, but also by the Teamsters and the construction unions.

"Little Davis-Bacon Acts"

Currently, thirty-seven states and the District of Columbia have enacted prevailing wage legislation that pertains to construction financed by an agency of the state.[12] Of these states, four have so-called "Little Davis-Bacon Acts," which apply to highway and heavy construction and which mandate that the collectively bargained rates be established as the prevailing ones;[13] nine others choose the collectively bargained rate in practice but still hold hearings, make surveys, or "ascertain and consider" the situation;[14] eighteen states and the District of Columbia set rates by using a survey and/or formula, or use the federal Davis-Bacon rate;[15] and three do likewise but also have minimum wages.[16]

These laws apply concurrently with the federal law, but the states usually bow to federal jurisdiction except in the rare situation where the state prevailing wage is higher than the federal one—then the state law is applied. Administration of the laws reflects the historic political strength at the state level of the construction unions. As a result, if the federal Davis-Bacon Act were to be repealed, union strength in highway and heavy construction would still be bolstered in most of the heavily populated states by state legislation and its administration.

The Steelworkers

Rival unionism is especially significant in highway construction in Pennsylvania, West Virginia, Kentucky, Ohio, and in other eastern coal-producing states. There many highway construction workers were unionized by District 50, which was originally organized by the late John L. Lewis as a division of the United Mine Workers

[12] The material in this section is based upon Armand J. Thieblot, Jr., "Prevailing Wage Laws of the States," *Government Union Review,* Vol. 4 (Fall 1983), pp. 3–65.

[13] Massachusetts, Montana, New Jersey, and Ohio.

[14] California, Connecticut, Hawaii, Illinois, Louisiana, Oregon, Pennsylvania, Rhode Island, and Wyoming.

[15] Alaska, Delaware, District of Columbia, Indiana, Kentucky, Maine, Maryland, Minnesota, Missouri, Nebraska, Nevada, New Hampshire, New Mexico, New York, Oklahoma, Tennessee, Washington, West Virginia, and Wisconsin.

[16] Colorado, Texas, and Vermont.

(UMW) for coal and coke by-product workers but which later became a catchall conglomerate union. District 50 agreements have always been of an industrial type, permitting the contractor to employ whom he chooses and to align the work as he desires, but providing for layoffs and recalls by seniority. Wage rates, if not set by Davis-Bacon standards, are likely to be lower than those of the AFL-CIO crafts. District 50 operated independently for many years, first under an affiliation with the UMW, then without any ties to other unions.

Open shop contractors found District 50 policies closer to their usual methods of operation than those of the standard crafts and frequently operated doublebreasted with District 50 contracts where the latter was a factor. In 1973, however, District 50 merged with the United Steelworkers of America. Several open shop contractors told us that during the first five years under Steelworker aegis, the former District 50's requirements became more stringent and its seniority-layoff procedures and pension-welfare requirements became onerous. Discussions with Steelworker officials, moreover, clearly indicated that, although they had no plans to disband the former District 50's construction locals, they planned no expansion in this industry and expected the membership therein to decline and eventually disappear. Obviously, the Steelworkers initially believed that their construction locals were an unnecessary source of friction among the standard construction unions. Hence, contractors were finding that the former District 50 local unions were less available as a sort of halfway house for union and open shop operations.

More recently, the Steelworkers and the construction unions have had a parting of ways. There was always friction between the two over the issue of whether maintenance and construction in steel plants should be performed by the in-house industrial union or contracted out to the construction trades. This dispute was aggravated by what the Steelworkers felt was the failure of the construction unions to permit ex-District 50 locals of the Steelworkers to operate without being harassed by construction unions attempting to take over their work. Accordingly, the Steelworkers abrogated their agreement to cease organizing activities in construction. Once again, the former District 50 is a real rival to the construction unions in highway and heavy construction work in the Appalachian coal areas.

The tension between the Steelworkers and the construction unions came to a head in March 1982, after Neshaminy Constructors, Inc., (Easterville, Pennsylvania) won a contract to build a waste-water treatment plant for the City of Philadelphia. The violence-prone Building and Construction Trades Council of Philadelphia (whose activities are examined in some detail in Chapter VIII) demanded that Neshaminy, which had been dealing with District 50 and the Steelworkers for a decade, subcontract much of the work to contractors employing its member unions. When Neshaminy refused, the building trades unions boycotted and harassed Neshaminy's suppliers, engaged in mass picketing and violence against Neshaminy's employees and property, and filed a charge with the City of Philadelphia that Neshaminy was violating state and/or federal labor standards. The National Labor Relations Board (NLRB), after initially refusing to issue a complaint, sought and won an injunction against the coercive and unlawful acts of the Philadelphia construction unions. After considerable legal maneuvering, the Building and Construction Trades Council of Philadelphia agreed to a stipulated settlement which required the unions to cease their mass picketing, "blocking the ingress or egress of Neshaminy employees or others from any business site of the company, threatening bodily harm ... placing broken glass under the tires of vehicles ... damaging or attempting to destroy trucks or other property of Neshaminy ... or ... other individuals ...." This settlement also prohibited boycotting or picketing companies doing business with Neshaminy.

The NLRB approved this settlement on February 16, 1983.[17] The secretary of the Building and Construction Trades Council called the result a "serious threat" to the Philadelphia building unions.[18] Neshaminy continues to work in Philadelphia, and the Steelworkers—whose membership was reduced from 1.4 million in 1979 to 700,000 in mid-1983 by the decline of industries under its jurisdiction—seems certain to continue to welcome members and affiliates in highway and heavy construction.[19] Thus, the Steelworkers are likely to continue as an alternative to the standard construction

---

[17] Building and Construction Trades Council of Philadelphia and Vicinity, AFL-CIO et al., 265 N.L.R.B. No. 147, December 16, 1982 and February 16, 1983; Hirsch, Regional Director, NLRB v. Metropolitan District Council of Philadelphia and Vicinity, United Brotherhood of Carpenters, No. 82–1771 (E.D. Pa. April 23, 1982) (Consent Decree); Hirsch, Regional Director, NLRB v. Building & Construction Trades Council of Philadelphia and Vicinity, No. 82–2685, (E.D. Pa. June 29, 1982) (Order granting temporary injunction); Hoeber, Acting Regional Director, NLRB v. Building and Construction Trades Council of Philadelphia and Vicinity, No. 82–3042, (E.D. Pa. July 20, 1982).

[18] "Steelworkers Called Bigger Threat Than Open Shop," *Builder & Contractor,* March 1983, pp. 7, 11.

[19] "Steelworkers' Union Considers Staff Cuts," *Philadelphia Inquirer,* June 26, 1983, p. B-6.

unions in such states as West Virginia, Kentucky, Pennsylvania, and Ohio.

EXTENT OF OPEN SHOP SURVEYS—NATIONAL

The only national studies of the extent of open shop and/or unionized segments in the highway and heavy construction sectors are the surveys conducted by the *Engineering News-Record,* by the Associated General Contractors (AGC), and by ourselves. Other national studies, noted in the previous chapters, did not cover these sectors. It is quite clear from the already noted work of Mangum and from our discussions in the field that, prior to the 1950s, highway work was not well unionized except in and around major cities. The passage of the Federal Highway Act, the administration of the Davis-Bacon Act and similar state laws, and the success of the Operating Engineers and other unions changed this rapidly. By 1958, Mangum could report twenty-eight states completely unionized, seven states partially unionized, and thirteen states largely open shop.[20]

Because of the pervasive impact of the Davis-Bacon Act, we can hypothesize that the heavy sector is as unionized as the highway sector. This means that by about 1960, although both the highway and heavy sectors were considerably more unionized than the home-building sector, they were both much less unionized than the commercial sector. Daniel Mills, writing about 1970, concurs with this broad generalization.[21] The surveys conducted by the Industrial Research Unit confirm this for more recent years and show a growing open shop trend.

The Engineering News-Record

The Engineering News-Record's compilation for 1983 of the 400 largest contracting firms supports the finding that the open shop has significantly increased its share of highway and heavy construction in recent years. It reported:

> Open shop contractors apparently found another plus in their relative preference for heavy and highway construction. . . . Over 40% of the open shop contractors surveyed are involved in this market, compared to 20% of the sample of union contractors.[22]

---

[20] Mangum, *The Operating Engineers,* p. 258.

[21] Daniel Quinn Mills, *Industrial Relations and Manpower in Construction* (Cambridge: The MIT Press, 1972), p. 17.

[22] "Open Shop Makes Gains on Survey," *Engineering News-Record,* May 5, 1983, p. 58.

Thus, the growing prevalence of open shop highway and heavy construction contractors is an obvious indication that they have gained in their share of work.

Associated General Contractors—Highway Sector Data

The 1969 and 1973 surveys of the AGC utilized in the previous chapters also included questions of union status for highway and heavy construction contractors. Thirty percent of the 3,044 respondents in 1969 reported that they were engaged in highway construction, compared to 39 percent of the 5,938 respondents in 1973.

Before examining the AGC data, it should be noted that the AGC has a rival association among highway contractors, the American Road & Transportation Builders Association (ARTBA), formerly the American Road Builders Association (ARBA). Some highway contractors belong to both the AGC and the ARTBA, but most belong only to one or the other. Thus, a survey by the AGC is inconclusive in some regions (for example New England) where the ARTBA is strong.

Table VI-1 shows the union status of AGC members engaged in highway construction by region as reported by the two surveys as well as the percent union by dollar volume in 1969, the only year in which it was available. Table VI-1 indicates that open shop highway construction among the various regions of the country tended to follow much the same pattern revealed in our previous discussions of other sectors. Open shop operations predominated in the South generally, while union representation remained highest in the eastern, east central, and western states. Since the rate of response in 1973 was apparently much higher than in 1969, any assessment of change over these four years must be made with caution. Nonetheless, the overall growth of open shop highway building seemed to have been concentrated in those regions where the unions have traditionally been strongest. In other words, the surveys reflect an expansion of open shop activity into areas of union strength, rather than a continued erosion of union representation in regions where the building trades have historically met substantial open shop competition.

The union percentage undoubtedly reflected not only the contractors that dealt with the standard AFL-CIO crafts, but also those that had Steelworker agreements. Without the inclusion of the Steelworkers, the union percentage in the Middle Atlantic region and also in the East North Central region could be less. Likewise, the decline in union strength in these areas could also be attrib-

TABLE VI-1
Union Status of AGC Members Engaged in Highway Construction by Region, 1969 and 1973

Region	Percent Union[a]	Number of Contractors					
		1969			1973		
		Union	Nonunion	Percent Union	Union	Nonunion	Percent Union
New England	73.0	26	18	59.1	18	47	27.7
Middle Atlantic	98.1	53	4	93.0	185	44	80.8
E.N. Central	91.6	122	14	89.7	345	72	82.7
W.N. Central	24.3	83	127	39.5	149	280	34.7
S. Atlantic	20.6	6	47	11.3	50	217	18.7
E.S. Central	55.3	12	20	37.5	28	41	40.6
W.S. Central	5.1	8	123	6.1	37	255	12.7
Mountain	78.9	73	46	61.3	138	94	59.5
Pacific	99.1	135	5	96.4	312	38	89.1
Total	44.0	518	404	56.2	1,262	1,088	53.7

Source: Associated General Contractors of America.
[a] Calculated by using the dollar volume of contracts granted in 1969.

utable to the decisions of open shop contractors to abandon Steelworker contracts as well as contracts with the crafts.

According to the AGC surveys, changes from union to open shop operations between 1969 and 1973 were less pronounced for highway contractors than for all contractors covered. In 1969, 43.8 percent of the highway contractors were open shop as compared with 46.3 percent in 1973, whereas 31.1 percent of all contractors were open shop in 1969 compared to 41.7 percent in 1973. This would confirm our previous findings of major open shop gains in commercial and industrial construction and would indicate a more stable status in the highway sector during this period. Federal funding and the impact of the Davis-Bacon Act could then account for maintenance of union strength in the highway sector. Although the highway sector remained less unionized than the commercial sector, the differences had considerably narrowed for AGC members by 1973. The 1969 survey showed that 77.2 percent of the commercial contractor respondents were unionized, as compared with 56.2 percent of the highway contractors (see Table IV-4). In 1973, however, the percentages were 63.9 for commercial and 53.7 for highway.

In most regions of the country, union contractors appeared to operate on a larger scale than their open shop competitors. The 1969 survey (but unfortunately not the 1973 survey) solicited information on the volume of work performed by respondents in the various industrial categories. Among those doing some highway construction, the average volume for the highway sector was usually substantially higher for union shops in all regions except the West North Central and West South Central. In fact, the large total number of open shops reported in the survey was attributable, to a significant degree, to the number of responses from these two regions (amounting to almost two-thirds of the open shop total). The responses from union shops were considerably more dispersed.

Associated General Contractors—Heavy Construction

Only 18 percent of the 3,044 respondents to the 1969 AGC survey reported that they were engaged in heavy construction as compared with 38 percent of the 5,938 respondents in 1973. The trends in Table VI-2, which contains the results of the two surveys for heavy construction, follow those for other types of construction. Thus, we find that the overall union ratio for heavy construction declined from 67.6 percent in 1969 to 59.8 percent in 1973, and that the open shop was especially strong in the South Atlantic, East South Central, and West South Central regions. Moreover, between 1969 and

TABLE VI-2
Union Status of AGC Members Engaged in Heavy Construction by Region, 1969 and 1973

Region	Percent Union[a]	Number of Contractors					
		1969			1973		
		Union	Nonunion	Percent Union	Union	Nonunion	Percent Union
New England	96.5	21	11	65.6	32	38	45.7
Middle Atlantic	96.1	47	3	94.0	174	21	89.2
E.N. Central	98.5	55	2	96.5	351	48	88.0
W.N. Central	33.6	51	41	55.4	141	166	45.9
S. Atlantic	47.6	16	26	38.1	63	201	23.9
E.S. Central	71.4	9	13	40.9	25	58	30.1
W.S. Central	6.6	21	63	25.0	63	275	18.6
Mountain	90.3	31	12	72.1	147	77	65.6
Pacific	55.2	122	8	93.8	383	42	90.1
Total	43.4	373	179	67.6	1,379	926	59.8

Source: Associated General Contractors of America.
[a] Calculated by using the dollar volume of contracts granted in 1969.

1973, the open shop gained in all regions including those most highly unionized.

District 50 was historically much less a factor in heavy work than in the highway sector. Thus, changes in the heavy sector's degree of unionization between 1969 and 1973 probably do not reflect to a significant degree any gains or losses other than those of the standard crafts.

More heavy construction respondents to the AGC questionnaire were unionized than were those in highway construction, not only nationally but in virtually every region, for both 1969 and 1973. One reason for this is that heavy work is concentrated in metropolitan areas (for example, subways and airports) or in remote locations where only the largest contractors, or a consortium thereof, are qualified to work and where union recruiting can supply manpower (for example, dams or mountain tunnels). On the other hand, since many highway projects are either in rural areas or are small in scale, smaller open shop contractors can provide competition. Although more unionized than highway construction, according to the AGC data, the heavy sector was less unionized than commercial construction in the early 1970s. Overall, the percentage of open shop heavy construction contractors in the AGC was very close to that of all contractors who answered the surveys for both 1969 and 1973. Thus, in 1969, this percentage for heavy construction contractors was 32.4 and for all contractors, 31.1. In 1973, the figures were 40.2 percent for heavy construction contractors and 41.7 percent for all contractors.

In terms of project type, the union contractors had the largest, most costly projects in 1969 in most regions, but not in the highly open shop areas. Again, the heavy responses from the southern regions affected the overall results.

The AGC was unable to provide an update of its 1969 and 1973 surveys, but a knowledgeable staff member estimated the union shop-open shop situation in each of the organization's highway-heavy chapters. These estimates are shown in Table VI-3 and indicate a tremendous erosion of union control since 1973. Indeed, if these estimates are correct, unionized companies are in the majority only in the Middle Atlantic and Pacific states, plus Massachusetts, Rhode Island, and Montana; work is split rather evenly between open and union shop concerns in Minnesota, Missouri, Northern Virginia, and West Virginia. These estimates are particularly surprising for Wisconsin, Minnesota, and Missouri (all 50 percent union), California (60 percent union), New Mexico (20 percent union), Idaho (100 percent open shop), and eastern Washington (50 percent union), all of which we believed to have a higher union penetration. We shall present more concrete data for some of these states later in this chapter. It is, however, noteworthy that all the states in which the unionized portion is estimated in Table VI-3 to be a majority have "Little Davis-Bacon Acts" that are administered in a manner that supports unionization.

The 1982 Industrial Research Unit questionnaire returns from AGC members do not support this overwhelming movement to the open shop that is indicated in Table VI-3. Of the 1,373 AGC members who responded to the questionnaire, 591 stated that they were highway construction contractors and 572, heavy construction contractors. Of the 591 highway contractors, 63.1 percent reported that they operated union, and 36.9 percent, nonunion; for the heavy sector, 64.2 percent were union, and 35.8 percent were nonunion. It is still possible that these questionnaire data and Table VI-3 are not in conflict, because some of the largest AGC highway and heavy sector chapters—those in Massachusetts and in the Middle Atlantic,

TABLE VI-3
Estimate of Union Shop-Open Shop Ratio
Highway and Heavy Chapters
Associated General Contractors
by State, 1983

Region and State	Percent Union	Percent Open Shop
New England		
Maine	0	100
Massachusetts	75	25
New Hampshire	0	100
Rhode Island	100	0
Middle Atlantic		
New Jersey	100	0
New York	90	10
Pennsylvania[a]	80	20
E.N. Central		
Illinois	95	5
Indiana	100	0
Michigan	100	0
Ohio	95	5
Wisconsin	50	50
W.N. Central		
Iowa	20	80
Kansas	0	100
Minnesota	50	50
Missouri	50	50
Nebraska	0	100
North Dakota	0	100
South Dakota	10	90

TABLE VI-3 (continued)

Region and State	Percent Union	Percent Open Shop
S. Atlantic		
Delaware	5	95
Florida	45	55
Georgia	5	95
Maryland	20	80
North Carolina	5	95
South Carolina	5	95
Virginia (Northern section)	50	50
(Balance of state)	0	100
West Virginia	50	50
E.S. Central		
Alabama	10	90
Kentucky[a]	60	40
Mississippi	0	100
Tennessee	50	50
W.S. Central		
Arkansas	0	100
Louisiana	15	85
Oklahoma	0	100
Texas	0	100
Mountain		
Arizona	25	75
Colorado[b]	25	75
Idaho[c]	0	100
Montana	75	25
New Mexico	20	80
Nevada[a]	80	20
Utah[a]	30	70
Wyoming	5	95
Pacific		
California[d]	60	40
Washington (Western section)	100	0
(Eastern section)[c]	50	50

Source: Estimates of Associated General Contractors staff official.
[a] Author's estimates
[b] See Table VI-6.
[c] See Table VI-8.
[d] See Table VI-7.

East North Central, and Pacific regions—remain union; and of course, respondents to the questionnaire may have been disproportionately union. Nevertheless, Table VI-3 may well overestimate open shop penetration.

American Road and Transportation Builders Association

The ARTBA has not made any surveys similar to those by the AGC, but at our request, the managing director of the ARTBA Contractors Division provided information on the union status for

1974 and 1981. In New England, the ARTBA Council was composed of union contractors in 1974; by 1981, that region's ARTBA membership was overwhelmingly open shop. The organization was open shop in the Southeast, except for Kentucky, which nevertheless had gone from strong union shop to only 50 percent union between 1974 and 1981, and the District of Columbia, which remained about 90 percent union. In Delaware, where the AGC chapter has gone open shop, the ARTBA has remained union; while in West Virginia it is largely open shop. Ohio, Indiana, Illinois, Wisconsin, and Iowa were still unionized in 1981, but Texas, Arkansas, and Oklahoma were open shop.

In the area between Pittsburgh and Philadelphia, ARTBA members were heavily involved with District 50 in 1974. When the Steelworkers took over District 50, many contractors went open shop, but the unionized ones still predominated in the state. In California, the ARTBA represents an open shop that is small but growing, as data presented later in this chapter will document. Thus, the ARTBA information supports that of the AGC in finding a strong open shop trend in the 1970s.

Federal Highway Administration

The Federal Highway Administration, pursuant to its enabling legislation, requires that highway contractors provide training for a percentage of their labor forces. Such training may be on-the-job, or through apprentice programs. We shall examine this training and its impact on minorities in Chapters X and XII. Here we are interested in the training data as an indication of union status of contractors.

Training provided by open shop contractors is mostly on-the-job. The unions generally prefer, and usually insist upon, apprenticeship training over which they exercise considerable control. The Operating Engineers and other unions, however, do sponsor some on-the-job training; for example, training provided under Steelworker agreements is usually the on-the-job type. As we shall see in Chapter X, open shop operators are increasingly sponsoring apprenticeship training. Nevertheless, an examination of the relative importance of on-the-job training in comparison with apprenticeship training under Federal-Aid Highway Projects does give a reasonable indication of union status. Table VI-4 presents such data by region for 1973, 1976, 1979, and 1981, all as of July 31 of the year.

The data for 1979 and 1981 are less valuable because of the severe depression in the highway repair and construction industry which began in the late 1970s and continued during the first few years of the 1980s. Nevertheless, these data do confirm the general picture already drawn of the extent of the open shop in highway and heavy sectors. Thus, in Region 1 (New England), nearly 70 percent of all trainees are on-the-job, and this is an area that we have found to be largely and increasingly open shop. Region 4 (the Southeast), Region 6 (the Southwest), and Region 7 (Iowa, Kansas, Missouri, and Nebraska) are heavily open shop, except for St. Louis, Kansas City, eastern Iowa, and parts of Missouri.

The high percentage of trainees in Region 3 reflects the open shop status of Virginia, Maryland, and the rural areas of Pennsylvania and West Virginia. Moreover, the Steelworkers, which is strong in this region, has also accepted trainees and has not stressed apprenticeship programs. This more than offsets the unionized states of Pennsylvania, Delaware, and West Virginia, as well as the District of Columbia.

On the other hand, Region 9 (California, Arizona, and Nevada) is much more unionized; two of the states are under a single Operating Engineers' local jurisdiction. This local has refused to accept on-the-job trainees, which comprised only a tiny percentage of the total trainees.

The presence of on-the-job trainees does not mean the absence of unions, nor does the presence of apprentices mean that the work is all union shop. Nevertheless, such classifications are strong indicators of trends. It is, therefore, probably significant to compare the 1973 and the 1981 data, especially since the numbers for 1979 are too small to be useful. We found that on-the-job trainees predominated in both years in Region 1 (New England), Region 3 (Middle Atlantic), and Region 4 (Southeast). In Regions 5 (Upper Midwest), 7 (Iowa, Kansas, Missouri, and Nebraska), 8 (Dakotas and Northern Rocky Mountains) and 10 (Alaska and Northwest), on-the-job trainees greatly expanded their share of the total training population. In Region 9, (California, Arizona, and Nevada) apprentices were virtually the only trainees, and surprisingly, in Region 6 (Southwest) apprentices also dominated, perhaps partly because large open shop companies there also have such programs. To the extent that these data reflect open shop-union shop trends, they indicate gains for the open shop. This conforms with our field investigations and with the results of the regional studies cited in the next section.

REGIONAL SURVEYS

A number of regional studies made during the last several years, principally by contractor associations or agencies, have included the highway and heavy sectors. In addition, our field interviews gar-

TABLE VI-4
Percentage Apprentices and On-the-Job Trainees as Indicators of Union Status by Region Under Federal Highway Projects Selected Years, 1973–1981

	Year and Types of Trainees					
	1973			1976		
Region	Total Trainees	Percent Apprentices	Percent On-the-Job Trainees	Total Trainees	Percent Apprentices	Percent On-the-Job Trainees
1	520	34.0	66.0	479	31.5	68.5
3[b]	704	31.4	68.6	725	37.5	62.5
4	629	10.7	89.3	468	13.5	86.5
5	1,054	55.2	44.8	1,117	58.5	41.5
6	583	13.4	86.6	371	19.9	80.1
7	366	14.5	85.5	245	40.8	59.2
8	321	49.2	50.8	299	55.9	44.1
9	568	99.9	(c)	317	99.1	(c)
10	194	50.5	49.5	289	59.2	40.8
All regions	4,939	40.4	59.6	4,310	45.6	54.4

Region	1979 Total Trainees	1979 Percent Apprentices	1979 Percent On-the-Job Trainees	1981 Total Trainees	1981 Percent Apprentices	1981 Percent On-the-Job Trainees
1	91	11.0	89.0	309	34.0	66.0
3[b]	88	18.2	81.8	388	34.5	65.5
4	67	14.9	85.9	384	12.0	88.0
5	161	41.0	59.0	813	63.8	36.2
6	67	1.5	98.5	468	9.3	90.7
7	62	27.4	72.6	246	58.1	41.9
8	61	31.1	68.9	205	53.7	46.3
9	39	97.4	2.6	219	98.6	1.4
10	59	42.4	57.6	286	76.9	23.1
All regions	695	29.1	70.9	3,312	46.4	53.6

Source: Office of Civil Rights, Federal Highway Administration

[a]Regional definitions:
(1) Maine, New Hampshire, Vermont, Massachusetts, Rhode Island, and Connecticut
(3) Delaware, District of Columbia, Maryland, Pennsylvania, Virginia, and West Virginia
(4) Alabama, Florida, Georgia, Kentucky, Mississippi, North Carolina, South Carolina, and Tennessee
(5) Illinois, Indiana, Michigan, Minnesota, Ohio, and Wisconsin
(6) Arkansas, Louisiana, New Mexico, Oklahoma, and Texas
(7) Iowa, Kansas, Missouri, and Nebraska
(8) Colorado, Montana, North Dakota, South Dakota, Utah, and Wyoming
(9) Arizona, California, and Nevada
(10) Alaska, Idaho, Oregon, and Washington.

[b]There is no Region 2.

[c]Less than 1 percent.

nered information on this subject. Both sources are discussed in this section.

Northeast, South and Midwest

The data in Table VI-3 are in basic accord with our field interview findings. Thus, northern New England—Maine, New Hampshire, and Vermont—is heavily open shop, but Massachusetts, Rhode Island, and Connecticut are heavily unionized. New York and New Jersey are union bastions. The Steelworkers are strong in the central areas of Pennsylvania, but the construction unions still have a clear majority of the work. Except for the District of Columbia and adjacent areas in Virginia, about one-half of the work in West Virginia and Florida, and a few parts of Texas, we found the South and Southwest to be almost completely open shop. The Midwest, east of the Mississippi, is one of the strongest union areas; however, union strength varies west of the Mississippi, as noted in Table VI-3.

Rocky Mountain

The open shop has made gains in the Rocky Mountain region also. In New Mexico and Arizona, where highway and heavy work were union-dominated during the early 1970s, large open shop contractors from Texas (which is overwhelmingly open shop) are now operating and have gained a majority of the business. Oklahoma, which was already heavily open shop, has become more so. As a result of these developments, highway and heavy construction contractors in Arizona and New Mexico persuaded the Operating Engineers and other unions to agree to a forty-four-month contract, which runs to mid-1985, that allows union contractors to field a labor force in which 20 percent of the craft personnel are subjourneymen and 30 percent are preapprentices. This has apparently made the unionized companies more competitive, particularly for the smaller projects "that are normally the stronghold of nonunion firms."[23] Nevertheless, the open shop contractors are still dominant in these states.

Table VI-5 shows the awards to union and nonunion highway and heavy construction contractors in Colorado for the first six months of 1973, 1979, and 1982, as found by the union-management–sponsored study made by the University of Colorado. Work in these sectors in Denver was once all union, but as early as 1973, the

---

[23] "Innovative Labor Pacts Scoring Market Gains," *Engineering News-Record,* July 7, 1983, p. 59.

TABLE VI-5

Percent Distribution of Highway and Heavy Construction by Work Category and Area, Colorado 1973, 1979, and 1982[a]

Area	Highway and Bridge Construction						Earthwork Construction						Utility Construction					
	Percentage No. of Awards			Percentage Dollar Volume			Percentage No. of Awards			Percentage Dollar Volume			Percentage No. of Awards			Percentage Dollar Volume		
	1973	1979	1982	1973	1979	1982	1973	1979[b]	1982	1973	1979[b]	1982	1973	1979	1982	1973	1979	1982
Denver Metro																		
Union	81	25	22	91	42	25	55	0	19	92	0	26	57	31	27	58	61	49
Open Shop	19	75	78	9	58	75	45	100	81	8	100	74	43	69	73	42	39	51
State Except Denver Metro																		
Union	54	25	34	60	19	33	33	25	0	17	99	0	44	17	12	42	52	2
Open Shop	46	75	66	40	81	67	67	75	100	83	1	100	56	83	88	58	48	98
All Colorado																		
Union	65	25	29	68	34	30	42	13	13	57	92	20	49	27	20	54	58	15
Open Shop	35	75	71	32	66	70	58	87	87	43	8	80	51	73	80	46	42	85

Source: David E. Moore, *Union and Open Shop Construction in the State of Colorado* (Denver: Construction Industry Management Council and Colorado Building and Construction Trades Council, 1983), Tables 6 and 7.

[a] First six months of 1973, 1979, and 1982.

[b] Too few awards for statistical reliability.

national AGC study (presented earlier in the chapter) found that 59 percent of the highway work outside of Denver was nonunion. Now the work both in Denver and throughout the state appears very heavily open shop, with both heavy and highway construction at least 70 percent nonunion.

California

The survey covering eleven counties of Southern California published by the Construction Industry Research Board (CIRB) of Los Angeles found that the highway and heavy construction sectors are the most unionized in that area. "The union-contractors' market share of this sector stood at 75% in 1981 down from 83% in 1977. This represents a 9.6% loss in union-contractors' market share, the smallest decline of any sector."[24] (See Table VI-6.) The CIRB also

TABLE VI-6
Union Contractors' Market Share[a]
Highway and Heavy Construction
Eleven-County Southern California Region
1977 and 1981

Area (No. of Responses by Area)	1977	1981
Regional Total[b] (284)	83	75
Los Angeles (159)	84	76
Orange (37)	82	72
Ventura (22)	90	90
San Bernardino (16)	77	70
Riverside (15)	83	80
Santa Barbara (15)	88	83
Kern (8)	88	79
San Luis Obispo (6)[c]	76	50
Mono (4)[c]	30	25
Imperial (2)[c]	70	50
Inyo (0)	n.r.	n.r.

Source: *A Survey of the Construction Industry Trends in the Eleven-County Southern-California Region* (Los Angeles: Construction Industry Research Board, January 31, 1983), Table 6.

[a] All percentages are frequency weighted averages of the midpoint of 20 percent intervals.

[b] The regional total for each category of construction is the weighted average of all responses. Responses were weighted by the actual county share of each type of construction for 1977 and 1981.

[c] Results for small-response counties are statistically unreliable. Variations as large as 20 percentage points could result with very small changes in the number of responses.

n.r. = no response

[24] *A Survey of the Construction Industry Trends in the Eleven-County Southern-California Region* (Los Angeles: Construction Industry Research Board, January 31, 1983), p. 13.

found that nonunion firms were underrepresented in the survey responses, and that firms that left the union ranks after 1976 tend to be larger and more likely to engage in public works than are firms that went nonunion in prior years.[25] Moreover, the newly turned nonunion firms "include more highway and heavy construction firms than do earlier departers."[26]

Despite their loss of market share between 1977 and 1981, union highway and heavy work contractors in Southern California increased their total dollar volume of work because of increases in construction. Moreover, union firms averaged a much larger volume of work than did their nonunion competitors. None of the nonunion firms did over $2 million of business while two-thirds of the union firms exceeded that figure. The greatest concentration of nonunion firms was found in relatively rural San Bernardino County, but, surprisingly, the second-largest grouping was in Los Angeles County, where they obviously have been successful in both private and public work.

The perception that open shop contractors have gained a moderate share of work in Southern California is shared by union and nonunion contractors. Table VI-7 shows that 62.5 percent of the union firms and 75 percent of the nonunion firms responding to the CIRB questionnaire believed that the union share of work had "decreased somewhat."

The CIRB's study confirms our general findings that the open shop is gaining in highway and heavy work, but it is doing so more slowly in California than in other regions, possibly because of legal and political impediments. California is one of the states in which its "Little Davis-Bacon Act" uses the union rate as the prevailing one, and during the period covered by the survey, a very prounion administration was in power.[27] Moreover, the Operating Engineers are very powerful in California, evidenced by the almost total lack of on-the-job trainees shown for the state in Table VI-4.

The ARTBA, however, has provided an organization for the open shop highway and heavy construction contractors in Southern California, and, by every indication, they are growing in both the number and volume of jobs. This growth, although modest, in such a union stronghold, signifies the beginning of a more widespread trend. In the rest of California, however, unionized contractors are clearly dominant in highway and heavy construction, aided by the administration of the state's Little Davis-Bacon Act.

---

[25] *Ibid.*, p. 18.
[26] *Ibid.*
[27] Edmund G. (Jerry) Brown, Jr., was governor at this time.

TABLE VI-7

Contractors' Responses Regarding Local Competition: How Has the Percentage of Work Going to Union Highway and Heavy Contractors Changed? January 1977–December 1981

	Share of Work Going to Union Contractors Percent of Responses in Each Category					
	Decreased		Stayed	Increased		Average
No. of Firms	Substantially (−2)	Somewhat (−1)	the same (−0−)	Somewhat (+1)	Substantially (+2)	Value (−2 to +2)
Union firms (24)	29.2	62.5	8.3	0	0	−1.21
Nonunion firms (8)	0	75.0	12.5	12.5	0	−0.62
All firms (projected)[a]	19.0	66.8	9.8	4.4	0	−1.01

Source: *A Survey of the Construction Industry Trends in the Eleven-County Southern-California Region* (Los Angeles: Construction Industry Research Board, January 31, 1983), Table 12.

[a] Values for all firms were calculated based on the number of responses from union and nonunion firms, adjusted for the underrepresentation and smaller size of nonunion firms. The number of nonunion firms responding was increased by a factor of 1.6 for calculation purposes.

Pacific Northwest

As discussed in previous chapters, the Pacific Northwest, which was once a construction-union citadel, has seen steady gains by the open shop in recent years. Western Washington, particularly the Seattle area, however, remains heavily unionized. Open shop highway and heavy firms are gaining in other parts of the state, despite some lag behind other construction sectors. Thus, as early as 1977, the Oregon-Columbia chapter of the AGC, covering Oregon and southwestern Washington, found that although 90 percent of the highway and heavy contractors who answered the questionnaire were union contractors, 18 percent stated that they were "seriously contemplating either going nonunion, or creating a separate company to operate nonunion."[28] By 1982, this AGC chapter was warning members that the "nonunion construction movement will continue to spread . . . to certain other markets such as paving. . . ."[29]

The growth of the open shop share of highway and heavy work in Oregon is further illustrated by Table VI-8, which shows the 1981 Oregon Highway Division awards and nonfederal Oregon state and local government public contract awards for the first four months in 1982. As in Southern California, the larger awards were won by unionized contractors. One large award for $40 million tipped the scales, however, because without it the union dollar volume percent for the 1981 highway awards would have shrunk from 76 to 45. On the state and local levels, the open shop contractors received two-thirds of the jobs but only one-third of the dollar volume. Since 1981, open shop contractors have won an increasing share of Oregon's highway and heavy work, including the larger jobs, as noted below.

Further evidence of the inroads of the open shop in heavy and highway work in the Pacific Northwest is provided in a study by the Inland Empire Chapter of the AGC, which draws members from eastern Washington and northern Idaho. Table VI-9 shows the dollar volume for contracts let in heavy and highway work in these regions by category of work from 1976 to 1983. Between 1976 and 1983, open shop contractors increased their share of the work in the various categories from 11.9 percent to 39 percent. Of special importance is that the open shop segment broke the monopoly of

[28] *Analysis AGC Labor Relations Questionnaire, November 1977* (Portland, Ore.: Oregon-Columbia Chapter, AGC, 1977), pp. 1–2. Questionnaire completed by heavy/highway contractors.

[29] *Construction Industry Profiles* (Wilsonville, Ore.: Oregon-Columbia Chapter, AGC, 1982), n.p.

TABLE VI-8
Oregon Highway Division Awards, 1981 and Oregon State and Local Non-Federal Public Contract Awards by Union Status, January–April 1982

	Oregon Highway Division Awards, 1981						Oregon State and Local Non-Federal Public Contract Awards, January-April 1982					
	Total		Union		Open Shop		Total		Union		Open Shop	
	No.	Percent	No.	Percent	No.	Percent	No.	Percent	No.	Percent	No.	Percent
No. of jobs	143	100	90	63	53	37	205	100	78	38	127	62
$ Volume of jobs (millions)	$129.0	100	$97.6[a]	76	$31.4	24	$108.3	100	$72.5	67	$35.8	33

Source: *Construction Industry Profiles* (Wilsonville, Ore.: Oregon-Columbia Chapter, AGC, 1982), n. p.
[a] If one job of $40 million is removed from the union totals, its dollar volume falls 31 percent.

TABLE VI-9

Dollar Volume of Contracts Let, Highway and Heavy Work, Eastern Washington and Northern Idaho by Sources and Category of Work[a] *Union and Open Shop, 1976–1983*

Source and Category	Total Volume $	Union Volume $	Percent Union	Open Shop Volume $	Percent Open Shop	Total Volume $	Union Volume $	Percent Union	Open Shop Volume $	Percent Open Shop
	1976					1977				
Washington Department of Transportation	17,749,357	17,749,357	100	—	00.00	34,744,522	34,744,522	100	—	00.0
Idaho Department of Transportation	13,973,531	13,973,531	100	—	00.0	14,013,129	13,939,846	99.4	73,283	00.5
Streets, bridges, & structures	986,912	663,686	67.2	323,226	32.7	5,678,894	5,395,630	95.0	283,264	04.9
Roads-crushing-paving	18,815,951	16,057,709	85.3	2,758,242	14.6	29,152,884	24,374,322	83.6	4,778,562	16.3
Utilities	17,937,439	13,732,319	76.5	4,205,120	23.4	37,647,842	26,664,534	70.8	10,983,308	29.1
Miscellaneous	64,396,986	55,739,244[b]	86.5	8,657,742	13.4	45,161,051	36,727,977[c]	81.3	8,433,074	18.6
Total	133,860,176	117,915,846	88.0	15,944,330	11.9	166,398,322	141,846,831	85.2	24,551,491	14.7
	1978					1979				
Washington Department of Transportation	64,026,760	64,026,760	100	—	00.0	54,824,977	54,401,029	99.2	423,948	00.7
Idaho Department of Transportation	11,105,512	10,220,073	92.0	885,439	07.9	25,124,997	25,066,739	99.7	58,258	00.2
Streets, bridges & structures	2,870,086	2,357,809	82.1	512,277	17.8	29,401,605	14,272,481	48.5	15,129,124	51.4
Roads-crushing-paving	20,375,709	13,667,882	67.0	6,707,827	32.9	20,662,035	16,377,412	79.2	4,284,623	20.7
Utilities	26,455,272	22,542,415	85.2	3,912,857	14.7	34,275,183	29,470,886	85.9	4,804,297	14.0
Miscellaneous	37,909,780	30,625,160	80.7	7,284,620	19.2	12,019,533	5,124,341	42.6	6,895,192	57.3
Total	162,743,119	143,440,099	88.1	19,303,020	11.9	176,790,536	144,712,888	81.8	32,077,648	18.1

TABLE VI-9 (continued)

Source and Category	Total Volume $	Union Volume $	Percent Union	Open Shop Volume $	Percent Open Shop	Total Volume $	Union Volume $	Percent Union	Open Shop Volume $	Percent Open Shop
			1980					1981		
Washington Department of Transportation	63,809,496	62,998,573	98.7	810,923	01.2	55,927,956	53,288,478	95.2	2,639,478	04.7
Idaho Department of Transportation	5,911,499	4,828,034	81.6	1,083,465	18.3	16,728,247	16,526,983	98.7	201,264	01.2
Streets, bridges & structures	2,789,736	1,777,232	63.7	1,012,504	36.2	7,800,150	5,299,318	67.9	2,500,832	32.0
Roads-crushing-paving	32,285,625	20,356,757	63.0	11,928,868	36.9	23,064,049	12,962,060	56.2	10,101,989	43.7
Utilities	23,682,520	11,271,824	47.5	12,410,696	52.4	48,534,584	30,645,724	63.1	17,888,860	36.8
Miscellaneous	138,066,385	114,318,709	82.7	23,747,676	17.2	109,479,196	94,981,035	86.7	14,498,161	13.2
Total	266,545,261	215,551,129	80.8	50,994,132	19.1	261,534,182	213,703,598	81.7	47,830,584	18.2
			1982					1983		
Washington Department of Transportation	96,495,893	82,729,649	85.7	13,766,244	14.3	94,525,458	74,442,855	78.7	20,082,603	21.3
Idaho Department of Transportation	14,495,151	13,022,524	89.8	1,472,627	10.2	17,303,265	14,279,891	82.5	3,023,374	17.5
Streets, bridges & structures	4,958,000	2,573,475	51.9	2,384,525	48.1	5,798,133	4,254,167	73.4	1,543,966	26.6
Roads-crushing-paving	27,396,926	10,917,455	39.8	16,479,471	60.2	28,187,208	13,124,828	46.6	15,062,380	53.4
Utilities	49,400,103	34,076,910	69.0	15,323,193	31.0	77,607,689	33,753,554	43.5	43,854,135	56.5
Miscellaneous	86,190,045	45,543,775	52.8	40,646,270	47.2	45,711,690	24,387,323	53.4	21,324,367	46.6
Total	278,936,118	188,863,788	67.7	90,072,330	32.3	269,133,443	164,242,618	61.0	104,890,825	39.0

Source: Inland Empire Chapter, Associated General Contractors, Annual Survey.

[a] 1. Washington Department of Transportation
Includes all contracts issued by that agency in eastern Washington.

2. Idaho Transportation Department
Includes all contracts issued in Districts #4 and #5 in northern Idaho, north of the 46.00 parallel.

3. Streets, Bridges and Structures Classification
Includes bridges, retaining walls, and other flood control structures let by cities, counties, highway districts, etc., in Washington and northern Idaho.

4. Utilities Classification
Includes all sewer, watermain, and treatment plants and well-drilling contracts in eastern Washington and northern Idaho.

5. Miscellaneous Classification
This classification includes demolition projects, projects on military bases, projects on the Hanford Reservation (requiring permits) and the Pullman campus, as well as landscaping and large Corps of Engineers and Bureau of Reclamation projects.

[b] Bacon Siphon, 1966.

[c] Grand Coulee Feeder Canal, 1977.

union contractors in work contracted by the departments of transportation of the two states and by 1983 had 21.3 percent of such work in Washington and 17.5 percent in Idaho for the area covered by the data in Table VI-9. This was done in spite of the existence of "Little Davis-Bacon" laws in both states that are administered to support unions and unionized contractors. The net effect of this change has been to make the open shop contractors effective competitors whom the Inland Empire Chapter, AGC, expect to have a majority of the work in 1984, or shortly thereafter, as discussed in Chapter VII (See Figure VII-2).

Further evidence of the decline in union control of the highway and heavy sector was provided by our field interviews. In addition, the Operating Engineers in Oregon have agreed to a twenty-seven months wage freeze for Oregon and southern Washington, in one case after first rejecting it.[30] An open shop contractor won a $29 million contract from the Oregon Department of Transportation in March 1983, the largest awarded an open shop firm by that state agency. The firm, R. A. Hatch Company, which operated union before 1980, has been the victim of mass picketing and violence. According to a company spokesperson, "union wage rates are not an issue because the project is covered by the Davis-Bacon Act. But, 'we don't go along with the union work rules.' "[31]

THE INDUSTRIAL RESEARCH UNIT QUESTIONNAIRES

The 1974 and 1982 Industrial Research Unit surveys provide further evidence of gains by the open shop in highway and heavy construction during the 1970s. Both surveys included within their coverage all members of the AGC and the ARTBA, as well as those of other associations. As in the case of other construction sectors, highway and heavy construction contractors who operated union differed markedly from those who operated nonunion in their estimate of the extent of open shop activity in the highway and heavy sector. Table VI-10 shows that about one-half of the nonunion contractors in the 1974 survey who were able to offer an estimate put the open shop share of highway work at between 61 and 100 percent, but more than one-half of the union contractors responding to the question estimated it at 20 percent or less. Table VI-11 shows ap-

[30] "Oregon Engineers Reconsider, Agree to Concession Agreement," *Construction Labor Report,* Vol. 29 (June 8, 1983), p. 506.

[31] "Union Spat Takes Odd Twists," *Engineering News-Record,* May 26, 1983, p. 205; and "Union Pickets Continue at Job Site," *Construction Labor Report,* Vol. 28 (June 8, 1983), pp. 506–07.

TABLE VI-10
Percent Distribution of Respondents' Estimates of the Proportion of Open Shop Work in Highway Construction by Region and Union Status

Region	Number	NR/DK[a]	0–20	21–40	41–60	61–80	81–100
Union							
Not reported	89	53.9	34.8	6.7	1.1	2.2	1.1
New England	27	66.6	18.5	0.0	3.7	0.0	11.1
Middle Atlantic	125	39.2	29.6	12.0	11.1	4.8	3.2
E.N. Central	147	29.2	50.3	8.8	5.4	2.0	4.1
W.N. Central	60	38.4	21.7	10.0	6.7	6.7	16.7
S. Atlantic	68	36.8	8.8	7.4	11.8	4.4	30.9
E.S. Central	32	28.1	31.3	12.5	6.3	3.1	18.8
W.S. Central	45	31.2	11.1	4.4	6.7	6.7	40.0
Mountain	52	36.5	36.5	15.4	1.9	5.8	3.8
Pacific	133	48.1	44.4	4.5	0.8	0.8	1.5
Total	778	40.1	33.3	8.4	5.5	3.3	9.4
Nonunion							
Not reported	83	44.6	15.7	8.4	9.6	3.6	18.1
New England	63	39.7	12.7	14.3	11.1	9.5	12.7
Middle Atlantic	134	47.0	23.1	6.0	9.0	6.7	8.2
E.N. Central	75	33.3	25.3	12.0	13.3	8.0	8.0
W.N. Central	45	20.0	6.7	8.9	6.7	17.8	40.0
S. Atlantic	220	45.5	6.8	3.2	5.0	10.0	29.5
E.S. Central	45	37.7	8.9	8.9	4.4	8.9	31.1
W.S. Central	92	33.7	8.7	2.2	5.4	7.6	42.4
Mountain	21	33.3	14.4	14.3	9.5	9.5	19.0
Pacific	6	16.6	66.7	16.7	0.0	0.0	0.0
Total	784	40.2	13.8	6.9	7.7	8.5	23.0

Source: Industrial Research Unit Questionnaire, 1974
[a] NR/DK = Not reported or do not know.

proximately the same relationship for the heavy sector, with both union and open shop contractors indicating a slightly higher proportion of union strength in the heavy sector than in the highway sector.

To the extent that respondents saw a change in open shop highway work over the years prior to 1974, that change was predominantly in the direction of greater open shop competition. On the whole, however, both union and nonunion respondents perceived less overall change in highway and heavy construction in the late 1960s and early 1970s than in other branches of the industry, as summarized in Table VI-12 and Table VI-13. This finding is largely consistent with the results of the 1969 and 1973 surveys by the AGC.

Since a relatively small proportion of the respondents could be classified as highway or heavy construction contractors, we at-

TABLE VI-11
Percent Distribution of Respondents' Estimates of the Proportion of Open Shop Work in Heavy Construction by Region and Union Status

Region	Number	NR/DK[a]	0–20	21–40	41–60	61–80	81–100
Union							
Not reported	89	59.6	34.8	0.0	2.2	2.2	1.1
New England	27	69.2	22.2	7.4	0.0	0.0	11.1
Middle Atlantic	125	40.0	37.6	7.2	8.8	2.4	4.0
E.N. Central	147	34.7	51.7	5.4	2.0	1.4	4.8
W.N. Central	60	36.6	30.0	8.3	3.3	6.7	15.0
S. Atlantic	68	39.7	13.2	8.8	14.7	2.9	20.6
E.S. Central	32	31.2	31.3	9.4	6.3	6.3	15.6
W.S. Central	45	37.8	13.3	2.2	6.7	4.4	35.6
Mountain	52	42.4	38.5	9.6	1.9	3.8	3.8
Pacific	133	46.6	47.4	3.8	0.0	0.0	2.3
Total	778	42.5	36.8	5.7	4.4	2.4	8.4
Nonunion							
Not reported	83	49.4	12.0	9.6	4.8	7.2	16.9
New England	63	38.1	23.8	11.1	1.6	9.5	15.9
Middle Atlantic	134	49.2	24.6	7.5	6.0	5.2	7.5
E.N. Central	75	36.0	33.3	12.0	8.0	4.0	6.7
W.N. Central	45	26.7	11.1	4.4	8.9	15.6	33.3
S. Atlantic	220	47.8	7.7	4.5	4.1	9.5	26.4
E.S. Central	45	46.7	13.3	2.2	4.4	4.4	28.9
W.S. Central	92	39.1	12.0	1.1	2.2	10.9	34.8
Mountain	21	38.1	19.0	14.3	4.8	14.3	9.5
Pacific	6	16.7	50.0	33.3	0.0	0.0	0.0
Total	784	43.5	16.5	6.8	4.7	8.3	20.3

Source: Industrial Research Unit Questionnaire, 1974
[a] NE/NK = Not reported or do not know.

tempted to determine whether the differences between the perceptions of union shops and open shops could be attributable to a lack of direct knowledge on the part of many respondents. If anything, however, union and nonunion highway and heavy work contractors (those with at least one-half of their volume in such construction) were even more inconsistent in their views on the open shop share of highway construction in 1974. For example, about 58 percent of the union highway firms felt that open shops had captured no more than 20 percent of the work by 1974, whereas a like proportion of the open shop firms put the percentage at 80 or more.

The responses to the 1982 questionnaire (Table VI-14) again showed a considerable discrepancy between the union and open shop contractors. Thus, 39.6 percent of the union highway contractors but only 18 percent of the open shop ones put the open shop share at 0–20 percent, whereas 24 percent of the union contractors and 43.8 percent of the open shop ones thought open shop penetration was above 40 percent. The 80–100 percent estimate was remarkably similar for 1974 (23 percent) and 1982 (21.6) on the part of open shop contractors and very close for union shop ones (9.4 percent in 1974 and 6.8 percent in 1982). In general, a comparison of Tables VI-10 and VI-14 does not indicate great gains for the open shop between 1974 and 1982; indeed there is some diminution of the open shop estimates for 1982. Whether this is attributable to sampling errors, a greater proportion of unanswered replies in 1974, or other reasons, a comparison of these tables yields results somewhat at

TABLE VI-12
Respondents' Estimates of Change in Open Shop Share of Highway Construction Over Few Years Prior to 1974 by Region and Union Status[a]

Region	Total	Total Indicating Change	Lower by 3	Lower by 2	Lower by 1	Higher by 1	Higher by 2	Higher by 3
Union								
Not reported	89	4	0	0	1	3	0	0
New England	27	0	0	0	0	0	0	0
Middle Atlantic	125	17	1	0	4	8	4	0
E.N. Central	147	10	0	0	2	7	1	0
W.N. Central	60	4	1	0	1	1	1	0
S. Atlantic	68	6	1	0	1	2	2	0
E.S. Central	32	5	0	0	1	3	0	1
W.S. Central	45	5	0	0	0	3	1	1
Mountain	52	4	0	0	2	2	0	0
Pacific	133	2	0	0	0	1	1	0
Total	778	57	3	0	12	30	10	2
Nonunion								
Not reported	83	5	0	0	1	4	0	0
New England	63	7	0	0	0	7	0	0
Middle Atlantic	134	10	1	2	0	5	2	0
E.N. Central	75	10	0	1	0	9	0	0
W.N. Central	45	8	0	0	2	5	0	1
S. Atlantic	220	9	0	0	2	6	1	0
E.S. Central	45	7	0	0	1	4	2	0
W.S. Central	92	8	0	0	0	3	3	1
Mountain	21	2	0	0	0	1	0	1
Pacific	6	1	0	0	0	1	0	0
Total	784	67	1	3	7	45	8	3

Source: Industrial Research Unit Questionnaire, 1974.
[a] Change expressed in terms of movement between ranges of 20 percentage points (i.e., 0–20, 21–40, etc.).

TABLE VI-13
Respondents' Estimates of Change in Open Shop Share of Heavy Construction Over Few Years Prior to 1974 by Region and Union Status[a]

Region	Total	Total Indicating Change	Lower by 3	Lower by 2	Lower by 1	Higher by 1	Higher by 2	Higher by 3
Union								
Not reported	89	4	0	0	1	2	1	0
New England	27	0	0	0	0	0	0	0
Middle Atlantic	125	15	0	1	3	8	3	0
E.N. Central	147	7	0	0	2	4	1	0
W.N. Central	60	4	1	0	1	1	1	0
S. Atlantic	68	6	1	0	2	2	0	1
E.S. Central	32	4	0	0	1	2	0	1
W.S. Central	45	3	0	0	0	2	0	1
Mountain	52	4	1	0	2	1	0	0
Pacific	133	1	0	0	0	1	0	0
Total	778	48	3	1	12	23	6	3
Nonunion								
Not reported	83	4	0	0	0	3	1	0
New England	63	6	0	0	0	6	0	0
Middle Atlantic	134	8	1	1	1	3	2	0
E.N. Central	75	8	1	0	1	6	0	0
W.N. Central	45	7	0	0	1	5	0	1
S. Atlantic	220	9	1	0	0	5	3	0
E.S. Central	45	7	0	0	1	4	2	0
W.S. Central	92	5	0	0	0	3	2	0
Mountain	21	1	0	0	0	0	0	1
Pacific	6	2	0	0	0	2	0	0
Total	784	57	3	1	4	37	10	2

Source: Industrial Research Unit Questionnaire, 1974.
[a] Change expressed in terms of movement between ranges of 20 percentage points (i.e., 0–20, 21–40, etc.).

variance with the other findings presented in this chapter. We do believe that the estimates in the middle ranges (21–40, 41–60, and 61–80 percent)—which all show a higher proportion of open shop penetration in 1982 for both union and nonunion respondents—represent a more accurate picture than the extremes (0–20 and 81–100 percent), and that these do conform with the data and information heretofore presented.

Much the same can be found in comparing the 1982 questionnaire responses for heavy construction contractors (see Table VI-15) with those of the 1974 questionnaire (see Table VI-11). The extreme categories yield very similar responses; the middle ranges show an increased open shop penetration.

TABLE VI-14
Percent Distribution of Respondents' Estimates
Proportion of Open Shop Work in Highway Construction
by Region and Union Status, 1982

Region	Union							Nonunion						
	Number	NR/DK[a]	0–20	21–40	41–60	61–80	81–100	Number	NR/DK[a]	0–20	21–40	41–60	61–80	81–100
Not reported	44	29.5	25.0	15.9	13.6	6.8	9.1	49	30.6	10.2	14.3	10.2	10.2	24.5
New England	62	19.4	32.3	21.0	21.0	3.2	3.2	111	32.4	22.5	9.9	11.7	7.2	16.2
Mid Atlantic	275	22.9	36.7	15.6	15.3	6.9	2.5	228	30.7	28.1	16.7	10.5	5.7	8.3
E.N. Central	474	17.1	61.6	12.7	5.1	1.9	1.7	205	24.4	35.6	18.5	11.7	5.4	4.4
W.N. Central	220	20.9	31.4	15.9	19.5	6.4	5.9	157	19.1	15.3	12.1	17.8	14.6	21.0
S. Atlantic	109	22.9	14.7	6.4	14.7	19.3	22.0	550	29.8	7.5	5.6	9.5	11.8	35.8
E.S. Central	69	27.5	11.6	14.5	13.0	13.0	20.3	111	21.6	9.0	5.4	15.3	18.0	30.6
W.S. Central	105	16.2	10.5	12.4	12.4	19.0	29.5	284	27.5	10.9	7.7	9.5	12.0	32.4
Mountain	108	23.1	17.6	22.2	17.6	6.5	13.0	184	19.0	17.9	18.5	20.7	14.1	9.8
Pacific	362	23.5	48.6	18.5	5.2	1.9	2.2	143	23.1	40.6	21.7	9.1	2.1	3.5
Total	1,828	21.1	39.6	15.3	11.2	6.1	6.8	2,022	26.5	18.0	11.7	11.9	10.3	21.6

Source: Industrial Research Unit Questionnaire, 1982.
[a] NR/DR = Not reported or do not know.

TABLE VI-15

Percent Distribution of Respondents' Estimates Proportion of Open Shop Work in Heavy Construction by Region and Union Status, 1982

Region	Union							Nonunion						
	Number	NR/DK[a]	0–20	21–40	41–60	61–80	81–100	Number	NR/DK[a]	0–20	21–40	41–60	61–80	81–100
Not reported	44	22.7	27.3	15.9	18.2	9.1	6.8	49	34.7	14.3	12.2	10.2	8.2	20.4
New England	62	21.0	46.8	17.7	3.2	6.4	4.8	111	33.3	34.2	5.4	5.4	6.3	15.3
Mid Atlantic	275	24.0	47.6	12.0	9.1	4.7	2.5	228	31.6	37.3	12.3	6.1	4.9	7.9
E.N. Central	474	16.5	67.7	8.6	3.8	1.3	2.1	205	25.4	43.9	12.2	8.8	3.9	5.9
W.N. Central	220	20.5	39.1	17.7	10.9	5.5	6.4	157	17.8	21.7	9.6	17.2	11.5	22.3
S. Atlantic	109	24.8	15.6	11.1	14.7	16.7	17.4	550	30.5	9.5	6.2	8.9	13.8	30.2
E.S. Central	69	26.1	15.9	13.0	13.0	8.7	23.2	111	19.6	16.2	9.2	7.2	15.3	32.4
W.S. Central	105	20.0	11.4	13.3	13.3	16.2	25.7	284	29.2	13.4	8.8	9.2	12.3	27.1
Mountain	108	23.1	23.1	17.6	14.8	9.5	12.4	184	21.2	26.1	17.9	16.3	10.9	7.6
Pacific	362	23.7	57.1	10.5	3.6	2.2	2.2	143	24.5	49.7	15.4	4.2	2.8	3.5
Total	1,828	21.3	46.7	12.2	7.9	5.4	6.6	2,022	27.6	23.8	10.1	9.3	9.9	19.3

Source: Industrial Research Unit Questionnaire, 1982.

[a] NR/DK = Not reported or do not know.

TABLE VI-16
Respondents' Estimates of Changes in Open Shop Share of Highway Construction by Region and Union Status 1978–1982

	Union								
Region	Number	NR/DK[a]	0–10	11–20	21–30	31–40	41–50	51–75	76–100
Not reported	44	43.2	20.5	13.6	2.3	2.3	6.8	4.5	6.8
New England	62	33.9	40.3	8.1	8.1	3.2	3.2	3.2	0.0
Mid Atlantic	275	45.5	29.8	11.6	5.1	1.5	1.5	2.5	2.5
E.N. Central	474	43.5	46.6	4.0	2.3	0.4	0.6	0.8	1.7
W.N. Central	220	44.5	31.4	10.5	3.6	2.7	1.8	2.3	3.2
S. Atlantic	109	45.9	9.2	5.5	4.6	7.3	10.1	5.5	11.9
E.S. Central	69	36.2	17.4	8.4	2.9	4.3	5.8	13.0	11.6
W.S. Central	105	37.1	9.5	13.3	7.6	3.8	5.7	4.8	18.1
Mountain	108	36.1	27.8	15.7	6.5	4.6	3.7	2.8	2.8
Pacific	362	41.4	46.4	4.1	2.2	0.0	0.2	0.5	5.0
Total	1,828	42.2	34.8	7.8	3.8	1.9	2.3	2.5	4.7

	Nonunion								
Region	Number	NR/DK[a]	0–10	11–20	21–30	31–40	41–50	51–75	76–100
Not reported	49	79.6	6.1	4.1	0.0	0.0	2.0	6.1	2.0
New England	111	70.3	9.0	7.2	1.8	2.7	0.9	2.7	5.4
Mid Atlantic	228	59.6	18.4	6.1	3.5	2.6	0.9	2.2	6.6
E.N. Central	205	55.6	28.3	7.3	0.5	1.5	0.9	1.5	4.4
W.N. Central	157	56.1	17.2	6.4	2.5	2.5	3.8	5.7	5.7
S. Atlantic	550	77.1	4.4	2.9	2.5	1.8	2.0	4.4	4.9
E.S. Central	111	62.2	7.2	5.4	5.4	2.7	3.6	5.4	8.1
W.S. Central	284	65.1	5.3	2.5	3.9	2.5	2.5	6.3	12.0
Mountain	184	46.7	16.3	10.3	8.2	3.8	6.0	2.7	6.0
Pacific	143	44.1	37.1	10.0	1.4	2.1	0.0	2.8	2.8
Total	2,022	63.4	13.4	5.5	3.1	2.3	2.2	4.0	6.2

Source: Industrial Research Unit Questionnaire, 1982.
[a] NR/DK = Not reported or do not know.

Tables VI-16 and VI-17 present the respondents' estimates of changes in the union shop penetration of highway and heavy construction from 1978 to 1982. As in the case of contractors in all sectors, a sizable proportion did not answer the question. For highway construction, 42.2 percent of the union respondents did not reply and 34.8 percent estimated only a 0–10 percent change; 63.4 percent of the nonunion respondents did not reply and the balance was widely scattered, with the largest group also in the 0–10 percent category. To the extent that these data yield significant results, they point to a small, increased penetration of the open shop.

The data for respondents' estimates of change in heavy construc-

TABLE VI-17
Respondents' Estimates of Changes in Open Shop Share of Heavy Construction by Region and Union Status, 1978–1982

	Union								
Region	Number	NR/DK[a]	0–10	11–20	21–30	31–40	41–50	51–75	76–100
Not reported	44	45.5	25.0	9.1	2.3	0.0	6.8	9.1	2.3
New England	62	33.9	46.8	6.5	6.5	1.6	1.6	3.2	0.0
Mid Atlantic	275	46.9	35.5	8.0	3.3	1.5	1.5	1.8	1.8
E.N. Central	474	43.2	47.9	3.4	1.5	0.4	0.4	0.8	2.3
W.N. Central	220	44.1	35.5	6.4	4.5	0.9	2.3	2.7	3.6
S. Atlantic	109	45.9	11.9	5.5	5.5	6.4	6.4	9.2	9.2
E.S. Central	69	34.8	20.3	7.2	4.3	4.3	2.9	17.4	8.7
W.S. Central	105	37.1	14.3	13.3	4.8	2.9	7.6	3.8	16.2
Mountain	108	36.1	34.3	9.3	6.5	4.6	3.7	3.7	1.9
Pacific	362	42.8	47.8	3.0	0.6	0.0	0.6	0.3	5.0
Total	1,828	42.6	38.0	5.8	3.0	1.5	2.1	2.8	4.3

	Nonunion								
Region	Number	NR/DK[a]	0–10	11–20	21–30	31–40	41–50	51–75	76–100
Not reported	49	79.6	8.2	2.0	0.0	0.0	4.1	4.1	2.0
New England	111	71.2	14.4	2.7	1.8	1.8	0.9	2.7	4.5
Mid Atlantic	228	59.2	20.2	6.1	1.3	2.6	0.9	2.6	7.0
E.N. Central	205	55.6	33.7	2.4	0.5	0.9	0.9	0.9	4.9
W.N. Central	157	56.1	18.5	5.7	2.5	3.2	1.9	6.4	5.7
S. Atlantic	550	77.0	5.1	2.4	2.7	1.5	3.3	4.4	3.8
E.S. Central	111	62.2	11.7	4.5	3.6	4.5	2.7	4.5	6.3
W.S. Central	284	65.1	6.3	2.8	3.2	2.1	3.2	7.7	9.5
Mountain	184	47.8	21.7	8.2	7.6	3.3	2.7	3.3	5.4
Pacific	143	44.1	41.3	7.0	1.4	0.7	0.0	2.1	3.5
Total	2,022	63.5	15.9	4.1	2.7	2.0	2.2	4.1	5.5

Source: Industrial Research Unit Questionnaire, 1982.
[a] NR/DK = Not reported or do not know.

tion presented in Table VI-17 are similar to those in Table VI-16, with concentrations in the 0–10 percent category and a large percentage of the respondents failing to answer the question.

CONCLUSION

There can be little doubt that the open shop gained ground in highway and heavy construction during the 1970s, but the extent of its gains are not clear. Based upon all the evidence presented, it appears that the gains made were significant, but less than those made in commercial construction. This may be attributed to the fact that highway and heavy construction have always been less unionized than commercial construction.

A key element in the future will be the impact of new Davis-Bacon Act regulations. The elimination of the "30 percent rule," described in Chapter II, is certain to make it more difficult for unionized contractors to win federally funded highway and heavy work, and this will further diminish the unionized sector. This is well illustrated by the letting of a Davis-Bacon covered contract in Alvin, Texas (Brazoria County). After both AGC and ABC chapters supplied data, the original union rates set by the U.S. Department of Labor for the construction of a municipal waste-water treatment plant were modified downward by as much as 43 percent per job classification, as shown in Table VI-18. Sixteen contractors, all non-union, submitted bids. Several probably would not have bid if they were compelled to pay the union rates—as much as $7 per hour higher than their regular rates—because of the difficulty of reducing rates for nongovernment jobs.[32] It is also likely that if the original rates had been maintained, union contractors would have bid. The lower rates will apparently save the government agencies involved between $300,000 and $500,000 in labor costs, or as much as 13 percent of the job's total cost.[33]

Despite changes in the Davis-Bacon Act's administration, "Little Davis-Bacon Acts" in Massachusetts, Rhode Island, New York, New

TABLE VI-18
Union vs. Open Shop Prevailing Rates[a]
Brazoria County, Texas, September 1983

Work Category	Union Contract Rates $	Open Shop Survey Rates $	Percent Decrease
Carpenters	18.52	11.96	35
Electricians	20.39	12.49	39
Iron workers	18.96	12.60	34
Laborers	13.72	7.84	43
Painters	17.92	12.16	32
Pipefitters	20.08	12.85	36
Truck drivers	14.04	12.38	12
Backhoe operators	18.78	11.97	36

Source: *Engineering News-Record,* October 13, 1983, p. 66.
[a] All rates include base wages and fringe benefits required under the Davis-Bacon Act.

---

[32] "Wage Protest Cuts Job Cost 13%," *Engineering News-Record,* October 13, 1983, p. 66.

[33] *Ibid.*

Jersey, Pennsylvania, Ohio, Indiana, Illinois, and California will probably continue to bolster union power and help them to maintain a large majority of highway and heavy work. One may expect, therefore, considerable variation from state to state in open shop penetration of highway and heavy work, with the probability of the open shop continuing to gain and to maintain a clear majority within the next decade, provided that the Davis-Bacon regulations are not once again revised and slanted in the unions' favor.

One other factor that has already aided open shop contractors, and is likely to continue to do so, is the widespread prosecution and conviction of companies that engage in bid rigging and conspire to divide up work in the highway sector. The net effects of these prosecutions have been to reduce job costs, to eliminate older, more inefficient companies, and to encourage new companies to enter the business.[34]

This is not to imply that such collusion was confined to unionized concerns. Nevertheless, the emphasis on improved productivity and lower costs inherently favors the open shop companies, which have lower labor costs, fewer featherbedding practices, and greater operating flexibility. The leading union in the highway and heavy sector, the Operating Engineers, is also one of the leaders in pushing restrictive and featherbedding work practices. Careful scrutiny of management bidding practices, great pressure on reduced costs, and the Davis-Bacon Act's administration rules that are less favorable to unions, clearly give the overall advantages to growth to open shop contractors. As is the case in other sectors of construction, this advantage remains despite concessions and new agreements, such as the one covering sixteen states signed in late 1983 between a National Joint Heavy and Highway Construction Committee (composed of the Operating Engineers and five other crafts) and contractors willing to work under it. This agreement bars strikes, eliminates travel time to jobs, reduces overtime to time and one-half in most instances, and provides for work assignment "on the

[34] A number of contractors have been jailed and many more fined. See, e.g., Steven Flax, "The Crackdown on Colluding Roadbuilders," *Fortune,* Vol. 108 (October 3, 1983), pp. 79–85; "Contractors Go to Jail for Highway Bid-Rigging," *Engineering News-Record,* September 29, 1983, pp. 42–43; and "How the Justice Dept. is Bagging Highway Bid-Riggers," *Business Week,* July 4, 1983, p. 89. Unfortunately, such antisocial action is confined neither to the highway sector nor to this country. See, e.g., "WPPSS to File Bid-Rigging Suit Against 6 [electrical] Firms," *Wall Street Journal,* August 15, 1983, p. 4; "Electrical Firms Indicted for Rigging Bids on Plants," *Engineering News-Record,* January 20, 1983; Selwyn Raab, "Investigators Charge Contractors Formed Club for Tax Fraud," *New York Times,* July 6, 1983, pp. A1, B10; and "German Bid-Riggers Nailed," *Engineering News-Record,* October 13, 1983, p. 67.

basis of traditional work jurisdictional lines," but permits the use of "composite crews" based upon "the amount of work involved for each union."[35] In short, it provides some relief for unionized contractors but still leaves the workforce less efficiently organized than that under open shop contractors.

[35] The text of this agreement is found in *Construction Labor Report,* Vol. 29 (November 30, 1983), pp. 1137–42. See also "Highway Labor Pact Signed," *Engineering News-Record,* December 15, 1983, pp. 58–59.

CHAPTER VII

Regional Variations

Having discussed the extent of open shop penetration by industry sector, we now turn to its geographic variations.

BACKGROUND AND EARLY STUDIES

Patterns of open shop construction activity tend to vary along two major dimensions. The first, discussed at length in Chapters III through VI, is the size and character of the construction project. The second is its location. Although these two factors are not completely unrelated, they do appear to exert independent influences on the ability of open shop contractors to capture construction work. Thus, certain branches of the industry exhibit relatively less unionization throughout the country (homebuilding is the prime example), and certain geographic regions are more hospitable to open shop activity across most or all categories of building (such as the Southeast). Although union strength in the aggregate has ebbed over the years, rankings along both these dimensions have been stable.

Geography affects open shop competition in two ways. The first is through variations in the degree of unionization among regions and states. The major concentrations of union strength have historically been found in two regional swaths: one ranging from southern New England southward through Pennsylvania and westward along the Great Lakes, and one along the West Coast. Open shops, on the other hand, have traditionally prospered most in the southern and border states, from Maryland to Texas. These generalizations must be viewed as statements of aggregate trends, however, not as unvarying patterns across multistate regions. Even in the South, for example, there are still islands where the challenge to union dominance has been relatively limited (e.g., the Beaumont, Port Arthur, and Orange areas, called the "Golden Triangle" of Texas).

The second effect of geography concerns urbanization. As a general proposition, open shop penetration varies inversely with the size of a metropolitan area and proximity to it. (There is, of course, a connection between the regional patterns and urbanization; key states in the Northeast, Midwest, and West have relatively high levels of population concentration.) Even in heavily unionized metropolitan centers, nonunion contractors are often active in their fast-growing suburban and exurban communities. What building work exists in rural areas is predominantly open shop. And in the cities themselves, open shops have won even the largest projects in small and medium-sized communities and have recently successfully invaded such major cities as Baltimore, Washington, and Atlanta.

Historical Findings

The variations just described were evident nearly four decades ago and, although open shop contractors have made discernible inroads across the board, the basic configurations do not appear to have changed until recently. In 1936, the U.S. Bureau of Labor Statistics (BLS) survey, discussed in previous chapters, found that the proportion of union workers in building construction ranged from 41 percent in the West South Central states (Texas, Oklahoma, Louisiana, and Arkansas) to 82 percent in the Middle Atlantic states (New York, New Jersey, and Pennsylvania). Similarly, the proportion of union workers was significantly higher in larger cities than in smaller ones. The results of the BLS survey are summarized in Table VII-1.

These basic geographic configurations of open shop construction have been regularly acknowledged in the literature. Thus, Haber and Levinson state that "geographically, the centers of greatest [union] strength are in the larger cities of the North, East, and Far West; the smaller cities, particularly in the South are very weakly organized."[1] Mills, writing sixteen years later, echoes the point: "Geographically, union organization is greatest in major cities, least in suburban and rural areas and the South."[2] The literature does not give an explicit analysis of the reasons for these patterns. Yet, because the patterns themselves have been remarkably stable over time, the relative success of open shop contractors in certain types of localities and regions is surely not randomly determined. Such an analysis is attempted later in this chapter, but it will be preceded by a detailed examination of the current state of the building trades

---

[1] William Haber and Harold M. Levinson, *Labor Relations and Productivity in the Buiding Trades* (Ann Arbor: University of Michigan, Bureau of Industrial Relations, 1956), p. 34.

[2] Daniel Quinn Mills, *Industrial Relations and Manpower in Construction* (Cambridge: MIT Press, 1972), p. 17.

TABLE VII-1
Union and Nonunion Workers, Building Construction, by Region and Size of City, 1936

Region	Union Workers	Nonunion Workers	Percent Union
United States	126,014	60,131	67.7
New England	8,832	4,390	66.8
Middle Atlantic	37,783	8,123	82.3
E.N. Central	30,796	10,384	74.8
W.N. Central	8,817	4,840	64.6
S. Altantic	12,223	13,455	47.6
E.S. Central	3,743	4,955	43.0
W.S. Central	4,609	6,669	40.9
Mountain	3,288	1,457	69.3
Pacific	15,923	5,858	73.1
City size			
1,000,000 and over	44,042	12,231	78.3
500,000-1,000,000	34,738	8,515	80.3
250,000-500,000	21,741	13,555	61.6
100,000-250,000	14,969	15,214	49.6
50,000-100,000	6,928	7,031	49.6

Source: Edward P. Sanford, "Wage Rates and Hours of Labor in the Building Trades," *Monthly Labor Review,* XLV (August, 1937), Tables 1 and 5.

unions across the nation. The discussion will be based, as in the previous chapters, on the findings of various surveys, on interviews conducted by the Industrial Research Unit staff in the periods 1972 to 1974 and 1981 to 1983, and on the 1974 and 1982 questionnaire results.

OPEN SHOP ACTIVITY IN THE NORTHEAST[3]

In the Northeast, perhaps more than elsewhere, an especially sharp distinction must be drawn between the populous, highly industrialized areas surrounding major cities and the large stretches of territory between. Union strength diminishes as construction moves farther from large population centers. Perhaps even more important, the open shop has penetrated into the perimeters of union strongholds, particularly since 1975.

New England

In sparsely populated northern New England, the preponderance of construction work has been taken over by open shop contractors.

[3] For purposes of this discussion, the Northeast includes the New England states, plus New York, New Jersey, and Pennsylvania.

In New Hampshire, for example, only six of the forty-seven members of the Associated General Contractors (AGC) were signatory to labor agreements by the summer of 1973[4]; since then, this state has gone almost completely open shop, except for atomic power and a few paper mill projects. In Vermont, the unions have become so weak that their apprenticeship programs could not be sustained; contractor members of the AGC there have been obliged to initiate and develop unilateral training programs. Vermont open shop contractors were also able to mount successful challenges to Davis-Bacon wage determinations before the recent administrative rule change by showing that the union rates were by no means "prevailing," even under the 30 percent rule. The picture for the unions has been equally bleak in Maine. Even the largest projects are likely to go nonunion unless the buyer, for his own purposes, insists on running the job with union workers. Two major paper mills were built union in the early 1970s, but in the 1980s the paper mills in Maine have been built nonunion.

In southern New England, union strength is concentrated in Boston and Providence. The Boston labor market is ringed by two highways: Route 128 about 10 miles from the city and Route 495 about 25 miles away. In the area between, perhaps the fastest growing section of the metropolitan region, a number of large projects have been won by open shop contractors. Some open shop contractors confided that they are reluctant to take on jobs within the Route 128 perimeter because of anticipated confrontations with unions, but the Associated Builders and Contractors (ABC) has targeted Boston as a city that can be opened.[5] Open shop contractors have performed several major nursing home, hospital construction, and other jobs in Boston since 1970.

Nevertheless, construction within the Route 128 perimeter remains heavily union, as it does in the city of Providence, Rhode Island. Many unionized contractors operating in Boston and Providence have doublebreasted affiliates elsewhere. A survey of construction in Rhode Island from 1983 to 1984 by the state chapter of the AGC found that approximately 24 percent of the work was being performed open shop—"a far greater percentage than ex-

[4] It is important to remember here that even jobs performed by open shop general contractors may include the services of one or more unionized subcontractors. In some areas of New Hampshire, for example, the Bricklayers union has maintained much of its strength with masonry firms, which themselves contract with nonunion general contractors. Occasionally elsewhere, but often in the South, unionized general contractors subcontract to open shop specialty contractors.

[5] "ABC Urban Push Hits Boston," *Engineering News-Record,* March 18, 1982, pp. 33–34. See Chapter IX for the union reaction to this push.

pected"—at a time when state building construction work had declined by 72 percent.[6] In Connecticut, significant gains by open shop builders induced the construction unions and unionized contractors in 1982 to form "a cooperative group fostering more union construction work through improved productivity and reduced jurisdictional disputes . . .,"[7] but results were not discernible by early 1984. Two large open shop contractors are headquartered in New England: Pizagalli Construction Company, South Burlington, Vermont, No. 128 on the 1983 *Engineering News-Record* 400 list, with contracts totalling $129.2 million; and Cianbro Corporation, Pittsfield, Massachusetts, No. 148, with contracts totalling $109.4 million.[8] The ABC has long had a large and flourishing chapter in New England, and several AGC chapters which once served only unionized contractors now generally admit open shop contractors and provide services for them.

The growth of open shop construction in New England is reflected in the 1969 and 1973 membership surveys of the AGC, as well as in the Industrial Research Unit 1974 and 1982 questionnaires. Over 76 percent of the AGC respondents in the region identified themselves as unionized in 1969; the percentage fell to less than 60 in 1973. (The vast majority of nonunion general contractors do not belong to the AGC.) In the Industrial Research Unit 1974 survey, the median estimate by unionized contractors of the open shop share of all construction was about 47 percent; the median estimate by nonunion contractors was about 65 percent. Twelve percent of the respondents felt that the open shop share had grown significantly in recent years; only about 2.5 percent believed that it had declined.

The 1982 survey responses are set forth in Table VII-2. For New England, only 62 unionized and 111 nonunion contractors provided estimates of the proportion of open shop work in all construction. Of the union respondents, 59.7 percent put nonunion construction above 40 percent; of the nonunion respondents, 67.5 percent made the same estimate, with much higher percentages in the 61 to 80 and 81 to 100 categories. The small response and the absence of homebuilders make it difficult to draw conclusions from these responses, which appear to understate the extent of open shop penetration as our field interviews found it.

---

[6] Associated General Contractors of America, Rhode Island Chapter, *Special Bulletin,* April 4, 1984, p. 2 and Exhibit E.

[7] "Statewide Labor-Management Promotion Group Is Being Formed in Connecticut," *Construction Labor Report,* Vol. 28 (March 10, 1982), p. 33; and "First Statewide Cooperative Accord Signed in Connecticut," *ibid.,* Vol. 29 (November 2, 1983), p. 1173.

[8] "The Top 400 Contractors," *Engineering News-Record,* April 28, 1983, pp. 75–76.

TABLE VII-2
Percent Distribution of Respondents' Estimates of the Proportion of Open Shop Work in All Construction by Region and Union Status, 1982

Region	Union							Nonunion						
	Number	NR/DK[a]	0–20	21–40	41–60	61–80	81–100	Number	NR/DK[a]	0–20	21–40	41–60	61–80	81–100
Not Reported	44	34.1	4.5	15.9	36.4	9.1	0.0	49	30.6	0.0	4.1	20.4	26.5	18.4
New England	62	12.9	6.5	21.0	51.6	8.1	0.0	111	25.2	2.7	4.5	34.2	18.9	14.4
Middle Atlantic	275	18.9	18.9	17.5	34.9	9.5	0.4	228	27.6	7.0	12.3	26.3	20.2	6.6
E.N. Central	474	12.7	26.6	28.5	27.0	4.6	0.6	205	21.5	5.9	13.7	33.2	20.5	5.4
W.N. Central	220	16.8	19.5	24.1	28.2	7.7	3.6	157	16.6	6.4	13.4	15.3	26.1	22.3
S. Atlantic	109	17.4	11.0	6.4	23.9	30.3	11.0	550	22.5	4.2	2.0	9.5	26.5	35.3
E.S. Central	69	14.5	4.3	8.7	33.3	29.0	10.1	111	17.1	2.7	4.5	13.5	30.6	31.5
W.S. Central	105	9.5	1.9	15.2	36.2	31.4	5.7	284	20.4	3.2	5.3	14.8	37.0	19.4
Mountain	108	16.0	4.6	13.9	38.9	20.4	2.8	184	15.8	2.7	6.5	28.3	34.2	12.5
Pacific	362	16.0	17.4	29.8	31.2	4.7	0.8	143	18.9	5.6	9.8	37.1	20.3	8.4
Total	1,828	15.9	17.1	22.3	31.5	10.9	2.4	2,022	21.4	4.4	7.0	20.5	26.7	20.0

Source: Industrial Research Unit Questionnaire, 1982.
[a] NR/DK Not reported or do not know

The National Electrical Contractors Association (NECA) data, cited in the previous chapters and set forth by region in Table VII-3, also emphasize the declining state of unionized construction in New England. In 1973, 16 percent of the 2,839 inside electrical employers in New England were unionized by the International Brotherhood of Electrical Workers (IBEW); by 1980, only 11 percent of the 2,996 employers were so unionized. The largest electrical employers are the most likely to be unionized. In 1973, 40 percent of the 20,339 inside electrical employees were IBEW members; by 1980, only 30 percent of a total of 19,492 were in the IBEW.[9]

Middle Atlantic

The state of New York is heavily nonunion. Between unionized New York City, where repair work and some housing outside Manhattan are likely to be open shop, and Buffalo, where open shop contractors have carved a niche, the open shop clearly dominates. This domination began in the late 1960s and early 1970s. Unionized employers formed the Construction Employers Labor Relations Association, which vainly sought to stem the tide and, failing, went defunct. Areas like Syracuse, Rochester, and Albany went open shop, except for state-financed work built under the state's "little Davis-Bacon Act," which was made even more restrictive and pro-union by 1983 amendments.[10] Open shop construction also made a significant entry into Westchester County, north of New York City, beginning in 1974. The picture drawn of New England—union predominance only in a few strongholds—is equally true of New York. As a result, the ABC has grown very strong in the state and AGC chapters admit and service nonunion contractors.

The union situation in New Jersey varies with the distance from New York City and Philadelphia. In southern New Jersey, the picture is much the same as in upstate New York, with much open shop activity. The major locus of union strength is in the counties which form part of the metropolitan Philadelphia area, although even there open shop activity is significant. The unions are stronger in the northern part of the state, especially in the counties included in the New York City metropolitan area. Yet open shop firms are sufficiently numerous in northern New Jersey to have formed an active chapter of the ABC. Spokesmen for unionized contractors in

---

[9] National Electrical Contractors Association, *Inside Electrical Construction Survey for the Years 1973–1979 and 1980,* March 2, 1981. All electrical data (See Table VII-3) in this chapter is from this source. See Chapter IV for discussion of these data.

[10] See Armand J. Thieblot, Jr., "Prevailing Wage Laws of the States," *Government Union Review,* Vol. 4 (Fall 1983), pp. 3–65.

TABLE VII-3

IBEW and other Inside Electrical Employers and Employees By Region, 1973 and 1980

Region	Inside Electrical Employers						Inside Electrical Employees					
	1973			1980			1973			1980		
	Total	Percent IBEW	Percent other	Total	Percent IBEW	Percent other	Total	Percent IBEW	Percent other	Total	Percent IBEW	Percent other
New England[a]	2,839	16	84	2,996	11	89	20,339	40	60	19,492	30	70
Middle Atlantic[b]	5,749	29	71	6,076	22	78	56,768	64	36	54,047	61	39
Midwest[c]	8,965	44	56	11,557	32	68	85,065	58	42	100,745	54	46
South[d]	8,685	21	79	11,864	13	87	104,802	30	70	126,765	27	73
Border[e]	1,811	22	78	2,256	16	84	21,245	38	62	24,417	34	66
Rocky Mountain[f]	1,878	40	60	3,009	21	79	19,680	54	46	26,549	40	60
Pacific[g]	3,960	75	25	5,492	45	55	36,305	66	34	55,253	48	52

Source: National Electrical Contractors Association, Annual Survey.

[a] Maine, Vermont, New Hampshire, Massachusetts, Connecticut, and Rhode Island

[b] New York, New Jersey, and Pennsylvania

[c] Ohio, Michigan, Indiana, Illinois, Wisconsin, Minnesota, Iowa, Missouri, Kansas, Nebraska, North Dakota and South Dakota

[d] Virginia, North Carolina, South Carolina, Georgia, Florida, Tennessee, Alabama, Mississippi, Arkansas, Louisiana, and Texas

[e] Delaware, Maryland, West Virginia, Kentucky, and Oklahoma

[f] Arizona, New Mexico, Utah, Nevada, Colorado, Wyoming, Montana, and Idaho

[g] California, Oregon, and Washington. (No data included for Alaska and Hawaii).

New Jersey have repeatedly called attention to the threat of open shop competition.

In Pennsylvania, union strength is greatest in the metropolitan centers of Pittsburgh, Philadelphia, and, to a lesser extent, Harrisburg. Although low-rise residential construction in and around major cities is almost entirely open shop, nonresidential building activity within the cities has remained largely union-dominated. As in other areas, open shop competition varies directly with distance from the central cities. Despite violence perpetuated on the Philadelphia construction scene, open shop contractors are now building major projects in the nearby suburbs and performing smaller work within the city. Schnabel Associates, headquartered in the northeast Philadelphia suburb of Kulpsville, ranked No. 263 on the 1983 *Engineering News-Record* 400 list with contracts totaling $64.5 million.[11]

The ABC has several flourishing chapters in the state, and even the Philadelphia AGC chapter was changing its bylaws in early 1984 preparatory to enrolling open shop members. While much highway work was going union shop, or to the Steelworkers, Lancaster and Berks counties, seventy miles west of Philadelphia, have become almost completely open shop. The only union work there appears to have been generated by Davis-Bacon, or "little Davis-Bacon" regulations; this will change with the new U.S. Department of Labor regulations. Again the big population centers are union, but the open shop is very strong in the less populated areas. The union shop companies also do the large industrial jobs in the Middle Atlantic region.

Despite the fact that the large AGC chapters in Philadelphia and Pittsburgh were all union, the 1973 survey of AGC members showed more than 10 percent of the Pennsylvania respondents reporting themselves as open shops, compared to only 2.5 percent in 1969; these surveys did not include most of the members of the several ABC chapters based in the state.

Thus, although the Middle Atlantic states have historically constituted one of the most strongly unionized areas of the nation, they have not been immune from the trend of erosion of union strength. The 1973 AGC survey found almost 20 percent of the respondents in New York and New Jersey to be open shop. In the 1974 questionnaire study by the Industrial Research Unit, more than one-half of the union respondents from the Middle Atlantic area estimated open shop work at more than 40 percent of the total.

[11] "The Top 400 Contractors," p. 83.

Among nonunion respondents, this proportion was greater than 80 percent. On the question of the growth of open shop activity in the years before 1974, about one-fifth of union respondents felt that open shop work had increased significantly, compared to only about 3 percent who believed it had declined. It seems clear that by 1974 the building trades had received a stiff challenge even in geographic and industrial sectors where they had been relatively secure.

Responses to the 1982 Industrial Research Unit questionnaire, shown in Table VII-2, support this conclusion. Of the 275 unionized respondents, 44.8 percent placed open shop penetration at 40 percent or more in the three Middle Atlantic states, as did 53.1 percent of the nonunion respondents; again, many more nonunion respondents believed that the open shop controlled a much higher proportion of the work. As homebuilders were excluded from the survey, these are significant results for a historically union region.

The NECA survey (Table VII-3) for the Middle Atlantic region shows that in 1973, only 29 percent of the 5,479 inside electrical employers in the Middle Atlantic region had been IBEW organized. By 1980, this ratio had dropped to 22 percent of 6,076 employers. On the other hand, the IBEW had unionized 64 percent of 56,768 electricians in the study in 1973, and 61 percent of 54,047 in 1980. These latter data reflect the strong union position in large-city high-rise office building, power plant construction, and other huge projects in the three states.

In the sheet metal craft, a study prepared for the Sheet Metal and Air Conditioning Contractors National Association (SMACNA) found unions to be strong but losing ground in Pennsylvania. Figure IV-8 shows that unionized contractors had about 75 percent of the work in 1977, but only about 58 percent in 1982. In the Northeast, which was defined to include Delaware and Maryland in addition to the New England and Middle Atlantic states, the Mechanical Contractors Association of America (MCAA) found a steady decline in the unionized contractors' share of mechanical work in all sectors of construction and predicted that the open shop would have a clear majority of all work in the region by 1985 (see Table IV-7).

OPEN SHOP ACTIVITY IN THE MIDWEST[12]

The Midwest is a markedly heterogeneous economic region, ranging from the agricultural states of the nation's heartland to the

[12] The Midwest includes the following states: Ohio, Michigan, Indiana, Illinois, Wisconsin, Minnesota, Iowa, Missouri, Kansas, Nebraska, North Dakota, and South Dakota. These are the East North Central and West North Central states in the U.S. Census regions.

heavily industrialized states along the Great Lakes. In the rural and farm states, open shop construction has historically been substantial; as shown in Table VII-1, the East North Central region, from Ohio to Wisconsin, was appreciably more unionized in 1936 than were the less industrialized West North Central states. Later membership surveys by the AGC tell much the same story: the vast majority of respondents from Ohio, Michigan, Illinois, Indiana, Wisconsin, Minnesota, and Missouri were union contractors, but those from Iowa, Kansas, Nebraska, North Dakota, and South Dakota were preponderantly open shop. The AGC data, summarized in Table VII-4, also show a swing to the open shop between 1969 and 1973 in both unionized and open shop states. The sole exception was North Dakota, from which the number of respondents was too small to be meaningful.

East North Central

Given the national trends outlined in earlier chapters, the building trades unions have not made significant inroads into areas where their strength has long been modest. Our concern here is with states and areas where unions have traditionally been strong. Our most detailed information comes from Ohio and Michigan, but field work in 1973 and 1982 was conducted in Indiana, Illinois, Iowa, and Missouri. Although it is always hazardous to attempt to generalize from a limited sample, these states and the major metro-

TABLE VII-4
Proportion of Union Respondents in AGC Membership Survey, 1969 and 1973

State	1969 Total Number	1969 Percent Union	1973 Total Number	1973 Percent Union
Ohio	80	98.8	267	86.9
Michigan	114	89.5	151	84.1
Indiana	85	95.3	135	84.4
Illinois	161	98.1	273	90.5
Wisconsin	60	96.7	101	90.1
Minnesota	99	82.8	172	74.4
Iowa	125	34.4	220	28.6
Missouri	79	92.4	173	85.5
Kansas	81	49.4	173	31.2
Nebraska	40	42.5	n.a.	n.a.
North Dakota	38	7.9	59	11.9
South Dakota	33	12.1	68	10.3

Source: Associated General Contractors of America.

politan areas within them are probably not significantly different from other states in the Midwest whose populations are largely concentrated in one or a few major urban centers. Here is where the unions have exercised their strongest influence. Thus, a nascent erosion of union strength in Detroit almost certainly signifies a broader expansion of nonunion competition in the rest of Michigan. And if open shop contractors in Detroit are beginning to make their presence felt, it seems quite likely that the process is being repeated (or is about to be repeated) in Minneapolis, Milwaukee, St. Louis, and Kansas City.

There are seven important metropolitan centers in Ohio: Cleveland, Columbus, Cincinnati, Toledo, Akron, Dayton, and Canton. In general, open shop construction has made its greatest inroads in the central and southern portions of the state, especially in Columbus, the capital and second largest city. In the northern areas, particularly Cleveland and Toledo, nonunion competition is more modest but the open shop has made considerable gains since the mid-1970s. The residential sector, long a union province through agreements with the local homebuilders' associations, has experienced more open shop building as the population moves farther from the city and its immediate suburbs. As a result, the Cleveland homebuilders no longer deal with unions, although some of their contractor members do. There is also an increasing amount of residential construction at and beyond the limits of Cuyahoga County, which encompasses Cleveland. As we have seen in earlier discussions, this trend has usually been, and continues to be, accompanied by a burgeoning of open shop activity in ancillary commercial building, such as shopping centers, generated by residential development. There has been enough nonunion building in the Cleveland area for the ABC, headquartered in Columbus, to have opened first a small satellite office and then an ABC chapter there. Cleveland itself, however, remains predominantly union.

The case is different in Columbus. The ABC in Columbus has become an important force in the industry, with about 260 contractor members; 50 to 60 of them have union contracts. All of the union contractors, however, operate on a doublebreasted basis. The ABC estimates that a little more than one-half of all open shop contractors in the Columbus area are members of the association. As is generally the case, open shop contractors in Columbus have made substantial inroads in residential and small-scale commercial construction, although there is no universal agreement on just what constitutes "small scale." According to an ABC spokesman, the only specialties which open shops have not penetrated are glazing and elevator construction. Both open and union shop groups now concede that the open shop has one-half of the commercial work in the Columbus area.

As the open shop has expanded, the ABC has also opened a chapter for southwestern Ohio, covering Dayton and Cincinnati, which is now larger than the Columbus chapter. In these areas, as well as in Columbus, the growth of open shop construction has led construction unions and unionized contractors to join in cooperative arrangements to reduce costs and improve productivity. These efforts have ameliorated the situation somewhat from the union point of view, but the open shop sector continues to expand, particularly in commercial construction.

In other areas of northern Ohio, the situation is similar to that of Cleveland. The unions, bolstered by the administration of the state's "little Davis-Bacon Act," remain strong in highway construction; they also dominate major industrial construction, although the open shop has built power plants in Ohio.

Michigan, unlike Ohio, has only one true major metropolitan center, Detroit. Detroit is the quintessential "union town," with an economy which relies heavily on durable goods manufacturing related to automobile production. Outside the Detroit Standard Metropolitan Statistical Area (SMSA), open shop construction is widespread. The labor relations director for the Detroit chapter of the AGC estimated in 1973 that the nonunion share of construction work "outstate" was perhaps one-half of the total. Today it is closer to 75 percent. Some unionized general contractors in Detroit have open shop arms which operate outside the jurisdiction of the Detroit-based unions. The rapid growth of the open shop share in Michigan in recent years has sparked considerable violence by the construction unions. Detroit remains a union town, and the Big Three automobile companies always build union. Nevertheless, homebuilding in the suburbs has turned open shop. The Detroit labor market includes four outside counties in which there is no union homebuilding. As in most metropolitan areas, there is much more residential building in the outer reaches than in economically depressed Detroit itself; homebuilding in the Detroit area is therefore heavily nonunion.

A growing ABC chapter reflects the changes in Detroit. A sizable portion of its membership comes from specialty contractors, particularly in the electrical industry, which has seen significant gains by open shop contractors.

Illinois remains a heavily unionized state. In the Chicago SMSA, including the outlying counties, even residential construction is unionized, but there is now some open shop homebuilding about thirty miles from the city's boundaries. In the southern part of the state, the unions have moved out from their East St. Louis base to maintain strong control, although recently Schnabel Associates built open shop subsidized housing in East St. Louis with Davis-Bacon wage rates. Peoria, in central Illinois, and Moline, farther north and west, are completely union towns, including residential construction. Unions and union contractors in Illinois are prime supporters of the state's Democratic political organization, including the Chicago machine; they contribute heavily to it and in turn receive its strong support. Throughout the state, highway and heavy construction is highly unionized.

Indiana presents a mixed picture. Gary and other communities near Chicago are heavily union, as is Evansville on the sourthern border. Indianapolis, however, which was once union for commercial and industrial work, but open shop for homebuilding and garden apartments (under four stories), has seen open shop gains in commercial work and high-rise apartment building. Many small towns and rural areas are entirely open shop. Highway and heavy work is divided, with the major jobs traditionally union but with recent significant open shop inroads.

Iowa's union condition varies with the distance from Illinois. Mississippi River towns like Davenport and Burlington are largely union, but as construction moves west union strength declines heavily in all branches of work.

Missouri boasts two of the country's strongest union cities, Kansas City and St. Louis, but the area between has seen a rising open shop movement. The ABC has established a chapter in Kansas City, but it is still small in numbers and job size. To offset potential open shop gains, unions and unionized contractors in Missouri have established a joint organization aimed at reducing costs and improving productivity. The effort seems to have reduced the number of strikes in the state's construction industry.

West North Central

The Dakotas and Nebraska (primarily agricultural except for Omaha) are also largely open shop. The installation of a missile site or similar major government construction project brings in a union job, but union influence is temporary. Highway work is usually open shop in these states. The huge Great Plains synfuels project in North Dakota is being built union; whether this will change the area remains to be seen. Synfuels work in Colorado has been done open shop.

Questionnaire and Survey Results—Midwest

In the Midwest generally, the respondents to the 1974 Industrial Research Unit survey were divided on the extent of open shop gains. The median estimate by unionized firms of the nonunion share of construction was about 29 percent; among open shop respondents the median estimate was 60 percent. Among both groups, most of those who perceived any significant trend (about 12 percent of the total) saw it moving in the direction of a larger open shop share in the years before 1974.

The 1982 questionnaire results give evidence of this trend. Only 32.2 percent of the 474 unionized respondents in the East North Central region estimated nonunion gains at 41 percent or higher, but 53.1 percent of the 228 nonunion respondents put the open shop share above 40 percent, including 20.2 percent of respondents who felt that the penetration was in the 61–80 percent category. For the West North Central region, both unionized and nonunion respondents agreed that the open shop penetration was higher. Forty percent of the 220 unionized respondents and 63.7 percent of the 157 nonunion ones put the open shop share at 41 percent or more, and 11.3 percent of the union respondents but 48.4 percent of the nonunion ones put the open shop penetration at 61 percent or more. The divergency of replies between the union and open shop respondents is probably based upon the location of the contractors involved. In any case, the Midwest has seen a fundamental diminution of union strength in recent years. (See Table VII-2.)

The NECA electrical trade data (Table VII-3) provide an interesting view of the mixed open shop/union shop situation in the Midwest. In 1973, 40 percent of the 9,772 inside electrical employers were under IBEW contracts; by 1980 this figure had slipped to 29 percent of 12,553 employers. The NECA employee data, however, demonstrate that union electrical construction for large projects is still strong in the Midwest. In 1973, 45 percent of the 94,524 employees covered by the NECA data were IBEW members. This percentage rose to 55 in 1974 and stayed above that mark until 1980, when 49 percent of the 110,090 employees were IBEW members. This is the only statistic in the NECA data that showed increased union strength, but even here the trend was downward by 1980.

In Indiana's sheet metal trade, the SMACNA study found that unionized firms dropped from 80 percent in 1977 to less than 65 percent in 1982, as shown in Figure IV-8. The data in Table IV-7 show that the union mechanical contractors had a substantial majority of all work except residential and garden apartments in the six heavily industrialized states of the Midwest in 1977.[13] By 1982, they had lost this lead in service and maintenance, and rehabilitation and remodeling. The MCAA originally predicted that by 1985 nonunion contractors would have a majority of commercial work, as well as 44 percent of industrial work, 46 percent of institutional work, and 48 percent of federally funded work in these states. The second MCAA study, however, found that these predictions were too conservative. By 1983, the open shop had won 46 percent of industrial work, 49 percent of institutional work, 63 percent of commercial work, and 50 percent of federally funded work, while increasing its hold in other categories. The specialty contractor data thus support the conclusion that open shop contractors have made gains in all sectors in the Midwest, but large-city high-rise office buildings and the largest industrial projects are still built union.

OPEN SHOP ACTIVITY IN THE SOUTHERN AND BORDER STATES

The South is the most open shop area in construction.[14] Union influence is weak, and the South is the home of most of the largest open shop contractors—Brown & Root, Daniel International, Austin, BE&K, Blount, J. Ray McDermott, H. B. Zachry, and others—who have grown strong in that region and then branched out elsewhere. The border states present a more mixed picture.

The South

There are union enclaves in the South. Central city office building in most cities, the "Golden Triangle" of Beaumont, Port Arthur, and Orange in Texas, and some other centers remain union. The unionized contractors, however, have seen these enclaves steadily diminish in the 1970s. The once all-union area of Galveston-Texas City, Texas, was breached when Brown & Root won major contracts to rebuild and modernize an Amoco refinery. The once all-union areas of Lake Charles and Baton Rouge, Louisiana, have also been successfully entered by open shop contractors, despite violence.[15] Atlanta, which was described as a union construction city in the

[13] Minnesota, Wisconsin, Michigan, Illinois, Indiana, and Ohio.

[14] The South is defined as Virginia, North Carolina, South Carolina, Georgia, Florida, Tennessee, Alabama, Mississippi, Arkansas, Louisiana, and Texas. Oklahoma, Kentucky, West Virginia, Maryland and Delaware are "border states." These states comprise the Southeast, East South Central, and West South Central states in the U.S. Census regions.

[15] See Chapter IX for details concerning the Lake Charles assaults.

1975 edition of this book, was an open shop success story by 1983. As late as 1981, ABC members had built only three projects in the Atlanta metropolitan area; by the end of 1983, forty-three were being built by its members, and the unionized share of construction volume had fallen from 70 percent to 10 percent.[16] Atlanta was the last stronghold of the unions in Georgia. In Savannah, the Sheet Metal Workers' union was eliminated after a long strike over wages and work rules, although the union workers apparently have found work in power plant construction outside the city.[17]

Construction in Florida has been a mixture of open shop and union shop. Before the construction of Disney World, Sea World, and Circus World, the central Florida area around Orlando was heavily open shop. Disney brought in the unions under a project agreement.[18] Because of the large amount of work, union workers migrated to the area, open shop firms signed union agreements, and the area became heavily unionized. Once these major jobs were completed, however, union work began to wane and the open shop regained its ascendancy. Then came Disney's EPCOT, and another union project agreement, but again a decline in unionism when the project ended. The changeovers have been violent; the decline in union work in central Florida was exacerbated by the precipitate fall in heavily union space work at nearby Cape Canaveral.

The west coast of Florida, including Tampa, is primarily open shop in the highway and residential sectors, including condominiums and apartments, and also in commercial work. Heavy industrial and power plant and large commercial work is likely to be union, although it has not been exclusively so.

Southeast Florida has seen considerable open shop encroachment on once strong union territory. Unions have lost most of the condominium and apartment work, as well as most of the highway and heavy work (except in the Miami area); and they have probably lost most of the commercial work as well.

The ABC has made some of its strongest inroads in Florida. In southeast Florida, the ABC has led the fight against union violence and has benefited considerably from this activity. Not only have open shop contractors joined the ABC, but others have decided to

---

[16] "Atlanta Building Trades' Project Phoenix Is Response to ABC's 'Open Shop War,' " *Construction Labor Report,* Vol. 29 (November 30, 1983), pp. 1326–27.

[17] "Savannah-Area Sheet Metal Workers May Drop Union Representation," *Construction Labor Report,* Vol. 29 (September 14, 1983), pp. 959–60; and "Savannah Sheet Metal Workers Vote Out Union in Decert Election," *ibid.,* Vol. 29 (October 12, 1983), p. 1072.

[18] Project agreements are discussed in Chapter VIII.

turn open shop as a result of the ABC's efforts. The ABC has two other Florida chapters, one on the west coast, and one in the central area. Both are growing rapidly and stimulating open shop activity in their areas.

Alabama was once one of the strongest unionized states in the South, but this has changed dramatically. Open shop contractors, supported by an enterprising ABC chapter, have made deep inroads in the once all-union Birmingham area; they now work in the central city as well as in the heavily open shop outlying areas. The takeover of Mobile construction by the open shop contractors was discussed in Chapter V. The repeal of the state's "little Davis-Bacon Act" has aided these developments. Mississippi has never been a union stronghold.

Tennessee unionism owes much to the Tennessee Valley Authority (TVA) which brought construction union strength to that state and has nurtured it with high wages ever since. Knoxville and Chattanooga are union towns, but Nashville and Memphis have seen unions lose heavily in recent years. Key contractors have broken with unions and the ABC is now strong in the state. Daniel built the new Nissan plant near Nashville, the only automobile plant built open shop in memory. The ABC has established an office in Knoxville, indicating that inroads are being made there for the open shop. Highway and heavy construction is largely union, less so in the central and western sector. Arkansas is largely nonunion and is becoming more so since Memphis on its eastern border has shifted in that direction.

As noted, the once all-union Lake Charles and Baton Rouge areas of Louisiana have seen major open shop inroads. New Orleans is a mixed situation, with the unions holding central city high-rise and government work, but losing in the suburbs and other sectors. Texas, particularly Houston and the Gulf Coast, was extensively discussed in Chapters IV and V. The Dallas-Fort Worth area has continued to move strongly toward the open shop since 1975; high-rise office buildings have been built in central Dallas by open shop contractors. Highway and heavy work, homebuilding, and commercial and industrial work, outside of a few union enclaves, are all open shop. Oklahoma, a border state, is mostly open shop, with the unions losing in their former stronghold of Tulsa. Power plants, refineries, and most major work in Oklahoma have been built open shop, particularly by Brown & Root and Zachry.

The Carolinas are the most completely open shop states in the United States. Moreover, the unions have suffered significant losses there in recent years. DuPont, which operates many plants in the

Carolinas, formerly built union but has turned construction over to open shop firms. This practice has been followed by other major companies, such as Firestone Tire and Rubber and General Electric. Two major utilities—Carolina Power and Light and South Carolina Light and Power—traditionally gave their work to Ebasco, a large National Constructors Association member, since both Ebasco and the utilities were once subsidiaries of the Electric Bond and Share Company, a utility holding company. In recent years, however, this tie has been broken and much work has gone to Daniel and other open shop concerns. Duke Power Company has divided its work on fossil fuel plants between its nonunion in-house construction crews and union contractors, but the nonunion in-house forces have performed virtually all work on nuclear plant construction.

Charlotte, the largest city in the two states, has four large union contractors for commercial and industrial construction; in the Carolinas, however, these contractors build with doublebreasted open shop firms. The highway and homebuilding sectors in these states are open shop. A sizable number of large electrical and mechanical contractors in the state operate open shop.

The militant open shop AGC chapter for the Carolinas extends its jurisdiction through both states and covers all sectors of construction. It works very closely with specialty trade associations and encourages users to build open shop. As one of the largest AGC chapters in the nation, it affects AGC national organization and has contributed materially to altering the policies of the national AGC, from service almost solely to union contractors to the present policy of service to both union and open shop contractors.

In Virginia, unions are strong in the Richmond and Norfolk areas and among roadbuilders in the Washington, D.C., region. The District of Columbia area is otherwise heavily open shop, even for large office and apartment buildings, as is Washington itself, as described in Chapter IV. As in most southern states where recent industrial progress has been rapid, manufacturers building plants sometimes have insisted on union contractors, but high costs in recent years have prompted them to turn to open shop contractors. Virginia Electric Power Company, in which the large National Constructors Association member Stone and Webster has an ownership interest, opened up its work to open shop constructors in the 1970s, ending one area of strong union control. DuPont, however, has maintained a union contractor in Richmond primarily because of its satisfaction with his work, and Philip Morris builds union.

Small commercial construction, homebuilding, and highway work are largely open shop throughout Virginia. Our 1973 field work and all available data pointed to a decline in an already low union influence, and our field work in 1982 found this decline moving at an accelerated pace.

The Border States

West Virginia and Kentucky have both highly unionized and open shop sectors, as well as Steelworker strength in highway work. Strong mining and industrial unions support the construction unions (or the Steelworkers); yet the open shop is significant, particularly in the less populated areas and in small construction projects, and appears to be gaining strength. Paper manufacturing plant construction, once all-union, has now seen BE&K and Daniel win major work in both states.

The ABC was spawned in the border state of Maryland. Baltimore is less unionized than any major city to the North; large buildings in its center have been ABC built, although unions have a big share of this work. Throughout the state, which has three ABC chapters, the open shop is dominant in all but a few major commercial and industrial jobs. Delaware, on the other hand, was until recently unionized in the northern sectors bordering the Philadelphia SMSA, but open shop in its southern region. The failure of the unions to stop picketing a DuPont project, as described in Chapter V, and a long strike by the Sheet Metal Workers in 1983, which delayed work on one of the last all-union projects in the state and which was one of several strikes in the area characterized by violence and property damage,[19] could well mean continued losses for, if not the virtual elimination of, union construction in this small state.

Questionnaire and Survey Results—Southern and Border States

The 1974 Industrial Research Unit questionnaire did not field usable results concerning the open shop situation in southern and border states. The results of the 1982 questionnaire, as set forth in Table VII-2, clearly reflect open shop strength. Thus, for the South Atlantic region, 65.2 percent of the 109 unionized respondents placed open shop penetration at 41 percent or higher, and 41.3 percent at 61 percent or higher; corresponding figures for the 550 nonunion respondents were 71.3 percent and 61.8 percent. The fact that so large a percentage of respondents were nonunion—83 percent—further indicates the strength of the open shop in this region.

The responses of contractors in the East South Central and West

[19] "Delaware Sheet Metal Workers Win Work Preservation Clause in 11-Week Strike," *Construction Labor Report,* Vol. 29 (September 21, 1983), pp. 987–89.

South Central regions were similar. Of the 69 unionized East South Central respondents, 72.5 percent estimated that the open shop had at least 41 percent of the work and 39.1 percent estimated an open shop share of 61 percent or better; 75.6 percent of the 111 nonunion respondents were in the 41 percent or better category, 62.1 percent in that of 61 percent or better. For the West South Central group, the figures were equally strong: 73.3 percent of the 105 unionized respondents put open shop penetration at 41 percent or higher, and 37.1 percent at 61 percent or above; for the 284 nonunion respondents, the respective percentages were 71.2 and 56.4.

For the southern states (see note 15), the NECA data (Table VII-3) found that only 20 percent of the 8,985 inside electrical employers were unionized by the IBEW in 1973. By 1980, this figure had declined to 13 percent of 11,864. Again demonstrating that larger jobs tend to be union, even in the South although to a lesser degree than elsewhere, 30 percent of the 105,728 employees were IBEW represented in these states in 1973, and 27 percent of the 126,792 in 1980. For the border states (see note 15), 22 percent of the 1,811 employees had IBEW contracts in 1973; 16 percent of the 2,256 in 1980. Following the same trend, 38 percent of the 21,245 employees in these states were IBEW represented in 1973; 34 percent of 54,417 in 1980.

Figure IV-8 provides data from the SMACNA study in regard to unionized sheet metal contracting in Texas and Florida for 1977 and 1982. For Texas, unionized contractors had 40 percent of the work in the former year, but only 30 percent in the latter. For Florida, the data were almost 43 percent in 1977, 35 percent in 1982.

For mechanical work, Table IV-7 summarizes the regional results of the study by the Mechanical Contractors' Association for the Southeast and Southwest.[20] In the Southeast, unionized contractors had a majority of all mechanical work in 1977, except residential and garden apartments and service and maintenance. By 1982, the unionized group had lost this majority in high-rise apartments, commercial, institutional, and federally funded jobs as well, leaving industrial as the only sector where the union group controlled a majority; this sector went 57 percent open shop in 1983. This study estimated that by 1985 the open shop group will control the great majority of all the work in other categories in the Southeast. For

[20] The Southeast is defined as Arkansas, Louisiana, Kentucky, Tennessee, Alabama, Georgia, Mississippi, North Carolina, South Carolina, and Florida. The Southwest is defined as Arizona, New Mexico, Oklahoma, and Texas. We, of course, place Arizona and New Mexico in the Rocky Mountain region.

the Southwest, a similar trend was found toward clear open shop domination in all sectors by 1985; this trend was achieved in all but the institutional sector by 1982, and this sector became 53 percent open shop in 1983. These studies of specialty contractors confirm our findings that the open shop has continued to make gains in the regions where it is the strongest as well as entering areas where it had little influence.

OPEN SHOP ACTIVITY IN THE WEST[21]

The western states, especially those along the Pacific Coast, are among the most strongly unionized in the nation. Even residential building, which tends to be almost entirely open shop in most other regions, still has an appreciable union component in some parts of the West. The strength of the unions here was reflected in the Bureau of Labor Statistics study nearly four decades ago, and again in the AGC membership surveys in 1969 and 1973. During this four-year period, however, the AGC surveys set forth in Table VII-5 show some diminution of union representation in most of the states in the region, although the differences in the rates of response for the two surveys preclude attaching much significance to this change. This diminution increased substantially in the 1970s.

Union strength in many of the mountain states tends to be concentrated in a single metropolitan area (Albuquerque in New Mexico, Phoenix in Arizona, Denver in Colorado, Las Vegas in Nevada) in which the preponderance of the state's population resides, although there are smaller communities where the building trades also exercise influence (Colorado Springs, Pueblo, and Fort Collins, Colorado, for example). As a result, building construction is generally more apt to be unionized than highway construction, much of which is carried out at some distance from metropolitan centers. In the 1973 AGC survey, 71.6 percent of the building contractors in the region were union, compared to 59.5 percent of those doing highway work. The Colorado studies presented in Chapters IV through VI showed highway construction less unionized in that state than either commercial or industrial construction. (See especially Table VI-5.)

Nevertheless, the highway and heavy work sector is also strongly unionized in key western states. As pointed out in Chapter VI, Local No. 3 of the Operating Engineers holds sway in California, Hawaii,

[21] The West includes the mountain states (Arizona, New Mexico, Colorado, Utah, Wyoming, Montana, Idaho, and Nevada) and the Pacific states (California, Oregon, Washington, Alaska, and Hawaii).

TABLE VII-5
Union Status of AGC Members in Western United States, by State, 1969 and 1973

State	Union Contractors		Nonunion Contractors		Percent Union	
	1969	1973	1969	1973	1969	1973
Mountain States	168	281	65	153	72.1	64.7
Arizona	44	64	2	16	95.7	80.0
Colorado	46	63	8	30	85.2	67.7
Idaho	18	43	7	17	72.0	71.7
Montana	27	31	5	10	84.4	75.6
Nevada	6	29	2	3	75.0	90.6
New Mexico	11	28	8	22	57.9	56.0
Utah	11	17	12	21	47.8	44.7
Wyoming	5	6	21	34	19.2	15.0
Pacific States	425	727	22	80	95.1	90.1
Alaska	17	30	0	1	100.0	96.8
California	220	386	3	23	98.7	94.4
Hawaii	3	15	2	3	60.0	83.3
Oregon	56	91	7	23	88.9	79.8
Washington	129	205	10	30	92.8	87.2

Source: Associated General Contractors of America.

Utah, and Nevada; it has been strong enough to preclude the use of trainees on Federal Highway Aid projects in its vast territory. The excellent study of open shop penetration in the eleven-county Southern California region which was utilized in Chapters III through VI showed that highway and heavy construction had the least gains for the open shop of any sector. (See Table VI-6.)

In the previous four chapters, we examined the extent of open shop construction in Colorado, California, Washington, Oregon, and Idaho in considerable detail, and need not repeat our findings. Suffice it to note that in Colorado the union situation substantially eroded in the 1970s so that, even in Denver, the unionized firms were on the defensive by 1983, with the open shop having taken a portion of commercial office buildings as well as much other work. In California, unionized firms still clearly hold a majority of the work, but the open shop contractors have made substantial gains in all sectors, especially in the outlying areas and to a lesser extent in Southern California. In the Pacific Northwest, these open shop gains have been greater than those in California. What once was a totally unionized area is now moving toward an open shop volume majority; Oregon reached that stage in late 1983.[22] As Figure VII-

[22] This was conceded by union officials. See ". . . As Northwest Unions Fight Back," *Engineering News-Record,* January 5, 1984, pp. 74-75.

1 shows, open shop construction is gaining the majority in eastern Washington and northern Idaho.

New Mexico has turned heavily open shop since 1970. Power plants and industrial construction have gone open shop, as have highway and residential construction. Albuquerque is no longer overwhelmingly union, and is heading toward an open shop majority. Arizona has made an even more remarkable switch. Residential construction, which was unionized in the early 1970s, is now overwhelmingly open shop. Highway and heavy construction have likewise gone from a union shop to an open shop volume majority. Open shop contractors, bolstered by a strong ABC chapter, have made significant progress in commercial and industrial construction.[23] Both union officials and unionized contractors agreed in 1982 that "the number of nonunion employees in the state's construction industry is on the rise" and that work "formerly done by union members is increasingly going to their nonunion counterparts."[24] One Phoenix union business agent summed it all up: " 'We have priced ourselves out of the market', he conceded."[25]

Utah has been heavily open shop for several years. Our estimate is that in terms of number of commercial and industrial projects, the state is 75 to 80 percent open shop; in terms of volume, 55 to 60 percent open shop. Highway work is, however, about 50 percent union, and residential construction is all open shop. The Intermountain Utility Project, which when completed could be the largest coal-fired generating facility in the world, selected Bechtel, the large union shop contractor, as construction manager over Daniel, which allegedly submitted a lower bid. The Bechtel arrangement includes a wage stabilization agreement and other arrangements that could force employees to join unions in violation of Utah's right-to-work law. Allegedly, Los Angeles political interest groups pressured for union construction. Los Angeles agencies are part of the consortium building the plant.

Utah is one of the few states that has repealed its "little Davis-Bacon" law. It is therefore likely that highway construction will be less unionized in that state in the future when the new Davis-Bacon regulations commence to show results.

Nevada is heavily unionized in the Las Vegas and Reno areas. Unions have, however, seen their position erode. Mine preparation

---

[23] "Arizona Crafts Charge Builders with Deliberate Attempt to Destroy Unions," *Daily Labor Report,* No. 165 (August 25, 1982), pp. A-2-3.

[24] "Nonunion Construction in Arizona on the Rise, Conference Is Told," *Construction Labor Report,* Vol. 28 (March 17, 1982), p. 82.

[25] *Ibid.*

FIGURE VII-1
Unionized and Nonunion Construction Trends
Eastern Washington and Northern Idaho[a]
1976–1983

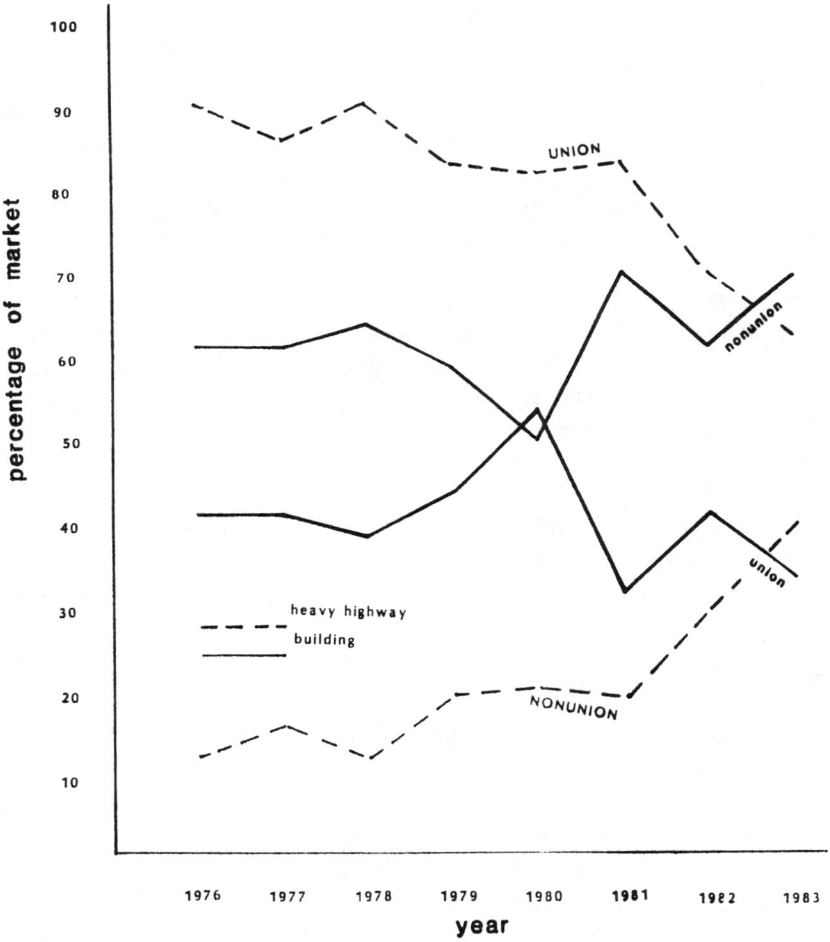

Source: Inland Empire Chapter, Associated General Contractors, Spokane.

[a] For geographic and construction sector coverage, see Tables IV-22 and VI-9.

in Tonopah was done open shop in the late 1970s for the first time, and other major jobs have gone nonunion, but Nevada's construction unions remain strong.

Historically, Montana has been a unionized state. It is still heavily union. Thus highway, industrial, and commercial work remains heavily union. Yet small nonfederal highway work, homebuilding, and small commercial projects have long been nonunion, and the union position in water and sewer work has eroded to 50 percent. When a $32 million coal-fired heating plant at Malmstrom Air Force Base near Great Falls was awarded to a California open shop contractor, the Montana unions precipitated violence in which equipment was destroyed and nonunion workers were assaulted. No adequate police protection was provided, despite a federal court injunction; instead, the governor appointed some local business leaders as mediators. They helped to induce Brenderson Construction Company, the contractor, to agree to do the project under a union contract.[26] Another project built nonunion, a pipeline project by Haines Pipeline Company near the town of Warm Springs, sustained $50,000 in damages by union construction workers. In this instance the company refused to accede to union threats, and the governor threatened to call out the national guard unless the violence subsided, which it has.[27] Obviously, the high construction costs in Montana have threatened the union position in the Rocky Mountain state, in which construction unions are stronger than anywhere in the region.

Wyoming, in contrast to Montana, is a strong open shop state. Some construction work around the coal mines is, however, unionized. Highway, residential, commercial, and most industrial work remains nonunion, with union work concentrated in the Cheyenne area.

Questionnaire and Survey Results—West

The responses to the 1974 Industrial Research Unit questionnaire from the Mountain and Pacific states were too few to permit any conclusions to be drawn as to the extent of open shop activity in those areas. Among the few responses, however, once again there was a marked difference between the views of union and nonunion respondents. At the same time, there appeared little indication of a major change in the open shop share over the few years prior to 1974 among either group of respondents, a picture which seemed roughly consistent with the results of our interviews. In short, there had for some time prior to 1974 been an appreciable open shop construction component in the Rocky Mountain area, especially in residential and (to a lesser extent) highway building, but the years

---

[26] "Violent Protests Against Open Shop Disrupt Projects in Mich., Montana," *Construction Labor Report.* Vol. 29 (September 21, 1983), pp. 983-85; and Gary Svee, "Public Loses to Politics." *The Billings Gazette,* February 27, 1983.

[27] Svee, "Public Loses to Politics."

before 1974 did not see a dramatic diminution of union strength, although there were some indications of a change. In the Pacific states in 1974, the unions seemed secure and open shop construction did not appear to be a threat to their position.

Of course, things have changed. The 1982 Industrial Research Unit questionnaire found that in the Rocky Mountain region, 62.1 percent of the 108 unionized respondents put the open shop penetration at 41 percent or more, 23.2 at 61 percent or more; in contrast, 75 percent of the 184 open shop contractors estimated the open shop share at 41 percent or greater, 46.7 percent at 61 percent or greater. Even the estimates of the unionized contractors indicate a decided loss for the unions.

The NECA inside electrical data for the Rocky Mountain and Pacific areas (Table VII-3) are also interesting. Forty percent of the 1,878 Rocky Mountain inside electrical employees had IBEW contracts in 1973; only 21 percent of 3,009 in 1980. In 1973, 54 percent of the 19,680 inside employees were IBEW represented; by 1980, only 40 percent of 26,549 were so represented.

In 1977, approximately 50 percent of the sheet metal contracting in Colorado was done by union contractors, according to the SMACNA study; by 1982, this ratio had declined to about 35 percent (Table IV-8).

The study of the Mechanical Contractors Association combined most Rocky Mountain states with those of the West North Central, and put Arizona and New Mexico with the Southwest. In the "Plains and Rockies" area, the union shop contractors were predicted to hold on to a majority of the commercial and industrial work by 1985 but to lose out in all other sectors. In fact, by 1983, the open shop had 57 percent of the commercial work but only 35 percent of the industrial. The Southwest, as noted, was expected to be overwhelmingly open shop by 1985. (See Table IV-7.)

For the Pacific states, the 1982 Industrial Research Unit questionnaire provided very significant data. In what was probably the strongest union regional bastion, 36.7 percent of the 362 unionized contractors and 65.8 of the 143 open shop ones estimated that the open shop had captured 41 percent or more of the work. Only 13.3 percent of the unionized contractors, but 46.7 of the nonunion ones, put the open shop penetration at 61 percent or more. Given the absence of homebuilders from these data and the difficulty of reaching small contractors, who are more likely to be open shop, with a mailed questionnaire, these results surely reinforce our findings that the Pacific Coast states are undergoing significant change in the degree of open shop penetration. (See Table VII-2.)

The NECA survey (Table VII-3) showed that 75 percent of the 3,986 inside electrical employers in the three Pacific Coast states dealt with the IBEW in 1973, but surprisingly only 45 percent of the 5,492 did so in 1980; for employees represented by the IBEW, a similar decline occurred from 66 percent of 36,305 to 40 percent of 55,253.

For Southern California, about 80 percent of the sheet metal contracting was unionized in 1977, according to the SMACNA study; by 1982, this had declined to approximately 66 percent (Figure IV-8). The Mechanical Contractors study found that the unionized contractors had a majority of all plumbing and mechanical work in 1977 in the three Pacific coastal states; by 1982 they had lost that majority in residential and garden apartments, high-rise apartments, service and maintenance, and rehabilitation and remodeling. This study predicted that in these states the open shop would have a majority of all sectors except industrial by 1985 (Table IV-7). Substantial gains by 1983 for the open shop indicate that this prediction may well have been conservative.

CONCLUSION

Although the preponderance of the evidence presented in the foregoing pages shows a broad diminution of union strength across the nation, there are still substantial differences among states, regions, and individual labor markets as to the magnitude of open shop activity in the construction industry. In part, the broad geographic patterns have remained constant since the mid-1930s. The South has historically been, and remains, the most completely open shop region. It is the home of most of the large open shop firms, which prospered there and now work far beyond the South's borders. The Pacific Coast states, the East North Central area of the Midwest, and the Middle Atlantic states have, on the other hand, long been the centers of construction union power, and in the center cities of these areas, this is still largely the case. Whereas the open shop has not only held its own in its southern stronghold, but has successfully overrun such union enclaves as Atlanta, Mobile, Baton Rouge, and Lake Charles, the union citadels in the North and West have, on the other hand, been materially breached. The Pacific Northwest has seen the open shop seize a large portion of the work; California remains strongly union, but that strength has been lessened by steady open shop gains; the East North Central States have seen the open shop take over the work outside most metropolitan areas except in strongly union Illinois; construction unionism in

New York is more and more confined to New York City, and to a lesser extent to Buffalo; in New Jersey, union strength wanes outside the New York and Philadelphia metropolitan areas; in Pennsylvania, it is weak between Pittsburgh and Philadelphia. New England, except Providence and Boston, is heavily open shop. These changes, which began in the late 1960s, grew rapidly in the late 1970s. In our first edition of this book we reported that open shop penetration of these once highly unionized regions was still more the exception than the rule, but by 1983 such a statement was true only for specific cities and states, certainly not for an entire region.

Many of the factors which affect the strength or weakness of the building trades unions among regions clearly go beyond the specific characteristics of the construction industry. The areas in which the building trades exercise the greatest influence tend also to be areas where unionization generally is relatively high. For example, union membership in the East North Central region, California, New York, and other industrial states has been relatively high; in these states, construction unions have been relatively strong. Thus in our original book, we found that the rank-correlation coefficient between union members as a percentage of all nonagricultural employment in 1970 (the latest year then available) and the proportion of union firms in the 1973 AGC membership survey was .798.[28] A breakdown of these data by states and regions is given Table VII-6, in which the parallel regional differences, with respect both to unionism in general and to construction unionism in particular, stand out starkly.

One must not, of course, draw too many inferences from such a high correlation. Nonunion contractors are often not represented in AGC chapters. Homebulders are also outside of the AGC's pale. On the other hand, the AGC does not include highly organized specialty contractors. Thus, the data in Table VII-6 must be considered indicative, not conclusive, for the early 1970s.

We have no comparable data for the 1980s, but it may well be that the high correlation found in Table VII-6 would also hold fast in the early 1980s. Figure VII-2 shows union membership as a percentage of the nonagricultural labor force for the South and Southwest, the Middle Atlantic and southern New England, and the East North Central states for 1980. Certainly, it is clear that the lack of unionization generally correlates highly with the low

---

[28] Union membership data from U.S. Bureau of Labor Statistics, *Directory of National Unions and Employee Associations 1971,* Bulletin 1750 (Washington, D.C.: Government Printing Office, 1972), p. 84.

TABLE VII-6

Union Membership as a Percentage of Total Nonagricultural Employment, 1970, and Proportion of Union Firms in 1973 AGC Membership Survey, by State

State	Percent Union Members	Percent AGC Union Contractors	State	Percent Union Members	Percent AGC Union Contractors
New England	26.3	59.6	North Carolina	9.4	6.3
Maine	22.0	10.0	South Carolina	11.6	7.7
Vermont	20.9	4.3	Virginia	18.9	11.7
New Hampshire	21.2	24.1	W. Virginia	46.8	57.9
Massachusetts	27.2	83.7	E.S. Central	24.0	36.0
Connecticut	27.5	88.9	Alabama	22.6	28.8
Rhode Island	28.0	87.1	Kentucky	32.2	56.0
Middle Atlantic	38.5	85.2	Mississippi	14.9	34.9
New York	40.2	81.0	Tennessee	24.0	21.6
Pennsylvania	40.0	89.4	W. S. Central	17.1	29.1
New Jersey	31.2	83.5	Arkansas	19.5	15.9
E. N. Central	38.6	87.5	Louisiana	19.3	56.7
Illinois	27.3	90.5	Oklahoma	18.6	31.8
Indiana	37.5	84.4	Texas	15.7	22.0
Michigan	43.5	84.1	Mountain	25.1	64.7
Ohio	38.9	86.9	Arizona	21.4	80.0
Wisconsin	33.3	90.1	Colorado	25.0	67.7
W. N. Central	29.2	47.1	Idaho	22.1	71.7
Iowa	24.5	28.6	Montana	34.3	75.6
Kansas	21.1	31.2	Nevada	36.5	90.6
Minnesota	31.9	74.4	New Mexico	18.8	56.0
Missouri	37.5	85.5	Utah	26.2	44.7
Nebraska	21.0	n.a.	Wyoming	24.8	15.0
North Dakota	21.5	11.9	Pacific	36.7	90.1
South Dakota	14.7	10.3	Alaska	34.4	96.8
S. Atlantic	18.6	23.5	California	35.7	94.4
Delaware	25.7	55.0	Hawaii	30.3	83.3
D.C.-Maryland	25.2	72.9	Oregon	36.7	79.8
Florida	16.2	38.2	Washington	45.3	87.2
Georgia	17.5	19.4			

Source: The union membership percentages were calculated from U.S. Bureau of Labor Statistics, *Directory of National Unions and Employee Associations, 1971,* Bulletin 1750 (Washington, D. C.: Government Printing Office, 1972), Table 18; and *Manpower Report of the President, 1974* (Washington, D.C.: Government Printing Office, 1974), Table D-1. The percentages are not entirely consistent with those reported in the BLS Directory. Separate calculations were necessary to obtain regional totals. AGC data are from the 1973 membership survey.

n.a. = not available.

FIGURE VII-2

Union (and Association) Membership As a Percentage of the Nonfarm Labor Force, by States, 1980

(Lightly Organized States)

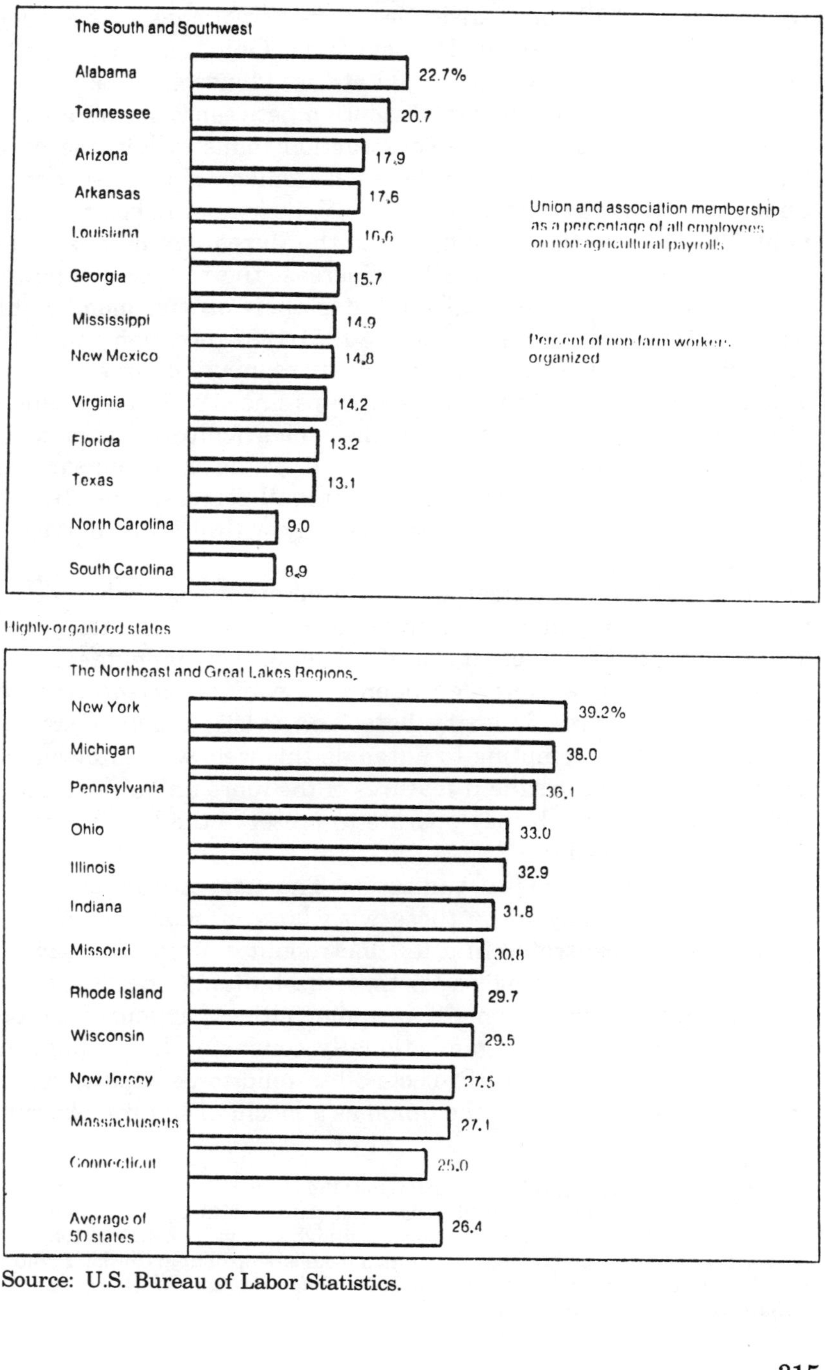

Source: U.S. Bureau of Labor Statistics.

construction union penetration in the South and Southwest. Yet the takeover of construction outside New York City and outside Detroit by open shop builders in New York and Michigan, the strides made by the open shop in Pennsylvania, Ohio, and other states listed as "highly organized," would at first glance seem to indicate that there is today much less correlation between a state's general unionization and that in the construction industry. On the other hand, anyone familiar with trade union membership statistics must understand that the data which the U.S. Bureau of Labor Statistics publishes are raw figures supplied to the Bureau by unions. For a variety of reasons, but especially to increase their illusion of power, most unions have traditionally inflated their official membership figures.[29] Moreover, as a result of the 1981-1982 recession and structural changes in mass-production industries, many unions have lost a substantial portion of their memberships since 1980. Today, unions represent less than 20 percent of the nonagricultural labor force, and perhaps as little as 15 percent.[30] It would then appear that increased open shop penetration might well correlate highly with declining membership in unions generally within a given state or region.

The broader question raised by the observations—why certain regions are more generally hospitable to unionism than others—goes beyond the scope of this study. The answer undoubtedly embraces a complicated interdependence of political orientation, demographic structure, industry base, and even random historical events. Without attempting to untangle this web, we can nevertheless try to outline the salient features of the more unionized regions and the ways in which they operate to preserve the holding power of the building trades.

One obvious variable is urbanization. The most broadly unionized states tend generally to be those with large proportions of their populations concentrated in a few metropolitan areas. We saw repeatedly in the foregoing pages that the building trades, despite some exceptions, are strongest in major cities, particularly older ones. Larger cities are organizationally congenial to the building trades for several reasons. Projects often tend to be expansive, enhancing the importance of the union as a mechanism for mobilizing

---

[29] A notable exception to this generalization is the United Steelworkers of America which publishes accurate membership data and continued to do so even when the membership declined in 1983 to 720,000 from a high four years before of 1,400,000.

[30] See the discussion and references to union membership data in Gordon F. Bloom and Herbert R. Northrup, *Economics of Labor Relations,* 9th ed. (Homewood, IL.: Richard D. Irwin, Inc., 1981), pp. 68–76.

large numbers of workers. Their concentration, moreover, facilitates the unions' ability to identify and combat nonunion operations as well as to monitor those undertaken with union labor. Construction unions have also long been a major political force in cities, which aids both to ensure that projects are built union and to deny police protection for open shop operations, as is detailed in Chapter VIII. It is no accident, therefore, that the centrifugal expansion of urban areas has frequently been associated with a diminution of union strength. It is also no accident that the three regions where unions have historically been most influential—Middle Atlantic, East North Central, and Pacific—are the three in which historically the largest proportions of the populations lived in urban areas.

The passing reference in the above paragraph to the timing of urbanization deserves a bit more comment. Certain southern states are becoming much more urbanized: Virginia, the Carolinas, Florida, and Texas are cases in point. In these states, nevertheless, open shop construction remains clearly dominant and has even gained as urbanization has advanced. Throughout the South, increasing urbanization (and industrialization) was initially accompanied by a growing proportion of the unionized workforce generally. In every southern or border state except Maryland and Arkansas, union membership as a percentage of total employment rose between 1964 and 1970.[31] Much of this growth was clearly attributable to the movement of national market manufacturing firms, which were already unionized, into the metropolitan areas of the South. Since construction is predominantly a local market industry, this "transfer effect" is largely absent; with open shop contractors firmly established in the region, the building trades unions do not appear to have benefited significantly from the changes in the South's basic economic and demographic character during this period. Since 1970, companies moving South—with the notable exceptions of a few, such as General Motors and Philip Morris—have attempted, largely successfully, to operate nonunion. This has been a factor, elsewhere as in the South, in inducing these firms to build open shop, as well as in materially slowing union growth generally in these states.

The implication here is that the level of industrialization in a state or region plays a greater role in maintaining union strength where it already is high than in augmenting it. Large, unionized manufacturing establishments may be reluctant to incur the enmity both of the building trades and of the industrial unions with which

[31] U.S. Bureau of Labor Statistics, *Directory of National Unions and Employee Associations, 1965 and 1971,* bulletins 1493 and 1750 (Washington, D.C.: Government Printing Office, 1966 and 1972), Tables 9 and 18, respectively.

they deal. This pressure, moreover, can operate even in relatively open shop areas. In one case in the early 1970s, Daniel International built a major paper mill in Maine with union labor because of pressure from the user. In union strongholds, industrial and commercial purchasers are sometimes willing to absorb higher costs in order to ensure labor peace. At other times, they may have no choice, as there may be no open shop contractors in the area equipped to handle large or highly specialized building projects.

There has, however, been a limit to such pressures, particularly as union wage costs have risen. As noted in Chapter V, new facilities in the highly unionized paper manufacturing industry, which were built union in the 1970s, were being constructed open shop in the 1980s, both in the South and in Maine. It has always been true that where the unions are not strong, pressures against building open shop are not effective. The decline in general union membership and power, and the increased economic pressures in the 1980s, both decreased attempts to force companies to build union and increased management's determination to resist actions which appease unions but add to their costs. These changing circumstances have prevented unions from making significant organizational inroads into the open shop's domain, and have strongly assisted open shop penetration into previously heavily unionized areas.

A related factor in regional differences is the degree of political influence exercised by the labor movement, which may be seen as both a cause and an effect of union power. A political environment favorable to unions can affect the fortunes of the building trades on at least three levels: through the promulgation of specifications for awarding public contracts; through laws and executive decisions either encouraging or hampering union activity; and through the enforcement of these policies. Where local political leaders have considerable labor support, their aggressiveness in dealing with mass picketing, demonstrations, and even violence has often been tempered. They may also be instrumental in persuading private construction purchasers to build union in the interest of "stability." In areas where local authorities have historically been antagonistic to union organization—the South is the best example—the forces which nationally have contributed to a weakening of the unions' base are not checked by favorable political alliances.

Public policy formulation affects open shop competition in a number of areas. Elsewhere in this volume we discuss the impact of prevailing wage laws and state apprenticeship regulations. Here we may add still another factor: right-to-work laws. These state statutes, which outlaw the negotiation of union security agreements, have special significance for the construction industry, in which, as explained in the following chapter, relatively little organization is effected through representation elections, and in which the unions historically enforced their control over labor supply through the closed shop and later through exclusive hiring halls. Of the twenty states which have right-to-work laws,[32] only Nevada has a strong union construction sector. Again, of course, the existence of a right-to-work law is both the cause and the consequence of relative union weakness. Such laws exist only where unions are weak, but they may well help to maintain the status quo.

It must also be understood that these factors cannot entirely explain the various regional and marketwide conditions found around the country. The variables here, either because of their conceptual basis or because of lack of data, do not admit of such neat quantification. Furthermore, some cities, some local unions, some crafts, and some individual projects defy the general patterns outlined here, and many of the explanations are undoubtedly idiosyncratic. The building trades have continued strong in Peoria, Illinois, a highly unionized industrial city; but in highly unionized Syracuse, New York, construction unionism is weak. Many such examples could be cited.

A final point should be stressed. The open shop has gained throughout the country because of various factors, including the movement to the suburbs and the increasing industrialization, and therefore rising construction volume, of the South. But the big gains have come since the late 1960s and the early 1970s. The overall, underlying reason seems clear: The unions have been pricing themselves out of the market. Economics remains a powerful force, and the market will exert itself despite many obstacles.

[32] Alabama, Arizona, Arkansas, Florida, Georgia, Iowa, Kansas, Louisiana, Mississippi, Nebraska, Nevada, North Carolina, North Dakota, South Carolina, South Dakota, Tennessee, Texas, Utah, Virginia, and Wyoming.

CHAPTER VIII

The Union Reaction

As the studies and comments quoted in the previous chapters clearly demonstrate, construction union officials have been well aware of the increasing expansion of open shop construction. They have moved to offset these open shop gains by three general approaches: (1) by economic moves, (2) by utilizing the legal weapons of conflict; and (3) by resorting to violence. This chapter surveys these three approaches.

ECONOMIC RESPONSES

The most direct way in which unions can combat nonunion operations is by narrowing the cost differentials between the two groups through wage and related concessions.

Wage and Related Concessions

Traditionally, unions have been reluctant to reduce wage rates,[1] but in the early 1980s, this policy changed. Despite high unemployment in recent years and the continuous increase in the market share of open shop construction since the late 1960s, it was not until the severe recession of 1981–83 that the construction unions negotiated widespread and substantial reductions in wages and benefits. Prior to that such concessions granted by the unions were moderate, localized, and generally informal so that basic rates and conditions could easily be reinstituted when it suited the unions.[2] After 1981, when wage increases approached $1.32, or about 12 percent per hour, a downturn began and gained momentum. In 1983, the Construction Labor Research Council (CLRC) found that wage and benefit settlements reported to its office and covering 719,586 workers, averaged $0.40, or 2.2 percent, per hour—the smallest hourly rate increase recorded in more than two decades

[1] See John T. Dunlop, *Wage Determination Under Trade Unions* (New York: MacMillan Company, 1944), Chapter IV, for the classic statement on this point.

[2] *Ibid.*, p. 67.

FIGURE VIII-1
Construction Settlements
First Year Increase in Wage Plus Benefit Rate
Dollars and Percents
1979 through 1983

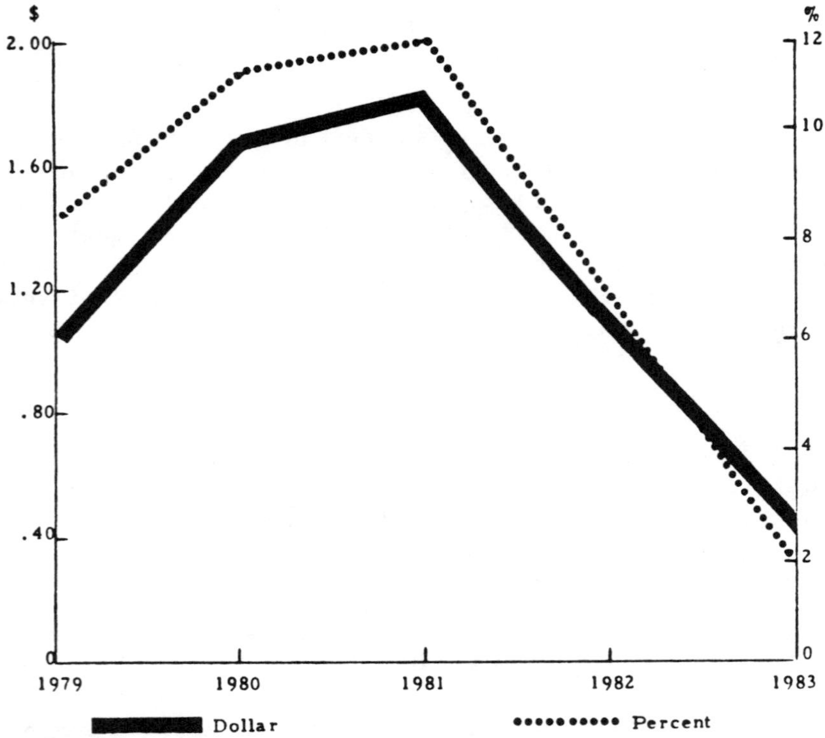

Source: "Wage and Benefit Settlements in Construction 1983," *Construction Labor Research Council,* December 28, 1983, p. 2.

(see Figure VIII-1). This was less than one-half the $1.10, or 6.7 percent, per hour raise negotiated in 1982. In addition, it is most significant that in 1983 over one-third of all negotiations studied by the CLRC resulted in wage and fringe benefit freezes or rollbacks.[3]

In previous chapters, we have recounted many cases in which unions agreed not only to reduce wages and benefits, but also to eliminate various practices which were decreasing productivity or increasing costs, such as time paid for but not worked, doubletime instead of time and one-half for overtime, and the insistence that

[3] "Wage and Benefit Settlements in Construction 1983," *Construction Labor Research Council,* December 28, 1983, p. 1; "Wage and Benefit Settlements—1984," *ibid.,* July 16, 1984, p. 1.

skilled workers perform unskilled work at journeymen wage rates. Although such concessions, together with wage reductions and/or moderation, certainly aid union builders in their competition with open shop contractors, they often fail to narrow the gap sufficiently to achieve the desired results. The basic reasons are that the wage and benefit costs of union construction remain high, and as will be discussed in Chapter IX, the craft structure of union organization carries with it an uneconomic system of labor deployment and utilization.

The cost advantages of open shop construction are significant and can apparently be overcome only when unionized contractors have experience, management expertise, or other special skills that open shop contractors do not. For example, as long as only a few open shop builders have the experience and capability to handle the industrial megaprojects, or in many areas, large high-rise office buildings, the unionized builders can offset the open shop contractors' lower costs. The whole thrust of the previous chapters, however, is that, as open shop builders gain this experience and capability, they win an increasing share of such work because their costs are lower. This means that despite some wage moderation, givebacks of restrictive rules, and other such measures, construction unions are finding it difficult to agree to sufficiently large cost reductions to meet the competition of open shop builders.

Union wage advances during 1983 were not always concessionary. Table VIII-1 shows that during 1983, the average wage increase in the agreements studied by the CLRC was, as noted, $.44, or 2.2 percent, per hour for the first year of the agreements. For the second year of the agreements, however, the average increase was $.95, or 4.6 percent, per hour. In New England, moreover, the average wage increase was $1.13, or 6.3 percent, per hour during the first year, and in the Middle Atlantic states, $1.12, or 5.8 percent, per hour during the same year, with both figures higher for the second year. Yet New England is fast becoming open shop territory, and the Middle Atlantic states have witnessed a steady erosion of union control since the mid-1970s.

In addition, not until 1983 did wage settlements in unionized construction decline to levels that corresponded with those in other segments of the economy. In 1981, agreements received by the CLRC contained an average wage increase of $1.82, or 12 percent, per hour, approximately one-third greater than that year's rate of inflation.[4] For the period from 1976 through 1982, the CLRC found

[4] "Contract Duration—Impact of Negotiating Three Year Agreements," *Construction Labor Research Council,* September 1982, p. 1.

TABLE VIII-1
Wage and Benefit Settlements
United States, Regions, and Selected States[a]
Construction, 1983

	Coverage		Average Increase[b]			
			Dollar		Percent	
	Settlements	Workers	1st Year	2nd Year	1st Year	2nd Year
Total U.S.	1,007	719,586	.44	.95	2.2	4.6
New England	74	46,812	1.13	1.17	6.3	6.1
Connecticut	22	10,523	1.42	[c]	7.9	[c]
Massachusetts	30	27,462	.96	1.17	5.0	6.0
Middle Atlantic	121	104,308	1.12	1.23	5.8	6.0
New Jersey	24	14,628	1.53	1.45	7.7	6.9
New York	22	19,472	2.05	[c]	9.6	[c]
Pennsylvania	47	37,899	1.30	1.49	7.4	7.8
Southeast	149	70,362	(.01)	.64	(0.1)	3.7
Alabama	21	4,873	.27	[c]	1.9	[c]
Georgia	23	12,652	(.53)	[c]	(3.6)	[c]
Kentucky	26	13,171	.08	[c]	0.4	[c]
Tennessee	22	11,686	(.18)	[c]	(1.2)	[c]
Virginia	26	11,401	-0-	[c]	-0-	[c]
E.N. Central	296	193,988	.37	.78	1.9	4.0
Illinois	112	99,053	.38	.58	2.0	3.0
Indiana	46	19,021	.22	[c]	1.1	[c]
Michigan	38	17,269	.22	[c]	1.1	[c]
Ohio	27	28,220	.56	[c]	2.9	[c]
West Virginia	20	12,467	.31	[c]	1.8	[c]
Wisconsin	46	14,998	.32	[c]	1.8	[c]
W.N. Central	136	52,794	.05	.80	0.3	4.1
Iowa	67	8,986	(1.47)	[c]	(9.8)	[c]
Missouri	51	36,104	.65	.80	3.5	4.1
S.C.	88	40,400	.14	.32	0.6	2.1
Louisiana	24	8,683	.06	[c]	0.2	[c]
Oklahoma	23	10,307	.02	.28	0.1	1.9
Texas	36	20,122	.27	[c]	1.4	[c]
Mt.-Northern Plains	50	19,083	.21	[c]	1.1	[c]
Montana	25	8,369	.45	[c]	2.8	[c]
Southwest Pacific	49	114,595	.38	[c]	1.5	[c]
California	42	110,748	.38	[c]	1.5	[c]
Northwest	44	77,244	.23	[c]	1.2	[c]
Washington	25	47,971	.21	[c]	1.1	[c]

Source: "Wage and Benefit Settlements in Construction 1983", *Construction Labor Research Council,* December 28, 1983, p. 5.

[a] States with a sufficient number of settlements to show a meaningful average.

[b] Weighted by number of workers covered.

[c] Too few agreements to show a meaningful average.

that wage and fringe benefit increases for construction workers averaged $6.21, or 50 percent, per hour, bringing the average hourly rate to $18.62.[5] Table VIII-2 shows average percent changes in wages and compensation in major collective bargaining settlements by quarter, from the fourth quarter of 1979 through the second quarter of 1983, as compiled by the U.S. Bureau of Labor Statistics. Not until the second quarter of 1983 did construction union-management settlements fall below the general industry average. Prior to that, construction settlements generally exceeded all other categories. Obviously, concessionary policies of construction unions must not only be very substantial if they are to succeed, but they must also be given with greater alacrity.

The extent to which the various signs of union restraint are caused by open shop competition is difficult to ascertain, although union spokesmen clearly blame open shop contractors. In Chapter XI, we shall demonstrate that union wage rates are substantially above those of open shop workers and that the union-open shop wage-rate differential has probably widened since the mid-1960s. This corresponds to the period of the open shop's greatest growth. Obviously these two factors must be related to the unemployment problems of unionized workers.

Restraint seems to require rather dire conditions before it is initiated; and even when those conditions are present, wage increases, such as those shown in Table VIII-1 for the New England and Middle Atlantic regions, indicate that the construction unions often do not believe that the demand for their services will be improved by wage restraint or moderation. Rather, union wage policy has been predicated, for the most part, on the assumption that construction employment will not be radically affected by wage costs, except for short periods, and therefore, that wage reductions in the long run prove more costly than helpful. Where difficult situations ensue, some restraint may occur, or off-rate adjustments may be made on an informal or temporary basis. When a large construction project is at stake, then a formal, but temporary, project agreement may solve the immediate problem for the unions without prejudicing future gains, as described below. Or, if the open shop is a continuing threat, then a well-propagated joint union-management organization may be established. Only when, as in the early 1980s, the unemployment rate is very high (for construction it was twice that of the labor force as a whole) and when open shop contractors are

---

[5] "1983 Construction Labor Rate Trends and Outlook," *Construction Labor Research Council,* February 1983, p. 2.

TABLE VIII-2
Average Percent Changes in Wages and Compensation in Major Collective Bargaining Settlements[a] *During Four-Quarter Periods*

Measure	Four quarters ending														
	1979	1980				1981				1982				1983	
	IV	I	II	III	IV	I	II	III	IV	I	II	III	IV	1p	11p
Compensation adjustments:[b]															
First year	9.0	9.4	9.2	10.3	10.4	10.2	10.5	9.7	10.2	8.0	4.4	4.1	3.2	2.7	3.1
Over life of contract[c]	6.6	6.7	6.5	7.0	7.1	7.1	7.8	8.4	8.3	6.2	3.0	2.8	2.8	2.9	3.4
Wage-rate adjustments in—															
All industries															
First year	7.4	7.7	7.7	9.1	9.5	9.2	9.9	9.7	9.8	8.4	5.1	4.4	3.8	3.0	2.8
Contracts with COLA clauses[c]	6.2	6.5	6.2	7.7	8.0	8.2	8.7	7.2	8.0	4.3	2.8	3.0	2.2	1.3	.8
Contracts without COLA clauses	9.1	9.3	10.3	11.2	11.7	10.3	11.0	10.7	10.6	11.5	9.0	7.5	7.0	6.1	5.0
Over life of contract[c]	6.0	6.0	6.1	6.9	7.1	7.0	7.6	8.0	7.9	6.7	4.2	3.7	3.6	3.6	3.7
Contracts with COLA clauses	4.6	4.5	4.3	4.9	5.0	5.1	5.6	5.7	5.5	3.0	2.0	2.1	2.1	2.2	2.4
Contracts without COLA clauses	8.0	8.1	9.0	9.9	10.3	9.2	9.4	8.9	8.8	9.5	7.9	7.0	6.6	6.2	5.0
Manufacturing															
First year	6.9	6.9	6.4	6.8	7.4	7.0	7.3	7.4	7.2	5.5	2.9	2.9	2.8	1.6	1.3
Contracts with COLA clauses	5.7	5.6	5.6	5.9	6.5	6.8	7.5	7.1	6.2	3.0	1.7	2.2	2.3	1.0	.4
Contracts without COLA clauses	9.4	9.7	9.7	9.7	9.7	7.5	7.1	7.7	7.6	9.4	7.8	7.1	6.3	5.4	4.9
Over life of contracts[c]	5.4	5.3	4.8	5.0	5.4	5.4	5.7	6.1	6.1	4.8	2.7	2.5	2.6	2.3	2.6
Contracts with COLA clauses	4.0	4.0	3.9	4.0	4.4	4.5	5.1	5.3	4.8	2.8	1.7	2.0	2.1	1.8	2.0
Contracts without COLA clauses	8.0	8.3	r8.4	8.5	8.6	6.7	6.3	6.7	6.6	8.0	7.4	6.1	5.8	5.1	4.8

Nonmanufacturing															
First year	8.0	8.7	10.2	10.7	10.9	10.8	11.2	11.4	11.2	9.4	7.1	6.0	4.8	5.1	4.1
Contracts with COLA clauses	7.0	8.5	9.1	9.7	9.7	9.7	9.4	7.3	8.9	5.1	4.5	4.3	1.9	2.5	1.7
Contracts without COLA clauses	8.8	9.0	10.7	11.7	12.2	11.9	12.7	12.3	12.1	11.9	9.2	7.6	7.2	6.3	5.0
Over life of contract[c]	6.8	7.0	8.4	8.2	8.3	8.3	8.6	9.4	8.8	7.3	5.6	4.9	4.7	5.5	4.6
Contracts with COLA clauses	5.5	5.7	6.3	5.9	5.7	5.7	5.8	6.5	5.8	3.1	2.5	2.4	2.1	3.5	3.4
Contracts without COLA clauses	7.8	8.0	9.4	10.4	10.8	10.6	10.8	10.1	9.9	9.8	8.1	7.3	6.9	6.4	5.0
Construction															
First year	8.8	8.9	11.6	13.4	13.6	13.7	13.9	13.6	13.5	13.2	9.7	6.7	6.5	5.3	2.8
Contracts with COLA clauses	8.7	8.7	13.5	13.9	13.9	13.9	12.0	8.8	8.8	8.6	6.2	−1.4	−1.4	−2.2	−2.2
Contracts without COLA clauses	8.8	8.9	11.4	13.3	13.6	13.7	14.3	14.0	14.0	13.6	9.8	7.1	6.8	6.2	3.5
Over life of contract[c]	8.3	8.4	9.9	11.3	11.5	11.6	11.9	11.4	11.3	11.1	8.5	6.6	6.3	5.5	3.3
Contracts with COLA clauses	4.8	4.8	6.9	8.8	8.8	8.8	9.2	8.4	8.4	8.4	4.4	−.2	−.2	.8	1.0
Contracts without COLA clauses	8.4	8.5	10.2	11.8	12.0	12.1	12.4	11.6	11.6	11.3	8.6	6.9	6.6	6.1	3.6

Source: *Current Wage Developments,* Vol. 35 (October 1983), p. 41.

[a] Limited to private nonfarm industry settlements affecting 1,000 workers or more (5,000 for compensation) and to production and related workers in manufacturing and to nonsupervisory workers in nonmanufacturing industries. Includes all settlements, regardless of whether wages and compensation were increased, decreased, or not changed. Data exclude possible adjustments in wages under cost-of-living clauses (except increases guaranteed by the contract).

[b] Limited to settlements involving 5,000 workers or more.

[c] Annual rate of change.

COLA = Cost of Living Adjustment.

increasing their share of work, do wage reductions and other cost modifications become common. Even then, concessions in construction have not usually been as common or as severe as in industry generally.

Union-Management Cooperation

In a number of communities, programs have been initiated by unionized contractors and construction unions which are designed to offset gains made by open shop contractors. These programs feature no-strike provisions, reductions in restrictive rules, and cooperative arrangements to improve productivity and to utilize joint efforts to seek business.

The National Construction Employers Council (NCEC), the organization that attempts to coordinate the activities of the various construction employer associations, has been a major sponsor of such programs. In November 1982, the NCEC and the AFL-CIO Building and Construction Trades Department signed an agreement to establish a "Market Recovery Program for Union Construction." Essentially, it provides ground rules for local implementation. Local groups are encouraged to assess open shop competition, identify nonproductive practices and work rules in local construction, review employment trends, determine the strengths and weaknesses of union construction, and institute a public relations program.[6] In 1983, the NCEC reported that twelve local groups had been assisted in establishing union construction recovery programs and that the National Joint Labor-Management Committee of the Market Recovery Program for Union Construction, co-chaired by representatives of the NCEC and the construction unions, had met each month in 1983 "to refine market recovery strategies," and to discuss other matters, including a proposed handbook which is designed as "an attitude adjustment tool."[7] Among the areas in which progress was reported are western Pennsylvania; Syracuse, New York; Tulsa, Oklahoma; and Cincinnati, Ohio.[8] Long-term results, however, remain to be determined. In the words of one construction labor-relations executive from upstate New York, the program may prove to be "a day late and a dollar short" because union contractors "are beyond the point of no return" and because unions "take a Nean-

[6] "Building Trades—NCEC Sign Market Recovery Agreement," *Construction Labor Report,* Vol. 28 (November 17, 1982), pp. 1257, 1313–14; "NCEC, Unions Sign Market Recovery Plan," *Engineering News-Record,* November 25, 1982, p. 88.

[7] "NCEC Board Charts Contractor Unity, Market Recovery Plans," *Construction Labor Report,* Vol. 29 (December 14, 1983), p. 1374.

[8] *Ibid.,* pp. 1374–75.

drathalic [sic] view of market recovery," which concentrates on "Draconian local ordinances" and resorts to violence "to restrict competition."[9] Most likely, the results will be somewhere between the extreme pessimism of this New York state executive and the extreme optimism of the program's proponents.

Other union-management cooperative plans have been initiated, some with NCEC-national union assistance. One of these is Operation Most—Management and Organized Labor Striving Together—which was organized in 1976 in Columbus, Ohio. Proponents claim that it has virtually eliminated jurisdictional disputes, improved productivity, and prevented open shop contractors from taking more than a 50 percent share of the business volume. Officials of the Ohio Building Chapter of the Associated General Contractors (AGC), which is a party to Operation Most, feel that it has been essential to the survival of unionized construction in central Ohio.[10] Similar programs have been adopted in Cincinnati and Dayton as a result of the Columbus experience. Yet an official of the Ohio Contractors Association Chapter of the AGC feels that the program is more talk than action, although he concedes that it creates a good image and therefore is helpful with business prospects.[11]

Another well-known cooperative plan is Pride in St. Louis, organized in 1972. It is directed by a ten-man, evenly split board that meets monthly. Proponents claim that it has brought labor peace, restored management rights, and increased productivity.[12] Solutions, of course, are strictly within the union craft structure. For example, an article in *Fortune* details how a contractor had ironworkers install forty-foot-high metal doors although the carpenters claimed that it was their work. They settled it by "a compromise in which the ironworkers completed the track installation and millwrights hung the doors."[13] The article did not state how much extra cost was involved in this deployment of labor, only that because of Pride it was settled without a jurisdictional strike.

Pride has been the prototype for similar programs in Denver, Toledo, Beaumont-Orange-Port Arthur, Texas (the "Golden Trian-

---

[9] "NCEC, Unions Sign Market Recovery Plan," p. 88.

[10] Interview with Leslie H. Jones and William J. Aver, Central Ohio Division, Ohio Building Chapter, AGC, Columbus, Ohio, August 20, 1981.

[11] Interview with Barry Connors, Ohio Contractors Association Chapter, AGC, August 20, 1981. Operation Most employs a public relations firm, Paul Werth Associates, Inc., to promote its concept.

[12] Irwin Ross, "The New Work Spirit in St. Louis," *Fortune,* Vol. 104 (November 16, 1981), pp. 92–106.

[13] *Ibid.,* p. 103.

gle"), and Indianapolis.[14] Such programs have had varied success. Again, they generally appear to have improved union-management relations, reduced jurisdictional disputes, and in some cases, raised productivity. Where the political situation permits open shop construction to operate, such as in Denver and Indianapolis, the programs do not seem to have stemmed its growth, although they may have slowed its gains.

Project Agreements

Project agreements are usually entered into by large industrial contractors and unions to cover such major construction as power plants, petrochemical complexes, etc. In a few cases, a project agreement is worked out for an area's or even an industry's projects. Not all project agreements are spurred by open shop competition. The need for uninterrupted construction, special problems, etc., can all result in a project agreement. There is no question, however, that open shop competition has been a decided spur in recent years for project agreement consummation.

Daniel International will operate union only under project agreements. The fact that this company has the capability to work open shop (and does so 90 percent of the time) on the largest of projects, places unions in a frame of mind to make a deal.

Project agreements do not normally affect wage rates, although in recent years special deals on wages have been made. Usually project agreements purport to reduce labor costs by banning jurisdictional dispute interruptions; by including provisions for no strikes and arbitration of disputes arising from a contract; by reducing the number of nonworking union stewards; by reducing or eliminating coffee break periods; by permitting management to select its supervisors; by reducing the overtime rate from doubletime to time and one-half; and by eliminating travel time, standby crews, etc. Most project agreements do not, however, make changes in basic craft rules—that is, who performs what work—and therefore do not permit one craft to do incidental work in another craft. On the other hand, in recent years, a number of project agreements have widened the right of management to utilize helpers and other subjourneymen on jobs.

Without a doubt, project agreements improve the economic situation for contractors and users. They protect them from local pressures to alter terms and conditions of employment and they

[14] *Ibid.*, p. 98.

provide the means to settle disputes. Moreover, both contractors and users usually have better control of costs and less interference with management prerogatives under project agreements. Finally, some recent project agreements also contain union wage and rule concessions. For example, the Houston Ship Channel Agreement, discussed in Chapter V, includes substantial reductions in wage rates, the use of subjourneymen for up to 40 percent of the workforce, and other major concessions.

Nevertheless, project agreements do not reach some key factors which give open shop contractors significant advantages. Foremost among these is the union craft structure and the problems and costs which are incurred because of the need to adhere to union jurisdictional lines in the deployment of labor. This factor is discussed in detail in Chapter IX.

Project agreements also may not provide relief in a key area that adds to contractor costs. For example, because of the opposition of the leadership of the International Brotherhood of Electrical Workers (IBEW) to a ten-hour day, four-day workweek,[15] the Houston Ship Channel Agreement did not permit this schedule without overtime after eight hours of work in one day. Again, as noted in Chapter V, this was one of the reasons why Brown & Root, rather than a unionized contractor, was awarded the 1983 Exxon refinery construction contract.

Project agreements can also be whittled down. Job actions on the part of unions can force contractors to give back what they have gained before construction commences. The Walt Disney World Epcot Center construction was done under a project agreement in Orlando, Florida, which is otherwise largely nonunion. The *Engineering News-Record,* however, quoted a contractor who was signatory to this contract as follows:

> "I have never seen worse labor relations. . . . We had sabotage, outbreaks of fighting, constant jurisdictional disputes, slow productivity, you name it." He says the project's labor agreement "read beautifully" but was impossible to enforce. "It gave us a reduction of 5 to 7%" for laborers, he explains. "That lasted about a week. After that, we had to pay a premium."
>
> Says one [other contractor]: "As it stands now, I wouldn't hire those [union] guys again no matter how bad the open shop [workers] are."[16]

Deals, protocols, or side letters made after construction commences

[15] See Chapter IV for a discussion of IBEW policy in this regard.

[16] ". . . as Contractors Sour on Hiring Them," *Engineering News-Record,* September 8, 1983, p. 60. See also, Allan Dodds Frank, "Comes the Mouse," *Forbes,* August 15, 1983, pp. 54–55.

can be very costly and have the effect of nullifying key sections of a project agreement. If unions refuse to live up to the special project arrangements, it is often difficult for the contractor or user to compel adherence. Another example involved a project agreement for the construction of a Florida electric power facility. It provided for the ten-hour day, four-day week schedule without penalty overtime pay after eight hours per day. It was signed by the IBEW representative but was repudiated by the IBEW national president. As a result, the contractor could only employ some electrical workers on a daily 8 a.m. to 4 p.m. schedule and others on a 10 a.m. to 6 p.m. schedule. This, of course, added to fringe benefit costs. Moreover, the pipefitters then demanded the same deal that the electricians had won.[17]

Daniel International built a second Florida power project, also under a union project agreement. When the IBEW refused to work the four-day schedule, Daniel employed an open shop contractor to do the electrical work. Similarly, Daniel built a facility in the St. Louis area under a project agreement. When the pipefitters refused to adhere to the agreement, an open shop contractor did the work.[18] A builder who works only union, however, does not have such freedom to seek alternatives in order to ensure the effectiveness of a project agreement.

Another problem is that contractors do not always make use of what they have gained under project agreements. The National Industrial Construction Agreement (NICA) of 1978, between the National Constructors Association (NCA) and eight construction unions, allows the use of up to 33.3 percent of the craft workforce as subjourneymen and/or apprentices with no apparent restriction. Yet a study of the Business Roundtable, published in 1982, found that "very little use has been made of this provision on jobs performed under NICA in the years since the agreement was signed."[19] Whether this is because of local union pressure or abdication of managerial responsibility is unclear. Managers who learn their work under a system which does not use subjourneymen are often insufficiently innovative to take full advantage of deploying labor more efficiently when the opportunity is presented. A study by the Mechanical Contractors Association of America (MCAA) also found

---

[17] See note 15, above; and field interviews, Tampa, Pensacola, Ft. Lauderdale, and Miami, January 21, 22, and 26, and March 14, 1982.

[18] Field interviews, Greenville, South Carolina, March 2, 1982, and St. Louis, Missouri, October 17, 1981.

[19] *Subjourneymen in Union Construction,* Construction Industry Cost Effectiveness Project, Report D-1 (New York: Business Roundtable, 1982), p. 6.

that, although the United Association of Journeymen and Apprentices of the Plumbing and Pipe Fitting Industry (UA) had permitted helpers in many contracts, local unions successfully resisted their use in a majority of cases.[20]

We may thus conclude that project agreements undoubtedly can improve the competitiveness of unionized contractors. Such agreements can return to management many important rights, which have been negotiated away or usurped by unions over the years, to manage more effectively. It is often up to management, however, to make sure that union concessions are used to the best advantage.

Project agreements also do not always fulfill their purpose because despite ironclad rules, jurisdictional strikes do occur. It takes just one ambitious business agent "to pull the men." Although such walkouts may be quickly settled, time and money are lost.

The project agreement is likely to reduce the economic differential between union and open shop operations, but at the same time, some very fundamental economic advantages remain with the latter. It is possible, through excellent management, high productivity, and skilled manpower, for union operations to eliminate this differential.

Other Special Contractual Arrangements

In a few locations, unions have established special companies or associations to compete. The Buffalo Building Trades Council Construction Corporation, a nonprofit construction corporation, was organized in 1984 by the unions in that city. They claimed that this corporation was necessary to compete with contractors which dealt with them but subcontracted most of their work to their double-breasted subsidiaries. The nonprofit corporation will be available on a craft-by-craft basis, to handle jobs on which users will be charged only for wages and fringe benefit payments set forth in local agreements. The Buffalo unions believe that this is the only way to offset the open shop competition from doublebreasted companies. A National Contractor Association official believes that this move is in the category of "rearranging the deck chairs on the Titanic."[21]

In Oregon and southern Washington, the construction unions developed a new agreement with an organization, the Union Con-

[20] *The Unionized Mechanical Construction Industry Continues in Crisis* (Washington, D.C.: Mechanical Contractors Association of America, 1983), p. 5.

[21] "Buffalo Crafts Create Nonprofit Company to Combat Pervasive Double-Breasting," *Construction Labor Report,* Vol. 29 (February 1, 1984), pp. 1509–10.

tractors Association, which apparently was established either by the unions or in consultation with them. This association was set up to compete with the Oregon-Columbia Chapter, AGC, the members of which have become increasingly open shop or double-breasted. Under the special standard agreement with the Union Contractors Association, the unions agree to have 20 percent of the wages of their members deducted and placed in an interest-bearing account. If the project utilizes union labor, the funds accrue to the contractors. If open shop contractors are used except where agreed to by the unions, the funds revert to the employees. A number of contractors have agreed to this arrangement, allegedly including some that originally planned to build open shop. Those that have announced agreement are building under Davis-Bacon regulations.[22] How wages in such circumstances can be reduced 20 percent, unless even with that reduction the net wages are prevailing, is not clear.

Pensions and Fringe Benefits as Organizing Devices

Because pensions are more common for union members than for those who work for open shop contractors, as will be discussed in Chapter XI, construction unions point to pensions and to other fringe benefits as reasons why employees and the public should support unionized construction. Moreover, pensions for unionized construction workers are generally negotiated between an employers' association and one or more of the craft unions. Such pensions usually cover all unionized contractors in the area and are portable; that is, a union member can change jobs and employers without losing pension credits or coverage of welfare plans. Most pension programs of open shop employers apply only to one contractor's employees.[23]

Claims of the superiority of union conditions are, of course, a standard union organizing device. In recent years, however, fringe benefits, and particularly pensions, have been seized by unions as a device both for encouraging unionization and for providing work for union members. Three major policies have been followed in this campaign: 1) to direct pension investments into sectors—for example, high-rise apartments—on condition that the construction

[22] "New Construction Labor Pact in Oregon Offers Several Innovative Features," *Construction Labor Report,* Vol. 29 (December 14, 1983), pp. 1371–72, 1383–91; "New Union Contractors Group in Ore. Is Beginning to Sign Up Members," *Construction Labor Report,* Vol. 29 (February 1, 1984), pp. 1511–12.

[23] The ABC and some other open shop groups do have portable welfare plans, as will be discussed in Chapter XI.

be union built; 2) to cause severe financial penalties to be levied against employers who withdraw from multiemployer pension plans; and 3) to attack nonunion pension plans as illegal. In these areas, the AFL-CIO Building and Construction Trades Department has a leading role.

The use of pension funds for social, union, or other purposes has often been advocated,[24] but the huge increase in pension assets since 1950 has made such attempts both more common and more insistent.[25] Investment of pension funds has been utilized as a tool to pressure companies which have defense contracts, which have investments in South Africa,[26] which do not recognize unions, or which otherwise offend some group. Likewise, pension funds have been utilized to help save New York City from bankruptcy, to provide employment in a particular state or area, or for other divergent purposes. In all such cases, the primary purpose of pensions—income for the retired—has been being reduced to a secondary objective. This raises many serious questions of policy and of law, particularly in regard to Section 404(a) of the Employee Retirement Income Security Act of 1974 (ERISA). This section governs fiduciary responsibility under the act by imposing stringent standards upon fiduciaries, who must bear in mind that when considering the ERISA legislation, Congress rejected various proposals for broadening the fiduciary standards of the bill in order to allow for the ramifications of divergent investing.[27]

The AFL-CIO is leading a drive to broaden investment opportunities for pension funds, so that these funds can be utilized to finance construction and other activities favorable to organized labor. Several state pension funds, for example, those of California and New York, have acceded to this idea. Of course, the unions want funding

---

[24] See the discussion from varying points of view in James T. Bennet et al., "Symposium: Union Use of Employee Pension Funds," *Journal of Labor Research,* Vol. II (Fall 1981), pp. 181–309; and Herbert R. Northrup and Richard L. Rowan, eds., *Employee Relations and Regulation in the '80s,* Labor Relations and Public Policy Series, No. 22 (Philadelphia: Industrial Research Unit, The Wharton School, University of Pennsylvania, 1982), pp. 409–64.

[25] As of 1979, total private and public pension plans, exclusive of the Old Age, Survivors Disability and Health Insurance (OASDHI), had assets of $619.1 billion. State funds accounted for $178.9 billion of this total, noninsured risk plans, $236.8 billion, and insured private, $20.4 billion. See Bennet et al., "Symposium," pp. 188–89, 221, 223–24.

[26] Interestingly, no equal opprobrium seems to be directed against companies or financial institutions which deal with the Soviet bloc.

[27] See James P. Northrup, "Investing Pension Assets for Social or Union Purposes: Management Analysis" in Northrup and Rowan, *Employee Relations,* pp. 441–42, for a capsule analysis of fiduciary requirements under ERISA.

only for unionized construction, for they see this as an opportunity to reclaim lost territory.

The potential impact of pension fund investment, however, can easily be overstated. Union control of pension funds, although relatively high in the construction industry, remains limited in most other industries. Union investment in housing developments or in other construction projects has had varied results. Overall, it is not believed that such investments could significantly alter the union-open shop balance.[28] Nevertheless, it is likely that the use of pension funds to create jobs for union members will increase.

In 1980, Congress enacted the Multiemployer Pension Plan Amendments Act, which was supported by a diverse coalition that included both employers and unions, especially the AFL-CIO construction unions. This law was designed to correct various problems which arose for multiemployer plans under ERISA. In particular, there was concern that many employers would withdraw from financially troubled multiemployer plans in order to avoid future liabilities that might arise if the plan were terminated, leaving those employers who remained within the plan and taxpayers to pay for a major share of multiemployer plan liabilities pursuant to ERISA.

To correct these perceived deficiencies, a number of changes were made in the ERISA requirements. Among the most important for this analysis were provisions to make employers withdrawing from multiemployer plans liable for the unfunded vested liabilities occurring during their period of contribution to the plan. The theory was to make all participating employers, not just those who stayed with the plan or the taxpayers, pay an appropriate cost.

The 1980 multiemployer pension law defined withdrawal from a multiemployer plan in general, but also included special definitions for the trucking, entertainment, and construction industries. According to this law, withdrawal in general occurs if the contractor either permanently ceases to have an obligation to contribute under the plan, or permanently has ended all covered operations under the plan. Withdrawal from a multiemployer plan in the construction industry, however, is defined as occurring only if the contractor ceases to have an obligation to contribute to the plan and continues to perform the same type of work in the jurisdiction of the collective bargaining agreement, or resumes such work within five years without then renewing the obligation to contribute.

To thwart doublebreasted companies that maintain a small operation subject to the plan in order to avoid withdrawal liability

[28] On this point, see particularly, James P. Northrup, "Investing Pension Assets."

assessments, the 1980 law also provides that a partial withdrawal occurs when "the employer's obligation to contribute under the plan is continued for no more than an insubstantial portion of work in the craft and area jurisdiction of the collective bargaining agreement of the type for which contributions are required." The construction unions were responsible for including this section in the law.[29]

Once an employer is found to have totally or partially withdrawn from a plan, his liability is determined by formulae written into the law.[30] The idea is to make the withdrawing employer pay his share of the plan's unfunded liabilities for the years in which he was required to contribute. The amounts which the sponsors of pension plans—in effect unions and employers remaining in the plans—have assessed on withdrawing employers is sometimes staggering and has prompted considerable litigation. Moreover, the law required withdrawal liability payments by contractors who terminate their plan affiliation retroactive to April 28, 1980. The constitutionality of this provision was upheld by the United States Supreme Court in June 1984, but in the tax bill signed four weeks after the Supreme Court decision, the effective retroactive date for pension liability was changed to September 26, 1980, the date that the Multiemployer Pension Act was signed. This not only released some companies from liability, but also required that many multiemployer funds refund several million dollars to those that paid the liability charges.[31]

The extent to which this law has achieved the objectives of the construction labor unions is difficult to determine. A study by the AGC found that 48.7 percent of the more than 650 multiemployer pension plans in construction were fully funded and therefore had no withdrawal liability. On the other hand, the other 51.3 percent were not fully funded, and for many, the liabilities were substantial. The funding of pension plans may have been understated in this study because these data were for 1978 and 1979, and following the

---

[29] The booklet prepared and distributed by the Associated General Contractors, Inc., *How the 1981 Multiemployer Amendments to ERISA Affect the Construction Industry. A Practical Guide to Understanding the Multiemployer Pension Plan Amendments Act of 1980* (Washington, D.C., 1981), has been very helpful in writing this section. See also William C. Martucci and John R. Utz, "Special Construction Industry Provisions under the Multiemployer Pension Plan Amendments Act," *Labor Law Journal,* Vol. 35 (May 1984), pp. 289–300.

[30] See the ABC booklet, note 29, pp. 19–27.

[31] Pension Benefit Guaranty Corporation v. R.A. Gray & Company, — U.S. — (June 18, 1984); *Daily Labor Report,* No. 118 (June 19, 1984), pp. D-1-6; Stephen Wermiel, "Tax-Bill Pension Provision is Apt to Switch Burden of Costs Within Several Industries," *Wall Street Journal,* July 18, 1984, p. 6.

enactment of the 1980 multiemployer pension law, many unionized contractors went to considerable lengths to improve the funding of such plans.[32] Although there are no data as to whether the law restrains contractors from leaving collective bargaining units, or conversely, whether the specter of being saddled with increased liabilities induces contractors to decline to join such units and to build under union contracts, the liabilities have obviously locked some contractors in and kept others out of bargaining units. The National Constructors Association is so certain that the law's impact has kept more contractors out than in, that it has employed former Secretary of Labor Ray Marshall, a close associate of the unions, to study the problem.[33] The elimination of retroactive liability for some contractors by the 1984 law further adds to the costs of those remaining in the multiemployer pension plans and could be an additional disincentive to contractor participation, thus subverting the unions' goals in sponsoring the legislation.

It is clear, however, that unionized contractors have worked energetically to improve the funding of multiemployer pension programs since the passage of the 1980 law, and that this can put them at a competitive disadvantage. Contractors have been insisting that a larger portion than heretofore of negotiated economic benefits be directed to reduce unfunded pension liabilities. This has both reduced these liabilities and increased costs that cannot easily be adjusted downward to meet competition.

Open shop contractors have also found that they can pay benefits at less costs than are necessary to operate union contractor programs. This is especially important where Davis-Bacon wage rates are established. Historically, the open shop builders competing on a Davis-Bacon, or state Little Davis-Bacon–controlled job, put the cost of fringe benefits into wages as required for companies without pension or benefit plans. This, however, put open shop builders at a disadvantage because social security, workers' compensation, and other government-mandated fringe benefits must be paid on the basis of total wages. Where, however, the open shop builder can obtain a benefit package at a lower cost than that of the negotiated multiemployer plan, the advantage shifts to his company, as shown in Figure VIII-2. Such open shop plans can be secured from the various associations or insurance companies, or from other carriers, as discussed in Chapter XI. The National Western Life Insurance

---

[32] "Factors May Cut Pension Liability," *Engineering News-Record,* March 3, 1983, p. 61.

[33] "Contractors Find Cure for Withdrawal Liability," *Engineering News-Record,* March 29, 1984, p. 168.

FIGURE VIII-2

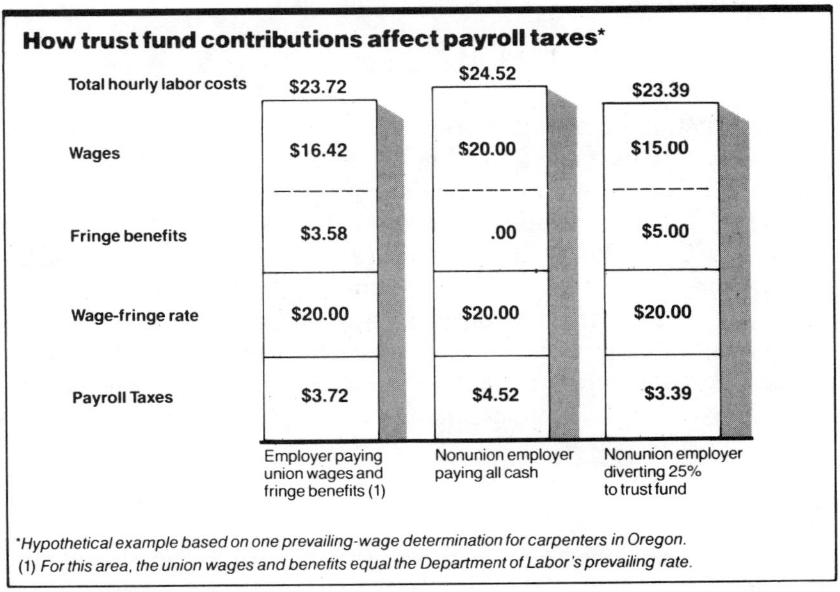

Source: Reprinted from *Engineering News-Record,* August 18, 1983, copyright, McGraw-Hill, Inc.

Company, Austin, Texas, established the Builders, Contractors, and Employees Retirement Trust and Pension Plan in 1977 for this purpose. It has been sufficiently successful for the AFL-CIO to attempt to restrict its activities, and by February 1984, it had agreements with 900 contractors.[34]

Pensions are thus a two-sided weapon in the open-union shop conflict. Divergent investing of pensions has created jobs for union members and may well increase the union job market in the future. It involves, however, a less prudent utilization of pension resources that could risk the welfare of retired union members. The Multiemployer Pension Plan Amendments Act of 1980 has kept contractors both in and out of the union fold. The act has also caused contractors to insist that the funding liabilities of pension plans be reduced by channeling a great proportion of economic benefits to that end. The effect has been to improve the viability of plans, but also to raise their costs; this has decreased the ability of unionized contractors to compete. Meanwhile, nonunion pension and benefit

[34] "Nonunion Pensions Draw Union Ire," *Engineering News-Record,* August 18, 1983, p. 74; Interview, National Western Life Insurance Company, Miami Beach, Florida, February 20, 1984.

plans appear to be lower cost and, consequently, may be expected to increase in use, as discussed in Chapter XI.

LEGAL RESPONSES

Construction unions have tremendous bargaining power. Because lost time is so expensive to a contractor, striking often forces contractors to accept union demands. By keeping materials from entering a job and preventing employees from working, picketing can shut down a job. Boycotting can thus cause severe losses to suppliers, contractors, and owners by preventing a product from being used on a job. Unions are also often able to coerce contractors to sign agreements that protect union interests. This section analyzes some union approaches and the laws under which they can be challenged, discusses the attack on doublebreasted operations, and assesses union political power.

Secondary Boycotts and Picketing

The nation's labor laws attempt to regulate union power in a manner that balances the contending forces in the interest of the public good. The Norris-LaGuardia Act of 1932 gave union strikes, picketing, and boycotts almost complete protection and license. The original National Labor Relations (Wagner) Act of 1935 greatly enhanced union power without restraining any of the freedom granted to unions by the Norris-LaGuardia Act. Twice since then, however, Congress has attempted to bring the nation's labor laws, and particularly unions' use of strikes, picketing, and boycotts, into better balance.[35] In the National Labor Relations Act amendments of 1947, known as the Taft-Hartley Act, and the Labor and Management Reporting and Disclosure (Landrum-Griffin) Act of 1959, sections were included which were designed, in part, to restrict union power in the use of strikes, picketing, and boycotts. Interpretive difficulties, inconsistent rulings by the National Labor Relations Board (NLRB), and courts favorable to union interests have given construction unions considerable latitude to utilize these weapons of conflict in their drive to restrict open shop builders.[36]

---

[35] For a layman's discussion of these and other labor laws, see Gordon F. Bloom and Herbert R. Northrup, *Economics of Labor Relations,* 9th ed. (Homewood, Ill.: Richard D. Irwin, Inc., 1981), Part Six.

[36] This discussion is based upon Ralph M. Dereshinsky et al., *The NLRB and Secondary Boycotts,* rev. ed., Labor Relations and Public Policy Series, No. 4 (Philadelphia: Industrial Research Unit, The Wharton School, University of Pennsylvania, 1981).

Thus unions are allowed to publicize a dispute by keeping pickets before an open shop construction site, but they must claim no organizing interest or the picketing becomes organizational picketing and can be enjoined in due course. Since the unions would be most happy to remove pickets if the job becomes union, the distinction between publicity picketing and organizational picketing is, to the layman, often more fictional than factual.

The consequences of publicity picketing vary from area to area and job to job. In certain highly unionized areas, it can seriously retard deliveries, cause customer hardship, and induce employees to leave. In areas dominated by the open shop, its impact is slight at most. Certainly, the potential impact of publicity picketing is a factor considered by users and open shop builders in determining whether it is feasible to build in an area.

To mitigate the effect of publicity picketing, employers have devised the "reserved gate" doctrine: a user can establish a separate entrance to the work site which is utilized only by contractors engaged in the construction. Picketing must then be confined to this particular gate and is an unfair labor practice if extended to other facility areas. What constitutes a reserved gate and what is considered legitimate conduct have not yet been firmly established by the courts.[37]

Secondary boycotts are an attempt to apply pressure on one party by economic action against a second party. The second party is thus induced to pressure the first party to accede to the unions' demands in order to relieve himself of a dispute in which he (the second party) is not directly involved. In a construction setting, for example, a union that picketed a supplier, or refused to handle that supplier's materials on construction sites because the supplier sold materials to an open shop builder, would be engaged in secondary boycott activity.

Whether such boycotting is deemed illegal, however, depends on numerous factors. The supplier's relationship to the open shop builder, various other relationships, proof of union-directed actions and motives, among others, are subject to interpretation by the NLRB and the courts. Moreover, the inconsistency of these decisions often tends to induce third parties to refrain from dealing with open shop builders in order to avoid such disputes. This encourages unions to inflict damage on open shop builders while the law is being tested. In the words of the most complete study of secondary boycotts:

---

[37] *Ibid.*, Chapter II.

> Congress first enacted a statute proscribing secondary boycotts in 1947. In the thirty-four years that have intervened since the passage of this statute, no clear definition of "secondary activity" has yet emerged. No consensus has been reached as to the scope of section 8(b)(4)(B), and there is no clear delineation of the types of conduct prohibited by the statute or the activity a labor organization may legitimately pursue to advance its economic interests.
>
> The language of section 8(b)(4)(B) itself does little to clarify the meaning of the elusive term "secondary boycott." Nowhere in the statute are definitions provided for such critical phrases as "primary strike" or "primary picketing" as those terms are used in the proviso to the secondary boycott provision. The NLRB and the courts have thus been forced to turn to the common law and legislative history to solve difficult questions arising under section 8(b)(4)(B). Neither source, however, has provided much additional guidance, frequently leaving the Board and the courts to their own devices in adjudicating complex issues.
>
> The case law under section 8(b)(4)(B) has thus been characterized by frequent reversals of policy and disagreements between the NLRB and the courts, and among the courts themselves. The Supreme Court has rendered very few decisions in the secondary boycott area, with each decision generating more confusion and leaving unsettled more questions than were resolved.[38]

Construction unions are encouraged by such inexact interpretations of the law to engage in extensive picketing and boycotting of open shop employers, their suppliers, and their users in the hope that they can sufficiently damage them economically so that unionized contractors will be utilized in their place, or at least on future projects. This is one reason why the authors of the study of secondary boycotts believe that the law should "clearly identify and enumerate the specific conduct that is deemed unlawful" and that it should "give stronger weight to the rights of neutrals that it originally set out to protect."[39] Until that happens, construction unions will continue to utilize the weapons of conflict to the maximum extent possible, causing open shop builders, at least, extensive and expensive litigation and, at worst, harassment and interference with their work.

Subcontracting Clauses

A potent weapon of the construction crafts is the subcontracting clause, which requires that work subcontracted by a unionized general contractor be given to a unionized firm. As explained in Appendix D, which deals with antitrust litigation, the courts have

---

[38] *Ibid.*, p. 331.
[39] *Ibid.*, p. 332.

ruled that such clauses are legal. They are being strongly pushed in some areas. For example, the carpenters' union in Southern California struck for four months in 1983 to obtain such a contract.[40] Subcontracting clauses are obviously potent in preventing subcontracts to nonunion firms, but of course, they may also encourage contractors and users to go completely open shop when feasible. Such clauses are most effective for unions in regions like Southern California, where unions are very strong, but have perhaps a reverse impact in an area like Baltimore, where open shop builders are readily available to take over all the work.

In late 1983, stronger versions of subcontracting clauses were developed by the Boston construction unions, the Asbestos Workers in Philadelphia, and the Sheet Metal Workers in Delaware. These clauses prohibit signatory contractors from working outside the labor agreement (for example, with a doublebreasted concern) as construction managers on a nonunion project, and prohibit them in their capacity as construction managers from awarding nonunion contracts. The most stringent was that obtained by Asbestos Workers Local No. 14 in Philadelphia, which precludes any direct or indirect interest of the contractor in any doublebreasted entity—whether financial, operational, or in any way involving one's associates or family. This extraordinary contract also includes a penalty for damages against the signatory contractor even if an arbitrator, the NLRB, or a court found that the nonsignatory firm was not subject to the agreement![41]

Serious questions about the legality of such clauses have been raised. For example, Peter G. Nash, former NLRB general counsel, believes that such clauses are in violation of Section 8(e) of the Taft-Hartley Act, which outlaws so-called hot cargo agreements that force a company to cease doing business with another company and that do not come within the exemption of Section 8(e). Even if legal, Nash believes that these agreements involve permissive rather than mandatory subjects of bargaining and, therefore, cannot be enforced by striking or picketing.[42] Edward B. Miller, former chairman of

[40] "Los Angeles Carpenter Negotiations Move Closer to Contract Settlement," *Construction Labor Report,* Vol. 29 (November 16, 1983), pp. 1239–41.

[41] For reports on these agreements, see "Boston Carpenters, Bricklayers Demand Work Preservation Language," *Construction Labor Report,* Vol. 29 (July 20, 1983), pp. 689–90; "Boston Bricklayers and Carpenters Come to Terms with Contractor Groups," *ibid.,* Vol. 29 (August 10, 1983), p. 813; "Philadelphia Asbestos Workers Win Work Preservation Language," *ibid.,* Vol. 29 (September 21, 1983), p. 983; and "Delaware Sheet Metal Workers Win Work Preservation Clause in 11-Week Strike," *ibid.,* Vol. 29 (September 21, 1983), pp. 987–88.

[42] Mr. Nash's opinion is set forth in *Cockshaw's Construction Capsules,* Vol. 13 (October 1983), No. 10.

the NLRB, expressed a similar opinion which is found in Appendix D. Nevertheless, the general counsel of the NLRB, whose term expired in April 1984, refused to issue a complaint against such clauses, and the Taft-Hartley Act provides no appeal from his decision. Many observers believe that a less pro-union successor may well reverse this decision.

These new work preservation clauses have initiated considerable concern. Open shop and doublebreasted companies are understandably worried as to their possible reach, especially since the AFL-CIO Building and Construction Trades Department has provided detailed instructions in its *Organizers Handbook* on how to negotiate and defend a clause modeled on that of the Philadelphia Asbestos Workers.[43] As noted in Appendix D, it is likely that antitrust cases will be brought against the unions, but their outcome is uncertain. Another reaction to subcontracting clauses, according to a telephone survey in late 1983, was a movement by some AGC and specialty contractor associations' members to withdraw from their organizations, either in order to go open shop or to bargain as individual firms.[44] The National Insulation Contractors Association (NICA), whose membership until that time was 90 percent unionized specialty firms dealing with the Asbestos Workers, following the signing of the agreement in Philadelphia announced that it was establishing "an open shop committee in the association to accommodate members who want to operate as nonunion contractors."[45] Although this association said that it would continue its union relationship, it noted that it had been losing members and income as members resigned and operated open shop. According to the executive vice president of this association, NICA was "one of the last associations to take this step," and changed only because of pressure from members and dwindling income. The NICA "could not hold out any longer in acknowledging the open shop. . . . we waited so long, we're playing catchup now."[46]

Use of NLRB Election Machinery

The NLRB representation procedure is not well suited to the construction industry. Often by the time the procedure is completed,

---

[43] See, e.g., Wilfrid C. Rodgers, "Boston's Carpenters in Tentative Accord," *Boston Globe,* July 30, 1983. A copy of the April 1984 edition of the *Organizer's Handbook* is in the author's possession.

[44] Memorandum from contractor association, November 23, 1983, in author's possession.

[45] "Insulation Contractors Approve Creation of Open Shop Committee," *Construction Labor Report,* Vol. 29 (December 7, 1983), p. 1355.

[46] *Ibid.*

TABLE VIII-3
National Labor Relations Board
Representation Elections in the Construction Industry
Fiscal Years 1977–1982

Year	Number of Construction Industry Elections	Number of Eligible Voters	Percent Elections Won by Unions	Percent of Votes Cast for Unions	Percent of Total Eligible Voters in Units Won by Unions
1977	126	2,776	43.7	39.8	40.6
1978	186	4,514	51.1	42.7	44.7
1979	246	6,563	50.8	46.9	49.8
1980	240	6,531	46.3	40.9	44.3
1981	187	5,894	47.1	40.9	38.0
1982	134	8,006	33.6	20.8	12.7
Total	1,134	35,243	46.6	37.5	36.3

Source: Wharton Industrial Research Unit data bank of NLRB elections.

a construction job may be finished, particularly if there is litigation over the bargaining unit, election conduct, or other matters. Thus, unions are likely to attempt to organize nonunion employees by this procedure only where the project is a long-term one, or where some special conditions exist.

In fact, the construction unions are unlikely to use the NLRB representation procedure in any case. Their objective in attempting to maintain or to increase their share of work is most often not to organize nonunion employees but to displace them with union members or permit holders. Thus, it comes as no surprise that construction unions do not see the NLRB representation procedure as an important adjunct to their campaigns against open shop contractors and have even suggested repeal of the National Labor Relations Act. This, they believe, would benefit unions by eliminating restrictions on strikes, picketing, and boycotts.[47]

When construction unions do utilize the NLRB representation procedure, however, they achieve better-than-average results. Table VIII-3 shows NLRB election results in the construction industry for fiscal years 1977–1982. Table VIII-4 compares the results in the construction industry with those in industry generally for 1977–1981. Construction union elections never represented more than 3.2

[47] See the remarks of officials of the Bricklayers' and Carpenters' unions summarized in "Deterioration in Collective Bargaining Relationships Threatens Labor Statute," *Construction Labor Report,* Vol. 29 (September 28, 1983), pp. 1019–21; and the note on the speech of Professor John T. Dunlop, *ibid.,* Vol. 29 (August 10, 1983), pp. 810–11.

TABLE VIII-4

Comparison of National Labor Relations Board Election Results, Construction and All Industry Fiscal Years 1977–1981

Year	Construction NLRB Elections as Percent of Total Elections	Eligible Voters in Construction NLRB Elections as Percent of All Eligible Voters	Percent Elections Won by Unions		Percent Votes Cast for Unions		Percent Total Eligible Voters in Units Won by Unions	
			All Industry	Construction	All Industry	Construction	All Industry	Construction
1977	1.3	.05	46.0	43.7	42.9	39.8	39.2	40.6
1978	2.3	1.1	46.0	51.1	42.2	42.7	37.6	44.7
1979	3.1	1.1	45.0	50.8	42.6	46.9	36.7	49.8
1980	3.2	.02	45.7	46.3	47.8	40.9	37.7	44.3
1981	2.5	1.8	43.7	47.1	41.2	40.9	36.8	38.0

Source: Table 16 in Annual Reports of the National Labor Relations Board, 1977–1981; and Table VIII-3.

percent of the total elections and covered less than 2 percent of eligible voters. During most years, however, the construction unions won a higher percentage of elections and won representation over a larger percentage of eligible voters participating in elections than was the case in industry generally. The sharp drop-off in construction union election victories in 1982 is probably a recession-induced phenomenon, but the NLRB election procedure is not a significant aspect of the union reaction to open shop gains.

The Attack on Doublebreasted Operations[48]

Unions are understandably concerned about doublebreasted operations and anxious, if possible, to prevent companies that operate union from establishing a second company that operates open shop. They have attacked doublebreasted arrangements within the legal framework by recourse both to the Taft-Hartley Act and to antitrust laws. The Taft-Hartley approach is discussed in this section and the antitrust approach in the following one.

The most direct attack on the legality of doublebreasted operations is made by maintaining that the separation between commonly owned union and open shop companies is artificial and, therefore, the union contract applies to both. One may also argue that the open shop company has violated Section 8(a)(5) of the Taft-Hartley Act by not bargaining with the unions about wages and conditions of employment for employees of the open shop company. If the union charge is sustained, the company will be found guilty of an unfair labor practice and ordered to bargain, provided that the open shop firm is found to be a separate bargaining unit. If the open shop and unionized companies are ruled to be one bargaining unit, the contract covering the union shop firm presumably applies to the open shop firm as well.

In determining whether doublebreasted companies should be treated as a single entity or two separate ones, the NLRB has long followed the policy of examining the degree of integration especially in regard to:

> Interrelation of operations, centralized control of labor relations, common management, and common ownership or financial control. No one of these factors has been held to be controlling, but the Board opinions have stressed the first three factors, which go to show "operational integration," particularly centralized control of labor re-

[48] A. Samuel Cook, Esq., of Venable, Baetjer, and Howard, kindly provided the author with a memorandum that has guided him in this discussion. It is reproduced in Appendix C.

> lations. The Board has declined in several cases to find integration merely upon the basis of common ownership or financial control.[49]

Although these criteria have been approved by the Supreme Court and are followed in each case, each case is wholly decided on the merits.[50] Companies which have been successful in operating doublebreasted have moved through this legal sea very carefully, with excellent legal advice, for there are numerous cases on the subject.[51] It is unlikely that any large company which goes doublebreasted for the first time will not be challenged. For this reason, and in order to obtain a continuous organization, many large unionized companies have established their doublebreasted concerns by purchasing an already existing open shop firm. Usually in these cases, existing management continues to operate the open shop concern at its separate headquarters, and careful separation is maintained in all management functions, particularly labor relations and personnel.

Another method of attack on doublebreasted operations, which has received NLRB and court approval, is to demand that unionized companies disclose to the unions whether they are engaged in doublebreasted operations in order for unions to determine whether this is being done in a manner that results in companies evading their contractual obligations toward unions. In a key case, the Associated General Contractors of Louisiana, New Orleans District, refused to agree to a clause that would apply the contract to doublebreasted companies controlled by the association's members. The unions filed an unfair labor practice suit charging that information on doublebreasted involvement was needed in order to prepare for future negotiations. Both the NLRB and the courts upheld the unions' case. The net impact of this union legal victory remains to be determined.[52]

Besides legal doublebreasted operations, our field interviews uncovered many allegations that small contractors often worked nonunion whenever they could, despite being parties to union contracts. No quantification of this phenomenon is possible, but the consistency of such reports during the 1981–83 field interviews leaves little

---

[49] *Twenty-first Annual Report of the National Labor Relations Board* (Washington, D.C.: Government Printing Office, 1956), pp. 14–15.

[50] Radio & Television Broadcast Technicians Local Union 1264 v. Broadcast Service of Mobile, 380 U.S. 255 (1965).

[51] See the cases cited in the memorandum by Mr. Cook, Appendix C.

[52] National Labor Relations Board v. Leonard B. Hebert, Jr. & Co., 696 F.2d 1120 (5th Cir. 1983); 703 F.2d 557 (5th Cir. 1983); *cert. denied,* 104 S. Ct. 76 (1983). "Supreme Court Upholds Order Demanding Disclosure by Double Breasted Firms," *Construction Labor Report,* Vol. 29 (October 12, 1983), pp. 1077–78.

doubt that it is widespread. Informants stated repeatly that these small contractors could not obtain work under union conditions and simply operated nonunion whenever possible. Often they apparently have had no legal advice and their knowledge of the legal problems involved has been superficial at best. Typically, they are not challenged because they are both too numerous and too small for the unions to afford to attack them. Larger companies, however, have found that, if they wish to operate an open shop doublebreasted concern successfully, they must have competent legal advice at all times. A detailed legal memorandum on the subject, prepared by A. Samuel Cook, Esq., and Associates, is found in Appendix C.

Impact of Antitrust Laws

Antitrust legislation, beginning with the Sherman Antitrust Act of 1890, was designed to prevent combinations of economic power from restraining interstate and foreign commerce. Section I of the Sherman Act states: "Every contract, combination in the form of trust, or otherwise, or conspiracy in restraint of trade or commerce among the several states, or with foreign nations, is hereby, declared to be illegal." In 1908, in the case of *Loewe v. Lawlor,*[53] commonly known as the "Danbury Hatters" case, the Supreme Court ruled that a union boycott was an illegal restraint of trade within the meaning of the Sherman Act. Unions then strove for a legislative exemption, believing that the Clayton Act of 1916 provided such exemption, but the courts ruled otherwise.[54] Finally, in 1932, Congress passed the Norris-LaGuardia Act, which, with some broad interpretations thereof,[55] seemed to give unions almost complete exemption from the antitrust laws. Beginning in 1965 with the *Pennington*[56] case, however, the Supreme Court began to limit the perceived union exemption. Today, antitrust legislation is significant in labor relations legislation, particularly in litigation pertaining to the construction industry. Both open and union shop adherents now look to antitrust suits to advance their cause. A memorandum analyzing the current state of antitrust law as pertaining to the open shop-union shop controversy, prepared by

[53] 208 U.S. 274 (1907).

[54] Duplex Printing Press Company v. Deering, 254 U.S. 443 (1920).

[55] See Apex Hosiery Co. v. Leader, 310 U.S. 469 (1940); United States v. Hutcheson, 312 U.S. 219 (1940); and Allen Bradley Co. v. Local Union No. 3, International Brotherhood of Electrical Workers, 325 U.S. 797 (1944).

[56] United Mine Workers v. Pennington, 381 U.S. 657 (1964). For a discussion of these and subsequent cases of general interest, see Bloom and Northrup, *Economics of Labor Relations,* Chapter 18.

Edward B. Miller, Esq., former chairman, National Labor Relations Board, is found in Appendix D. He concludes that the number of antitrust cases involving the open shop-union shop conflict is likely to increase in the years ahead.

Political Pressure

Throughout this volume many instances of successful construction union political pressure were noted. The very existence of the Davis-Bacon Act and state "little Davis-Bacon" laws are testimony to the political power of unions. The failure of the NLRB's general counsel to issue a complaint against questionable union demands for expensive work preservation clauses, as described earlier in this chapter, is another example. Investment by several construction union pension funds at below market interest rates in order to spur union construction employment resulted in a unanimous recommendation by the Department of Labor to sue the trustees of two funds in April 1983. By late January 1984, the secretary of labor, following intensive union pressure, had taken no action.[57] The cities of Rochester and Oswego, New York, have enacted ordinances requiring bidders on city construction projects to have apprenticeship programs approved by the New York State Department of Labor. Such laws are clearly designed to eliminate open shop contractors since that department is notoriously pro-union[58] and since many open shop firms do not train through apprenticeship programs or through other approved plans.[59]

Construction unions lobby at all governmental levels in an attempt to offset the economic advantages of open shop construction by inducing various public bodies to increase construction and to utilize only union contractors. Such lobbying and pressure are not always successful. For example, the construction unions have been unable to induce Congress and the President to agree to "common situs" legislation which would permit virtually unlimited picketing and boycotts.[60] Yet there can be no doubt that the construction unions are an effective political force at the municipal, state, and

[57] Randall Smith, "Use of Union Funds to Create Union Jobs Raises Issue of Legality," *Wall Street Journal,* January 17, 1984, pp. 1, 16.

[58] The New York State Commission of Investigation has accused the state Department of Labor of wasting huge amounts in the administration of the "Little Davis-Bacon" law by policies which favor unions. See "New York Wage Law Blasted," *Engineering News-Record,* January 19, 1984, pp. 66–67.

[59] "Open-Shop Contractors Fight 'Biased' Bid Rules," *Engineering News-Record,* January 19, 1984, p. 67.

[60] Congress passed such a proposed law in 1976 but President Ford vetoed it. In 1977, the House of Representatives voted down a similar bill.

federal levels and are likely to continue to be so despite the decline in the ratio of unionized construction. They have the money, personnel, and talent, are experienced in political action, and continue to work energetically in order to offset the open shop's economic advantages by political restraints on market forces.

VIOLENCE[61]

Violence has been a factor in construction labor relations for many years. At the turn of the century, officials of the Ironworkers' union were convicted of dynamiting nonunion construction jobs.[62] Montgomery's classic history of industrial relations in the Chicago construction industry makes repeated references to bombings and other violence.[63] Ryan recounts one strike alone in San Francisco in which "over three hundred cases of violence occurred."[64] Seidman's study of labor racketeering is replete with instances of violence perpetrated against construction union members and contractors.[65] The McClellan congressional hearings of the 1950s contain numerous similar accounts.[66] Even such a sympathetic historian of a construction union as Mangum recounts the violence accompanying union organizing drives.[67]

That the increased penetration of open shop firms since the mid-1960s has led to a resurgence of violence seems incontrovertible. In 1972, the *Engineering News-Record* reported:

> Violence is the gut reaction of unions in some areas to counter the open shop. In recent years violence and property damage have hit

---

[61] This discussion, which builds upon our original book, has been updated and expanded from Armand J. Thieblot, Jr. and Thomas R. Haggard, *Union Violence: The Record and the Response by Courts, Legislatures, and the NLRB,* Labor Relations and Public Policy Series, No. 25 (Philadelphia: Industrial Research Unit, The Wharton School, University of Pennsylvania, 1983), Chapter VI.

[62] Luke Grant, *The National Erectors' Association and the International Association of Bridge and Structural Ironworkers* (Washington, D.C.: United States Commission on Industrial Relations, 1915).

[63] Royal E. Montgomery, *Industrial Relations in the Chicago Building Trades* (Chicago: University of Chicago Press, 1927). In the index, the word "violence" is followed by twelve separate citations.

[64] Frederick L. Ryan, *Industrial Relations in the San Francisco Building Trades* (Norman, Oklahoma: University of Oklahoma Press, 1936), p. 191.

[65] Harold Seidman, *Labor Czars: A History of Labor Racketeering* (New York: Liveright Publishing Corporation, 1938).

[66] Useful summaries of the McClellan hearings, in which the Teamsters and Operating Engineers were featured, are found in John Hutchinson, *The Imperfect Union: A History of Corruption in American Trade Unions* (New York: E.P. Dutton & Co., Inc., 1970); and Robert F. Kennedy, *The Enemy Within* (New York: Popular Library, 1960).

[67] Garth L. Mangum, *The Operating Engineers: The Economic History of a Trade Union* (Cambridge: Harvard University Press, 1964), pp. 181–95.

nonunion projects in Florida, Maryland, Tennessee, Georgia, Ohio, and Pennsylvania.[68]

The extensive study of labor union violence, published by the Wharton Industrial Research Unit in 1983, noted 319 reports of violence which occurred between 1976 and 1981 and involved building trades unions or councils. Of these, fewer than 100 were directed against employers or involved strikes or contract disputes of the traditional type. The great majority arose out of protests against nonunion or open shop work being done in an area where work had traditionally been done by union shops.[69]

The geographic extent of violence in construction is wide, and it involves most of the traditional building crafts. During the mid-1970s, the NLRB issued complaints on charges of coercive violence against four international construction unions, the Construction and Laborers Union, the United Brotherhood of Carpenters and Joiners of America, the United Union of Roofers, Waterproofers and Allied Workers, and the International Association of Bridge, Structural, and Ornamental Iron Workers, plus state and local union bodies of the Operating Engineers in Ohio, Michigan, New Jersey, and Pennsylvania, Ironworkers in Michigan and Massachusetts, Laborers in Kentucky, Plumbers in Michigan, Roofers in Pennsylvania, and Building and Construction Trades Councils in Michigan and New Jersey.[70] In litigation, some of the unions charged—notably those in Michigan—agreed to refrain from further violence as part of consent decrees, and others were found guilty of violent and coercive behavior.[71]

Besides these NLRB cases, considerable other litigation in various parts of the country has involved the issue of violence in the construction trades. The use of violence is both common and extensive in the industry. As examples of its scope and character, summaries of events which have occurred in recent years in four localities (Denver, Philadelphia, Michigan, and the Gulf Coast) are set forth below. These summaries are followed by more brief descriptions of violent activities drawn from other places around the country.

---

[68] "Open Shop: A Growing Force and a Catalyst for Change," *Engineering News-Record,* November 2, 1972, p. 20.

[69] Thieblot and Haggard, *Union Violence,* Chapter VI.

[70] This summary is based on the NLRB's general counsel's press release and *Order Consolidating Cases, Consolidated Complaints* and *Notice of Hearing* in Cases Nos. 8-CB-2112, 22-CB-2598, 1-CB-2316, and 7-CC-757, 7-CB-2834.

[71] Laborers' International Union of North America, 219 N.L.R.B. 142 (1975), *affirmed,* 532 F.2d 749 (4th Cir. 1976).

Denver

One of the most frequently used tools of labor violence is fire. Buildings under construction by nonunion contractors, lumber yards supplying them, completed projects built by them, construction offices and trailers, and even the homes of nonunion contractors have been set on fire at one time or another as a lesson in the added costs of nonunion work. This sort of activity was especially dramatic in Denver during the 1960s and 1970s.[72]

The construction-related fires in Denver involved principally apartment houses and condominiums, but other buildings and lumber yards supplying them were also hit. During this period, about forty such fires were reported which involved open shop builders or contractors who were involved in disputes with construction trade unions. The fires usually occurred when they could do the maximum damage—after a building was closed in, but before internal walls and ceilings were in place, for example—and the total amount of destruction during the period was estimated to be over $10 million.

Investigation of these fires was stymied for a considerable amount of time while attorneys for the Northern Colorado Building and Construction Trades Council vainly sought to suppress evidence obtained through wiretapping. The council appealed all the way to the U.S. Supreme Court, but the wiretap procedures were found to have been legal under Colorado law, and the evidence was eventually admitted. In 1975 and 1976, this, along with other evidence collected by the Colorado Bureau of Investigation, resulted in arson convictions and jail terms for a business agent of the Colorado Building and Construction Trades Council, two business agents of the Cement, Lime and Gypsum Workers' International Union affiliated with that council, and the executive secretary of the New Mexico Building and Construction Trades Council. During the trials, there was considerable evidence that this "arson organizing" extended beyond the Denver area, with Arizona, Kansas, and California all mentioned as other sites where it may have occurred. There have been no related arson indictments or convictions in those other areas.

In December 1981, the Denver headquarters of the Colorado Contractors' Association, the highway and heavy chapter of the Asso-

[72] The Denver fires story appeared regularly in the press there. See *Rocky Mountain News*, July 27, 1972; May 13 and December 14, 1974; April 10, June 26, September 24, and October 29–30, 1975; and January 30, 1976; and *Denver Post*, July 28 and 31, August 8 and 28, September 7, October 10, and December 29, 1972; March 1 and 2 and June 13, 1973; December 15, 1974; and January 9 and April 14, 1975. See also C. Stevenson, "Arson-to-Order in the Building Trades," *Reader's Digest*, March 1976, reprint, n.p.

ciated General Contractors (AGC) in Colorado, was hit by three fires in three nights. Damages caused by these obvious arsons were estimated at $100,000. As discussed in Chapter VI, the highway and heavy sector of Colorado is about 70 percent open shop (see Table VI-5). Earlier in 1981, the state's union highway and heavy contractors were struck by their unions for four months, which resulted in major gains for open shop contractors. The Colorado Contractors' Association has training and referral programs for open shop contractors which further upset the unions. The arsonist has not been apprehended.[73]

Philadelphia

Philadelphia, like many of the older cities of the East Coast and Midwest, is, as described in Chapter IV, a union town. The construction unions there have been key supporters of several mayors and enjoy a close relationship with City Hall. Philadelphia is also a city where construction union violence has been endemic. One of the most blatant and brutal cases involved Altemose Construction Company, an open shop contractor.

Altemose Construction Company.[74] In the early 1970s, a dispute arose between Altemose and the Building and Construction Trades Council of Philadelphia over Altemose's refusal to sign a "subcontractor's agreement" which would have obligated it to let subcontracts exclusively to firms using only unionized labor. Although Altemose was willing to use union subcontractors if they provided the best bid, it was unwilling to deny bidding rights to otherwise qualified nonunion firms.

The Altemose dispute reached its climax during the summer of 1972 after Altemose had been awarded the general contract for the construction of a hotel and office building complex called the Valley Forge Plaza in the suburbs of Philadelphia. In the words of the U.S. Court of Appeals for the Third Circuit:

---

[73] "AGC Chapter Burned Out of Denver Headquarters," *Engineering News-Record,* December 17, 1981, p. 172.

[74] The material for the Altemose story, unless otherwise noted, is based on numerous press stories and articles collected over the years. See, particularly, for the Altemose case, the (Philadelphia) *Evening* and *Sunday Bulletin,* June 5, 11, 12, 14, 15, 18, 20, and 22, July 16 and 17, August 19, 20, 21, 22, 24, and 29, September 15, 18, 19, 20, 21 and 28, and October 18 and 24, 1972; February 18, June 28, July 19, 24, 25, and 27, September 11, October 23, November 8, 10, 11, 22, and 30, and December 1, 5, and 28, 1973; January 20, February 6 and 9, March 3, April 4, 7, 11, and 14, September 8, October 2 and 20, and November 1 and 29, 1974; and January 14, 17, 21, and 22, and March 12, 1975; and *Philadelphia Inquirer,* May 26, and June 7 and 14, 1972. See also Anthony Gnoffo, Jr., "The Men Who Built King of Prussia," *Philadelphia Inquirer,* May 21, 1984, pp. 2M–4M.

> In the morning of June 5, 1972, approximately 1,000 men from the Trades Council arrived at the Plaza construction site in buses chartered by one of the Council's member unions and proceeded to destroy vehicles, equipment and other property belonging to Altemose, causing damage in excess of $350,000. Although Altemose immediately obtained a preliminary injunction in state court prohibiting picketing by members of the Trades Council within one mile of any Altemose site or its office building, a second mob descended the following day on the office of Altemose Enterprises and threatened to burn it down. Violence was prevented only through the combined efforts of state and local police. Subsequently, on August 17, J. Leon Altemose was attacked and beaten by members of the Trades Council in broad daylight, while attempting to enter a bank in downtown Philadelphia. Members of the Trades Council also put pressure on the employees of the construction site to cease working for Altemose. Edward Fitzpatrick, a business agent for one of the member unions in the Trades Council, called one of the subcontractors on the site, Richard Czeiner, and told him that the unions would prefer that he not work there. Czeiner refused to quit, however, and subsequently several of his vehicles were destroyed or damaged by fire bombs. Despite strong security measures taken by Altemose, construction on the project was plagued by sabotage.
>
> As a result of the violence and threats, many subcontractors withdrew from the site and Altemose and Energy [Czeiner] had considerable difficulty in obtaining qualified workers willing to risk work there.[75]

About two weeks after the initial violence against Altemose, vandals registered at a Sheraton Hotel in Philadelphia—the same chain for which Altemose was constructing the Valley Forge Plaza—and seriously damaged a number of rooms. After that, other Altemose projects were bombed and other violence occurred, all of it apparently directed toward intimidating present and potential customers of Altemose.

Altemose's suppliers and creditors were also pressured to cease doing business with the company. Users were warned not to put the company on their bid list. Altemose completed the Valley Forge Plaza, but the company has faced difficulties ever since, perhaps partly as a result of the weak construction market of the late 1970s and early 1980s but also because the unions have maintained their pressures on all fronts. In 1976, for example, the unions induced the governor of Pennsylvania, the host for the Governor's Conference that year, to move it away from Valley Forge so that the Altemose-built hotel would not receive the patronage.

---

[75] Altemose Construction Company v. National Labor Relations Board, 514 F.2d 8 (3rd Cir. 1975). See also Altemose Construction Company v. Building and Construction Trades Council of Philadelphia and Vicinity, Civil Action No. 73-173 (E.D.Pa., 1983).

The union attack on Altemose has been total, apparently in an endeavor to drive the company out of business. Thus, two plumbers sought employment with Altemose, and after being discharged, filed charges with the NLRB. When the NLRB hearing was held, the hearing room was ringed with demonstrators who threatened the Altemose participants. Yet the NLRB regional director did not order a change of venue and the administrative law judge credited the discharged plumbers, discredited the company witness, and found Altemose guilty of an unfair labor practice, a decision with which the NLRB concurred.

In a decision that was about as critical of NLRB procedure as any ever written, the Court of Appeals for the Third Circuit reversed this decision and castigated the NLRB for permitting a hearing under such circumstances, pointing out numerous discrepancies which reflected upon the credibility of union witnesses, and questioning several key rulings of the administrative law judge.[76] One point made clear by the court decision was that the unions were not interested in organizing Altemose's employees; their objective was to secure jobs for their members by forcing Altemose or other nonunion companies to hire only union subcontractors in their own specialized trades.

The combined pressures on Altemose were telling. By 1980, the company was considerably reduced in size, and its future as a building contractor was greatly diminished, although in 1984 it appears to be making a comeback. If the results were a victory for the unions, however, they were not without cost.

The violence and destruction that resulted from the mass picketing at Altemose in 1972 was blatant and undisguised, and it gave rise both to NLRB cases and to criminal charges. The first criminal trial involved a battery of lawyers and cost the Philadelphia Building and Construction Trades Council an estimated $300,000. Nevertheless, in it and in two other trials on related crimes and violence in Norristown, Pennsylvania, eighteen unionists, most of whom were members of Local 30 of the United Union of Roofers, Waterproofers and Allied Workers, were convicted of multiple accounts of violence, and eleven of them served time in jail.[77] The attack on Mr. Altemose, however, brought no convictions.

---

[76] *Ibid.*

[77] Besides the eleven who were sent to jail, seven were acquitted, and five received probation after their original convictions were overturned on appeal and they later pleaded guilty. See Sara Schwieder, "11 Roofers are Ordered to Prison for Rioting," *Philadelphia Inquirer,* June 12, 1980, p. B-4, and David Harris, "5 Roofers Get Probation in '72 Riot," November 4, 1980, p. F-1.

Roofers Local 30. To those familiar with construction labor relations in southeastern Pennsylvania, it is not surprising that members of Local 30 were involved in the Altemose case, as that local was long involved in construction-related violence. Indeed, it so prided itself on its tactics that at one time its members had brutalized open shop builders and workers with baseball bats specially inscribed with the Local 30 insignia. In addition, Local 30's business agent had rented the bus used for the attack on the Altemose construction site.

Injunctions have been issued against Local 30; it has been charged in a number of NLRB complaints;[78] and it has been the subject of feature articles in Philadelphia newspapers and magazines which documented with remarkable specificity its involvement with violence.[79] Nevertheless, during the early 1970s, with the exception of particularly blatant examples such as the attack on Altemose, its violence and transgressions went largely unpunished.

There can be little doubt but that the political atmosphere of Philadelphia during this time contributed to this neglect. The pro-union stance of the former mayor was well known,[80] and union officials were represented on all key city agencies dealing with housing, zoning, or building codes.

An example of the degree to which union violence and intimidation was condoned in Philadelphia at that time is found in an account of a Democratic Party dinner in the city in 1974. The mayor had not been invited to sit on the dais for the dinner because of a dispute with a city party chairman. The following scene evolved:

> While over 4,000 Democrats were gathering at Civic Center to dine and hear speeches, a flying squad of roofers burst into the hall. They overturned tables, beat up a 60 year old . . . [opposition member] and a number of deputy sheriffs who were serving as ushers, while the police looked the other way.
>
> A few minutes after the roofers had set the tone, [the mayor] . . . pulled up in his city limousine. [The business agent of the Roofers, the president of the Building and Construction Trades Council, and the business agent of Teamsters Local 107] were riding with him.

---

[78] The Board found Local 30 guilty of violence in at least two cases and issued remedial orders. See Roofers Local 30 and Associated Builders and Contractors, Inc., 227 N.L.R.B. 1444 (1977); and Roofers Local 30 and Edward R. Kitson and Martin J. Sobol, 228 N.L.R.B. 652 (1977).

[79] See Jim Riggio, "The Hardhats' Holy War," *Philadelphia Magazine,* Vol. 63 (August 1972), pp. 55–61, 126–38; and the series by L. Stuart Ditzen, "The Roofers: A Study of Union Violence," (Philadelphia) *Evening Bulletin,* July 15–18, 1974.

[80] See, e.g., "Union Men Held in $60,000 Vandalism," (Philadelphia) *Evening Bulletin,* April 10, 1975, p. 6.

> Led by a fife and drum corps, . . . [the mayor] and 500 union toughs stormed into the Civic Center, completely upstaging the Democratic politicians and scaring hell out of everyone there. . . . They encircled the diners and dignitaries, chanting obscenities . . . and knocking people down along the way. The union chiefs formed a phalanx around [the mayor], who boasted that it was the greatest night in his life.[81]

With the advent of a new mayor, less beholden to the construction unions but equally anti-open shop, and the successful prosecutions in the courts and by the NLRB, the violent activities of Local 30 quieted somewhat during the late 1970s, at least insofar as they were directed against the nonunion construction industry. Local 30 then turned the focus of its attention to Atlantic City and the burgeoning new market there brought on by legalized gambling. It attempted to establish a guard employees' local subsidiary there. On December 16, 1980, John McCulloch, business agent of Local 30, was murdered in his home by Willard E. Moran, who was later convicted and sentenced to death. Moran then allegedly divulged information that implicated the business agent of the rival Atlantic City guards' union and a racketeering figure who had also been charged in the killing. Press stories linked McCulloch with the late Angelo Bruno, longtime alleged head of the Philadelphia mafia.[82] Racketeering and violence have often gone hand-in-hand in the construction industry.[83]

Schnabel Associates. With the death of its longtime business agent, Local 30 ceased to be at the cutting edge of Philadelphia

[81] Mike Mallowe, "You Got a Friend in the Business," *Philadelphia Magazine,* Vol. 65 (September 1974), pp. 118–23, 200–13.

[82] According to press stories, Angelo Bruno was murdered along with several of his alleged associates in what was termed a fight with their New York City counterparts over jurisdiction in Atlantic City. Control of the restaurant guards' and related union there is thought to be the key to racketeering control of that obviously lucrative site. See, e.g., the series of articles in the (Philadelphia) *Evening Bulletin,* June 7–10, 1981; *Philadelphia Inquirer,* December 17, 18, 21, 1980, and September 8–10, 1982; and also the series in the *New York Times,* August 29–31, and September 1, 1982.

[83] See Seidman, *Labor Czars.* In New York City, where violence and racketeering is apparently endemic in construction, this situation has been clear for many years. See, e.g., the series on the subject in the *New York Times,* April 25–27, 1982. These articles have led to an official inquiry by the state of New York. At one of the hearings, testimony was heard concerning violence at construction sites. See Selwyn Raab, "State Panel Told about Violence in Construction," *New York Times,* September 16, 1982, p. B-3. A few weeks later, a Teamster official implicated in both violence and racketeering in New York construction was found guilty of seven charges of labor racketeering and tax evasion. See Selwyn Raab, "Key Teamster Leader Is Convicted of Labor Racketeering by L. I. Jury," *New York Times,* October 9, 1982, p. 1. For other New York racketeering-labor convictions, see "Union Aide Accused of Million in Bribes," *New York Times,* June 8, 1979, p. B-2, and "Union Chief Facing Trial Feared Dead," March 25, 1982, p. B-8.

construction violence, but violence associated with the Philadelphia construction trades has, nonetheless, continued. Its most frequent target in the early 1980s has been the firm of Schnabel Associates, which grew from a small Philadelphia suburban concern to one operating all over the Eastern Seaboard, and which rose in 1983 to No. 263 on the *Engineering News-Record* list of the 400 largest building contractors. Schnabel was always operated open shop.

As it has grown, this firm has endured picketing and violence in a number of areas, but only since 1980 has it become a prime target of violence. In 1982, Schnabel was building at five sites in the Philadelphia area. The building trades unions picketed at all the sites, interfering with deliveries, harassing and threatening employees, and destroying and stealing equipment. According to the *Engineering News-Record,* the violence was particularly rough during the construction of a $3.3 million home for the elderly in Boothwyn and a middle school in Radnor, but pickets were also very much in evidence at the site of a Haverford condominium and elsewhere.[84]

When the problems began at the Boothwyn site, Schnabel began videotaping picketing activity and maintaining a journal of the day-to-day events which transpired at all five of the firm's active work sites around Philadelphia. After viewing hundreds of hours of the tapes, a reporter for the *Engineering News-Record* wrote:

> The tapes show pickets beating company employees, spiking and slashing tires, throwing rocks and even a stolen chain saw through the windows of cars and trucks entering the jobsite. In the films, union business agents involved in the picketing appear to be directing part of the activity, and local law-enforcement officials are seen standing idly by as pickets violate court injunctions against mass picketing and violence.[85]

The attorney for the construction unions, who had represented them throughout the period covered by this discussion, made the purpose of this harrassment very clear. Again as reported by the *Engineering News-Record:*

> Building trades attorney Bernard Katz says his clients are using the same tactics to fight Schnabel that they have been using against nonunion contractors for "more than 30 years. What we want to do, quite simply is erase the competitive advantage Schnabel has by virtue of his hiring workers at substandard wages."
>
> According to Katz, this is done by several means, all of which "make it difficult for him to do business here." Among the extra costs that

[84] See Kathy Boccella, "Middle School Project Halted after Union Pickets Close Site," *Ardmore* (Pa.) *Main Line Times,* November 12, 1981, p. B-2, 14.

[85] "Unions, Open Shop Clash Over Jobs," *Engineering News-Record,* May 27, 1982, p. 10.

> Schnabel has to bear, he says, are higher prices for building materials—"no driver wants to cross a picket line"—and a disproportionate amount of money spent on security and legal fees. Even if Schnabel completes the Boothwyn project, says Katz, "We are sure we will see less of him in the future" because of the building trades' efforts.[86]

In light of these comments, it is difficult to disagree with the claim of Schnabel's attorney that the violence perpetrated by the union is "planned and calculated."[87] Schnabel has filed a multi-million-dollar damage suit against the Building and Construction Trades Council of Philadelphia, four local unions, and six union officials, but Attorney Katz states that the construction unions "are not the least bit concerned. Believe me, we lost our lawsuit virginity a long time ago."[88]

Other Philadelphia Violence. In 1983, the Carpenters' union emerged as the leader in Philadelphia construction union violence. Small jobs and contractors were picketed, individuals were threatened and attacked, and construction work was destroyed.[89] In one incident "About 100 construction workers smashed windows, tossed bricks and pushed a piece of construction equipment into a hole" at the construction site of the Ukranian Self Reliance Federal Credit Union in Northeast Philadelphia, where the members of the credit union were building the edifice themselves. According to Attorney Katz:

> There was picketing, and there was an assault on the pickets by scab laborers. Periodically we are exposed to provocation of people who resent our picket lines and when that happens you have to defend yourselves.[90]

The "defense" consisted of $6,000 worth of damage against a self-help group which no other witness saw assault anyone. Indeed, the action so antagonized people in the area that a local of the Sheet Metal Workers had its apprentice class rebuild the damage and install a modern heating and cooling system without charge.[91]

Many other actions of violence have occurred: attacks on Neshaminy Constructors and the Steelworkers were described in Chap-

[86] *Ibid.,* p. 11.
[87] *Ibid.,* p. 10.
[88] *Ibid.,* p. 11.
[89] See, e.g., Lucinda Fleeson and Edward Colimore, "Nonunion-Labor Dispute Turns Violent," *Philadelphia Inquirer,* November 18, 1982, pp. A-1, 4; Lucinda Fleeson, "Restaurant Owner, Union Settle Dispute," *Philadelphia Inquirer,* September 27, 1983, pp. B-1–2; and *idem.,* "Carpenters Build Stress in Jobs Push by Union," *Philadelphia Inquirer,* December 5, 1983, pp. A-1, 4.
[90] Fleeson and Colimore, "Nonunion-Labor Dispute," p. A-1.
[91] Lucinda Fleeson, "Union's Labor of Good Will. Vandalism Prompted Free Work on Building," *Philadelphia Inquirer,* October 29, 1983, pp. B-1–2.

ter VI; because the Southeastern Pennsylvania Transportation Authority (SEPTA) has constructed various facilities with its in-house workers who are unionized, it has been picketed and its entrances blocked;[92] and both large and small companies have been harassed, with violence concentrated against the smaller ones by the Carpenters.[93] Police protection has been meager. Obviously, the penalties handed out to those who attempted to destroy the Altemose business has not deterred their successors. Pickets even assaulted Robert Wholsen, Sr., a leading open shop contractor, so severely that he was required to undergo brain surgery for a subdural hemorrhage. Wholsen's company took over from striking union members at the DuPont site discussed in Chapter V. The assault occurred at Chadds Ford, Pennsylvania, about thirty miles northwest of Philadelphia where the Wholsen company was building a museum.[94]

Michigan

Although Detroit is very much a union town, over the past few years, open shop construction has been gaining ground in other areas of Michigan, and there have even been some inroads into Detroit itself. The construction unions responded with violence which reached a crescendo in the early 1970s at Kalkaska and Midland, and resurfaced at Port Huron in 1983.

Kalkaska.[95] The town of Kalkaska is a community of 1,600 inhabitants in the northern sector of Michigan's southern peninsula. In July 1971, the Shell Oil Company announced plans to build a gas processing plant there. Shell awarded the contract to the Delta Engineering Company of Houston, a large open shop builder with considerable experience in this type of construction.

---

[92] Dominic Sama, "SEPTA Workers Enter Delco Construction Site," *Philadelphia Inquirer,* July 21, 1982, pp. B-1–2.

[93] Terry E. Johnson, "Union Rally Demands Sun Boycott," *Philadelphia Inquirer,* November 6, 1982, p. B-1.

[94] "Brain Surgery for Assault Victim," *Construction Labor Report,* Vol. 29 (September 21, 1983), p. 985.

[95] The Kalkaska story is based upon the following press stories: *Detroit Free Press,* February 20, 21, and 27, May 2 and 3, June 14, 15, 17, 19, 20, 21, 24, 26, 29 and 30, July 12, 17, and 18, 1973; *Detroit News,* June 16, 17, and 21, and July 17, 1973; *Bay City Times,* February 20, May 15 and 20, June 2, 19 and 24, 1973; *Midland Daily News,* June 15, 16, 20 and 28, July 13, 17 and 18, 1973; *Jackson Citizen Patriot,* June 18 and 19, 1973; *Traverse City Eagle,* July 11, 12, and 17, 1973; *Grand Rapids Press,* July 26, 1973; *New York Times,* March 25 and July 22, 1973; *Business Week,* July 7, 1973; *Construction Labor Report,* No. 929, July 25, 1973; *Solidarity,* June 1973; *Building Tradesman,* June 22 and 29, 1973.

Violence began in February 1973, when 400 union demonstrators did $100,000 worth of damage at the Kalkaska site and at another open shop construction site, injuring two state troopers in the process. In addition, the construction unions began a series of demonstrations at Shell service stations across the state, aided in some cases by members of the Teamsters and the United Automobile Workers.

A threatened injunction stopped the picketing of service stations, but 200 to 500 construction workers continued on a reign of terror in Kalkaska for the next several months. During the seige, the state of Michigan had to station as many as 350 state troopers—20 percent of its total force—in this small town to protect Delta Engineering's labor force from the demonstrators, most of whom were from Detroit. The cost to the state's taxpayers for this police protection was about $30,000 per day.

Repeated efforts by Shell and Delta to obtain enforceable injunctions were unsuccessful. Finally, in June 1973, the NLRB moved, in federal district court, for an injunction against the secondary boycott activities and coercion of individuals on the part of the unions and scheduled a hearing for the issues in July.

The district court judge involved was a former labor mediator. Instead of granting an order, he pushed the parties into mediation sessions. The unions were demanding that the nonunion Delta Engineering Company, although subject to no contracts or agreements with any of the unions involved, pay money into the unions' welfare funds as a condition of settlement. Delta refused to go along with this "legal" form of extortion, but did offer to employ fifty union men in its work force. The counteroffer was refused and negotiations continued.

After being prodded by the judge and by the governor of Michigan, Delta and Shell agreed to donate $250,000—the estimated cost of the union welfare programs—to the Kalkaska Township and to try to employ more workers from within a 100-mile radius. The unions were satisfied by this. The violence was called off, and the construction work remained open shop. As part of the settlement, however, Shell and Delta suspended all legal action, which provided the unions with both a reward and a victory for their use of violence. According to one press report:

> The union said they felt the publicity given to the often violence-marred dispute will reap long-range benefits for union members.
>
> "I don't think anybody else will try to do this [build a major job with nonunion work] again in Michigan," one union source said. "There's no doubt about that."

> He predicted "a hell of a lot of construction in the future in the oil industry in Michigan and I bet it's all going to be union."[96]

According to another union official, "There is no doubt it [the settlement] was good for the union movement. In the last three weeks we signed up 20 to 25 companies."[97] As one press article put it, neither Shell nor Delta were ever given the NLRB hearing or court injunction to which they were entitled. Instead they "were coerced by the federal judge and the state government into negotiating with a union organization not involved in their project,"[98] or it might be added, which did not represent their employees.

The judge's rationale for mediating instead of issuing the injunction (that one was warranted is indisputable) was that it would not be enforced, and that he always preferred mediation. Such an approach by a federal judge seems questionable.

> Somehow . . . we cannot help but have misgivings about a settlement which, in effect, has been arrived at because members of the Building Trades Council—in order to achieve their ends—were willing to commit acts of violence and vandalism. . . . That such repugnant methods should be rewarded in any way is troubling indeed.[99]

Since the settlement of this dispute, Shell has not attempted to build nonunion in the Kalkaska area and peace has reigned. Thus, the violence and the denial of due process to those upon whom violence was inflicted have saved jobs for union construction workers, and the unions have successfully used violence and the continued threat of violence to force a company to pay out a large sum of money which it otherwise would not have paid and for which it recieved no benefit other than the cessation of violence—"protection" money, in all but name. The unions thus achieved one of their aims: eliminating the economic advantage of nonunion construction by equalizing the costs of union and nonunion construction work.

Midland.[100] Having won a victory at Kalkaska, the Michigan construction unions turned their attention to Midland, Michigan, where the Dow Chemical Company is located. Dow lets contracts

---

[96] *Traverse City Record-Eagle,* July 17, 1973.

[97] Edward J. Doherty, "Union Trouble in Kalkaska, Court Bows to Extortion," syndicated story for North American Newspaper Alliance, August 1973.

[98] *Ibid.*

[99] Editorial, *Grand Rapids Press,* July 26, 1973.

[100] The Midland and related material, exclusive of the Kalkaska situation, is based upon stories in the *Midland Daily News,* September 4, 5, 6, 14, 15, 17, 18, 20, 21, 27, 28 and 29, October 2, 6, 8, 9, 10, 13 and 17, November 1 and 11, 1973, January 21, February 12, March 5, September 24, October 30 and 31, and November 1, 1974; and Michigan State Building and Construction Trades Council, NLRB Decision and Order, Cases No. 7-CB-2897 et al., February 7, 1974.

for construction on its huge complex to either union or open shop builders based solely on the bids and provides separate gates for each on construction sites. Between September 1973 and January 1974, the open shop gate at Dow's complex was picketed by 50 to 200 building tradesmen. Dow received fairly good police protection, but entering workmen were severely harassed.

Mass picketing continued until Dow filed charges with the NLRB and eventually won a consent decree in which the unions agreed to stop the use of violence and agreed to a stipulation which could be enforced by any court of appeals upon NLRB application and with defense waived in advance by the unions.

Although the picketing ended at Dow, violence continued. One of Dow's contractors, the Colinson Construction Company—an open shop contractor that had long been the butt of labor violence—received additional threats and had three of its construction projects not located at Dow damaged by bombs. This violence was stopped by another consent decree and a stipulation similar to that for Dow. The stipulation indicates rather conclusively the tactics that the Michigan trades have utilized:

> A. Respondents . . . shall:
> 1. Cease and desist from:
> (a) . . .participating in, directing, authorizing, ratifying, condoning or causing acts of property destruction and vandalism, mass picketing and obstruction of ingress and egress, threats of violence and bodily harm, and similar coercive conduct.
> (b) In any other manner, or by any other means, restraining or coercing employees in the exercise of their rights guaranteed by Section 7 of the Act.[101]

The stipulation also prohibited the unions from picketing at non-union gates and from engaging in sympathy strikes and secondary boycotts at the affected sites. After these stipulations were issued, construction violence in Michigan was reduced considerably, although some outbursts did occur, and twice, the Michigan State Building and Construction Trades Council was found in contempt for further violence.[102] In 1983, however, violence again broke out, this time at Port Huron.

Port Huron. In this southeastern Michigan city, Jerry Bartow, an open shop contractor, was building a $50,00 addition to a plant of the U.S. Manufacturing Company. Work began in June, 1983.

---

[101] Michigan State Building and Construction Trade Council, AFL-CIO, Settlement Stipulation, NLRB Case No. 7-CB-3030 (February 1975).

[102] National Labor Relations Board v. Michigan State Building and Construction Trades Council, AFL-CIO and Eugene D. Tolot, No. 75-1453 (6th Cir. April 22, 1977); 6605 F.2d 1093 (6th Cir., 1980).

Within one month, serious vandalism and violence began. Bartow claimed that he received no support from the state police, and that the vandalism came from construction workers employed on a Detroit Edison construction project. According to Bartow, whenever the unions wanted a show of force, they went out on a wildcat strike at the Edison project.[103] It is perhaps not coincidental that the resumption of violence in Michigan followed the election of a pro-union governor.

The Gulf Coast

The Gulf Coast, stretching from New Orleans to the Mexican border, is one of the strongest nonunion areas in the country, both in construction and in industry generally. Nevertheless, within this nonunion area, there are some very strong union enclaves. These include parts of Louisiana; the "Golden Triangle" of Beaumont, Port Arthur, and Orange, Texas; and the Galveston-Texas City area. All of these union strongholds have seen violence erupt when nonunion firms have attempted to handle construction projects.

Lake Charles. Lake Charles, Louisiana, is in Calcasieu Parish, near the Texas border. Through the years, it has witnessed its share of labor violence. The Southwest Louisiana Building and Construction Trades Council, which also coordinates construction union activities in several neighboring parishes, is powerful in Calcasieu. Nonunion workers, and those of a rival independent union which is headquartered in Baton Rouge, have been attacked periodically.

When Brown & Root, the country's largest open shop contractor, built a paper mill in the Calcasieu town of DeRidder, mass picketing, violence, bombings, and shootings occurred.[104] Brown & Root was forced to bring in personnel and materials by helicopter until order was restored by the state police. In 1975, the American Dredging Company sought an injunction to halt repeated acts of violence by the Laborers' and Operating Engineers' unions there. In one of the incidents, thirty persons wielding clubs, chains, wrenches, and other weapons attacked nonunion workers.[105]

The climax of the violence in this parish came in January 1976 when a mob of 100 persons, led by the head of the local Plumbers'

---

[103] "Violent Protests Against Open Shop Disrupt Projects in Mich., Montana," *Construction Labor Report,* Vol. 29 (September 21, 1983), p. 984.

[104] "Bomb Bridge in Louisiana Labor Dispute," *Chicago Tribune,* September 10, 1968.

[105] These incidents are recounted in an article in the *Orangeburg* (S.C.) *Times-Democrat,* April 19, 1976, which dealt with union violence. See also "Ex-Teamster Claims Attack Netted $1,000," *Baton Rouge Advocate,* February 4, 1971.

union and using a large forklift truck as a tank, overturned gates and shot their way into a construction site of the open shop Payne & Keller Company. One person was killed, several were wounded or injured, and extensive damage was done. This action occurred on the fourth day of violence at the same site. It was coordinated with an attack on another Payne & Keller site in Houston, where additional violence occurred.[106] In the indictments and trials which followed, the leader of the union attack was sentenced to twenty-two and one-half years in jail.

Beaumont-Port Arthur-Orange. In Texas, the area around Beaumont, Port Arthur, and Orange is known as the Golden Triangle. With its petroleum deposits, refineries, chemical plants, and other processing industries, it has long been a fertile source of employment and a strong union area. Two incidents illustrate the tactics of construction unions to keep it a union area.

In 1979, Brown & Root contracted to build part of a dock and shore base facility for offshore oil and gas operations of the Atlantic Richfield Company. The construction site was located on a road known as FM-3322 on the Sabine Ship Channel within the Golden Triangle. In the words of Brown & Root's subsequent suit for damages:

> On the morning of June 12, 1979, a mob of violent and unruly persons estimated at 1,000, more or less, appeared on FM-3322 at the entrance to the Sabine Site, blocking access to the construction site; blocking free and unrestrained travel on FM-3322; tearing down the fences separating the Sabine Site from the public road; stoning, beating upon and damaging the temporary office building of Plaintiff Brown & Root at the Sabine Site; yelling insults and profanity and threats of bodily harm to the employees of Plaintiff Brown & Root inside the Sabine Site and the temporary office building; and preventing other Brown & Root employees from traveling on the public road and gaining access to their place of work at the Sabine Site. The mob, varying in size and composition, continued such unlawful activities from day to day, from June 12, 1979 until or about August 1, 1979.[107]

In August 1979, an injunction was granted forbidding violence at

---

[106] The story of the attack was carried by the various presses in most areas. The *New York Times* and the *El Paso Times* stories from the Associated Press on January 15, 1976 are the primary sources used here. At Houston, buildings were set afire and considerable damage was done. Again noting the nexus between corruption and violence, the head of the Plumbers' union in Lake Charles allegedly had interests in contracting concerns, the Laborers' local there has been fined for accepting payments to let nonunion firms work, and the business agents of both the Electrical Workers and the Plumbers have been convicted of taking bribes.

[107] Wilson and Brown & Root, Inc. v. Sabine Area Building and Construction Trades Council, Complaint Before the U.S. District Court, Southern District of Texas, Galveston Division, 1979.

the site, but it continued nevertheless. Employees of Brown & Root were chased as they left the site and some were driven off the road; the company's equipment was damaged; and "interference, intimidation, blocking of access, riot, and gauntlet on the public road approaching the Sabine Site" continued. As a result

> Brown & Root was compelled to replace some of its employees . . . with pipefitters and piledrivers sent to the job by Defendant Sabine Area Council who worked with a substantially equal number of . . . Brown & Root employees in completing the construction work. Thereafter, Defendants discontinued their acts of intimidation at the Sabine Site.[108]

Brown & Root, unable to obtain sufficient police protection for its employees, gave in to the unions' demands for jobs as the only way to restore order and complete the job. It subsequently filed damage suits against the unions as a result of the violence described above, and several of its employees who were injured or intimidated have also filed suit; but regardless of the success of those suits (at the time of this writing, they had not yet been heard), the unions achieved their objective of forcing their will upon the largest open shop construction firm in the country.

Three years before Brown & Root's difficulties in the Golden Triangle, the A. W. Cross Construction Company was awarded an $8 million contract by the United States Army Corps of Engineers to erect the Alligator Bayou Pumping Station and Gravity Drainage Structure on the hurricane levee along Taylor's Bayou near Port Arthur. The Cross company is also an open shop contractor.

On January 15, 1976, the Sabine Area Building and Construction Trades Council held a meeting at which the nonunion status of Cross was discussed and plans for a "citizen's protest" meeting were mentioned. Two days after that meeting, a crowd of 300 persons assembled at the pumping station construction site. Mr. Cross and one of his supervisors attempted to induce the crowd to leave the site. In the words of the Court of Appeals for the Fifth Circuit:

> Several vehicles made brief forays up the access road, and their occupants confirmed with Cross and Scott [his supervisor] that they were at the Cross Construction Company jobsite. The crowd began to get unruly, pushing and shoving the remaining Cross workers as they arrived. Nevertheless, Cross's employees began work as usual. Then, shortly after 7:00 that morning, a group of four pickup trucks, each carrying between twelve and eighteen persons, emerged from the crowd gathered at the access road and drove onto the jobsite. Plaintiff Scott went out to meet the intruders and to request them to leave

---

[108] *Ibid.*

> the area, but one of them approached Scott and said, "Man, you all have got to be crazy . . . this is a union town."
>
> Scott told his interlocutor that they did not want any trouble, and he attempted to gather together the other employees and to leave the jobsite. However, before he could complete his mission, someone stepped out of the group and struck him on the head. Suddenly, the mob swarmed over the construction site, brutally beating Cross and his employees with iron rods and wooden boards, overturning and setting fire to the trailer that served as the construction site office, smashing automobile and truck windshields, and vandalizing company tools and equipment.
>
> The entire episode lasted only a few minutes, but the destruction was devastating. Cross and his employees were treated for their injuries at a local hospital, and work at the construction site did not resume for nearly three weeks. Some of Cross's employees, frightened by the possibility of repeated attacks at the jobsite, refused to return to work. In addition, the violence and vandalism delayed the completion of the project by about six months, ultimately causing the Cross Construction Company to default in its contractual obligation to the U.S. Army Corps of Engineers.[109]

Three persons were convicted of participation in the riot, two of whom were given jail sentences.[110] The company and some of its employees also sued individuals and the Sabine council and its constituents for conspiracy and won a judgment in the United States District Court and Court of Appeals applying the Ku Klux Klan Act of 1871, as a "conspiracy designed to deprive nonunion workers of the First Amendment right to freely associate with one another where that conspiracy does not occur in conjunction with legitimate union activity and is perfected by force and violence."[111]

A permanent injunction restraining the Sabine council from engaging in violence at the construction site was won—too late to solve the company's problem—and damages were awarded to the individuals and to the company. The case was, however, reversed by the United States Supreme Court in a five to four decision on the grounds that the 1871 law did not apply to this type of dispute.[112] Thus, the case did not provide a means to halt such violence.

Conclusion

The accounts of violence in Philadelphia, Denver, Michigan, and the Gulf Coast are illustrative of its prevalence. Table VIII-5 shows

---

[109] Paul E. Scott v. Bill Moore, 640 F.2d 708 (5th Cir. 1981), *aff'd,* 680 F.2d 979 (5th Cir. 1982), *rev'd,* 103 S. Ct. 3352 (1983).

[110] "Texans Guilty of Labor Violence," *Lake Charles American Press,* July 19, 1976, n.p.

[111] 680 F.2d at 1004.

[112] 103 S. Ct. 3352 (1983).

TABLE VIII-5
Selected Incidents of Violence Directed Against Nonunion Employees or Open Shop Contractors Involving Construction Trades Unions 1976–1981[a]

California	Several nonunion construction projects burned by arsonists; 108 company truck tires damaged.
Connecticut	Nonunion workers harassed by disorderly pickets; rock thrown at truck entering nonunion work site; nonunion entrance blocked by mass pickets; picket protesting nonunion contractor jumped into truck and broke window.
Delaware	Stones thrown; cars kicked and pounded; workers threatened; open shop contractor harassed; nonunion workers severely injured by thrown rocks.
Florida	Traffic blocked; cars damaged; workers threatened; stones thrown; trucks burned; power poles cut; rocks and debris thrown; workers injured; shots fired at company truck; tires slashed; rocks thrown; crane engines filled with sand; equipment damaged; cranes burned; plumbers' union official convicted of plot to burn many nonunion businesses.
Georgia	Employees' cars sprayed with paint; shots fired into truck carrying nonunion employees, two of whom were wounded.
Illinois	Bombs destroyed part of hotel being built nonunion; pickup truck overturned; fences torn down; bat-wielding union members severely beat two nonunion workers; equipment damaged.
Iowa	Cars and homes of nonunion employees vandalized; nails put on road; rocks and eggs thrown at employees; construction site blocked; building under construction damaged.
Louisiana	Four persons injured by a flash fire at nonunion construction site; five nonunion workers wounded, one critically, by club-wielding union members.
Maine	Rocks thrown; road blocked; newsmen and others threatened.
Maryland	Tires slashed; rocks thrown at police; police assaulted.
Massachusetts	Unruly picketing; cement wall torn down; equipment damaged; $2,000,000 worth of damage done to equipment of two nonunion companies by union members; police assaulted.
Michigan	Buildings built by nonunion contractor damaged by explosions; area around nonunion site vandalized.
Minnesota	Wiring of housing project cut and snapped; rocks thrown; company trucks damaged; tires slashed; owner of nonunion firm reportedly hit by union official.
Mississippi	Car windshield broken; nonunion employee injured.
Missouri	Open shop construction site bombed; nonunion workers threatened, harassed, and assaulted; rocks thrown at nonunion employees; thirty engines of heavy construction equipment damaged.
Montana	Shed housing nonunion company's records burned by mob; four nonunion workers assaulted.
Nebraska	Supervisor at nonunion construction site beaten by union members.
Nevada	Car windows smashed, office trailer damaged, employees threatened.

TABLE VIII-5 (continued)

New Jersey	Windows of company truck broken; bomb found on nonunion construction site; telephone lines cut and pieces of it taken; tires slashed; employees threatened; office trailer vandalized, telephone lines cut, building plans destroyed; office trailer burned by rioting union members at nonunion site; rocks thrown by pickets at cars and vans entering nonunion jobsite; three nonunion workers beaten by union members; nails put on road; human or animal feces thrown into office trailer; worker beaten; threats made.
New York	Construction site blocked; crane damaged; workers threatened; car windows broken.
Ohio	Car window broken; threats to do damage made; nonunion workers roughed up; security guards attacked; windows broken; entrances blocked; company equipment damaged; structure under construction damaged; workers threatened, assaulted, and injured; vandalism occurred; hotel firebombed with Molotov cocktails; windows of restaurant being constructed with nonunion help broken.
Pennsylvania	Bulldozer damaged; wires cut; windows broken; masked men attempting to bomb a nonunion construction site were caught by security guards, who were then kidnapped, taken to a remote area, and beaten; nonunion employees attempting to cross picket line were threatened and assaulted; $100,000 worth of machinery owned by open shop company damaged; tires slashed; truck windshield broken; driver injured; nails strewn on driveway; employees assaulted; thirty tires slashed and windows broken on employees' cars, company truck firebombed; nonunion contractor's home shot into, contractor threatened with death; employees and guests of hotel harrassed; five nonunion employees beaten with iron pipes and pick handles; owner of nonunion firm and his son beaten by union officials and members; equipment of twelve nonunion contractors damaged; picket with a two-by-four chased and threatened employee.
Tennessee	Entrance to construction site blocked; cars damaged; workers threatened; site closed down; hundreds of rocks thrown by pickets; $50,000 worth of equipment damaged; employees threatened with bodily harm and death; rocks thrown; tires slashed; windows smashed.
Texas	$1,000,000 worth of damage done to nonunion worksite by union members; construction site vandalized; seventeen cases of arson; much equipment damaged; employees threatened; mass picketing; windows of employees' cars smashed; employees threatened and harassed; nonunion employees run off road and threatened.
Washington	Objects thrown at vehicles; windows broken.
W. Virginia	Worker shot at when crossing picket line; tacks spread on road.
Wisconsin	Dynamite found on a crane belonging to a nonunion contractor.

Source: Newspaper clipping file of the National Right to Work Legal Defense Foundation, sorted by the Wharton Industrial Research Unit, and reproduced from Armand J. Thieblot, Jr., and Thomas R. Haggard, *Union Violence: The Record and the Response of Courts, Legislatures, and the NLRB,* Labor Relations and Public Policy Series, No. 25 (Philadelphia: Industrial Research Unit, The Wharton School, University of Pennsylvania, 1983), pp. 166–67.

[a] Unions involved include Building and Construction Trades Councils, Bricklayers, Carpenters, Electrical Workers, Ironworkers, Laborers, Operating Engineers, Painters, or Roofers.

that the use of violence against open shop builders and employees is widespread. Without doubt, it tends to increase as the market share of open shop contractors increases.

In many of these instances, the purpose of the violence is clear: to frighten contractors and employees from working open shop and to frighten construction users from awarding contracts to open shop builders. To the question of how they determine where they will bid on work, large open shop contractors readily reply: "Where we can be reasonably sure of police protection." Many users consider the same thing before letting contracts. Violence has thus been effective in many areas in maintaining the union shop, and such violence will undoubtedly continue until effective laws to control it are passed, and undoubtedly more important, until laws against labor violence are enforced as they are when violence emanates from any other source.

PART THREE

Labor Markets And Policies in Open Shop Construction

CHAPTER IX

Recruitment and Deployment

There are few industries in the United States whose labor market operations are as extensively influenced by unions as construction. Through both formal contractual mechanisms and informal pressures, the building trades unions share broadly in the rule-making process which regulates manpower flows and assignments both among and within employing units. There will be debate, to be sure, on whether these various regulations and restrictions are, on balance, socially benign, malignant, or neutral, but these questions need not detain us here. Let it suffice to say that they exist and, for better or worse, serve to constrict the decision-making latitude of construction managers in unionized operations with regard to the recruitment and deployment of their workers.

Prior to our initial study, no researcher had seriously asked how labor market decisions were being made by open shop construction contractors. How, for example, do open shop companies go about finding needed employees for replacement or expansion? How do they augment and contract their workforces in response to abruptly changing manpower requirements? How do they distribute the jobs to be done among workers of varying skills and capabilities? How, in sum, does a construction labor market operate where the guiding principles are the unilateral judgments of employers, rather than the provisions of union-management contracts?

The theoretical function of a factor market is to allocate resources to their most productive uses. If the market performs optimally, a given output will issue at the least possible cost—resources will be neither underutilized nor overutilized. As needs change, resources must be shifted, and in the process workers are displaced. In weighing the social benefits of efficiency against the individual costs of displacement, American unions, in particular the building trades, have traditionally emphasized the latter. The fear of job scarcity has, in many instances, led to bargaining goals which underscore the preservation of jobs, even though these jobs may be redundant in terms of the employer's production needs.

It is not universally accepted that the interests of society are invariably well served by the apotheosizing of economic efficiency. But the interests of the employer, however they may be defined, are certainly furthered by retaining flexibility in obtaining and allocating his workforce. Indeed, open shop contractors frequently argue that it is this flexibility, more than anything else, which gives them a competitive edge over their unionized counterparts. Moreover, even while pointing to competitive problems relating to high wages and strikes, unionized contractors often declare that loss of the right to manage is their most serious problem.

How, then, is the right to manage exercised by open shop contractors, and how are the results different from those obtained by union contractors? This chapter is concerned with a variety of labor market practices, all of which relate ultimately to manning the job. We begin with the task of recruitment, an area where unions, either directly through hiring halls or indirectly through hiring restrictions, play a significant role. Within this context, we will also explore the movement of workers between the union and nonunion sectors of construction.

The other major area of concern in this chapter is the allocation of workers within the employing unit, the deployment of labor at varying levels and of varying skills. A central question in this context is how flexibility in making job assignments is exercised in the open shop sector, in contrast to the union sector where the managerial role is circumscribed by such contractual guidelines as craft jurisdictions. A further question involves the redeployment of labor in response to changing seasonal manpower needs, a condition endemic, to a greater or lesser degree, to most branches of construction. Other issues relating to labor deployment which will be discussed in this chapter include: the effect of the absence of various job-creating requirements (e.g., crew sizes, journeymen-foremen ratios, and standby manning); remuneration for nonworking time (travel and show-up pay); and restrictions on the use of materials, equipment, and technology.

HUMAN RESOURCE RECRUITMENT

Human resource needs in construction are of both a short-term and long-term character. Over a period of years, the industry must provide both for the secular growth of the workforce and for vacancies created by death, retirement, and industry mobility. Long-term needs, especially those for skilled labor, are met predominantly by the various training programs (discussed in Chapter X). Training programs, however, require time, and the contractor who wins a large job or loses some key workers may not be able to afford the luxury of gradually upgrading his workers. He must enter the market to find new labor. "Entering the market," however, means different things to the union and open shop contractor.

For the union contractor, it most likely means a call to the union hiring hall. As Philip Ross has pointed out, the term "hiring hall" is often used loosely to refer to both a system of job assignments in which participants are given preference over nonparticipants and a routing referral service in which the union serves as a broker of information between employers and workers.[1] Whether contractors utilize the union hiring hall because it is convenient and useful or because its use is required by contract, the hiring hall is the primary source of labor for many union contractors.

According to a recent study by the Business Roundtable, hiring hall provisions are found in 60 percent of the union agreements in the construction industry. Only one-half of such agreements require exclusive use of the hall, a majority allow management to reject referrals, and about one-half permit management to request at least part of the workforce by name. "Reliance by [union] contractors on the hiring hall because of lack of an alternative, however, has resulted in exclusive use a much greater percentage of the time than is required by labor agreements."[2]

To the union contractor, the hiring hall is a mixed blessing.

> Many contractors over the years have considered the hiring hall to be the most convenient means of meeting the constantly changing labor force requirements. The hiring hall has served as a source of employment information for applicants. It also may perform the difficult role of prorating work among members when job opportunities are scarce. Taken as a whole, the hiring hall has relieved the contractor of the difficulty and expense of operating his own personnel function. That relief, however, has been attained at the price of creating a major base of union power and badly handicapping the ability of the contractor to manage his work force.[3]

The hiring hall does provide the union contractor with a reasonably reliable source of labor, usually on comfortingly short notice. It is also tied into a network of other halls in different areas, so that a manpower shortage in one locality can be met by drawing

---

[1] Philip Ross, "Origin of the Hiring Hall in Construction," *Industrial Relations,* Vol. 11 (October, 1972), p. 366.

[2] *Constraints Imposed by Collective Bargaining Agreements,* Report C-4, Construction Industry Cost Effectiveness Project (New York: The Business Roundtable, 1982), p. 16.

[3] *Ibid.,* quoting *Coming to Grips with Some Major Problems in the Construction Industry* (New York: The Business Roundtable, 1974).

upon a manpower surplus in another. Through the hiring hall, the employer is thus plugged into a well-organized flow of labor market information which gives him access to a broad range of job-seekers.

Although different hiring halls have varying degrees of flexibility, the system has its costs, in that it often places serious restrictions on the contractor's freedom to hire. Especially in times of high unemployment, he may be contractually obliged or informally pressured to hire a worker different from the one whom he wants, or he may be unable to obtain a craftsman with experience specific to his job needs. Moreover, in times of a tight labor market, the union may be either unable to supply personnel, or it may accept, or give permits to, unqualified workers whom the employer must try out at great cost. Finally, contractors reported to us that union locals have often exercised control over the labor supply to restrict an employer's labor requirements in order to stretch out work, force up wages, compel the utilization of overtime at double the regular wage rate, or even favor one contractor over another. The disadvantages of the hiring hall to the contractor and the user were starkly presented in a 1974 Business Roundtable report:

> The day-to-day operation of hiring hall administered by a union official can be ruthless. Referral lists can be manipulated, contractors enticed to order men by name or specialty skill in order to avoid the next man on the list, dissident members or "permit" men referred constantly to jobs of short duration, and many similar injustices perpetrated.
>
> Control over the referral process also may have a direct impact upon the layoff of employees. For example, instructions from a union official to quit a job are quickly honored if a member hopes to obtain a referral to another job in the future. In reality, contractors who are dependent on the hiring hall have substantially lost the right to hire or terminate their employees.
>
> Union control of the administration of the hiring hall results in unions being the repository of virtually all available information concerning the quality and quantity of the construction manpower pool within a labor market. The same information is usually inaccessible to contractors or construction users, except for general information about total membership in local building trades unions. This imbalance in the availability of construction manpower information prevails throughout the organized segment of the construction industry, regardless of the form of the hiring hall, and places contractors at a severe disadvantage in performing their role as employers.
>
> Union control over the work referral system has also had some profound economic implications on contractors and the users of their services. An increase in demand for labor encourages an increase in supply in most industries. Not necessarily so in construction! A union-administered work referral system establishes an environment conducive to artificial shortages. These shortages, in turn, have enabled

TABLE IX-1
Hiring Methods Employed Always or Often by Union and Nonunion Contractors, 1974

Method of Hire	Number of Contractors Reporting Use			
	Union[a]		Nonunion[b]	
	Number	Percent	Number	Percent
Ask current employees	298	38.3	464	59.2
Ask other employers	88	11.3	129	16.5
Contact past workers	393	50.6	262	33.4
Contact prior applicants	158	20.3	355	45.3
Hiring hall	517	66.5	4	0.5
Newspaper advertisement	53	6.8	330	42.1
State employment service	46	5.9	222	28.3
Vocational schools	31	4.0	126	16.0
"Off the street"	78	10.1	171	21.8

Source: Industrial Research Unit Questionnaire, 1974.
[a] N = 778.
[b] N = 784.

> the unions to achieve stronger bargaining positions, and thereby to obtain inflationary wage increases, excessively restrictive work practices, and other costly conditions of employment. In his role as hiring hall administrator, a union official is often torn between the longer term interests of the industry and the union members on the one hand and the political appeal of maintaining a surplus of job opportunities and additional income through overtime.[4]

With more latitude in hiring, the open shop employer, nevertheless, may have less access to the local labor supply. Since there is usually no central agency which performs the brokerage function of the union hiring hall, a nonunion employer is obliged to rely upon a variety of other methods to inform prospective employees of his manpower needs. Taken together, these methods may eventually get his message across to a large number of available workers within the labor market, but they are unlikely to work as expeditiously or thoroughly as a telephone call to a single source.

The broader use of diverse methods of hire by open shop contractors is illustrated in Tables IX-1 and IX-2. Table IX-1 is based on responses to the Industrial Research Unit's 1974 survey of both union and open shop contractors, and Table IX-2 is based upon a similar 1982 survey. Employers who participated in those surveys were given a list of various hiring methods used by both union and nonunion contractors and asked to indicate whether they use each

[4] "The Hiring Hall in the Construction Industry," in *Coming to Grips With Some Major Problems in the Construction Industry* (New York: The Business Roundtable, 1974), p. 17.

TABLE IX-2
Hiring Methods Employed Always or Often by Union and Nonunion Contractors, 1982

	Number of Contractors Reporting Use			
	Union[a]		Nonunion[b]	
Method of Hire	Number	Percent	Number	Percent
Ask current employees	713	39.0	1,155	57.1
Ask other employees	328	17.9	538	26.6
Contact past workers	634	34.7	550	27.2
Contact prior applicants	247	13.5	857	42.4
Hiring hall	1,252	68.5	33	1.6
Newspaper advertisement	90	4.9	648	32.0
State employment service	63	3.4	422	20.9
Vocational schools	41	2.2	253	12.5
"Off the street"	107	5.9	367	18.2

Source: Industrial Research Unit Questionnaire, 1982.
[a] N = 1,828
[b] N = 2,022

method "always," "often," "occasionally," or "never." Tables IX-1 and IX-2 show the number and percentage of respondents who reported using each of the methods listed "always" or "often." With the obvious exception of hiring halls and the less obvious one of contacting past workers directly, all other methods of hire were used more frequently by open shop contractors than by union contractors. Moreover, the percentages reported for 1974 and 1982 were not significantly different.[5] The reason union contractors tend to contact past workers to fill their job openings is because (apart from a normal desire to utilize known workers) of provisions in many hiring hall arrangements encouraging, or indeed requiring, employers to recall previous employees before soliciting new workers from the hall.

All the hiring methods listed in Tables IX-1 and IX-2, with the exception of taking on workers "off the street," are initiated by the employer. These methods may be generally divided into two groups: the "formal" approach and the "informal" approach. The formal approach includes methods which employ some outside agency which routinely performs some kind of placement function. The informal approach includes methods in which solicitation is limited to the employer's circle of contacts.

[5] The 1982 questionnaire contained a typographical error by having the second category listed as "ask other employees" instead of "ask other employers." This, however, does not alter the analysis between informal and formal means of recruitment.

Informal Approach

These methods generally involve the employer "letting the word out" that he is in the market for workers. He may ask his present workers if they know anyone who is looking for work. He may ask his business associates—subcontractors, suppliers, competitors—if they know workers who have recently been laid off. Among open shop contractors especially, workers are commonly "loaned" for a limited time. This practice benefits the borrower by providing him with a reliable worker, and the lender by enabling him to release a temporarily redundant worker while not losing him permanently. He may contact workers whom he has employed in the past to see if they are, or expect to be, available. Finally, he may keep a file on previous job applicants who could not be hired at the time of their application and contact them when work becomes available.

On the whole, open shop contractors appear to have utilized these informal methods more frequently than union employers. They also utilize informal methods more frequently than formal methods, suggesting that the formal methods available to nonunion contractors are not deemed wholly reliable or desirable, or at least that they will usually seek workers from close at hand before resorting to a broader search. Many union contractors (66.5 percent in 1974, 68.5 percent in 1982), by contrast, invariably use the hiring hall.

The informal hiring methods manifestly rely upon the grapevine to disseminate job information. Such an approach offers no assurance that information will either reach all those who may be interested in work, or that the information will be transmitted with accuracy. The grapevine will be successful only where the contacts of employers and workers overlap. Since these contacts are many and dispersed, the probability of information crossing between the supply and demand sides is less than that where all or most information is routed through one or a few points.

Formal Approach

The parallel to the hiring hall in open shop construction—to the extent that there is one—lies in the use of formal methods of advertising existent job openings. These may include newspaper want ads, an employment service or agency, or the placement office of a trade or vocational school. However, these alternatives are not relied on by open shop contractors to the extent that the union hall is used by organized firms.

As the figures in Tables IX-1 and IX-2 show, four to eight times the proportion of open shop than union shop respondents reported using newspapers, state employment services, or vocational schools on a regular basis to recruit their workers. Among respondents who answered "occasionally" or "never" to questions about their use of placement or referral services other than hiring halls, most of the union shops said they "never" used such services while most of the open shops said they used them "occasionally." Because union shops are often required by provisions set forth in labor contracts to use hiring halls, this finding is scarcely surprising. Moreover, previous research evidence substantiates the relatively infrequent use of placement and referral services by union contractors, as well as the greater, though far from preponderant, use of them by nonunion firms.[6]

The picture thus drawn of hiring practices may be summarized as follows: union contractors use hiring halls extensively, either because they are required by labor agreements or because they are convenient. Alternatively, they contact workers whom they have employed in the past or ask their present workers to recommend potential applicants. Open shop contractors, on the other hand, appear to prefer informal to formal methods of hiring, which may reflect a dissatisfaction with the formal methods available to them. Of these formal methods, more than one-third of the respondents reported using newspaper advertisements regularly and more than one-fifth state employment services.

Examples of Formal Open Shop Approaches

We have found few formal hiring systems in the open shop sector while conducting our field work in 1973–1974 and 1981–1983. Three or four chapters of the Associated Builders and Contractors (ABC), some chapters of the Associated General Contractors (AGC) that have enrolled a sizable number of open shop contractors, and a few open shop specialty contractor groups have established formal referral systems. In addition, many open shop associations or chapters act as referral services for their members, but only on a very limited and infrequent basis. Essentially, association staff members will inform contractors seeking workers about other contractors who are cutting back. They, however, arrange few direct meetings between employers and workers. Association spokesmen report that although contractors will occasionally seek to hire through associations, workers do not often approach them for employment.

---

[6] Howard G. Foster, *Manpower in Homebuilding: A Preliminary Analysis,* Manpower and Human Resources Studies, No. 3 (Philadelphia: Industrial Research Unit, The Wharton School, University of Pennsylvania, 1974), pp. 103–108.

In a few areas, however, contractors' associations are beginning to develop more sophisticated referral services for their members. Records are kept both on employer and worker applicants, with the goal of enabling the system to match supply and demand expeditiously. The outcome, in large measure, is a hiring hall without restrictions on the employer (or what Ross would simply call a "referral service"). In one case, such a service was established on a nationwide basis for contractors and workers in the homebuilding industry. With financial support from the U.S. Department of Labor, the National Association of Home Builders (NAHB) operated the National Construction Job Clearinghouse through its headquarters in Washington, D.C., from 1971 to 1976. For the first two years it handled returning veterans and thereafter any applicants. Contractors were invited to submit announcements of job openings for twenty-two occupational categories. Workers submitted a brief resume indicating their specific job interest, prior experience, date of availability, and their preferred geographic location. The clearinghouse sent the resumes to appropriate employers and helped to arrange interviews. The service was provided free of charge to both contractors and applicants. It grew steadily until over 5,000 referrals per year had being made. Then, as the housing market entered a recession in 1975–1976, demand slackened. By 1976 the NAHB found the service "too much of a burden" and discontinued it.[7]

As the open shop expanded in a number of areas, contractor groups began to explore avenues for centralizing the job-matching process. What is perhaps the first of these efforts was outlined at a 1974 meeting of AGC committees. The impetus for it was described as follows:

> Our chapter has always had about 25 to 30% of its members operating on an open shop basis. In the last years, we learned that more of our members would be operating open shop if they could locate skilled manpower. There has been no central place from which they could locate the skills necessary to man open shop operations. If they are on a union basis, they can call the union halls. Those who operate successfully open shop have developed a following of men over the years.
>
> To meet this open shop need, our chapter has decided as a part of its open shop services to set up an open shop Hiring Center. This is not a recruiting thing or a training thing. You would probably call it a locating service.[8]

The referral service, located in Houston, is provided through an

[7] Telephone interview with Philip Polivchak, Director of Manpower and Training, NAHB, August 4, 1983.

[8] "Construction Workers' Register," mimeo, n.d. (late 1974).

independent corporation called Construction Workers' Register, which is legally distinct from the local AGC chapter and which is used by both nonunion AGC and ABC contractors. The Register refers non-blue-collar personnel, such as engineers and office workers, as well as craftsmen and laborers. Workers are sought through newspaper ads and word of mouth. Employers who subscribe to the service register their openings by telephone and in return receive a list of workers in the trades they specify who are known to be available for work. All direct contacts and actual hiring are done by the employers themselves. Employers and workers are expected to inform the Register when their needs have been met, and the Register's lists are accordingly updated each day. The service is financed by its sponsors, and there is no charge to the individual contractor or the worker. The success of this service has led to its being copied by several open shop groups in other parts of Texas. Table IX-3 summarizes the Houston Construction Workers' Register activities from 1979 to 1982. The decline in 1982 reflects a lower business volume, not declining utilization.

Other programs operate with varying formality, several on the Houston Register model, but most as informal services offered through association chapters. Many remain scattered and experimental, but could well increase in size and formality as open shop contractors solidify their gains. The fact that open shop contractors are seeking to replicate some aspects of the service function of the hiring hall without becoming subject to its restrictions demonstrates that the damnation expressed by industry spokesmen about union halls should not go unqualified. Their undesirable features are not inevitable concomitants to their beneficial functions. As we shall see below, open shop contractors successfully compete with the union sector by improving upon its labor market operations; in other respects—in some forms of training as well as recruitment methods—they are beginning to do it by copying some features of the devices utilized by union shops while attempting to avoid aspects which they find unacceptable.

TABLE IX-3
Construction Workers' Register of Houston Activities, 1979–1982

	1979	1980	1981	1982
Job orders	636	391	797	486
Jobs available	1,466	927	1,951	1,318
Workers hired	1,244	850	1,796	1,258
Cost per worker hired	$24.66	$35.04	$19.13	$28.77

Source: The Register.

Bourdon-Levitt Findings

Bourdon and Levitt published a study in 1980 which stated that contractor interviews "reported very few significant differences in hiring practices between different types of firms, either union or open-shop."[9] They also claimed that when union contractors did use hiring halls, "they often requested journeymen by name, thus avoiding the personnel assignment function of the business agent."[10]

Bourdon and Levitt then reported several exceptions to this surprising generalization. According to their analysis, the hiring hall and/or the union business agent played a major role when large numbers of workers were being referred for a major job over a short period, when tradesmen crowd the hiring halls looking for work in periods of high unemployment, and "when the contract provisions on hiring were strictly enforced."[11] These authors admit that, in these three situations, the impact on management is disruptive and productivity is adversely affected.

Bourdon and Levitt state that "the hiring methods used by open-shop contractors are similarly informal and diverse," and therefore they conclude: "Since hiring occurs in similar fashions in both union and open-shop sectors of the industry, the overall impact of the union on referrals is much less than is usually assumed."[12] This conclusion is, of course, consistent with their general thesis that the economic advantages of the open shop and the differences in general between union and open shop operations have been much exaggerated.

Our surveys and field interviews (which were conducted both before and after Bourdon and Levitt's study); the work of the Business Roundtable; and the numerous national, craft, and regional surveys presented in Chapters III-VI contradict Bourdon-Levitt's basic thesis, which seems clearly unable to explain how the open shop could steadily increase its market share if open shop contractors handled labor so similarly to union shop contractors. We conclude, therefore, that the Bourdon-Levitt thesis is not supported by any other objective evidence that we have uncovered.

Union Background of Open Shop Workers

A worker hired by an open shop contractor may come from several sources. He may be a new entrant into the labor force taken on

---

[9] Clinton C. Bourdon and Raymond E. Levitt, *Union and Open Shop Construction.* (Lexington, Mass.: D.C. Heath and Company, Lexington Books, 1980), p. 62.
[10] *Ibid.*
[11] *Ibid.*
[12] *Ibid.*, pp. 62–63.

initially as a helper or trainee. He may be attracted from another industry such as manufacturing or farming, where his job provided him with some building experience. He may, of course, have quit or been laid off from a job with a different nonunion contractor. But perhaps not as readily expected, the worker may have come from employment in the union sector of construction.

Since (as will be shown in chapter XI) wage rates in union shops tend to be significantly higher than in open shops, it is tempting to view intersector migration as operating only in one direction. Union membership has, to be sure, been augmented by workers whose initial exposure to, and training in, construction work was with nonunion employers. These workers may become union members by soliciting employment during periods of high activity or simply by having their employer organized by a union. There has long been evidence to show that the movement from nonunion to union status is significant.[13]

At the same time, however, movement between sectors is decidedly not one way. For a variety of reasons, union workers do seek work in open shops. The transition may be temporary or permanent, depending upon the circumstances and the reasons given by the worker who opted for the change. We may estimate the extent of this mobility from our surveys of open shop contractors. Both in 1974 and in 1982 respondents were asked to indicate the approximate proportion of their personnel who were once or are currently union members. Their responses are shown in Tables IX-4 and IX-5. As these tables indicate, a majority of the respondents reported having some workers with union membership in their background. These percentages, however, may actually understate the true magnitude of the movement. Many union workers seeking employment in open shop firms do not make their union status known either because they feel that this would reduce their chances of being hired by a nonunion contractor or because they fear that this would later jeopardize their returning to work in the union sector. Thus, the percentages in Tables IX-4 and IX-5 include only those past or present union members of whom the respondents were aware.

Some union workers, of course, will migrate to the open shop only as a temporary expedient. Although we do not have the longitudinal data necessary to test this proposition, it is reasonable to conclude that the amount of movement varies directly with un-

---

[13] Howard G. Foster, "Labor Supply in the Construction Industry: A Case Study of Upstate New York" (Ph.D. diss., Cornell University, 1969), pp. 264–267.

TABLE IX-4
Percentage of Employees of Open Shop Contractors Presently or Previously Union Members, 1974

Percentage Union Members	Contractors Reporting	
	Number	Percent
Zero	206	26.3
1 to 10	241	30.7
11 to 25	128	16.3
26 to 50	92	11.7
51 to 99	25	3.2
100	5	0.6
Not reported	87	11.1
Total	784	49.4

Source: Industrial Research Unit Questionnaire, 1974.

employment among union workers. Whether the unemployment is occasioned by a general downturn in construction activity or by a shift of work from union to nonunion contractors, some union members will undoubtedly seek open shop work rather than remain unemployed while waiting for business conditions to improve. In the latter case, of course, actual migration is likely to be greater, for it is then that displacement in the union sector is more evenly matched by job vacancies in open shops. Nearly all large open shop contractors told us that many of their employees—including those of long service—maintain their union memberships, and that when

TABLE IX-5
Percentage of Employees of Open Shop Contractors Presently or Previously Union Members, 1982

Percentage Union Members	Contractors Reporting	
	Number	Percent
25	624	30.9
50	196	9.7
75	73	3.6
100	47	2.3
Do not know	898	44.4
Not reported	184	9.1
Total	2,022	50.9[a]

Source: Industrial Research Unit Questionnaire, 1982.

[a] 50.9 percent is calculated by 2,022 x $\frac{3,972}{100}$

2,022 is the total of nonunion contractor respondents
3,972 is the total of respondents: union, nonunion, and not reported.

union contractors are depressed, they receive many more applications from unionized tradesmen.

Another temporary circumstance under which union workers might seek employment in an open shop is a strike. The weakness of union contractors in collective bargaining is frequently analyzed in terms of the job alternatives available to union workers while they are on strike. Even though many open shop contractors are unwilling to take on a worker who will leave when the strike is over, others are satisfied to obtain a skilled craftsman (who, after all, requires little real investment by the employer) even for a limited time, and still others may hire a striking union worker unknowingly.

There is also permanent movement from union shop to open shop. One avenue is a direct parallel to the unionization of a firm, whereby some incumbent employees are automatically admitted to the union. Similarly, a union contractor may decide to go nonunion or to set up a doublebreasted operation and some of his workers may transfer to the new nonunion component, although this is not too likely. The result, of course, is much the same as when a worker changes employers in the course of moving from union to nonunion status.

Some moves may begin as temporary ones but become permanent. For example, union workers who are forced to seek employment in open shops because of a scarcity of union jobs may initially expect to return to work in the union sector when openings materialize, but may later decide to remain with the nonunion contractor. A commonly cited reason why some workers' favor the open shops is the relative stability of employment there.[14] Many open shop contractors assert that despite wide differences in wage rates, their workers earn as much or more than the average union member on an annual basis. Workers may also dislike the uncertainty engendered by the possibility of strikes, or the control union officials exert over their job opportunities. Further, the tendency for nonunion firms to be smaller and have less worker turnover may lead to stronger personal and social ties among employees or between employer and employees. On the other hand, workers who prefer union shops may regard the hiring hall as a desirable, indeed an indispensable service; higher wage rates worth the price of some unemployment; personal and social relationships irrelevant to job satisfaction; and the general protection provided by a strong union, through grievance procedures and other devices, as highly valuable. Our point here is not that employment conditions are objectively

---

[14] See discussion of seasonal employment in this chapter.

better in open shops than in union shops, but that different workers place greater or lesser importance on the various attributes of a job. As long as the open shop controls a majority of the job opportunities in construction, migration from the union shop to the open shop can be expected to continue.

DEPLOYMENT IN OPEN SHOP CONSTRUCTION

The most salient aspect of human resource deployment practices of open shop contractors is that they are managerial functions, subject only to the managerial judgment of the supervisor and various legal requirements. Conversely, unionized contractors must make job assignments within occupational classifications defined by the jurisdiction of the various building trades. In some instances, assignments must be made where the employer, if it were within his discretion, would not assign anyone at all. In other cases, workers may not be assigned, except at premium rates, at certain hours of the day, or on certain days of the week regardless of the total number of hours worked. These restrictions and requirements have appreciable cost implications, and their absence serves to reduce the labor cost of open shops even where they are obliged—as, for example, under Davis-Bacon or state prevailing wage laws—to pay the same wages as union contractors.

Restrictive practices are not necessarily limited to the matter of personnel deployment. Three broad categories of union work rules which serve to impede efficiency are:

1. Onsite rules, requiring certain work to be done on the premises and prohibiting or limiting the use of prefabricated products;
2. Restrictions against the use of certain tools and devices;
3. Requirements for excessive personnel on the job, including what appear to be irrational limits on the variety of work certain categories of workers may perform, and the requirement that skilled craftsmen perform semiskilled or unskilled work.[15]

Limits on Production Techniques and Methods

Research evidence suggests that contractual limitations on production techniques and tools are much less serious and widespread in the union sector than is commonly believed, and a considerably less onerous burden on the contractor than restrictions on manpower deployment. In a major study conducted over thirty years

[15] See Sumner H. Slichter, *Union Policies and Industrial Management* (Washington, D.C.: The Brookings Institution, 1941), Chapters VI-IX.

ago, Haber and Levinson concluded that "it . . . is reasonably clear that the building trades unions have been more receptive to new techniques than has been widely believed. In the summer of 1952, the overwhelming majority of new techniques were found to be in wide use in all or nearly all of the cities surveyed."[16]

A 1974 study by the U.S. Bureau of Labor Statistics suggested that restrictions on production techniques and tools were still the exception rather than the rule. The study encompassed 769 agreements covering 1.2 million workers, or about one-half of all unionized building tradesmen. Limitations on prefabricated materials and components were found in 15 percent of the agreements covering less than 12 percent of the workers.[17] Such restrictions were common only among plumbers and sheet metal workers.[18] Limitations on tools and equipment were found in only 8 percent of the agreements covering less than 12 percent of the workers. Such restrictions were prevalent among painters (about one-half of the agreements covering four-fifths of the workers) and involved limits on the use of spray guns and the widths of brushes and rollers.[19]

A 1982 Business Roundtable report, in the Construction Industry Cost Effectiveness Project series, found restrictions on off-site fabrication in one-half of the plumber-pipefitter contracts examined, which it declared had a very heavy impact on industrial and power plant construction. It estimated that these restrictions added $30 million to construction costs.[20] A second report in this series found additional prefabrication restrictions imposed on jobs by local union practices.[21]

---

[16] William Haber and Harold M. Levinson, *Labor Relations and Productivity in the Building Trades* (Ann Arbor: Bureau of Industrial Relations, University of Michigan, 1956), p. 153.

[17] U.S., Bureau of Labor Statistics, *Characteristics of Construction Agreements, 1972–73,* Bulletin 1819 (Washington, D.C.: Government Printing Office, 1974), Table 39, p. 27.

[18] Haber and Levinson *Labor Relations and Productivity,* pp. 128–130, 141–145. Haber and Levinson found widespread restrictions on precut, prethreaded, and preassembled pipe, but a common use of prefabricated sheet metal work.

[19] U.S., Bureau of Labor Statistics, *Construction Agreements,* p. 27. See also Haber and Levinson *Labor Relations and Productivity,* pp. 121–127. In practice, some restrictions on the use of prefabricated materials may have been enhanced by the controversial decision of the U.S. Supreme Court in *National Woodwork Manufacturer's Association v. NLRB,* 386 U.S. 612 (1967). As noted in Chapter VIII, this decision permits a union to "preserve work" by refusing to handle prefabricated materials, thus negating what appeared to many as a clear prohibition against such action in the Taft-Hartley Act.

[20] *Constraints Imposed by Collective Bargaining Agreements,* p. 12.

[21] *Local Labor Practices,* Report C-5, Construction Industry Cost Effectiveness Project (New York: The Business Roundtable, 1982), p. 8.

The foregoing observations suggest an important caveat with respect to a comparison of labor deployment practices in union shops: Construction is a large and diverse industry, and the union sector consists of many formal crafts; therefore few statements can be made regarding restrictive practices that would cover all crafts in all localities. Yet, there is an unfortunate tendency to assume that any restrictive practice occurs everywhere. One can be appropriately critical of chains on one contractor's arms and another's legs without asserting, hyperbolically, that there are chains on all eight limbs. Union restrictions, for example, are much more common in a union city like Philadelphia than in a heavily open shop area like Houston.

On the other hand, many restrictions exist which are not found in formal agreements. Construction unions can, and frequently do, force additional concessions on the job by direct actions which restrict the use of materials and equipment or add excessive manpower. These restrictions find their way into practice and can become as binding and onerous as restrictive agreements.

As a result of the recession and the gains of the open shop, the early 1980s saw a clear admission on the part of the unions of the existence of restrictive practices in their willingness to give up uneconomic rules in order to compete. The pages of the weekly *Construction Labor Report,* the reports of the Construction Labor Research Council and *Cockshaw's Construction Labor News + Opinion,* and the stories in *Engineering News-Record* have been filled with union concessions on this score, plus comments by contractors that restrictions still exist. Practices bred over the years are not easily ended, but the period since 1981 has seen a tremendous reexamination of some of the most overt restrictions.

Assignment of Workers

The question of how many workers are to be assigned to specific jobs or whether the employer should assign any at all are left to the following section. For the moment we assume that the work must be done and that the employer must assign workers to do it. There are, then, two broad areas of inquiry: (1) should the work be assigned to a skilled or to an unskilled worker; and (2) if assigned to a skilled worker, must he belong to a specific craft?

Under normal circumstances, it may be expected that the contractor will prefer to assign work to employees who are skilled enough to perform it. The assumption, of course, is that wage levels are tied fairly closely to skills, so that workers assigned to tasks which require less training or experience than they have had, constitute an unnecessary cost to the employer. There are, to be sure, circumstances under which skilled workers may understandably be given unskilled work, such as when employers have no other tasks to be done, yet are reluctant to lay off their skilled people for fear of losing them permanently. But in typical situations, routine, undemanding jobs will be assigned to workers in the lower occupational classifications.

The union contractor, however, may be required to use skilled tradesmen for some kinds of unskilled or semiskilled work. The jurisdictional claims of the various building trades are not typically grounded on the skill requirements of the work, but on its relationship to other functional components of the job. The jurisdictions of most skilled craft unions, therefore, include work which requires appreciably less training than that required of the average journeyman. Haber and Levinson provide the following examples:

> A large proportion of bricklayers' locals in the cities covered by the field survey required that only members of the union could wash down and point brick, a relatively unskilled task. The cement masons' locals often required that the cement finishing machine be operated by a skilled man, though a semiskilled worker would usually be satisfactory on work where the quality of finish was not of great importance. Most of the plumbers', electricians', and sheet metal locals demanded that the unloading and carrying of materials and fixtures be done by union members. In carpentry, a small number of locals insisted upon the right to strip wooden forms from concrete; in most cases, however, this task, as well as the unloading and carrying of lumber, was done by laborers. Finally, minor problems sometimes arose in the finishing trades, particularly in painting, because of the demand that cleaning up the job be done by skilled men.[22]

Since Haber and Levinson wrote this nearly thirty years ago, the takeover of unskilled work by the skilled craft unions has expanded. Thus, it is not infrequent to encounter jurisdictional disputes between the plumbers and the laborers as to whom will carry and lay underground pipe; between the carpenters and laborers over work pertaining to rough forms; between the laborers and any one of several crafts over the handling of materials, etc. If laborers can do such work satisfactorily, it is difficult to objectively sustain a reason why a craftsman must do it at skilled workers' wages.

Helpers and Subjourneymen. Part of the drive for union craftsmen to take over the work of less skilled personnel involved the elimination of helpers and subjourneymen and the substitution of the single-rated journeymen in all crafts. This drive occurred after World War II and has been very successful, although recently many

[22] Haber and Levinson, *Labor Relations and Productivity,* p. 179.

local unions have modified their practices because of unemployment and open shop competition.

The elimination of the helper has been a union objective in the construction crafts for many years. At the turn of the century, unions were demanding restrictions on how helpers could be utilized, and contract provisions which reserved semiskilled and unskilled work for craftsmen.[23] In 1941, the late Professor Sumner H. Slichter reported that the building trades unions worked energetically to prevent helpers from competing with journeymen.[24] Most crafts succeeded by the 1960s in eliminating helpers or subjourneymen entirely. Even though the work may be only semiskilled and the worker a "semicompetent craftsman," if he is used in the union jurisdiction to perform that work, he must receive the full craftsman's wage rate.

Davis-Bacon Impact. In the post–World War II era, major restrictions on the use of helpers, or in most situations the elimination of the helper entirely was aided, abetted, and institutionalized by the U.S. Department of Labor's administration of the Davis-Bacon Act. The job classifications required under Davis-Bacon procedures rarely recognized a helper classification. Open shop contractors bidding on Davis-Bacon work were thus not permitted to utilize the most economical deployment of their labor force, but were required to adhere to the union rule that anyone in a given union's jurisdiction must receive the union journeyman's rate. The only situations in which a helper classification was recognized by the U.S. Department of Labor and permitted under Davis-Bacon rules occurred when (1) the scope of the helpers' duties—meaning the physical tasks performed—was defined and differentiated from that of journeymen, and (2) the particular helper classification prevailed in the area. The use of helpers rarely prevailed under Davis-Bacon rules because the Department of Labor would usually find that the helper was absent in 30 percent of the employee groups. Once the helper classification was eliminated for a government job, all future jobs were required to follow the single journeyman rule. Thus, as noted in Chapter II, one large union job in an area could eliminate the helper classification or that of other subjourneymen for all Davis-Bacon projects in that area.[25]

---

[23] John H. Ashworth, *The Helper and American Trade Unions,* Johns Hopkins University Studies in Historical and Political Science, Series XXXIII, No. 3 (Baltimore: Johns Hopkins Press, 1915), esp. Chapter I.

[24] Slichter, *Union Policies,* pp. 46–47, 304–305.

[25] Armand J. Thieblot, Jr., *The David-Bacon Act,* Labor Relations and Public Policy Series, Report No. 10 (Philadelphia: Industrial Research Unit, The Wharton School, University of Pennsylvania), pp. 126–129.

When the Davis-Bacon regulations were changed by Secretary of Labor Raymond J. Donovan, the construction unions brought suit in a vain effort to prevent the change. Robert A. Georgine, president of the AFL-CIO Building & Construction Trades Department submitted an affidavit claiming that this action was discriminatory because collective agreements (forced on constructors by unions) precluded many contractors from taking advantage of the changes in the new regulations.[26]

Recession Impact. Union policy in regard to helpers has been somewhat modified in recent years as a means of attempting to gain work lost to the open shop. It was noted in Chapter IV that a number of crafts had permitted unionized contractors to utilize helpers and/or subjourneymen at below journeymen's rates. Failure to permit such personnel in crews adds greatly to costs, as illustrated below.

Excessive Crew Costs: Example and Findings. A good example of costs added by union contracts which can be avoided by operating open shop is found in the installation of electrical outlets and connections. A unionized contractor may utilize crews consisting of three to five persons, depending on the nature of the project, all paid journeymen's rates of $12 to $20 per hour or more. Yet much of the work is unskilled or, at best, semiskilled, as for example attaching the conduit; inserting and pulling the wires; and nailing, screwing, or bolting the conduit and boxes to the walls. In fact, the crew will usually have one or at most two first-class craftsmen who will direct it and make the actual connections and installations, particularly where they are especially complicated.

In contrast, the open shop contractor will do the same work with crews of one craftsmen or, perhaps on very large projects, two top-rated craftsmen, one or two helpers, and the balance laborers. The craftsmen may be paid as much or more than the union rate; the helpers are paid a lower rate, depending upon their skill and experience; and the laborers, still less, again varying according to their experience and abilities. Given the large number of crews required to complete a job and the differences in their composition and wage schedules, the savings of an open shop operation to the contractor and customer can thus be quite substantial on this single aspect of worker utilization.

In addition to information obtained from our field interviews, there have been some important findings on the extent of the varying use of skilled personnel among open and union shop contractors.

[26] Building & Construction Trades Department v. Raymond J. Donovan, Secretary of Labor, 712 F.2d 611 (D.C. Cir., 1983), cert. denied, 104 S. Ct. 975 (1984).

TABLE IX-6
Proportion of Unskilled Manpower Requirements by Region and Type of Project, Various Years

Type of Project	Year	Region				
		U.S.	North-east	North Central	South	West
1. School	1959	29.1	27.8	27.2	38.7	23.0
2. Federal office building	1959	32.5	28.3	31.6	33.7	25.1
3. Hospital	1959-1960	26.7	24.6	26.6	32.0	18.6
4. Public housing	1959-1960	30.9	23.3	25.3	40.8	22.0
5. One-family house	1962	23.3	25.2	14.9	31.0	14.8
6. College housing	1960-1961	31.8	24.9	29.5	40.5	24.7
7. Sewer works	1962-1963	38.8	31.7	34.9	44.3	29.5
8. School	1964-1965	30.9	24.6	24.3	42.7	22.3
9. Hospital / nursing home	1965-1966	19.6	17.8	18.1	28.3	10.8
10. Public housing	1968	30.2	25.4	26.7	36.4	20.4

Source: U.S. Bureau of Labor Statistics Bulletins:
1-1299 (1961); 2-1331 (1962); 3-1340 (1962); 4-1402 (1964); 5-1404 (1964); 6-1441 (1965); 7-1490 (1966); 8-1586 (1968); 9-1691 (1971); 10-1821 (1974).

One is the already noted willingness on the part of unions in recent years to permit the utilization of helpers in order to give unionized contractors a better opportunity to compete with their open shop competitors who obviously make effective use of helpers. A second is the series of studies done by the BLS in the 1960s and early 1970s of personnel requirements on various kinds of construction projects. These surveys have consistently found a higher proportion of unskilled labor being used on building work in the South, where, as we have seen, union contracts are distinctly less widely applicable than in the rest of the country. This point is illustrated in Table IX-6. Other factors may, of course, be responsible, but it does seem likely that the extent of the open shop in the South is the major point of difference in skilled craft utilization.[27]

More recently, the Business Roundtable, as part of its Construction Industry Cost Effectiveness Project, examined the use of helpers

[27] It could be charged, for example, that the southern data indicate the use of skilled blacks who are given unskilled designations and thereby discriminated against. In fact, as we shall make clear in Chapter XII, the takeover of unskilled work by skilled craft unions has involved the displacement of blacks by whites. The elimination of the helper classification has also eliminated many minorities.

and subjourneymen both by unionized and open shop contractors. Despite some agreement permitting their use, the report concluded after a field survey of unionized contractors:

> The results were disappointing in that comparatively few projects were found where subjourneymen were actually used. The broadest and least restrictive subjourneymen [contract] language occurs in the National Industrial Construction Agreement (NICA), negotiated to enable members of the National Constructors Association to compete with open-shop contractors. Though several jobs under the NICA were in progress, the study team was unable to find any substantial use of subjourneymen on any of them.[28]

The Business Roundtable study group did find the use of helpers greater on projects built under project agreements, but even then, their utilization was dwarfed by the extensive use of helpers which was found in open shop construction projects in the power plant, paper, petrochemical, refinery, and metallurgical industries. In these industries, the average use of helpers varied from 35 percent to 50 percent of the workers in each craft, while the ratio of helpers to subjourneymen varied to yield the maximum overall efficiency, depending upon the type of job and the quality of available helpers, and to some extent upon the craft and the contractor.[29] The Roundtable report concluded:

> Open-shop contractors have a significant economic advantage becuse of their ability to fully utilize helpers for unskilled and semiskilled work in all crafts, and to add manpower to labor pools. Despite the limited use of subjourneymen on union construction projects to date, enough experience has been observed to demonstrate the advantages to unions, contractors and owners.
>
> Using a high percentage of subjourneymen, properly screened, may bring labor cost savings, on a particular job, as great as 20%. The long-range benefit may be equally important because it offers the construction industry a new tool for meeting future manpower needs. Open-shop contractors have demonstrated an ability to train large numbers of workmen, including minorities and beginners, in a relatively short time by a variety of successful methods. These techniques should be equally successful in the union sector. Moreover, building trades unions and union contractors would win an increased share of the construction market by a more competitive performance.[30]

In sum, the open shop contractor is free to divide work among his employees according to their skill, rather than their organizational affiliation, and therefore enjoys a direct cost advantage over

---

[28] *Subjourneyman in Union Comstruction,* Report D-1, Construction Industry Cost Effectiveness Project (New York: The Business Roundtable, 1982), p. 8.

[29] Ibid., pp. 8, 12–13.

[30] *Ibid.,* p. 14.

his unionized competitor. He may not only utilize his unskilled labor to the fullest extent, but he is also at liberty to assign his skilled workers to whatever jobs need to be done when there is a temporary lull in their own work, thus preserving the stability of his workforce and increasing employee loyalty.

Craft Jurisdictional Diseconomies. A very significant difference in terms of the allocation of personnel between unionized and open shop contractors involves the manner in which union craft jurisdictional boundaries interfere with the economic assignment of work among the various skilled trades. Since differences in the wages between two crafts tend to be significantly less than the difference between skilled and unskilled labor generally, it might be of relatively little concern to the contractor as to which is assigned the work. Nevertheless, contract requirements which prescribe job assignments to specific crafts may entail several costly practices. Three examples are illustrated:

1. Some jobs, such as activating equipment or temporary heating and lighting devices, could be done in a short time by many craftsmen of average mechanical ability, but the contractor is required to bring in a specific type of journeyman and pay him for a full day's work even though there is no other work for him to do.
2. A small task associated with the work of one craft requires the temporary use of a worker in a different trade. Even if the workers at the site are capable of performing the task, they must delay until a member of the craft which "owns" the job is brought in to do it. Thus, drilling holes or affixing a holding board for pipe or electrical work must be done by carpenters, although electricians and plumbers who are on the job could easily do it.
3. Sometimes a relatively simple job overlaps the jurisdiction of several crafts, and workers from all crafts involved must be utilized for a brief time, but paid a full day's pay. Thus, it is possible that, for example, if a hospital table was moved which had fittings for gas, electricity, and water, a union contractor might have to utilize and pay a steamfitter, an electrician, and a plumber to make it operable.

The extent and impact of these jurisdictional restrictions today cannot be readily measured. But their existence was corroborated by interviews with contractors and other industry spokesmen during our field work in 1974 and 1982. It is apparent, moreover, that the open shop contractor enjoys at least some cost advantage by being free to allocate work among his skilled tradesmen with a minimum of interruptions and friction. This is not to say, to be sure, that nonunion workers are perfectly interchangeable. The average bricklayer does not simply drop his trowel and begin stringing wire or laying pipe when the masonry work is done. There is, however, considerable interchange of work among carpenters and cement finishers. The general mechanic or mechanics qualified in two or more crafts are also becoming increasingly common on the payrolls of open shop contractors, both increasing the job security of the workers and decreasing the costs of the employers.

The major savings for open shop contractors in labor deployment still occur when there is a temporary need for a worker to perform routine tasks out of his normal occupational class. The freedom of the open shop worker to work out of his occupational class manifestly allows the job to continue more smoothly and less expensively than if the work has to be done in bits and pieces by several craftsmen. Whether the savings result from routine task performance or the use of multiqualified craftsmen, they are substantial, not only in less direct wage costs, but also in fringe benefit costs, which substantially depend on the number of employees rather than the hours worked, and, perhaps most of all, in the elimination of pay for time not worked, which is a fundamental cost of rigid craft jurisdiction and job demarcation.

Finally, and also of major importance, the open shop contractor does not have to contend with time lost because of jurisdictional strikes. Despite national rules and detailed manuals, the question of which union has jurisdiction over a given task can be a fluid matter that varies from locality to locality, often depending on the power, prestige, and aggressiveness of local union business agents. Craft jurisdictions are also often essentially obsolete because of changing technology and, therefore, do not reflect work realities. Since jurisdiction determines who performs the work and how much a local union will receive in dues and permit fees, the interunion conflict can be fierce. As a result, year after year a significant proportion of the work stoppages in construction involve interunion disputes over jurisdiction, even though such strikes are forbidden by law and contract.[31] The cost of this lost time is a serious problem for unionized contractors who attempt to compete with open shop

[31] Most contracts forbid jurisdictional strikes, and various national compacts have been made to settle such matters, but the agreements are often not observed and the national compacts are neither binding nor effective. Jurisdictional strikes are an unfair labor practice under Section 10(k) of the National Labor Relations (Taft-Hartley) Act, and Section 10(K) provides injunctive relief and opens the way for damage suits in federal courts.

contractors, especially when it is noted that "construction workers are on strike more frequently and lose more working time than workers in any other industry."[32]

After studying the jurisdictional problem, the Business Roundtable Construction Industry Cost Effectiveness Project concluded as follows:

> Exclusive jurisdiction—which is based on the notion that every task can be performed only by members of a particular union—is a major source of inefficiencies in construction. It may well be the greatest handicap faced by union contractors as they attempt to avoid further losses in their shrinking share of the construction market. . . .
>
> Costly jurisdictional disputes still occur, but an *even greater cost* to the industry *results from inefficient work assignments routinely made to comply with precedents and jurisdictional agreements and to avoid disputes*. . . . jurisdiction imposes major costs on union contractors.
>
> Mechanisms for voluntary settlement of jurisdictional disputes have only increased the problem by developing ever more precise craft assignments of work that could be performed by any of several crafts. The national voluntary jurisdictional dispute settlement plan is in disarray and is not functioning. When the National Labor Relations Board (NLRB) has been utilized to settle jurisdictional disputes, its decisions usually have been based more logically on employer preference and efficiency considerations. But use of the NLRB is impractical in many disputes because the Board takes so long to act, if it chooses to act at all.[33]

Determination of Workforce Size

In addition to provisions requiring the assignment of certain kinds of work to specific crafts, union contracts often contain various clauses which, according to employer spokesmen, mandate the use of excessive manpower. Where the open shop contractor is free to ensure that all workers are engaged in productive labor, the unionized employer is in several respects constrained in the number of workers which he must assign to a particular function. It is difficult to evaluate the net effects of these provisions for at least

---

[32] David B. Lipsky and Henry S. Farber, "The Composition of Strike Activity in the Construction Industry," *Industrial and Labor Relations Review,* Vol. 29 (April 1976), p. 388. Professor Mills has noted that "construction is a strike-prone sector of the economy: . . . with only about 5 percent of the nation's total employment, contributed about 24 percent of all strikes and 19 percent of all workdays lost due to strikes." D. Quinn Mills, "Construction," in *Collective Bargaining: Contemporary American Experience* ed. Gerald G. Somers (Madison, Wisconsin: Industrial Relations Research Association, 1980), p. 77.

[33] *Exclusive Jurisdiction in Construction,* Report C-1, Construction Industry Cost Effectiveness Project (New York: The Business Roundtable, 1982), p. 1.

two reasons. First, it is difficult to determine the extent to which the contractual obligations actually differ from the practices which would be voluntarily adopted by the employer—or, for that matter, the practices which are in reality adopted by contractors not subject to union rules. Second, it is not always clear that the restrictions are socially unjustifiable in themselves. Arguments, for example, that certain manpower requirements are necessary for safety reasons cannot be dismissed out of hand. In many cases, however, safety is not a factor, and the alternate rationalizations are much less persuasive.

Charges of excessive manpower requirements may be grouped broadly into three related categories: those mandating minimum crew size; those mandating assignment of workers to unnecessary jobs; and those mandating a minimum number of foremen. Foremen requirements appear to be the most common. In the 1972-1973 BLS survey of 769 construction agreements, 65 percent of them (embracing about 60 percent of the workers surveyed) specified the assignment of a foreman for a given number of workers (in most cases, only four or fewer). About one-quarter of the contracts also prescribed a ratio of foremen to workers, a provision found most commonly among electricians and plumbers.[34] The study also found provisions in many agreements, although it did not specify the actual number, prohibiting foremen from working with tools once the crew reaches a certain size.

Recent studies by the Business Roundtable emphasize that such provisions are common. About 30 percent of the contracts examined by the Roundtable study group had restrictions on foremen not working and about 50 percent contained foremen-journeymen ratio requirements.[35] Moreover, first-line foremen and, in many circumstances, higher-level supervisors, were contractually required to be union members, were employed through the union hiring hall, and had their compensation set in negotiations.[36]

Supervisors are not "employees" as defined by the Taft-Hartley Act, and therefore there is no requirement that they be a part of the bargaining unit or that management bargain concerning their compensation. Union power has accomplished this in construction. As a result, supervisors can be disciplined by unions for carrying out management policies, for example, for attempting to increase productivity or for opposing wasteful allocation of labor.

---

[34] U.S., Bureau of Labor Statistics, *Construction Agreements,* p. 13; and *Constraints Imposed by Collective Bargaining Agreements,* p. 14.

[35] *Contractor Supervision in Unionized Construction,* Report C-3, Construction Industry Cost Effectiveness Project (New York: The Business Roundtable, 1982), pp. 3–4.

[36] *Ibid.,* p. 11.

Open shop contractors, of course, also utilize foremen, and it is not improbable that, in some instances, they would meet the union standards anyway. But not all situations are the same, and the use of foremen (especially nonworking foremen) should logically be predicated on the complexity of the work and the need for close supervision. A fixed formula does not distinguish between the routine job and the special case. The open shop contractor who is able to get along with fewer foremen saves money both in terms of saving the foreman's premium wage and obtaining more productive working time. On balance, however, this factor appears to constitute less of a cost advantage to the nonunion contractor than most other restrictive practices. When the open shop contractors and executives were interviewed, we asked them to outline the reasons for the growth of open shop construction. Foremen assignments were seldom mentioned either in 1973–1974 or in 1981–1983.

On the other hand, the unionized contractors interviewed clearly felt that their position was made more difficult by the unionization of supervisors. They were unanimous in their belief that this practice inhibited efficiency, proper management of the workforce, and above all, discipline.

Employment of unnecessary workers by union contractors was also very high on almost everyone's list. The Operating Engineers have traditionally come under the most criticism for this practice. One example is clauses which mandate the assignment of a worker to a limited number of small machines, such as pumps and compressors which, according to employer spokesmen, require no continuing attendance. Another is the specification that an oiler to work with the operator of each machine. Finally, many contracts limit the number of times an individual can be transferred from one piece of equipment to another. Although the operating engineers typically receive much of the attention in this connection, other crafts are not immune. The Business Roundtable has cited some illustrations in two reports, one in 1974 and the other in 1982:

> Examples of costly "standby" manning include: carpenters being required to stand by during the pouring of concrete even though wooden forms are not used; cement finishers in some areas have to be on hand when concrete is being poured even though they are not actually needed or used; the union steward who must be on the job at all times work is performed even though he is not required and may not be able to perform any of the available work.[37]

[37] "Restrictive Work Practices in the Construction Industry," mimeographed report prepared by the Construction Roundtable, New York, 1974, p. 14. The Construction Roundtable is now part of the Business Roundtable.

The 1982 report, in a "Summary of Jurisdictional Inefficiencies," listed in more than five pages cases where unnecessary workers were used. Electrical workers, iron workers, boilermakers, and others were found in many cases to rival operating engineers in demanding the employment of unnecessary workers.[38]

Unions claim that some of these situations involve safety practices. Moreover, none of the practices deplored by the Roundtable are either universal in the union sector or always necessarily in conflict with the prevailing norm in open shops. The evidence suggests, however, that they are sufficiently common to help explain the growing penetration of open shops into previously union markets. It is significant that complaints from organizations of union contractors and their customers—the National Constructors Association (ACA), AGC and Business Roundtable—even during the period of explosive wage settlements, did not center as much on wage rates as on what is broadly described as "featherbedding." The open shop contractor, facing no jurisdictional restrictions or disputes and no limitations on the size and allocation of his workforce (other than those prescribed by building codes and other laws), has found himself in a position of increasing competitive advantage simply because of his greater flexibility in operations. The extent to which this flexibility is differently exercised may obviously vary considerably, and the differences, in practice, may indeed have been exaggerated by spokesmen for embattled union contractors. But even a uneconomic practice that is exaggerated is still one that exists and, in a competitive industry like construction, may spell the difference between winning and losing an important contract. Other implications of this question will be explored below.

SEASONAL EMPLOYMENT

One of the more significant burdens borne by the construction labor market is the necessity of adjusting to seasonally changing manpower needs. Although the magnitude of seasonal change varies along a number of dimensions—geographic region, type of construction, and function of the contractor or subcontractor are the most prominent of these—the aggregate swings in the industry's employment over a twelve-month period are substantial. It is not unusual for employment to vary substantially—20 percent or more—between February and August. This means that, in good employ-

[38] *Exclusive Jurisdiction in Construction,* pp. 39–45.

ment years, the industry may recruit 800,000 or more workers between the winter low and the summer high, and then determine what to do with the surplus during the slack season.[39]

Only part of the answer is found in the unemployment statistics. Between February and August construction unemployment typically falls by only a fraction of the concomitant rise in employment. The major adjustment is made in the size of the labor force rather than simple fluctuations caused by the release of workers who go on the unemployment rolls and are absorbed again later. The latter alternative creates problems in both the union and nonunion sectors. The open shop contractor in laying off a regular employee runs the risk that the employee will be unavailable when he is needed again later and that a suitable replacement will be difficult to recruit. For the union contractor, this process is less troublesome, for the hiring hall usually provides a ready source of manpower during the upswing. However, large numbers of unemployed members constitute serious political problems for the union, which will, therefore, normally seek to maintain its regular membership at a level somewhat lower than that needed to meet all peak season needs.

Labor force adjustments occur at both seasonal extremes. During the upswing, some previously unemployed workers are recalled; in addition, other workers are recruited on a temporary basis to meet peak season needs. The open shop contractor is also free to upgrade workers; transfer them from one type of work or craft to another; restructure jobs to make better use of scarce skills; schedule overtime at time and one-half rather than double time, as is called for by most union contracts (see Chapter XI); have supervisors and managers do more manual work; have more work prefabricated; and in other ways take advantage of the flexibility which union rules deny unionized contractors.

During the downswing, three main phenomena can be broadly identified: the temporary seasonal workers return to their primary activity; some regular workers are laid off; and certain changes take place in the unionization of the remaining workforce. Although many of these adjustments are common both to union and open shops, the relative extent and importance of certain adaptations tend to be somewhat different among the nonunion contractors.

It is common for construction contractors to supplement their regular workforces with seasonal help. These temporary workers are often students or teachers taking summer jobs, although they

---

[39] Data from U.S. Bureau of Labor Statistics monthly compilations.

may also include moonlighters from other industries. Respondents to the Industrial Research Unit surveys were asked in 1974 and 1982 whether they "typically put on supplementary summer help (college students, etc.)." The responses, presented in Tables IX-7 and IX-8, demonstrate that this practice is utilized by both union and open shops, although it is used more frequently by open shop contractors.

Perhaps the most persuasive explanation for the greater use of summer help by nonunion contractors lies in the relative wages which these workers can be paid. Although some local unions permit students to be paid at the apprentice rate, others require a full journeyman's wage for all workers (other than bona fide apprentices) in the union's jurisdiction. The open shop contractor, however, is free to pay all temporary workers according to their skill and experience. Summer workers are likely to be untrained or at least less accomplished than the average worker with a full-time attachment to construction work. For the union contractor who must pay the going rate to such a worker, there is a real disincentive to utilize him. For the nonunion contractor who can peg the wage to the worker's productivity, these workers can be a satisfactory alternative to higher paid skilled workers who may, in any case, not be available when most needed.

Tables IX-7 and IX-8 may also demonstrate another characteristic of the industry's employment patterns. The percentage both of unionized and open shop employers utilizing summer supplementary employees was markedly less in 1982 than in 1974. Undoubtedly, the reasons include, above all, the more difficult economic situation in the latter year. With thousands of full-time construction workers on layoff, seasonal employees would naturally find fewer jobs.

Open shop contractors are also likely to respond differently during the downswing. The Industrial Research Unit 1974 questionnaire asked contractors if they "have had to lay off any workers in the

TABLE IX-7
Constructor Use of Supplementary Summer Employment Union and Nonunion, 1974

	Union		Nonunion	
Response	Number	Percent	Number	Percent
Yes	473	60.8	614	78.3
No	290	37.3	162	20.7
Not reported	15	1.9	8	1.0

Source: Industrial Research Unit Questionnaire, 1974.

TABLE IX-8
Constructor Use of Supplementary Employment
Union and Nonunion, 1982

	Union		Nonunion	
Response	Number	Percent	Number	Percent
Yes	862	47.2	1,298	64.2
No	850	46.5	636	31.5
Not reported	116	6.3	88	4.4

Source: Industrial Research Unit Questionnaire, 1982.

past three years because of the seasonal slowdowns." As shown in Table IX-9, open shop contractors were more likely to respond in the negative in 1974. The responses to the 1982 questionnaire precluded any firm conclusions about seasonal layoffs because it was impossible to determine from the data those layoffs that resulted from the usual seasonal slowdowns and those produced by the general economic downturn in that year.

The responses to the 1974 questionnaire were largely consistent with the assertions of spokesmen for open shop contractors to the effect that employment tends to be steadier in the nonunion sector. Since there is no reason to expect that union and nonunion employers differ in the extent to which their needs fluctuate over the seasons, the responses suggest that open shops make more concerted efforts to avoid layoffs. These efforts, to be sure, are usually not represented as gratuitous or altruistic. In our personal interviews with contractors, they were more often represented as necessary and prudent, in part to offset the relatively low wages paid and in part as a recognition of the open shop's relative difficulty and uncertainty in finding suitable replacements later. A typical response would take the followng form: "We keep our employees by offering them steady work, thus firming up their attachment and loyalty to

TABLE IX-9
Seasonal Layoffs by Union and Nonunion Contractors
1971–1974

	Union		Nonunion	
Response	Number	Percent	Number	Percent
Yes	600	77.1	430	54.8
No	156	20.1	345	44.0
Not reported	22	2.8	9	1.1

Source: Industrial Research Unit Questionnaire, 1974.

the company." In short, the conventional premise that the construction worker forms a bond with his trade rather than his employer is less true of nonunion workers.

The open shop contractor seeks to ensure that his employees can be kept on during slow periods in a variety of ways, some of which are not available to the union employer faced with jurisdictional lines. He may "beat the bushes" for work which he would not actively solicit in busier times, perhaps even performing it at little or no profit simply to keep his workforce occupied. He may construct, on speculation, rental apartments or homes. He may simply give some workers a paid vacation. Open shop highway contractors and others devote considerable manhours to machine rebuilding and preventive maintenance. Others utilize the slow months for intensive training activities; still others assign janitorial or any other duties available to "keep the men on." Several open shop contractors interviewed by the Industrial Research Unit staff in 1974 and 1982 emphasized that they purposely utilized summer help, particularly students, in order to minimize winter layoffs.

In the light of these differences in the labor market practices between union and open shop firms it would, nevertheless, be erroneous to assume that unionized workers are all job-hoppers and are lightly dismissed by their employers. A study of the records for 1981 of a large multicity carpenter's fringe benefit fund by the Construction Labor Research Council found that, as the unionized worker becomes older, he tends to stay with one firm and to work longer annual hours. Moreover, the longer the annual hours, the greater is the proportion of the total work performed by the longer hours personnel. Thus, in this study, the 25 percent of the workers with the most annual hours performed 50 percent of the work, and as a worker increased his annual hours over 1,000, he tended to stay with one contractor. The study concludes:

> It is readily apparent that there is a wide range of individuals with a wide range of work experiences in the construction industry. When examining these hours worked data, averages are likely to either mislead or obscure a situation. Some general conclusions can, however, be reached from the data.
>
> In an apparent contradiction, there is a sizeable cadre of construction workers who work close to or at a full work year and perform a significant portion of the industry's work while at the same time a good portion of those persons available for work are very much underutilized within the industry. The high hour workers have a strong attachment to the industry, are more likely to have a continuous relationship with one employer, are older than average and, quite likely, are the most productive workers in the industry. Generally, the opposite is true of the low hour workers. The large number

> of low hour workers with high turnover rates are probably more costly, but they also provide an available labor pool in periods of high demand.[40]

In times of severe recession, as in 1980-1982, all contractors, open shop and unionized, are compelled to lay off workers. Even in many recessions, however, the open shop contractor's flexibility and responses to changing seasonal needs are likely, in most cases, to provide a more stable workforce as a whole although some long service union craftsmen may have relatively steady work.

CONCLUSION

The open shop contractor recruits his workforce through more informal channels than does his unionized counterpart. This gives him none of the advantages of the union hiring hall, but frees him from its significant disadvantages. Because of the increasing need for more effective recruiting tools, open shop contractor groups are developing centralized recruitment mechanisms suited to their needs, but these are still in an experimental stage and are relatively rare.

Open shop contractors have significant advantages in the deployment of personnel on the job. Unrestricted by union rules, they can pay employees according to skill and productivity while assigning them according to need. Free of worry of jurisdictional disputes, free to determine the skill content of each job, and free to determine the extent of job manning needed, the open shop contractor can, and indeed does, utilize labor more efficiently. On the other hand, because of the absence of craft restrictions, skilled personnel can be retained on the payroll and used below skill levels when employment is down and it is desired to conserve such skills for the firm.

Finally, through greater use of supplementary summer help and a host of measures to contrive work in slow seasons, open shop contractors are both permitted and induced to maintain a more stable regular workforce, although a minority of union craftsmen are employed for a stable work year.

---

[40] "Annual Hours of Construction Workers. Analysis of Worker Characteristics," Construction Labor Research Council, August 1983, p. 7.

CHAPTER X

Training and Development

There are few industries which can function properly without ongoing efforts to generate trained human resources. The study of investment in human capital has itself become a specialty within the scholarly discipline of economics. Government, academe, the labor movement, and industry have all explicitly acknowledged the importance of employee development in terms of individual fulfillment, industrial prosperity, and aggregate economic growth.

The role of occupational training is especially critical in construction. Although all industries employ skilled workers, construction employs an exceptionally large number of them.[1] In 1982, 55.0 percent of all workers in the industry were classified as skilled. Among other industries employing a substantial manual workforce, the percentages were 22.7 for mining, 19.3 for manufacturing, and 21.0 for transportation and public utilities.[2] Thus, construction employs about one-quarter of the skilled blue-collar workers in the entire labor force.

Each year the labor force must be replenished to offset death, retirement, turnover, migration to other industries, and growth. Turnover alone can require 1.8 new workers for each separation, with new employees comprising the largest percentage of those who leave a company. To the individual firm, as well as to the industry as a whole, training is thus a vital component of efficient labor market adjustment. As one open shop contractor put it metaphorically:

> Compare yourself as a construction manager to the coach of a successful basketball team, or any other kind of team in the world of sports, and you will realize for a coach to have a winner, he must have the "horses" in the slang of sports. A successful coach must first recruit his players and then coach them in the fundamentals and in his method of playing the game. And finally, he must put them all

[1] The term "skilled" as used here relates roughly to the census classification "foremen, craftsmen, and kindred workers." Semiskilled and unskilled workers are officially designated as "operatives" and "laborers" respectively.

[2] U.S. Department of Labor, Bureau of Labor Statistics, *Handbook of Labor Statistics,* Bulletin No. 2175 (Washington, D.C.: Government Printing Office, 1983), p. 55.

> together as a team to become a winner. With good success in recruiting, coaching and teamwork, he will have a winner. The same is true in the construction industry. We must recruit our help, train our help and put all of the men together as a team in order to have a winner.[3]

Training for construction occupations takes place both inside and outside the industry. Some workers pick up certain basic skills in vocational or trade schools. Some workers have obtained training in branches of the military which perform their own construction work or through various programs designed to prepare servicemen for civilian life. Some pick up trades on their own with the help of friends and relatives. Finally, some learn skilled building occupations through training or experience in other industries, such as farming, manufacturing, or shipbuilding. On the whole, however, the industry itself must take primary responsibility for training, both formal and informal, if its employees are to acquire the necessary skills to perform construction work. Therefore, how open shop contractors train their workforce, and how this training compares with that in the union sector are extremely important issues, particularly in view of the extent and growth of the open shop. We shall examine both informal and formal training programs in this chapter, including the general picture and specific company and association efforts.

INFORMAL TRAINING IN THE OPEN SHOP

There is no universally acknowledged distinction between formal and informal training. Indeed, learning forms may be viewed as ranging along a spectrum from mere observation of the production process itself to highly structured programs involving both a set sequence of on-the-job experiences and rigorous off-the-job instruction. The point at which training becomes "formal" can be fixed in any number of ways: for example, the intent of the employer; the ordering of the trainee's assignments; the explicit role of the journeyman who is doing the training. Since these notions are difficult to deal with empirically, we shall regard in this chapter as "formal" only those approaches to training which include some form of supplemental instruction, either theoretical or manipulative or both, away from the workplace.

By this definition, most training which occurs in construction is informal. Such training is of course purposeful, valuable, and ef-

---

[3] Associated General Contractors, "Excerpts from the AGC Regional Open Shop Conference," Atlanta, February 12, 1972, (mimeographed), pp. 36–37.

fective, but most contractors when asked about formal training (as we define it) would agree that on-the-job learning is greatly facilitated by a grounding in the theory, terminology, and technology of the construction process—subjects most easily conveyed in a pedagogical setting. At the same time, related instruction can be expensive, so that our distinction between formal and informal training is also useful as an indicator of the employer's commitment to investing in his workers.

Informal training generally goes beyond mere observation by the trainee. When employers report that all their training is on the job, that may mean many things. Most commonly, it means that a new worker is assigned to work alongside a journeyman, performing first the simpler tasks associated with a particular job and later graduating to more complex operations. The journeyman is expected to explain the job while doing it and to give the trainee some practice under his direct supervision. The distinction between training and random learning is the role of the journeyman. When he is training his work pace is deliberately slower than when he is simply being observed in a random learning situation. Most contractors assert that their journeymen (or they themselves) actively engage in training, and that trainees are hired with the understanding that they will be taught aspects of their work on the job.

Such training, if successful, performs a special service for the contractor in teaching the trainee precisely what he is expected to do for the particular contractor, or particular type of construction work. It may not prepare the trainee for jobs with other contractors. Since a priority of open shop contractors is to hold their labor, whereas the union shop contractor depends upon the union hiring hall for labor supply, such on-the-job training is much more appropriate for the open shop.

In most informal training there is no predetermined ordering of job assignments which build upon each other. The trainee is assigned to a job that needs to be done at the time. If his prior experience is not sufficient to learn that job expeditiously, then the journeyman must use him only for less significant tasks. His role then becomes that of a helper rather than a trainee. The distinction between the helper and the apprentice, drawn by Haber and Levinson thirty years ago, is still appropriate: "The principal difference between the helper and the apprentice is that the former is given no formal training, but concentrates full time on his semiskilled function."[4] The result is "a more specialized labor force

[4] William Haber and Harold M. Levinson, *Labor Relations and Productivity in the Building Trades* (Ann Arbor: University of Michigan, Bureau of Industrial Relations, 1956), pp. 98–99, note 27.

capable of doing an acceptable job in a limited area, but with virtually no competence in other important aspects of the trade."[5] Of course, many helpers do in time become craftsmen, while many apprentices never reach first-class journeyman status. The distinction is, however, indicative of a basic difference in training concepts and aims.

Although an employer can do quite well with the bulk of his workforce specialized in very narrowly defined skill areas, some workers must have broad familiarity and expertise in the overall construction process. For example, there must be someone who can read blueprints, lay out a job or sections of a job, perform the more delicate and exacting tasks associated with a trade, solve unforeseen problems as they arise, and generally supervise and orchestrate the work of others. In short, any project requires a complement of all-around journeymen:

> Training for craft occupations in construction is designed to produce a journeyman who is able not only to perform certain tasks under supervision but to supervise himself to a large degree. The journeyman is trained to perform much of the supervisory and planning functions that in other industries are the role of management. . . . The journeyman is expected to understand the fundamentals of his craft as well as how to perform specialized tasks; he may be asked to lay out work from blueprints and to supervise his own performance on the job.[6]

The worker who is trained informally on a "catch-as-catch-can" basis is less likely to attain this level of journeyman status—or at least he is not likely to attain it as quickly as one whose training is more systematic. The fundamentals of his craft are not easily or efficiently taught on the job; the supervising journeyman is not apt to interrupt the work to deliver a comprehensive lecture on blueprint reading or the composition and properties of cement. The contractor may be able to function with a large number of semiskilled or narrowly trained specialists, but they cannot constitute his entire workforce. The open shop employer, in particular, unfettered by union jurisdictional restrictions, may be able to subdivide the work in order to maximize the output of his specialists and to minimize his overall labor costs, but this practice puts a premium on the role of the exceptionally skilled and well-rounded craftsman who can coordinate and direct the work.

Conversely, the unionized contractor who must utilize all journeymen and no helpers must often require skilled personnel to

---

[5] *Ibid.*, p. 99.

[6] Daniel Quinn Mills, *Industrial Relations and Manpower in Construction* (Cambridge: M.I.T. Press, 1972), pp. 183–84.

perform unskilled or semiskilled work. As set forth in Chapter IX, many such contractors frequently and rightly complain that they must pay the journeyman's rate to persons with a wide spectrum of capabilities, including some that are far from all-around craftsmen. Thus, the open shop contractor's use of the on-the-job trained helper or less experienced craftsman creates a workforce organization responsive to practical job specifications. This flexibility in assigning workers, as has been frequently noted in previous chapters, is a fundamental reason why open shop construction is more economical.

A significant advantage of on-the-job training in the open shop sector is that opportunity is afforded for intercraft mobility. Typically, employees with no experience are hired as laborers. If they show promise and openings are available, they can win assignments for cement work and/or rough carpentry, or they may progress from one craft to the other. With craft union barriers absent, a capable person can change his line of work and improve his income. Since open shop builders' wage rates depend on ability and performance, rather than on the particular craft involved, and since it is to the advantage of the open shop contractor that his employees have as many capabilities as possible, upward movement and line-of-work changes are encouraged.

Upward movement into the more specialized and difficult trades, such as the electrical, mechanical, and plumbing work, however, is less likely to occur for two principal reasons. First, these trades are typically subcontracted. A general contractor's employees are thus not likely to have an opportunity to move into such crafts. Second, such crafts require greater technical and/or classroom training and thus are more difficult to enter through on-the-job training. Nevertheless, some mobility into these crafts has occurred, and cases were cited during field interviews in which skilled craftsmen (particularly plumbers and electrical helpers) were former laborers and cement finishers. To be sure, most were not yet journeymen, but their mobility had been considerable and probably would not have occurred in the union sector because their formal education and age excluded them from systematic apprentice programs. As we shall discuss in Chapter XII, upward mobility through on-the-job training has been especially important for minorities who have been bypassed by the high educational, low age, and often discriminatory entrance requirements of union apprentice programs.

Despite the significance of on-the-job training, the critical role of the open shop journeyman requires an understanding and appreciation of the formal training undertakings which have been increasingly expended by open shop contractors' associations. While nonunion employers have long engaged in structured training programs,[7] the growth of open shop construction since 1960 has been matched by a marked expansion of such programs. Prior to our original study, however, there had been no comprehensive inventory or assessment made of formal training in the open shop.

FORMAL TRAINING IN OPEN SHOP CONSTRUCTION

Formal training in open shop construction is of two types. The first is the traditional apprenticeship approach; the second is the task method.

Apprenticeship

The most common type of formal training in the construction industry is apprenticeship. Apprenticeship has been described as combining "a schedule of work processes in which an apprentice is to receive training and experience on the job" with "organized instruction designed to provide the apprentice with knowledge in technical subjects related to his trade."[8] It also normally includes a wage progression throughout the indenture (usually calculated as a percentage of the journeyman's wage) and formal recognition upon completion of the program, which must take place over a stipulated number of years.

Apprentices may be indentured either to an individual employer or to a special multiemployer organization. To gain certain benefits under public statutes (such as apprenticeship wage rates under prevailing wage laws or payments under the GI Bill) programs must be registered with the U.S. Bureau of Apprenticeship and Training (BAT) or with a state apprenticeship council. Open shop employers have historically had difficulty in securing registration for their programs. The staffs of BAT and most state apprenticeship agencies have been drawn largely from union ranks,[9] and such agencies once

---

[7] See Haber and Levinson, *Labor Relations,* p. 98, for an analysis of the balance between formal and informal open shop training at the time of their study.

[8] U.S. Department of Labor, Manpower Administration, *The National Apprenticeship Program,* 1968 ed. (Washington, D.C.: Government Printing Office, 1968), p. 6. The Manpower Administration is now known as the Employment and Training Administration. The definitions, however, are still current.

[9] Writing in 1968, Professor Mangum summed up the extent of union control of BAT: "The choice of apprentices, training methods, and the costs of training were totally matters decided by local unions and employers. Its power bases were the Building Trades, Metal Trades, and Printing Trades, departments of the AFL-CIO, with a potent assist from the fact that the chairman of the Labor Department-HEW Appropriations Subcommittee was a former Building Trades man. The BAT Administrator was selected after consultation with the craft unions and most of the field staff were former union officials or journeymen." Garth L. Mangum, *MDTA: Foundation of Federal Manpower Policy* (Baltimore: The Johns Hopkins Press, 1968), pp. 45–46.

considered union participation as a sine qua non of a plan's acceptance for registration.[10] Moreover, prior to 1970, BAT would not register an apprentice program in an area where another was in existence if the new program's requirements were not as rigorous. The twin pressures resulting from the growth of the open shop and discrimination by unions and contractors against minorities led to a change in these policies during the Nixon administration. Guidelines now in effect substitute national apprenticeship standards for the system of local standards previously applied.[11] Thus, the programs of the Associated Builders and Contractors (ABC) and other open shop organizations, including the open shop segment of the Associated General Contractors (AGC), have been required to conform to the national standards, but they can vary from local ones where the definition of "more rigorous" might be regulations which extend training periods or set entrance requirements that limit the number of trainees.

Scholars and the public alike tend to regard the provision of apprenticeship training programs as a union or union-management function. Although the existence of construction craft training in the nonunion sector was briefly noted by Haber and Levinson in the 1950s,[12] it was conspicuously unacknowledged in the literature prior to the first edition of this book. Typically, even such a knowledgeable observer as Mills, as late as 1972, wrote about apprenticeship administration solely in terms of joint labor-management committees.[13]

The attitudes of BAT have been reinforced by the Department of Labor's administration of the Davis-Bacon Act (prior to the 1984 change of regulations) and by the administration of state Little

[10] In a 1974 interview published in ABC's journal, BAT's then Director, Hugh Murphy, a one time official of the Bricklayers' union and brother of a longtime president of that union, acknowledged that "the programs submitted by ABC have at times experienced inordinate difficulties in obtaining approval," but he believed that this is becoming "less the case, as more people have accepted that ABC is sponsoring and developing extremely fine apprenticeship programs. . . ." *The Contractor,* June, 1974, p. 33.

[11] See the Court's opinion in *Clayton Daugherty v. United States,* No. 70-H-1096 (S.D. Tex. July 22, 1974); summarized in *Construction Labor Report,* No. 981 (July 31, 1974), pp. A-1-2. In this case, a local of the Sheet Metal Workers attempted to halt an ABC apprentice program because it followed national, not local, standards. Promulgation of such standards has been ruled the discretionary right of BAT director, provided that they are reasonable. See also *Gregory Electric Co. v. U.S. Department of Labor,* 268 F. Supp. 987 (D.S.C. 1967). Of course, new regulations could also be issued to harm open shop programs as well as to further them. Basically, the reasonableness of the regulations and their administration depends upon the extent of union control of the Department of Labor, which was very great under the Carter Administration, and of course much less so under the Reagan one.

[12] See Haber and Levinson, *Labor Relations,* p. 98.

[13] Mills, *Industrial Relations,* pp. 186–189.

Davis-Bacon laws. Particularly by excluding helpers, trainees, and subjourneymen from the wage categories set in regulations under these laws, not only were these classifications virtually eliminated from unionized construction, but also open shop contractors who desired to bid for work controlled by Davis-Bacon or similar state standards were required to attempt to adopt apprentice training if they wished to use and compensate employees at other than prescribed journeymen's rates. According to a Business Roundtable study, the impact of BAT rules, combined with Davis-Bacon administration, has been widespread:

> Since roughly a third of all construction is at least partly financed by federal funds, subject to similar restrictions, union practices have come to prevail on matters well beyond wages particularly in the realm of training. As a result, the use of lesser skilled workers on federally financed construction work is usually limited to those already enrolled in union apprentice programs. This has imposed serious limits on the use of innovative training techniques, notably those that let individuals advance to higher levels of classification and pay in line with skills acquired rather than time served.[14]
>
> The long term impact of apprenticeship training reaches much further than is commonly supposed. As Dr. Quinn Mills observes in his book, *Industrial Relations and Manpower in Construction:* "Apprenticeship appears to supply a core of key journeymen, foremen, supervisors, and even contractors in most trades. In this sense, it can be as much a management training mechanism as a means of training the workforce itself." Thus, it tends to build in construction management a strong support for traditional programs and resistance to innovation.[15]

Another important facet of BAT policy is a reluctance to register apprentice programs for new skills or combinations of crafts. It did, however, approve the informal classification "metal building assembler" in late 1981,[16] and this has encouraged open shop organizations to push for approval of various multicraft programs which depart from the traditional union craft structure. It is not likely, however, that the tradition-bound BAT and other similar state agencies will support such innovations either sympathetically or widely, especially since the proponents of change are users and open shop contractors, and the opponents are primarily the construction unions.

---

[14] *Government Limitations on Training Innovations,* Construction Industry Cost Effectiveness Project, Report D-2 (New York: The Business Roundtable, 1982), pp. 3–4.

[15] *Ibid.,* p. 6.

[16] *Ibid.,* p. 7, note 4; and "BAT Has Approved Open Shop Apprenticeship Classification of Metal Building Assembler," *Construction Labor Report,* No. 1359 (December 30, 1981), p. A-5.

Nevertheless, the new regulations promulgated for the Davis-Bacon Act should encourage a drift away from formal apprenticeship by some open shop contractors. During recent years, the number of apprentices registered in BAT-approved programs has varied from 275,000 to 320,000. Usually about 50,000 to 60,000 completed their indenture in any one year, indicating a dropout rate of about 80 to 85 percent. Construction apprentices are by far the largest group in the BAT data.[17]

Although the BAT and the U.S. Department of Labor have demonstrated increasing flexibility toward nontraditional open shop training programs, many state apprenticeship councils remain inflexible and discriminate against open shop training. In fact, the situation remains so bad that in early 1984 the U.S. Department of Labor sent a memorandum to the twenty-eight states which have apprenticeship councils advising them not to reject training programs just because the councils may not favor open shop training.[18] Yet it is not likely that some of these councils will change their position unless forced to do so, and the U.S. Department of Labor does not have the power to pressure them into making such a change. As a result, litigation has commenced in at least two states—Washington and Nevada—seeking orders against state apprenticeship councils to require approval of plans of open shop associations and / or contractors, and bills have been introduced in state legislatures to accomplish the same purpose. Unions are fighting these changes because, as already explained, under Davis-Bacon (prior to the new Davis-Bacon rules) and similar state regulations, the failure to have an approved apprentice program has meant that an open shop contractor would be required to compensate all employees at the journeyman's rate.[19]

Task Training

Task training has been defined by the already noted Business Roundtable report as follows:

> An important difference between labor-management apprenticeship programs and the training done by open-shop contractors and their associations is the greater use of task training by non-union employers. Like many other industries, construction has become increasingly specialized. Training of workers merely to accomplish specialized tasks has, in many instances, proven to be more efficient and eco-

---

[17] Data from Bureau of Apprenticeship and Training, U.S. Department of Labor.

[18] "States Urged to Approve More Open-Shop Training," *Engineering News-Record*, February 16, 1984, p. 14.

[19] *Ibid.*

> nomical for specific projects than the broader training offered by traditional apprenticeship programs. Trainees can reach necessary levels of skill much faster when they concentrate on specialized tasks. The broad knowledge and abilities learned more slowly through traditional apprenticeship are not required by every member of the workforce.[20]

Task training, moreover, does not confine the learner to a narrow skill education. By using the "building block" method, which will be described in greater detail below, an individual can acquire all-around knowledge of construction (particularly in programs such as those developed by the ABC, Brown & Root, Daniel, or other associations and major contractors) through a combination of school work and study. The individual can progress as far as he or she desires and at a speed which is required to master the task and subject. Thus, whereas a sheet metal apprenticeship may take four years, the same skills and competence under task training may be acquired in two to five years or at intervals over a longer period.

There are other varieties of formal training besides apprenticeship. Preapprenticeship training is often used in connection with affirmative action programs to help minorities prepare themselves for entrance into apprenticeship training. These programs are usually brief in duration (seldom exceeding six weeks) and concentrate upon improving literacy skills and providing exposure to the fundamentals of a trade. We shall again note them in Chapter XII when discussing minority representation in open shop construction. There are also programs to upgrade journeymen that are designed both to update the skills of established craftsmen and to coach them in techniques for teaching these skills to on-the-job trainees. In addition, there are management courses for supervisors and others interested in management training, many of which have been developed by associations, particularly the ABC and the AGC. All of these programs are much less widespread than apprentice training in both the union and open shop sectors, but the rapid expansion of the open shop firms has increased the need for supervisory and other managerial programs.

ASSOCIATION PROGRAMS

Until the early 1970s, most formal training efforts within the open shop portion of the construction industry were sponsored by individual contractors for their own employees. Unlike unionized contractors, whose training is administered by joint union-manage-

[20] "Government Limitations," p. 6.

ment apprenticeship committees (which are financed by contributions from all employers to an apprenticeship fund), open shop contractors are usually obliged to underwrite their own programs. Now, however, most contractors' associations with substantial open shop representation have organized formal training programs for their members. As these programs grow and are introduced into more and more areas, it may be expected that an increasing number of contractors—both those who have provided training on their own and those who have not—will participate in association-sponsored training activities. Because of their broad backing, multiemployer programs can offer training in a wider range of crafts and do it more efficiently than programs run by individual contractors.

Association open shop training programs were in effect in the early 1980s in chapters of the ABC, the AGC, the National Association of Home Builders (NAHB), the Independent Electrical Contractors (IEC), the National Association of Plumbing-Heating-Cooling Contractors (NAPHCC), and other organizations. An examination of the principal programs follows; training undertaken by individual open shop contractors is reserved for the next section.

Associated Builders and Contractors

The most extensive network of organized apprenticeship programs in the open shop sector has been developed by the ABC, which has employed a full-time training staff since the early 1970s and encourages its chapters to do likewise. The ABC moved heavily into training in the late 1960s and pushed hard to obtain approval of its programs by the BAT and by state apprenticeship agencies. Largely because of opposition from building construction unions and because of the already noted relationship of organized labor to these agencies, the ABC chapters first experienced great difficulty in gaining approval of its programs. As noted, this was overcome in the early 1970s, and progress has moved forward rapidly. Table X-1 shows the programs operating in various ABC chapters through 1974, as compiled by the then director of training activities for the association. The most common trades for which training was given were carpenter, bricklayer, air conditioning mechanic, electrician, plumber, and sheet metal worker. In addition to the programs indicated in the table, there were scattered programs for asbestos worker, precast concrete specialist, lather, painter, and welder. Table X-1, moreover, understates the proliferation of ABC training programs actually underway in 1974, apparently because of a lag in reporting from chapters to ABC headquarters. For example, our 1973–1974 field interviews with association executives uncovered

TABLE X-1
Approved Apprenticeship Programs
Conducted by the Associated Builders and Contractors,
by Trade and Association Chapter, 1974

Trade	Chapter[a]																		
	1	2	3	4	5	6	7	8	9	10	11	12	13	14	15	16	17	18	19
Air conditioning mechanic	X	X		X		X	X	X	X			X	X	X	X	X	X		
Bricklayer					X	X	X				X	X				X		X	X
Carpenter	X	X	X	X	X	X	X	X		X	X	X	X	X	X	X	X	X	X
Cement mason					X							X						X	
Electrician	X	X	X	X	X	X	X	X	X			X	X	X		X	X	X	X
Heavy equipment operator		X										X							
Plumber / pipefitter	X	X	X	X	X	X	X	X	X	X	X	X	X	X		X	X	X	X
Sheet metal worker	X	X			X	X	X					X	X	X		X			X
Stonemason							X					X							

Source: Associated Builders and Contractors.
[a] Key to chapters:
1. Anne Arundel, Md.
2. Baltimore Metro.
3. Central Florida
4. Central Ohio
5. Central Pa.
6. Cumberland Valley
7. Delaware Valley
8. Florida Gold Coast
9. Florida Gulf Coast
10. Georgia
11. Kentuckiana
12. Keystone
13. Metro Washington
14. South Lousiana
15. South Texas
16. Tennessee
17. Texas
18. Western Michigan
19. Yankee

nascent programs in western Michigan (Detroit) and western Pennsylvania (Pittsburgh), as well as two programs (for bricklayer and heating and air conditioning mechanic) in Central Ohio.

Table X-2 shows the enrollment in ABC-sponsored apprenticeship programs by craft and by association chapter. These data are based on a survey conducted by the national office of the ABC in 1982 to 1983. Because of the severe depression in the construction industry at that time, the number of apprentice programs and apprentices enrolled in such programs were less than the previous year, but the total still approximated 4,800. Of particular interest is the large number of electrical programs; 48 percent of the total enrollees were in this craft. A total of 722 apprentices graduated from the ABC programs in 1983. This comprised 15 percent of the enrollees and is about equal to the average completion ratio for all BAT-approved programs.[21]

ABC Response to Industrial Research Unit Questionnaire. Further analysis of training by members of the ABC can be gained from the 1982 Industrial Research Unit questionnaire, which was sent to every other member of this association. Table X-3 shows the training background of new employees as revealed by ABC respondents. A very small percentage of employees with formal training of any kind were hired by the respondents. Those with on-the-job training from previous employers were employed often (65.5 percent) or always (5.7 percent), but union apprentice graduates were employed always or often only a total of 11.7 percent, and open shop apprentice graduates were employed always or often 24 percent. These figures certainly do not support the standard union allegation that open shop contractors have unions and unionized firms do their training. Whether the small percentage of new hires from open shop apprentice programs is attributable to their lack of availability or to the belief of open shop contractors that other forms of training are more helpful (such as the Wheels of Learning discussed below), or both, cannot be ascertained.

Table X-4 examines the participation of ABC member firms in training programs. Slightly less than a majority answering the questionnaire participate in association-sponsored programs. (We assume ABC members are not involved in union-management programs.) More than two-thirds (68.3 percent), however, provide workers with their own off-job training or instruction, with again more than two-thirds (68.7 percent) providing their own instruction.

---

[21] The completion rate on BAT programs varies from 20 percent in more prosperous times to figures that are considerably lower in recessions. A rate of 15 percent is probably a good average.

TABLE X-2
Apprentice Training Program Enrollment
Associated Builders and Contractors
By Chapter and Craft, 1982–1983

Chapter	Craft															
	Carpenter		Electrician		Plumber		HVAC		Sheet Metal		Painter		Pipefitter		Other	
	Enrolled	Completed 1983	Enrolled	Completed 1983	Enrolled	Completed 1983	Enrolled	Completed 1983	Enrolled	Completed 1983	Enrolled	Completed 1983	Enrolled	Completed 1983	Enrolled	Completed 1983
Alabama Mid Gulf	28	0	156	18	42	0	19	0	42	5						
Arkansas	7	6	62	8					31	6	2	0				
Golden Gate (Cal.)			40	0	17	0	22	0					27	0		
Baton Rouge-Pelecon (LA)			79	16												
Anne Arundel So. Md.	9	2	103	27	27	8			17	1						
Baltimore Metro (Md.)	29	2	146	55	110	28			45	7						
Florida Gold Coast	47	2	197	28	11	0									8[b]	0
Florida Gulf Coast	71	28	244	35	83	10	14	3	49	7						
Georgia	19	0	27	0	17	0			7	0	4	0			4[c]	0
Indiana	22	0	15	0												
Iowa	22	8	75	14	21	3	8	0	10	1	9	0				
New Orleans-Bayou	22	4	63	10					16	4					6[d]	0
Cumberland Valley	25	2	92	31	21	6	12	2	7	3					13[e]	6

Metro-Washington	23	8	14	3	40	12	19	12	18	3					1[e]	1
Massachusetts-Rhode I	6	0	31	1	2	0	12	0	25	0	2	0	2	0	11[d]	1
Western Michigan	18	11	67	15	14[a]	3[a]			11	0			[a]	[a]		
Mississippi			24	0												
Heart of Amer. Missouri			20	0												
Northern New England	1	0	52	0	13	0									1[f]	0
Empire State (N.Y.)			68	0			21	0								
Central Ohio	21	2	54	10	16	2			17	3						
Southwest Ohio	9	0	32	3	8	0			14	0						
Oklahoma	38	11	79	8	12	0										
Keystone	46	8	77	18	19	8			14	3					1[c]	0
Southeast Pennsylvania	26	4	3	3	8[a]	1							[a]	[a]		
Tennessee	11	2	66	11	11	0										
Central Texas			43	5												
Greater Houston, Texas			66	6	15	0										
San Diego, California			48	6					14	0						
Rocky Mountain	53	11	187	13	31	0	19	0	74	7	23	0				
Central Florida	49	9	168	16	61	7	2	0	26	1	1	6			12[b]	0
															1[e]	0
															12[g]	5
															89[d]	18
															22[h]	0
Texas Gulf Coast	14	0	156	37			13	0			7	1	80	12	22[i]	2
															10[j]	1

See footnotes at end of table.

TABLE X-2—Continued

Chapter	Carpenter Enrolled	Carpenter Completed 1983	Electrician Enrolled	Electrician Completed 1983	Plumber Enrolled	Plumber Completed 1983	HVAC Enrolled	HVAC Completed 1983	Sheet Metal Enrolled	Sheet Metal Completed 1983	Painter Enrolled	Painter Completed 1983	Pipefitter Enrolled	Pipefitter Completed 1983	Other Enrolled	Other Completed 1983
Inland Pacific									7	1					3[k]	0
Wisconsin			2	0	4	0									1	0
Wyoming	9	0	23	0	8	0									3[l]	1
Totals	625	130	2,579	397	611[a]	104	161	17	444	52	48	7	80[A]	12	219	34

Source: Associated Builders and Contractors

[a] Plumbers and pipefitter combined; totals only pipefitter
[b] Ironworker
[c] Insulation mechanic
[d] Sprinkler fitter
[e] Bricklayer
[f] Cement mason
[g] Roofer
[h] Glazer
[i] Operating engineer
[j] Instrument fitter
[k] Auto mechanic
[l] Diesel mechanic

TABLE X-3

Training Background of Employees Hired by Contractor Members Associated Builders and Contractors, 1982

	Always		Often		Occasionally		Never	
Employed with prior training from:	No.	Percent	No.	Percent	No.	Percent	No.	Percent
A. Vocational school	17	1.7	197	22.4	547	62.4	115	13.1
B. Military	1	0.1	66	7.6	530	61.6	263	30.5
C. Union apprentice	14	1.6	88	10.1	365	42.1	398	46.0
D. Open shop apprentice	22	2.4	192	21.6	428	48.2	245	27.6
E. Apprenticeship in the industries	0	0	57	6.8	343	41.3	430	51.8
F. On-the-job training with former employer	53	5.7	606	65.5	236	25.5	30	3.2

Source: Industrial Research Unit Questionnaire, 1982.

TABLE X-4
Participation in Training Programs by Contractor Members Associated Builders and Contractors, 1982

	Yes		No	
	No.	Percent	No.	Percent
Participate in local training programs sponsored by associations	470	48.8	492	51.2
Provide workers with own off-job training or instruction	658	68.3	306	31.2
Instruction given by own people	524	68.7		
by vocational schools	239	31.3		
Length training lasts				
Up to 4 weeks	234	31.4		
4 to 8 weeks	111	14.9		
8 to 16 weeks	103	13.8		
Over 16 weeks	297	39.9		
No. of hours per week of classes				
Up to 5 hours	543	78.2		
5 to 10 hours	117	16.9		
10 to 15 hours	11	1.6		
15 to 20 hours	23	3.3		

Source: Industrial Research Unit Questionnaire, 1982.

The largest percentage of these programs last over 16 weeks (39.9 percent), with the second concentration four weeks or less (31.4 percent). Classroom instruction is usually five weeks or less (78.2 percent).

This record reveals that formal training has indeed a reasonable following among ABC members, and that claims concerning the dearth of training among open shop contractors, discussed in the concluding section of this chapter, are perhaps exaggerated. Of course, the Industrial Research Unit questionnaire provides no analysis of the quality of the training, but then most statistics on unionized training do not either. It would appear, however, that the data presented here clearly indicate that the commitment of the ABC leadership to the expansion of off-the-job training has its roots in the desire and need of its members for training. This provides a hopeful future for the development of the Wheels of Learning and the Open Shop Training Trust Fund.

Wheels of Learning. Open shop apprentice training is subject to some of the same criticisms that have been made against joint union-management apprentice training. In particular, apprentice training develops the all-around craftsmen who play a significant role, but it is time oriented, rather than task oriented, and attempts to require all enrollees to become all-around craftsmen whether such training is desired by the individual or required by the employer. Responding, therefore, to the needs of its members and impressed by the task training long utilized by such major companies as Brown & Root, Daniel International, Zachry, and others, the ABC instituted its Wheels of Learning program in 1980.

Wheels of Learning is described as "a competency-based, task-oriented training program ... covering 21 trades [that] provides a modular training format which allows employers to design their training program to suit their needs and their workers' interests."[22] Under this program, each trade is divided into skill modules. A worker receives a certificate upon completion of a module and can proceed at his own speed on to other modules. Modules provide both classroom and on-the-job training. Classes may be taken exclusively in one module, or modules can be combined to produce a formal apprenticeship program. Experienced workers can use the training to "brush up" or to learn new skills. Since the programs are offered by various ABC chapters, trainees can transfer jobs without disrupting their studies. Figure X-1 illustrates the modules for the multiskilled craft of sheet metal building erector, and for the single-craft ironworker apprenticeship.

To put Wheels of Learning into operation, the ABC made a special assessment of its members and worked together with the American Institute of Construction, which promotes construction as a profession, and with associations of two-and four-year colleges and universities that have construction programs. Several universities were contacted to develop training programs, including those for supervisors. Under the Wheels of Learning program supervisors and foremen are trained to provide instruction, and are therefore critical to the program. As a result the ABC has stressed this aspect and has developed considerable materials under its Training Orientation Program for Supervisors and Train the Trainer Program. The ABC has also developed modules in the Wheels of Learning series for foremen relating to employee relations, labor, and other aspects of the work of supervisors in a program known as Supervised Training Orientation Program.

Since the Wheels of Learning Program has operated for only a very short period, no definitive evaluation can be made. It does, however, appear to provide the vehicle for a sound task training approach, and by January 1983, it was being utilized by twenty-

---

[22] From an undated ABC pamphlet, "The Wheels of Learning. Training Today for Tomorrow."

FIGURE X-1

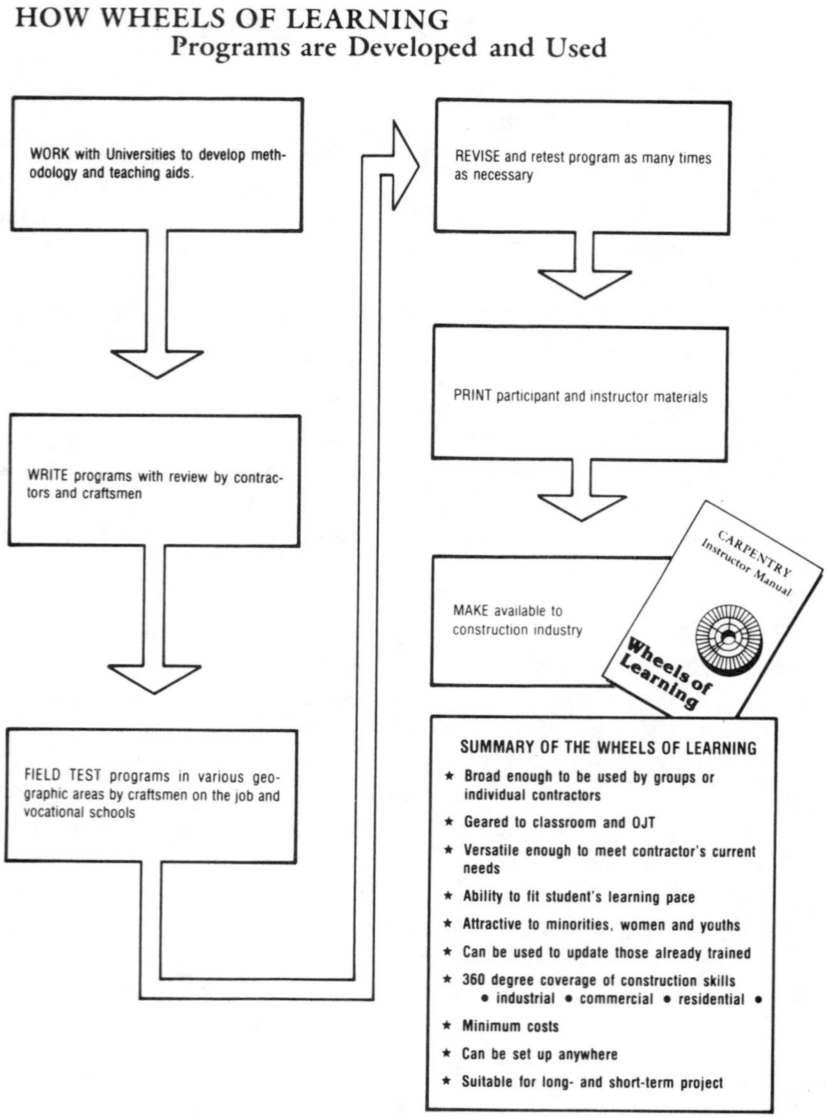

Source: Associated Builders and Contractors.

Curriculum Options

Creating a Multiskilled Craftworker

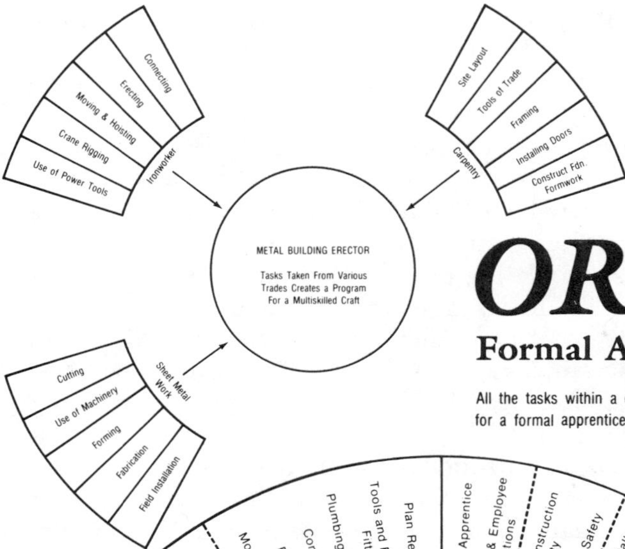

OR

Formal Apprenticeship

All the tasks within a craft can be used as the curriculum for a formal apprenticeship program.

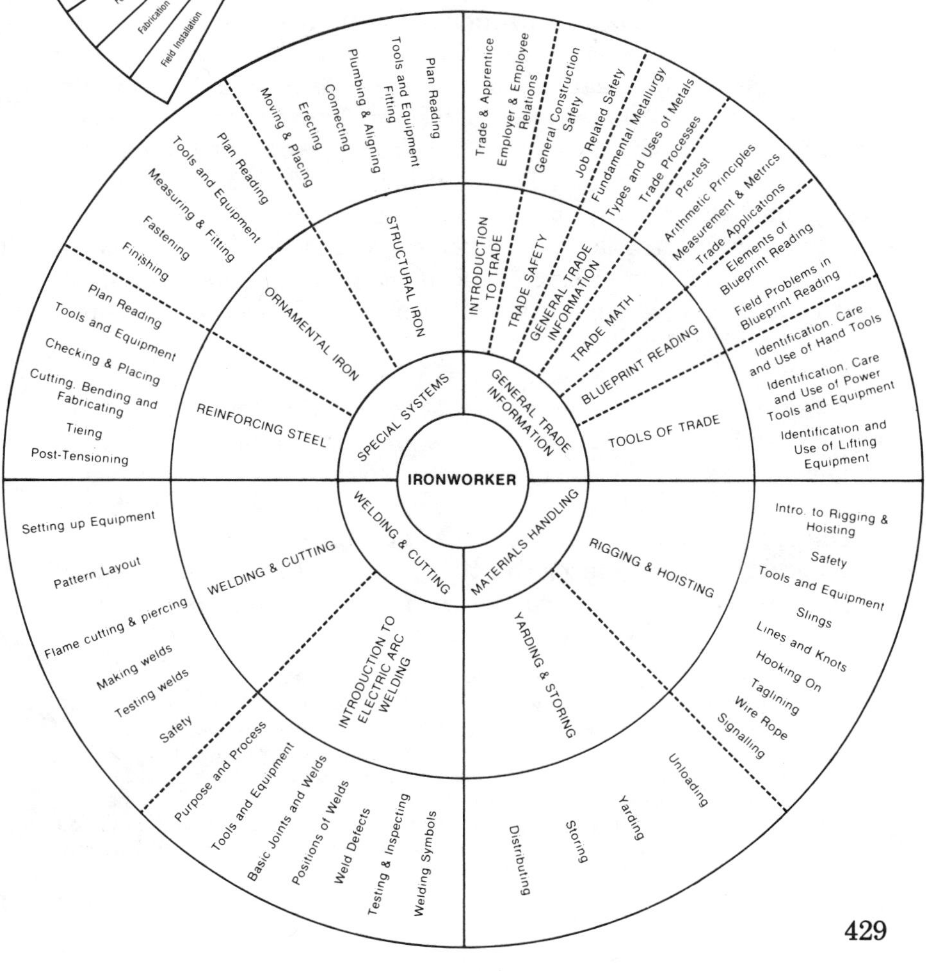

two ABC chapters to train approximately 2,000 workers.[23] Considering the depressed state of the industry at that time, this was an auspicious beginning. As more modules are developed and if trainees' experiences with the Wheels of Learning Program continue to be satisfactory, it can be expected to attract considerably more trainees in the future.

Open Shop Training Trust Fund. In September 1982, the Business Roundtable issued a report which declared that "less than 10 percent of all funds now going into construction craft training is directed toward open shop training. Likewise, less than 10 percent of those individuals completing construction craft training are in open shop programs."[24] The accuracy of this statement is discussed later in the chapter. What is pertinent here is that the ABC leadership welcomed the criticism and proceeded to put it to good use. It organized the National Open Shop Training Trust to fund "any legitimate training program in the construction industry" and invited other contractor organizations to participate. After the trust received formal approval of the ABC board of directors, a meeting of all interested contractors' organizations was held in Washington, D.C., in August 1983. Twelve contractors' associations were represented at the meeting, and by January 1984, several had committed their support to the trust.[25] The trust hopes to raise $50 million per year from users, as part of construction contracts, to support training. It plans to provide certificates to graduating trainees which will certify their completion of specified training. This would both give contractors assurance of their competency and aid the open shop training graduates to obtain jobs more easily.

The training trust is clearly designed as the capstone of ABC's training activities. It can also enhance cooperation between the ABC and a number of other associations which have significant open shop memberships. Its success will depend not only on its sponsors, but also upon the support of the major users, who through the Business Roundtable have been so critical of construction training in general and of the alleged dearth of training (to be assessed below) in the open shop sector in particular. If the users are in fact ready

---

[23] Data from ABC chapter telephone survey, January 1983.

[24] *Training Problems in Open Shop Construction,* Construction Industry Cost Effectiveness Project, Report D-4 (New York: The Business Roundtable, 1982), p. 1.

[25] The author attended this meeting at which, besides the ABC, the following were represented: American Concrete Institute, American Subcontractors Association, Association of the Wall and Ceiling Industries, Ceilings and Interior Systems Contractors Association, National Ready Mix Concrete Association, National Sand and Gravel Association, Painting and Decorating Contractors of America, and National Association of Plumbing-Heating-Cooling Contractors.

to fund the training as part of open shop construction costs as the Roundtable recommended, and as indeed they do when they give a contract to a unionized builder, the trust should stimulate a major advance in open shop construction training and, indeed, in construction training generally.

AGC Open Shop Training

The AGC has a long and distinguished record of participation in supervisory and management skills training programs. With the exception of the Carolinas Construction Training Council under the leadership of the Carolinas AGC, its skills training programs were basically joint union-management endeavors until the mid-1970s. This, of course, reflected the AGC's early heavily union construction orientation.

As the membership of the AGC became more open shop, its interest in unilateral, and particularly task-oriented, training increased. As set forth in the AGC Bulletin:

> Starting in 1973 and utilizing contractor review committees, a full time curriculum writer, and consultants, AGC began developing industry instructional systems for bricklaying, commercial carpentry, cement masonry, construction craftsman (multicraft), millwright, and equipment operator. The systems now complete are competency based and use the most recent advances in instructional technology. Participants can progress through these materials at a pace which matches their abilities. To ensure that these systems are utilized to the fullest extent possible, AGC employs a training program coordinator, who works with AGC chapter training programs and the vocational education community to see that AGC materials receive the maximum exposure.[26]

The AGC unilateral training program standards were finally approved on February 1, 1980 by the BAT. According to the AGC, "the approved Model, like AGC's curriculum, represents an improvement over many traditional training programs because it is competency based and permits trainees to advance through training at a pace commensurate with their skills."[27] As of April 29, 1983, ten chapters, all in the South and West, except one in New Hampshire, had such programs in effect.[28]

The AGC, like the ABC, has full-time staff devoted to open shop training. Technical assistance, literature, lessons with vocational institutions, and other services are provided by both organizations.

---

[26] Associated General Contractors of America, *AGC Unilateral Trainee Programs on Line in Eleven Chapters,* General Bulletin No. 8-83, April 29, 1983, p. 1.

[27] *Ibid.*

[28] *Ibid.,* pp. 2–4.

The AGC open shop committee, however, rejected invitations to the ABC's August meeting to discuss the National Open Shop Training Trust Fund, and declined to participate in the fund. The official AGC attitude on the subject is that it should support the joint implementation of training programs with other associations at the local level, but that the "National AGC's training role should be continued on its current basis, [and] the proposed national open shop training trust fund is not the means of fulfilling training requirements."[29] Whether the AGC believes that the training trust would be too dominated by the much larger ABC or might severely antagonize AGC unionized members could not be ascertained.

Response to Industrial Research Unit Questionnaire. The AGC has both unionized and open shop contractors in its ranks, with the former predominating at a slightly greater than 60 to 40 basis. The responses to questions relating to training in the 1982 Industrial Research Unit questionnaire returned by AGC members, as set forth in Tables X-5 and X-6, provide both insight into the extent of open shop training and a comparison between unionized and open shop firms within the same organization.

Table X-5 gives the training background of employees hired by the AGC contractors. Of the unionized employees hired by AGC contractors, 77.2 percent had always or often been trained under union apprentice conditions, but only 11.7 percent of the open shop hires fell into this category, and generally open shop AGC contractors did not employ persons with such a background—again evidence that claims of union training of open shop workers may be exaggerated. As in the case of the ABC contractors, open shop AGC contractors found the bulk of their employees (77 percent) had received on-the-job training by prior employers. Open shop apprentice courses were considerably less important in this regard, accounting for only 18.1 percent of persons always or often employed.

The differences between open shop and union shop approaches to training are further illustrated by Table X-6. Whereas 83.2 percent of the unionized contractors participate in an association or union-management training program, only 38.6 percent of the open shop contractors are so involved. On the other hand, 56.5 percent of the open shop contractors, and 45.9 percent of the unionized contractors provide workers with their own off-the-job instruction. The length of training and the hours per week of such training

---

[29] AGC, *Open Shop Committee Report to the Board* (duplicated), October 17, 1983, p. 2. The AGC board approved this position in October 1983. See "Training Trust Advances," *Engineering News-Record,* October 27, 1983, p. 65.

TABLE X-5
Training Background of Employees Hired by Contractor Members of Associated General Contractors by Union and Open Shop Status, 1982

Employed with prior training from	Always				Often				Occasionally				Never			
	Union		Open Shop		Union		Open Shop		Union		Open Shop		Union		Open Shop	
	No.	Percent	No.	Percent	No.	Percent	No.	Percent	No.	Percent	No.	Percent	No.	Percent	No.	Percent
A. Vocational school	10	1.4	1	0.2	66	9.5	58	11.9	456	65.6	371	76.3	163	23.5	56	11.5
B. Military	0	0	0	0	29	4.3	40	8.3	389	57.0	300	62.4	264	38.7	141	29.3
C. Union apprentice program	141	17.9	3	0.6	467	59.3	54	11.1	157	19.9	181	32.4	23	2.9	247	50.9
D. Open shop apprentice program	0	0	4	0.8	21	3.1	83	17.3	145	21.6	239	49.9	504	75.3	153	32.0
E. Apprenticeships in other industries	1	0.2	3	0.7	18	2.7	55	11.8	212	31.8	207	44.5	435	65.3	200	43.0
F. On-the-job training with former employer	27	3.6	31	6.1	370	49.6	358	70.9	290	38.8	104	20.6	60	8.0	12	2.4

Source: Industrial Research Unit Questionnaire, 1982.

TABLE X-6
Participation in Training Programs by Open Shop Contractor Members Associated General Contractors, 1982

	Union				Open Shop			
	Yes		No		Yes		No	
	No.	Percent	No.	Percent	No.	Percent	No.	Percent
Participate in local training program sponsored by association or union-management committee	683	83.2	138	16.8	200	38.6	318	61.3
Provide workers with own off job training or instruction	376	45.9	443	54.9	293	56.5	225	43.4
Instruction given								
by own people	312	66.0			242	68.9		
by vocational schools	161	34.0			109	31.2		
Length training lasts								
up to 4 weeks	216	47.3			125	37.6		
4 to 8 weeks	74	16.2			59	17.7		
8 to 16 weeks	43	9.4			50	15.2		
over 16 weeks	124	27.1			98	29.5		
No. of hours per week of classes								
up to 5 hours	346	83.4			239	79.6		
5 to 10 hours	50	12.0			44	14.7		
10 to 15 hours	8	1.9			5	16		
15 to 20 hours	11	2.7			12	4.1		

Source: Industrial Research Unit Questionnaire, 1982.

programs are quite similar for each group, and two-thirds of each group provide their own instruction.

The AGC data from the 1982 Industrial Research Unit questionnaire reveal considerable differences in training methods and employee training backgrounds between unionized and open shop contractors and employees. We shall examine the record from all responses to the questionnaire later in the chapter, when the need for more formal training in the open shop segment is further discussed. It is quite clear, however, that, given the extent of open shop penetration in the industry, the open shop contractors must increase their formal training without decreasing their on-the-job instruction. Before we analyze this need further, however, the training programs of other associations and of individual companies will be examined.

National Association of Home Builders

In 1967 the NAHB organized its Manpower Development and

Training Division for the purpose of establishing, operating, and evaluating programs to train skilled construction mechanics for the homebuilding sector. Entry-level programs are offered in eleven construction trades: carpentry, bricklaying, cement masonry, electrical work, plumbing, plastering, building and apartment maintenance, painting, heating and air conditioning, landscaping, and solar installation. Carpentry is by far the largest of the programs. Before organizing its manpower division, the NAHB had conducted a job training program at a Job Corps Center for three years.

Funding for the NAHB programs comes heavily from the U.S. Department of Labor, the Department of Health and Human Services, and the Department of Housing and Urban Development. According to the NAHB, "millions of dollars" have been contributed by these agencies for the association's various training activities, and "state and local home builder associations have assisted this effort by raising another $5 million" between 1967 and 1983.[30]

Through its manpower division, the NAHB assists local home builder associations (HBAs) to develop and implement their industrial programs. Federal and supplementary funding is arranged through contracts involving the NAHB and local HBAs. The manpower division of the NAHB provides technical assistance in obtaining approval for training programs from federal and state training and apprenticeship agencies. In 1983, the NAHB stated that over 100 affiliated state and local builders' associations offered training programs, in addition to twenty-six Job Corps centers and several special programs designed to aid minorities and other disadvantaged groups. With federal funding and NAHB interest and supervision, these programs were maintained during the severe housing depression of 1980–1983, although on a more limited basis.

Once a local program is established, a full-time coordinator is hired to oversee training activities. The contract provides funds for such expenses as the coordinator's salary, classroom instructors, books and materials. Thus, no direct expenses are incurred either by the trainee or by the individual builder. Classroom instruction is provided either through local vocational schools or through the program directly.

The program progresses in two phases. For carpentry, for example, there is first a six-week preapprenticeship period during which trainees are instructed in the rudiments of the carpentry trade. During this period, subjects include mathematics for carpen-

---

[30] This section is based on the NAHB's literature dealing with its manpower division.

ters, construction materials, blueprint reading, carpenters' tools, and safety. Upon completion of this initial phase, trainees usually enter a full-blown apprenticeship program combining on-the-job training (through indenture with a local builder) and 144 hours per year of related instruction.

According to the NAHB, overwhelming success has been achieved in placing the graduates of its training activities. The association claims that over 50,000 had been trained through 1983, and that "there is ample evidence to indicate at least a 90 percent placement record for all graduates of NAHB training programs."[31] Detailed placement statistics, however, do not seem to be available, nor is there information on the number or percentage of trainee graduates who remain in their trade after graduation. The NAHB program coordinators do, however, work very hard at placement. They often aid graduates in obtaining transportation to, and living accommodations in, the vicinity of jobs. They also assist graduates in relocating to localities where the demand for construction workers is strong if homebuilding is depressed in the area where they received their training.

The association also sponsors a journeyman upgrading program designed to prepare craftsmen for supervisory roles in construction. Conducted entirely within a classroom format, the program provides thirty-six hours of instruction in such areas as blueprint reading, job cost estimating, leadership and supervision, and new materials and technology. As many as 1,000 craftsmen per year have gone through this program, but it is not known how many have assumed leadership or supervisory positions.

In many areas, the NAHB training programs have greatly aided minority integration. The association coordinators have often been quite adroit in writing regulations that meet the requirements of the BAT but also permit disadvantaged youths to qualify for the programs. For example, it provides a preapprenticeship program to improve the candidates' reading and mathematics skills so that the minimum qualifications for journeyman training can be achieved.

Nationally, the NAHB does not work closely with the ABC nor with other associations in its training activities. In some areas, however, we found that local HBAs cooperated with the ABC training program and recruited graduates from it. In such instances, there was no NAHB training active or supported by the local homebuilders.

---

[31] National Association of Home Builders, Manpower Development & Training Division, "A Preview of NAHB Training Programs Today," Washington, D.C., undated.

Other National Associations

As their open shop membership has grown, a number of national associations have responded to their members' needs for training by developing open shop training programs. An official of the NAPHCC, whose membership is 60 percent open shop, stated that as of October 1983 there were "approximately 200 Open Shop Apprentice programs active," but he claimed specific information about these programs was not available.[32] The NAPHCC is supporting the ABC-initiated training trust. It has a contractor committee that oversees and promotes open shop training.

The National Utility Contractors Association (NUCA), whose membership is about evenly divided between unionized and open shop contractors, also has open shop training programs. For the most part, NUCA chapters work with local technical institutes, vocational schools, or other associations rather than sponsor the training themselves, and larger contractor members of NUCA do their own training.[33]

The open shop IEC, established in 1956, has from its inception emphasized formal training. Like the ABC, the IEC employs a full-time training director and staff and actively encourages its chapters to develop and to expand apprenticeship training. The IEC also supports the ABC-initiated training trust. Nationally promulgated apprenticeship standards serve as a guide to local chapters in developing programs which will be registered by state and federal apprenticeship agencies. These standards, which were approved by the BAT in 1972, do not appear to differ markedly from those used for union electricians. They cover such matters as minimum hours of on-the-job training and related instruction (2,000 and 144 per year, respectively), minimum qualifications for admission, wage schedules, affirmative action, subject requirements, and so forth.

Each program is administered by a local apprenticeship committee. In September 1983, the IEC had over 1,000 apprentices enrolled in its programs. At that time, it had 2,000 electrical contractor members in twenty-eight states.[34] It claims to have graduated 10,000 apprentices. Training programs are a principal, if not the principal, concern of the IEC.[35] Given the need for electrical workers to qualify for state and municipal licences and for their work to pass stiff

[32] Letter to the author from Russell J. Alessi, Technical Director, NAPHCC, October 7, 1983.

[33] Interview, NUCA officials, Washington, D.C., March 5, 1982.

[34] Letter to author from Mark Levin, Executive Vice-President, IEC, September 19, 1983.

[35] Interview, Mark Levin, Washington, D.C., March 5, 1982.

safety inspections, the IEC' stress on training is certainly understandable.

Carolinas Construction Training Council

One of the most extensive training efforts initiated unilaterally is led by the Carolinas AGC, which covers all of North and South Carolina, and is composed overwhelmingly of open shop contractors. In the late 1960s, the Carolinas AGC brought together five local training programs under the umbrella of a new organization called the Carolinas Construction Training Council (CCTC). Although the Carolinas AGC was instrumental in its formation, the CCTC is an independent unit with over 300 contractor members, fewer than one-half of whom are AGC members. The remaining participants belong to homebuilding and specialty contractors' associations throughout the Carolinas. The CCTC, which began operation on January 1, 1969, was formed to initiate and coordinate a variety of training activities in both building and highway construction, and it has accomplished this goal with the cooperation of numerous specialty contractor organizations.[36]

The council is administered by a twelve member board of trustees drawn from contractor members in North and South Carolina. Its day-to-day operation is handled by a five-person office staff in Charlotte, North Carolina, and eleven field coordinators based in various locations in the two states. The coordinators assist members in setting up and monitoring individual training programs, taking major responsibility for such tasks as recruiting and transferring trainees, arranging for related class instruction, and ensuring that the standards of apprenticeship and equal employment opportunity agencies are met. The coordinators are also responsible for keeping records on the activities of the various programs; as such, they are the primary source of information on the progress of trainees.

The council has had two major sources of funding. The first is an assessment on members based upon man-hours worked in selected trades. In addition, the council has received revenues under contracts with the NAHB, largely under its carpentry apprenticeship program, and with various federal agencies. The CCTC's dependence on federal funds has diminished in recent years, with membership contributions now accounting for the bulk of its budget.

The CCTC presently administers two major kinds of training programs, in addition to its upgrading and preapprenticeship pro-

[36] The discussion of the CCTC is based upon interviews with the staff, 1973 and 1983, and materials supplied by that staff.

grams and its special programs for disadvantaged persons. The first of these is its apprenticeship programs, which are designed primarily for building construction crafts and which, as of September 1974, enrolled over 1,000 apprentices in twelve trades. The Council also runs an on-the-job training program for highway construction, under the sponsorship of the Carolinas Branch of the AGC. The on-the-job-training program, started in 1969 with funds from the Federal Highway Administration, has developed standards for fifty-eight separate trade classifications. Its enrollment generally runs at less than one-tenth of the apprenticeship programs. These programs are discussed at greater length below.

The operations of the CCTC have, by all standards, grown markedly since the council's inception. Table X-7 summarizes the enrollment in the apprentice programs by craft for the period 1975 to 1982. The number of apprentices has averaged about 650 per year during this period. This represents a decline from the 1972–74 period, when approximately 1,000 apprentices per year were in training under CCTC auspices.

As Table X-7 shows, nine crafts, plus a miscellaneous group, are included in the training, with the great bulk of trainees concentrated in three crafts: carpentry, electrical, and plumbing/pipefitting. Like apprenticeship generally, the council's programs include both on-the-job and related training components. Related instruction is usually carried out in cooperation with local trade schools and technical institutes. Instructors may be provided either by the school or by the council itself. The CCTC also works with local high

TABLE X-7
Enrollees and Completions in Carolina Construction Training Council Apprenticeship Programs, 1975–1982

Trade	1975	1976	1977	1978	1979	1980	1981	1982
Bricklayers	56	28	21	33	31	23	15	16
Carpenters	282	201	155	137	173	191	177	139
Electricians	253	189	189	264	306	269	259	314
Plumbers/pipefitters	120	121	153	152	130	127	124	125
Cement masons	4	2	6	3	1	2	0	1
Sheet metal workers	65	65	66	66	71	85	93	69
Insulation workers	7	8	8	7	13	16	18	9
Ironworkers	1	3	2	1	1	2	0	0
HVAC	6	9	5	8	8	8	9	8
Other		3	4	5	3	2	1	0
Total	794	629	609	676	737	725	696	681
Completions	130	134	121	88	63	80	62	61

Source: Industrial Research Unit Questionnaire, 1982.

school vocational programs in recruiting apprentices: a majority of all apprentices enroll directly from high school, and some are given advanced standing because of their prior schooling. The balance are recruited from the unskilled ranks of member contractors, from the military services, or "off the street."

Council spokesmen estimate that 20 to 30 percent of its apprentices are black, although representation in the various trades is uneven. The electrical and plumbing trades, as in much of construction generally, show the least minority gains. The CCTC staff itself includes black administrators and field coordinators. There are also a few black firms among the council's membership. A small percentage of women are now also enrolled.

The major problem plaguing the CCTC's apprenticeship program is the attrition of the enrollees. Only about 10 percent of the enrollees complete the program. The key reason seems to be the opportunities which enrollees have had over the years to join non-AGC member firms at higher than apprentice wages or to be advanced to journeyman status prematurely, thus dulling their incentive to continue in the program. To the extent this is true, of course, the apprenticeship program has contributed to the trainees' stock of skills even if they fail to complete the program. According to CCTC administrators another factor that has contributed to the attrition problem is the long time it takes to complete the program, as set by federal apprenticeship standards based on union-negotiated standards. As a result, trainees either become impatient and drop out or finish the program and become supervisors, in either case subverting the intent of developing journeymen. Similar to many open shop programs, the CCTC may well move in the direction of shorter, more specialized training programs, especially in the basic construction trades.

In short, the evidence on the CCTC apprenticeship programs seems to call for mixed reviews. The willingness of open shop contractors in the Carolinas to commit themselves to training is impressive; the growth and direction of the programs in most respects demonstrates their commitment to such programs. At the same time, CCTC apprenticeship seems not yet to have contributed in a really major way to manpower development for open shops in the regions served. With over 300 firms voluntarily contributing money for training purposes, however, and with the demonstrated ability to recruit as many as 2,000 apprentices a year, (which the council had on its rolls in 1972 and 1973),[37] the council clearly has a base from which to address its attrition problem.

---

[37] See Table X-3 in the original edition of this book for data for 1969–1974.

On-The-Job Training Program. The CCTC's on-the-job training program for highway contractors established training standards for fifty-eight occupational classifications used on highway construction projects. The standards specify three kinds of employer obligations under the program: (1) a contribution to the CCTC based on total hours worked by all employees in classifications for which the contractor desires training; (2) a specified number of hours to be spent at various components of the trade; and (3) minimum wage rates at progressive stages of the program. The duration of the program varies with the different classifications, ranging from 520 working hours for several of the trades to 2,080 hours for equipment mechanics.

CCTC coordinators assist in the recruiting and enrolling of trainees into the program. Actual training is the responsibility of the contractor. Contractors are expected to file weekly reports on each trainee with the council, including information on hours worked within the training classification, an evaluation of the trainee's progress, and a statement of the wage rate paid. The contractor is also expected to assign a skilled journeyman or supervisor to oversee the training. Once the trainee is enrolled, the Council appears to have a minimal role in the training process, although coordinators are available to assist with any problems encountered by either employers or trainees.

The on-the-job training program, however, performs a function beyond that of developing skills, in that it is designed to meet the training and affirmative action requirements of the Federal Highway Administration, which will be discussed in Chapter XII. There is an avowed goal of 65 to 70 percent minority enrollment, and coordinators maintain ties with recruiting sources in the minority community. Furthermore, enrollment in the approved CCTC program ensures that trainees may be classified as apprentices under Federal Highway Administration and Davis-Bacon regulations, thus permitting them to be paid at a wage rate somewhat lower than the prevailing journeyman's wage. CCTC personnel also advise participating employers on proper record-keeping procedures to assure Davis-Bacon compliance.

The rather loose control exercised by the council over on-the-job training activities raises the question whether the training purpose has been secondary to that of formally meeting federal guidelines. There is, for example, no explicit testing mechanism to ensure that trainees completing the program have in fact mastered the skills taught in the program. Many subjects—such as safety, shop work, use and maintenance of tools and equipment, and, in some cases, even blueprint reading—are expected to be taught on the job by the contractor himself, with little practical opportunity for oversight by program staff. To be sure, the conscientious contractor will comply with the standards, but it is the conscientious contractor who would probably do that even without a formal training program. Thus the major beneficiary of the program may well be the contractor, who may be less interested in training than in a convenient way of meeting the obligations placed on him by the government. On the other hand, the CCTC program ensures that some training will occur and at the same time encourages the open shop highway contractors to engage in the training and employment of minorities.

Evaluation. Despite these criticisms of the on-the-job training program, it must be noted that the CCTC and its member contractors are responding to a real need in the open shop construction sector. The council's personnel acknowledge that the number of apprentices completing their programs is too small and are reevaluating the need for program changes, such as gearing the apprentice's advancement to individual learning and testing standards rather than to rigid time periods. The on-the-job training program must be viewed in light of the fact that training in highway construction has traditionally been less formal than in the building sector of the industry. Formal training by small open shop contractors generally is in a nascent period, and thus, any shortcomings are perhaps less important than the fact that such training is taking place at all. Much of the labor force in highway work, especially where labor markets have been tight as in the Carolinas, is marginal and/or casual and in need of training. Programs like that of the CCTC provide the means to meet this need in addition to federal requirements.

Association Training—Final Comment

The training programs discussed in this section are still largely in the formative stages and the number of trainees involved is not large. They are growing, however, and point to the fact that formal training in the construction industry is far from limited to joint apprenticeship programs. As open shop contractors continue to compete for larger construction projects, their dependence on formal training has grown, and the programs of their organizations must expand because of the obvious need. The major open shop associations understand this, as the proposed training trust clearly demonstrates. Meanwhile, individual open shop contractors have also been developing their own training mechanisms and experimenting with different forms of training.

TRAINING BY AVERAGE-SIZE INDIVIDUAL OPEN SHOP CONTRACTORS

Contractors may satisfy their need for skilled manpower in two ways. They may try to employ workers who have already been trained, or they may undertake to train present workers themselves. Our questionnaires to both union and open shop contractors in 1974 and 1982 sought to elicit information on each of these methods. One series of questions related to the prior training experience of workers; another dealt with training by contractors directly. Although employers may not always be aware of the backgrounds of their employees, their responses, nevertheless, shed some light on training patterns as well as on intersector mobility in construction.

The Industrial Research Unit Questionnaires

We asked contractors to indicate the training backgrounds of their workers. Training formats included vocational schools, military service, several kinds of apprenticeship programs or on-the-job training. The responses to these questions for the 1974 questionnaire are summarized in Table X-8; and for the 1982 questionnaire, in Table X-9.

In interpreting and analyzing Table X-8, we proceed on the assumption that a "no response" is likely to mean that the contractor was not accustomed to hiring workers with the particular training background in question, or at least that he was not aware of having hired workers with that experience. The most significant responses would seem to be "always" and "often," for these serve to designate the forms of training which make a substantial contribution to the employer's inventory of skills.

It seems clear from Table X-8 that structured training processes were seldom found in the backgrounds of open shop workers in 1974. Although over two-thirds of our open shop respondents identified on-the-job experience as an important form of prior training, only about one-fifth of them so designated open shop apprenticeship. Other forms of training were even less prevalent, although it is noteworthy that three out of ten respondents reported hiring workers, at least on occasion, who had been apprentices in the union sector. The finding that in 1974 only 21 percent of the open shop contractors often or always hired workers with open shop apprenticeship training, while almost 66 percent of the union contractors often or always hired workers with union apprenticeship in their backgrounds, suggests that formal training in the open shop sector was then considerably less extensive than in union construction.

TABLE X-8
Prior Training of Workers, Union and Open Shop Contractors by Type of Training, 1974

	Always		Often		Occasionally		Never		No Response	
Type of training	No.	Percent	No.	Percent	No.	Percent	No.	Percent	No.	Percent
Union[a]										
Vocational school	6	0.8	68	8.7	354	45.5	135	17.4	215	27.6
Military service	1	0.1	40	5.1	305	39.2	186	23.9	246	31.6
Union apprenticeship	149	19.2	355	45.6	171	22.0	47	6.0	56	7.2
Open shop apprenticeship	5	0.6	25	3.2	140	18.0	325	41.8	283	36.4
Other apprenticeship[b]	4	0.5	25	3.2	161	20.7	298	38.3	290	37.3
On-the-job	23	3.0	364	46.8	191	24.6	43	5.5	157	20.2
Open shop[c]										
Vocational school	4	0.5	125	15.9	487	62.1	98	12.5	70	8.9
Military service	2	0.3	68	8.7	411	52.4	195	24.9	108	13.8
Union apprenticeship	1	0.1	34	4.3	210	26.8	405	51.7	134	17.1
Open shop apprenticeship	8	1.0	160	20.4	317	40.4	202	25.8	97	12.4
Other apprenticeship[b]	2	0.3	74	9.4	287	36.6	264	33.7	157	20.0
On-the-job	32	4.1	499	63.6	184	23.5	15	1.9	54	6.9

Source: Industrial Research Unit Questionnaire, 1974
[a]N = 778.
[b]Apprenticeship in an industry other than construction.
[c]N = 784.

TABLE X-9

Prior Training of Workers, Union and Open Shop Contractors by Type of Training, 1982

Type of training	Always		Often		Occasionally		Never		No Response	
	No.	Percent	No.	Percent	No.	Percent	No.	Percent	No.	Percent
Union[a]										
Vocational school	37	2.0	191	10.4	818	10.4	385	21.1	397	21.7
Military service	0	0.0	56	3.1	702	38.4	629	34.4	441	24.1
Union apprenticeship	517	28.3	833	45.6	285	15.6	74	4.0	119	6.5
Open shop apprenticeship	2	0.1	50	2.7	351	19.2	969	53.0	456	24.9
Other apprenticeship[b]	1	0.1	33	1.8	370	20.2	949	51.9	475	26.0
On the job	63	3.4	652	35.7	610	33.4	195	10.7	308	16.8
Other	8	0.4	54	3.0	329	18.0	407	22.3	1030	56.3
Open shop[c]										
Vocational school	21	1.0	363	18.0	1172	58.0	265	13.1	201	9.9
Military service	1	0.0	138	6.8	1067	52.8	573	28.3	243	12.0
Union apprenticeship	8	0.4	143	7.1	688	34.0	930	46.0	253	12.5
Open shop apprenticeship	39	1.9	408	20.2	857	42.4	509	25.2	209	10.3
Other apprenticeship	6	0.3	133	6.6	715	35.4	858	42.4	310	15.3
On the job	120	5.9	1260	62.3	463	22.9	59	2.9	120	5.9
Other	21	1.0	151	7.5	591	29.0	238	11.8	1021	50.5

Source: Industrial Research Unit Questionnaire, 1982.

[a] N = 1828

[b] Apprenticeship in an industry other than construction.

[c] N = 2022r

Table X-9 demonstrates a similar pattern for 1982. The 1982 data are complicated by a very large number of no responses from both unionized and open shop concerns, but the responses received did show a paucity of formal open shop training. Given the different populations included in the two surveys, as noted in Chapter II, Tables X-8 and X-9 show a rather remarkable similarity in the responses for 1974 and 1982. Moreover, since larger contractors are generally considered more likely both to have formal apprenticeship training and to respond to questionnaires, it is possible that Tables X-8 and X-9 actually overstate the extent of open shop formal training. Our interviews revealed both more interest and activity by open shop contractors in formal training in 1982–1983 than in 1973–1974. To be sure, the interviewees tended to be the larger contractors, but since business conditions were far worse in the latter years than in the former and since it is much easier for contractors to maintain formal training in prosperous times than in depressed ones, such interest and activity are significant for the future.

Table X-9 also emphasizes a point made earlier in the discussions of the ABC and AGC contractors' responses to the 1982 questionnaire. The open shop contractors only occasionally find union apprenticeship training a significant source of trained personnel. The open shop segment clearly places its heaviest dependence on on-the-job training, with open shop apprenticeship the second most important source of training.

Finally, as the principle of task training had by then become more widely understood and utilized by 1982, perhaps that year's Industrial Research Unit questionnaire should have included a task training category. Because it did not include this category, we cannot be certain whether contractors using task training designated it as apprenticeship or on-the-job training.

Examples of Individual Contractor Programs

The bare recital of statistics, however, does little to depict the character and substance of the training activities undertaken by individual open shop contractors. It may be useful at this point to describe in detail (on the basis of our field interviews), the approaches of some employers to ensure an adequate supply of skilled manpower. These illustrations are not necessarily representative of training generally, but they should serve to impart some flavor to what has heretofore been a rather general discussion of training forms.

General Contractor A specializes in water pollution control facilities. Based in New Hampshire, he operates throughout the Northeast. He utilizes several approaches to training. When a project is located within the jurisdiction of an ABC chapter, trainees may be enrolled in an ABC program only for those portions of the curriculum applicable to *A*'s work. *A* also has a variety of carpentry programs with off-the-job components, some lasting only a few weeks and others four years. Some are highly specialized—for example, form building or blueprint reading. The company employs a full-time training director who fashions specific programs to meet immediate needs. The company president estimates that about 10 percent of his 400 current workers have undergone one of these programs.

General Contractor B operates exclusively in and around its New Hampshire base. Its training programs for carpenters had at the time of our interview 22 workers out of a total workforce of 150. Classes met once a week for two hours, with instructors drawn from among the company's own supervisory staff. Instruction ran for sixteen weeks and covered such subjects as blueprint reading and the care and use of tools. *B* also runs an informal program for equipment operators, in which workers are brought out to job sites during off hours and are taught the care and operation of heavy equipment used by the firm.

Electrical Contractor C employs ninety workers in Columbus, Ohio. At the time of our interview he was a participant in the local ABC apprenticeship program, but for some years before that he ran his own program. It met two nights a week for four years and proceeded to cover the electrical code book stage by stage. Instruction was led by the contractor himself or one of his journeymen and included both lectures and manipulative practice. The program was selective, with no more than six trainees at one time.

Electrical Contractor D, with twenty-five employees in Detroit, had just joined the newly established ABC apprenticeship program. The classes met one or two days a week for four years and were paid entirely by the firm. The company owner believed that he was the first nonunion electrical contractor in the area who provided formal training for his employees.

Plumbing and Heating Contractor E, employing ninety-five workers in Columbus, had a rather skeletal program for plumbers and mechanics for a short time prior to joining the local ABC apprenticeship program. The firm ran its own school in the shop for four or five sessions, according to a lesson plan it developed itself. Workers came on their own whenever they wanted to attend. Before the school was set up, all training was on-the-job.

General Contractor F is the largest in a central Texas city. This city was primarily union from World War II until the late 1960s. At that time, a series of jurisdictional disputes led to a determination by the city's largest contractors to go open shop. A series of strikes and decertification elections followed with the result that only the Carpenters maintained a union contract, and this union's agreement was significantly modified to eliminate restrictive and costly practices.

One strike involved the Bricklayers and shut off a source of labor supply that could not be supplemented from the open market. The contractors, therefore, set up a special training institute, recruited twenty black laborers who previously had no skill training, paid them the minimum wage to learn, taught them to lay brick for eight hours per day over a six week period, and then put them on the job. According to the town's largest contractor, "They may not be the best bricklayers in the world, but they can lay 500 to 850 bricks while the union guys were laying 400 to 500."

Electrical Contractor G is headquartered in another Texas city. He employs 38 persons permanently and his force occasionally goes up to 150 persons. This company does the electrical and sometimes the mechanical work on water and power plants. To train its employees, the company uses a manual prepared by a Texas AGC chapter, which it augments with correspondence courses. All superintendents and the three partners are journeymen electricians and they conducted the first training classes themselves. The employees pay for correspondence courses through payroll deductions, but the money is refunded with interest if the course is completed. Moreover, job preference is given to those who complete the courses. Since the company operates over a wide area of the Rocky Mountain and Southwest regions, it is always looking for trainees who might make journeymen. It thus encourages this type of training at all its locations.

Mechanical Contractor H is a large Texas organization capable of handling jobs of well over $1 million. Its employees handbook states:

> You earn advancement by excelling at your present job. Performance, trainability, and length of service are all weighed whenever an advancement is considered. It is your primary responsibility to prepare yourself for promotion. One of your supervisor's primary responsibilities is to encourage, develop, and recommend his employees for promotion. Make sure you tell him of any special training you have had or are taking so that it may be entered in your personnel record.

The after-hours education program includes a continuous effort to qualify personnel for journeymen and master mechanic ratings and to tutor them so that they can pass tests in order to obtain necessary licenses. Classes are held both after work and in early mornings. The company has about twenty master mechanics, although only one is required on each job. The company believes that maintaining a surplus of trained personnel is the key to continued expansion. Among the journeymen plumbers are several blacks; the company makes a special effort to recruit and train both blacks and Mexican-Americans. It also conducts extensive supervisory training programs in cooperation with a major university.

Most of these cases involve specialty contractors, especially electrical and mechanical contractors. The big shift of the highly skilled trades from union to open shop during the 1970s, which has been repeatedly noted in earlier chapters, seems certain to compel open shop specialty contractors to increase their formal training programs in the future. The above sketches indicate the wide variation in which formal training can successfully be accomplished. Undoubtedly, many other models for training are available and equally successful for the small open shop contractor.

TRAINING BY LARGE OPEN SHOP CONTRACTORS

The small size of most open shop contractors inhibits their developing elaborate training programs of their own. Broad scale training is usually feasible only through participation in a local multiemployer program sponsored by a contractors' association. The situation is different, however, in the case of those open shop concerns whose volume places them among the largest construction firms in the country. Their manpower needs are so extensive as to warrant a heavy investment in training, including the establishment of a separate department within the organization devoted exclusively to manpower development. Their size, moreover, enables them to experiment with diverse training forms which are often geared to the needs of specific jobs.

In this section, we examine the training efforts of three of the largest open shop contractors: Brown & Root, Inc., of Houston, Texas; Daniel International Corporation of Greenville, South Carolina; and H. B. Zachry Company of San Antonio, Texas. Brown & Root and Daniel, now a subsidiary of Fluor, are two of the largest contractors—union or nonunion—in the nation, each with gross volumes of several billion dollars per year. Zachry's contracts in 1983 totaled over $485 million, and it ranked forty-eighth on the *Engineering News-Record* 400 list. Although each large firm has a somewhat different approach to training, the case studies presented here are in line with those of other large open shop contractors.

BROWN & ROOT, INC.[38]

Founded in 1914, Brown & Root has grown to a multibillion-dollar corporation employing an average of over 50,000 persons worldwide in the last five years. The company specializes in large industrial and heavy construction projects, including petroleum refineries, chemical plants, manufacturing and process plants, power plants, mining facilities, smelters, offshore platforms, pipelines, dams, bridges, and highways. It also provides contract maintenance services to operating plants which it builds. Brown & Root is organized into several operating units, each reflecting the various kinds of construction in which it is engaged.

Faced with the challenge of developing a skilled construction workforce for its projects in remote locations, both domestic and international, Brown & Root pioneered a modular, competency-based training program. This program provides workers with a knowledge of the resources, techniques, equipment, and tools used in a specific construction craft, as well as skills basic to all crafts, such as mathematics, blueprint reading, and safety procedures. The company also offers supervisory training designed to develop skills in key management areas, such as leadership, communication, planning and scheduling, productivity, quality control, safety, timekeeping and payroll, and personnel practices. This training is strongly oriented toward internal company policies and procedures as well as project needs. Brown & Root's training program combines on-the-job training and, as needed, classroom instruction.

The foundation of the company's human resources development program is the Instructional System Design (ISD) concept. Using ISD, Brown & Root devised a series of craft skills inventories that list the general and specific skills needed by workers in order to demonstrate competence in a craft. These inventories are used by trainees to plot their career advancement from entry-level to craftsman status, and by supervisors to evaluate trainees' performance.

Brown & Root has also used ISD to develop modular classroom training texts for the various crafts. The modules in these texts are performance oriented and job related and, therefore, may be sequenced to meet specific project requirements. At the completion of classroom training, project managers and supervisors use the skills inventories to determine whether trainees' newly acquired skills match project needs; and, if necessary, the training modules

[38] The discussion in this section is based largely on information contained in the Annual Report of Brown & Root's Personnel Training and Development Department and the company's Employee Training Handbook, plus interviews and information supplied by company personnel, July 1973, November 1974, and August 1981.

are revised to better meet those needs as well as technological changes.

The objectives of Brown & Root's human resources development program are as follows:

1. To develop a local workforce capable of performing engineering and construction tasks that meet a project's quality-control standards, cost, and schedule.
2. To develop a local workforce capable of operating and maintaining a project's facilities when construction is completed.
3. To reduce the impact of a project on the community by developing and utilizing the local workforce.
4. To upgrade the skills of local workers, thereby enabling them to contribute positively to the growth of the regional economy.

During the planning stages of a project, management conducts a survey to determine the characteristics and availability of the workforce in the area surrounding the project. Management compares these findings with project manpower requirements and estimates the scope and duration of a project training program. Then a training schedule is submitted to the customer for approval as part of the overall project cost. In most cases, Brown & Root has found the customers willing to defray a portion of the training cost because of the benefits to be gained from project training.

To further ensure funding for project training, Brown & Root has established good working relationships with the vocational/technical education departments of a variety of states, including Texas, Oklahoma, Alabama, North Carolina, Virginia, Arkansas, and Colorado. These states have helped to develop their manpower resources by providing Brown & Root with training facilities, equipment, and wages for instructors.

A trainee's salary at Brown & Root is based on evaluations of his performance on the job. This is in contrast to traditional apprenticeship programs where trainees receive raises (usually expressed as a percentage of the journeyman's rate) at predetermined time intervals.

Career advancement is also based on performance, rather than on the completion of a training program. The expectation is that training will improve competence and productivity and in turn lead to faster promotion—although the correlation is not automatic.

Brown & Root maintains a computerized personnel records and management system to increase the effectiveness of its training program. This data base summarizes each employee's training and employment history. Supervisors seeking information about an employee may consult this data base. It facilitates the screening and referral process and helps to eliminate the time lags inherent in traditional hiring systems. It also provides a means for the company to contact former employees when moving back into an area on a new project.

One advantage of Brown & Root's training program is that it offers great flexibility in meeting training requirements for diverse projects. It allows trainees to develop skills relevant to specific project and community needs in a hands-on, performance-oriented environment. At the same time quality-control standards built into the program ensure that the construction meets owner's specifications and government regulations, particularly those having a direct impact on trainees' health and safety.

Another advantage of the program is its cost effectiveness. Based on its experience with projects not requiring BAT approval, Brown & Root has found that a one-to-one trainee-journeyman ratio can reduce project costs to the owner. (The BAT-approved ratio averages one to four nationally.) As the workforce gradually develops, local workers can fill more and more craftsman positions, and when the construction phase of a facility has come to an end, a skilled workforce is available to operate and maintain it.

Since Brown & Root initiated its training program in 1968, it has achieved the following milestones: (1) trained over 66,000 employees; (2) provided over 600 million manhours of on-the-job training and over 30 million manhours of classroom instruction; (3) invested over $200 million of company resources in training.

The company's training program has been so successful that its methodology was adopted by both the AGC and ABC. Moreover, it was approved by the U.S. Department of Labor as an alternative to traditional apprenticeship training.

In the late 1970s, having completed most of its training and development effort, Brown & Root's Personnel Training and Development Department was transformed from a corporate staff organization to a line organization. Major responsibility for training and development now rests with the various operating divisions, with corporate staff providing necessary supplemental support.

Daniel International Corporation[39]

Daniel's size rivals that of Brown & Root in terms of both volume and the size of its workforce. Its training activities, however, are

[39] Information on Daniel's training activities was obtained in interviews with various company personnel, Greenville, S.C., February 1974, July 13, 1981, and March 2, 1982, and Philadelphia, September 1974, amplified by materials generously provided by these specialists, and by Gerald M. Glenn, Senior Vice-President.

even more extensive than Brown & Root's. As a result, it has both trained more personnel and developed more extensive training procedures.

Formal training by Daniel began in 1971 and has progressed rapidly since then. Like Brown & Root, Daniel employs the great bulk of its specialty craftsmen and trains personnel in all crafts that the company has on its payroll. Approximately 25 percent of the trainees are black, and 5 percent female. Most training is conducted at the job site. Training needs are determined after alternate sources of recruitment are utilized. Comprehensive, elaborately computerized records are kept on all of its jobs in an area, with all employees entered by craft, wage rate, skill level, and so forth. When the company prepares to begin work in an area where it has previously completed a project, it consults these records and directly contacts workers whose past performance was satisfactory. After as many local skilled workers as possible are assembled, training programs are established to produce the remainder of those needed. If the local market cannot supply sufficient resources, additional manpower can be recruited from the company's three manpower centers in Greenville, South Carolina; Houston, Texas; and Salt Lake City, Utah.

Daniel's program, Basic Employees Skill Training (BEST), has three components: preemployment training, on-the-job training, and after-hours related instruction. The preemployment phase is made available predominantly to new hires with little or no prior experience at a trade and entails a basic craft. Trainees attend sessions on a full-time basis for varying durations during which they receive no remuneration. The length of this phase depends on the craft: insulating requires one week, carpentry two weeks, pipefitting two to four weeks, and welding eight weeks or more. (In the case of welding, this time is needed for the worker to obtain certification.) Thus, by the time the work and the formal training period starts, the worker has acquired an elementary familiarity with construction work in general and with the basic vocabulary, tools, and materials of his prospective trade in particular. The objective of preemployment training is to provide the trainee with entry-level skills that will enable him / her to be productive on the job immediately after being hired.

The preemployment phase of training is significant in at least three respects. First, it serves to screen out those workers whose interest in formal training is limited. About one-half of the workers entering preemployment training drop out during this phase of the program. Second, the substantive knowledge gained during preemployment training facilitates the rest of the training process. Among trainees who enter a regular program from the preemployment phase, only about 10 percent drop out before finishing; among workers who bypass the preemployment phase, the dropout rate is over 20 percent. Third, specially designed preemployment training can quickly generate marginally trained workers who are, nevertheless, capable of some productive work right away. This training may serve to carry the company over a temporary labor shortage on a particular project. In one instance, on a major chemical plant project in North Carolina, Daniel trained 215 pipe insulators with two weeks of intensive preemployment instruction. This feat was accomplished by breaking down the relatively simply functions of the trade and using an unusually large ratio of trainees and helpers to journeymen. A large percentage of the trainees were black women.

Related instruction is usually arranged through the facilities of local education centers or high schools. The instructors, however, are invariably drawn from Daniel's own ranks. After-hours classes are almost always conducted at the job site. Trainees attend classes for about three hours a week on the average (it varies by craft) and will normally complete the program in two years. Total class time in the program averages 275 hours. The company attempts to coordinate a trainee's job assignment with the material that he is learning in the classroom at a given point in time.

Advancement within the program is organized around blocks of skills into which the requisites of each craft have been divided. As each block of skills is mastered, the trainee becomes eligible to move to a higher skill level with a commensurate increase in his wage. Records are kept on the length of time each trainee spends at specific functions of the trade. Information on the trainee's activities, both on the job and in related instruction, is prepared weekly at the project site and submitted to company headquarters, where it is entered on the computerized records.

Each project has an Evaluation Board, usually consisting of the craft superintendent, general foreman, foreman, project personnel supervisor, and project training coordinator. As trainees complete the time requirements of each skill level, the company transmits their names to the Evaluation Board, which meets monthly to consider whether a trainee should be advanced to a higher skill level and wage rate. A trainee is assumed to have met the requirements for a skill level within the indicated time frame unless his foreman is able to specify detailed reasons why he should not be advanced. It is possible for a trainee to advance ahead of the recommended time intervals if the foreman can demonstrate his exceptional ability to the Evaluation Board. Promotion within the program is geared to meeting predetermined objective criteria. Thus, Daniel's program appears presently to place somewhat heavier emphasis on acquisition of craft-related subskills demonstrable to the Evaluation Board as a criterion for advancement than does Brown & Root's program, which leaves the question of promotion largely to the discretion and initiative of the supervisor.

In recent years, Daniel has established a certification procedure (Craft Certification) which attests to an individual's rating as a journeyman. Individuals who qualify receive craft certification cards that they can use for employment purposes elsewhere if necessary, but such rated journeymen usually find employment at Daniel sites. New employees must pass a written test, be employed for thirty days, and be evaluated relative to job performance skills before they can achieve this journeyman status. A multicraft group is included in the certified trades, and individuals can be certified in more than one trade. One Daniel employee has sixteen certifications.

A somewhat innovative aspect of the Daniel training philosophy is a formal upgrading program. The aim of this program is to upgrade selected hourly craftsmen to supervisory and management positions. Courses have been established and are being taught in presupervisory training (future foremen), foreman, and superintendent training. Management training specialists visit Daniel construction sites on a periodic basis to conduct these seminars. The upgrading program enhances management skills and also provides Daniel with a cadre of skilled supervisors and managers to meet its future needs.

The concept of "skill levels" is somewhat analogous to the modular approach utilized by Brown & Root. It provides an element of flexibility which can facilitate the training process in at least two important ways. First, a new hire can be slotted into the program at a particular skill level, depending upon his prior training and experience. Since participation in the program is voluntary, new employees do not invariably enter training at all. In any event, however, they are evaluated by the company's hiring agent and placed at a pay scale commensurate with that of trainees who have reached an equivalent level of development. Second, an employee may enter training programs for various trades throughout his tenure on the project.

In addition to project-related training as described above, Daniel also operates regional craft training centers located in strategic geographical locations to support area manpower requirements. The training centers may be opened or closed as manpower demands dictate. Daniel has been successful in cooperating with federal and state governmental agencies to provide training for qualified unemployed or disadvantaged trainees under federally sponsored programs through these training centers.

H. B. Zachry Company[40]

This large general contractor, specializing in major heavy and industrial projects, conducts formal training programs for a variety of trades as needed at its headquarters in San Antonio, Texas. In addition, when a project requires a rapid expansion of the workforce in a short time, special programs are developed to meet the specific needs of that job. The company has programs for all the crafts which it utilizes in its operations, except sheet metal and ornamental iron work. The company also conducts courses in management development for all its supervisors. Voluntary courses in management are offered to craftsmen when the need for foremen arises.

The company employs a full-time staff member on each job requiring training to supervise and coordinate each training program. This staff member assumes responsibility for recruiting and selecting trainees, placing them on particular jobs, arranging for related instruction, and monitoring the progress of trainees through the program. Overall responsibility and coordination is handled by the personnel department in the home office at San Antonio. On-the-job training is overseen by foremen or craftsmen designated by the home office as field training supervisors, who provide the basic orientation to company policies and safety regulations, ensure that the trainee is given experience in various facets of the trade, maintain weekly records of the trainee's progress, and determine whether the trainees are qualified to advance to successive stages of the program. Advancement is contingent upon successful completion of each phase of training regardless of the actual time spent learning the tasks. Progressive wage rates are attached to the various phases,

[40] Information obtained through interviews with Mr. E. M. Hammond, Vice-President and Personnel Director, H. B. Zachry Company, Houston, Texas, July 1973; San Antonio, Texas, November 1973; and Dallas, Texas, August 11, 1981; amplified by materials kindly supplied by Mr. Hammond.

and trainees may be placed at any level depending on their previous training and experience. The number of trainees in the program at any one time is determined through an evaluation of the human resource needs for the particular job under contract.

Zachry's training programs have three components: classroom instruction, shop work and on-the-job training. Classroom instruction is arranged through local educational institutions if adequate facilities are not found at the job site. It covers such subjects as the use of various tools and materials, blueprint reading, and basic construction mathematics. At the beginning of each phase of classroom work, the trainee is given a list of performance objectives, and the instruction is designed to prepare him to meet these objectives. After the trainee passes a qualifying test for each phase, he moves on to the next phase. Shop work involves manipulative training under conditions simulating actual job operations so that the trainee gains practice in the use of tools and equipment. On-the-job training is provided by the field training supervisor. Trainees are assigned to observe and practice specific aspects of the trade and are moved to successive functions as their competency is developed to the satisfaction of the supervisor.

As noted earlier, special programs may be established at particular locations to meet immediate needs. At a major chemical plant expansion in Texas, a program was designed to produce large numbers of semiskilled electricians, form setters (carpenters), pipefitters and welders, as well as to upgrade a number of electrician's helpers to junior craftsmen status. The electrician's program was broken into three blocks, or modules: high-voltage electricity for four weeks; control and hookup for six weeks; and instrumentation for six weeks. As each module was completed, the trainees went into the field and worked at the operation they learned. Many later entered another module and ultimately progressed to the three-week upgrading program. The other programs varied in length from four to eight weeks.

The components of these programs varied in relation to the needs of the trade being developed. Electricians and form setters spent two hours in class, two hours in the shop, and four hours on the job each day. The pipefitter's day was equally divided between classroom instruction and on-the-job training, while welders spent most of their time in the shop. The electrician's upgrading program was predominantly on-the-job training. These programs were used to develop a total of 900 helpers in blocks of time not exceeding eight weeks. Once finished, the helpers were assigned to work closely with craftsmen in the field.

Large-Company Training—Final Comment

Training programs in other large open shop companies use similar approaches as well as many variations. The commitment to training by these companies is obvious: on-the-job training is easier and faster when an employee has a foundation in the theoretical understanding of construction work.

Still, the training approaches discussed above differ in significant respects from the more traditional forms of apprenticeship in the building trades. Their hallmark appears to be flexibility. Programs are designed to make maximum use of a trainee's skills even while he is progressing through the curriculum. Material tends to be organized in blocks, with each block developing a set of skills which can be put to immediate use at the workplace. There is, furthermore, the idea that training need not be designed exclusively to develop all-around journeymen. It can, instead, be fashioned to turn out semiskilled employees who can work productively under the supervision of a journeyman. Indeed, open shop contractors are wont to stress the efficiency gained by utilizing work crews with fewer journeymen than those typical of a union project. This is what task training is all about, and it began in order to meet the manpower needs of the large open shop company.

This approach to training permits the short-term development of a large workforce needed to complete a specific project expeditiously. The examples given above of Daniel in North Carolina and Zachry in Texas are cases in point. Such training allows the firm to minimize its investment while still generating sufficient skills to complete the job, a highly desirable feature for a company which will be leaving an area once the project is finished.

Another distinct characteristic of these task-oriented programs in the three large firms is that they underscore individual advancement. In no case is it assumed that a trainee has mastered a particular block of skills simply because he has worked at it for a given time period. Formal testing and evaluation is ongoing throughout the duration of the programs. Traditional apprenticeship tends to be more rigid, with specific time periods and wage rates for each chronological stage of the program. Open shop task programs, at least in terms of their formal descriptions, tend more to emphasize the substance of learning rather than "time in rank."

These features of open shop training illustrate why they are at odds with some of the traditional apprenticeship standards—including the overall duration required to complete apprenticeship programs—promulgated by federal and state agencies for registration purposes. As a result, these programs tended not to be registered with an apprenticeship agency, and thus trainees could not until recently be used on work covered by the Davis-Bacon Act and state prevailing wage statutes. What is involved has been the already noted fundamental conflict in philosophies between the one which sees formal training as appropriate only for the development of the most highly skilled journeymen, and the other which views training as adaptable in different forms to various kinds of needs, including the need for more narrowly trained helpers and semiskilled "craftsmen." The business success of companies like the three discussed above emphasize our previous point that the philosophical approach historically adopted by the BAT needs continued and thorough review.

THE BUSINESS ROUNDTABLE STUDY

In September 1982, the Business Roundtable issued a report stating that, although the open shop accounted for 60 percent of the construction market, as compared with 40 percent for unionized firms, it spent only $11 to 12 million on training, as compared with $230 million spent by union firms. Giving some credit for open shop expenditures not uncovered, this report concluded that 90 percent of construction training funds were spent by the unionized sector, and only 10 percent by the open shop segment.[41] The Business Roundtable report listed the following reasons for the lack of support for training:

1. Fear of losing a bid due to added training costs.
2. Fear of training employees and losing them to a competitor.
3. Fear of working with a nonproven program.
4. Failure to recognize a need for training, since workers can be pirated from their competitors.
5. Lack of employee acceptance of traditional training programs.
6. Lack of appreciation of the improved productivity that can be realized through training of their workers.[42]

These reasons are in accord with our findings with one exception: users, including undoubtedly those who comprise the Roundtable, have themselves also been less than enthusiastic about paying for

[41] *Training Problems in Open Shop Construction,* pp. 14–15.
[42] *Ibid.,* pp. 10–11.

formal training when it can be avoided—that is, when open shop contractors are utilized.

Our findings also agree with a basic premise of the Roundtable report that formal training by open shop contractors must be increased. This is very apparent both from our field interviews and from the results of our questionnaires. It is, moreover, made more urgent by the tremendous expansion of the open shop in the higher skilled trades, such as electrical, mechanical, plumbing and sheet metal work. These are skills which, at least in their more advanced functions, require formal training, and the open shop segment must do more of this if the supply of skilled workers is to meet the demand in the future.

The Business Roundtable's estimates concerning expenditures for training in the unionized and open shop segments do not, however, accord with our findings. There are several reasons for this. First, our research covered the industry in much more depth and did not rely so much on data from the U.S. Department of Labor, associations, or large construction companies as did the Roundtable report. There is, for example, no record that the Roundtable researchers engaged in extensive field interviews or utilized a questionnaire as we did. Thus, their results exclude information which we obtained from more typical contractors.

Second, despite the existence of another excellent Business Roundtable report in the same series,[43] the report dealing with open shop training did not consider the impact of different worker deployment systems upon training. By their efforts to eliminate the helper and have the journeymen take over unskilled and semiskilled work, construction unions have eliminated a natural upgrading and on-the-job training system. The only recourse for training in union construction has been the apprentice system, which is almost universally conceded to be cumbersome and ill-suited for much of the training in the industry, although suited at least partially for some all-around type training, as still a third Roundtable report so well pointed out.[44] Unfortunately, the widespread utilization of helpers by open shop contractors and the tremendous amount of on-the-job training that results from this system escape the type of analysis found in the Roundtable's open shop training report. In addition, the use of helpers greatly reduces the need for apprentices who are utilized by unionized firms to do some of the work performed by

---

[43] See *Subjourneymen in Union Construction,* Construction Industry Cost Effectiveness Project, Report D-1 (New York: The Business Roundtable, 1982).

[44] See *Government Limitations on Training Innovations.*

helpers on open shop projects. Given the fact that only 15 percent of apprentice students actually complete their terms, one cannot be certain that the expenditure for such training is often more than an equivalent for the use of helpers, especially if the latter are receiving both valuable experience and sound on-the-job training.

Third, and of considerable importance, is the fact that paying for training and receiving it under union contracts are sometimes quite different things. Union-management training facilities are often quite elaborate with considerable amounts spent for buildings, equipment, administrators, and other overheads, some of which do not necessarily contribute either to the quantity or quality of training.

More important, unionized contractors can receive a lot less than they pay for, as a study of the Edison Electric Institute (EEI) clearly demonstrated. In 1979, the EEI surveyed ninety-three of its member companies in the electric utility industry who were constructing facilities. The EEI study, which is set forth in Table X-10, found that:

> Despite agreements which call for 20 to 30% apprentice to journeymen ratios on our members' projects, the average ratio is 7%. Unions and contractors tell us 7% isn't a bad ratio and the $8,436,281.27 we paid for apprentice training in one year isn't much when compared to the billions of dollars spent for power plant construction.
>
> Our answer is that $8,436,281.27 is a lot of money when you get little return on your investment.[45]

According to *Cockshaw's Construction Capsules,* which reported this study, "EEI member firms see 'few apprentices or trainees on our projects.'" By 1982, the EEI reported that the situation had not altered appreciably, and *Cockshaw's* added that "owner/users in the non-utility industrial segment have long been irked by the small ratio of apprentices on their projects."[46]

Finally, the Business Roundtable study does not give appropriate consideration to the institutional factors which have limited open shop training. Open shop training through cooperative efforts among associations is both relatively new and has historically been thwarted, rather than aided, by government. For example, the ABC's first apprenticeship program was approved in Baltimore in 1959 by the Maryland State Apprenticeship Council, but it was not until 1970 that the BAT gave the ABC approval of any of its apprenticeship programs.[47] The NAHB began its training in 1967; it

---

[45] *Cockshaw's Construction Capsules,* Vol. 12 (July 1982), No. 7.

[46] *Ibid.*

[47] *Training Problems in Open Shop Construction,* p. 8.

TABLE X-10

Apprenticeship-Journeymen Ratios—Negotiated and in Practice Electric Utility Construction Projects, 1979

	Percent of Apprentices To Journeymen Per Agreement	Journeymen (No. on 9-30-79)	Apprentices (No. on 9-30-79)	%	Average Cost Per Apprentice	Contributions for Apprenticeship & Training 1979
Asbestos workers	25	484	42	8.7	$1,038.35	$33,227.16
Boilermakers	20	4,007	387	9.7	848.05	401,125.38
Bricklayers	20	189	16	8.5	487.85	6,829.89
Carpenters	20	6,249	746	11.9	1,059.85	750,372.25
Cement finishers	20	958	152	15.9	344.66	51,354.82
Electricians	33⅓	10,322	652	6.3	2,067.83	1,327,547.58
Ironworkers	33⅓	5,634	651	11.6	1,307.79	818,675.15
Laborers		13,234	264	2.0	6,113.41	1,283,815.59
Millwrights	25	1,087	118	10.9	1,123.67	111,243.51
Operating engineers	20	4,590	280	6.1	3,817.08	1,049,698.20
Painters	20	871	78	9.0	726.80	48,695.45
Pipefitters/plumbers	20	12,756	953	7.5	2,542.06	2,328,529.95
Roofers	100	58	0		0	954.00
Sheet metal workers	25	1,140	43	3.8	3,026.75	99,882.78
Other		2,429	221	9.1	612.46	124,329.56
TOTAL		64,008	4,603	7.2	$1,794.04	$8,436,281.27

Source: Edison Electric Institute. Reproduced from *Cockshaw's Construction Capsules,* Vol. 12 (July 1982), No. 7.

was not, however, until the early 1970s that the AGC open shop group became significantly influential in that organization, and until 1980 that the AGC won approval from the BAT of its unilateral trainee program standards. Even today, many state apprenticeship councils and local governmental agencies discriminate against open shop training programs.[48]

Open shop contractors are also often smaller than their unionized counterparts. Such companies are unlikely to be able to afford the cost of formal training, but often do a considerable amount of informal instruction. It is impossible to measure, but equally impossible to ignore. This does not mean that on-the-job and formal training are equivalent. As most of the employers whom we interviewed acknowledged, the manipulative skills developed on the production line are more easily and more quickly inculcated in the trainee who has had some theoretical exposure to the fundamentals of his trade and to the materials, implements, and methods of construction generally. There is, however, need for several types of training, including both formal and on-the-job training.

CONCLUSION

Open shop contractors utilize on-the-job training as their primary training tool, but the extent of formal training has expanded, and there is evidence that it will continue to do so. It is now generally recognized that many union craftsmen learn their trades informally, despite that sector's much longer experience with widely organized apprenticeship programs. There will undoubtedly always be a place for the specialist who performs a simple, repetitive task which can be learned rather quickly by observation, as well as the journeyman who is less than an all-around craftsman but can do most jobs. As long as the building trades unions continue to jealously guard their jurisdictional lines, these specialists will play an even larger role in the nonunion workforce. Nevertheless, as open shop contractors compete for an ever larger share of the union market; as they challenge the union contractor for the larger jobs requiring sophisticated methods and coordination of complex activities; and as more union contractors opt for open shop or "doublebreasted" operations, the need for structured training forms becomes increasingly apparent. Multimillion and multibillion dollar projects are not built entirely either by specialists or by first-class craftsmen.

[48] See "States Urged To Approve More Open-Shop Training," for evidence that state apprenticeship councils restrict open shop training. Localities are also pressured to deny open shop builders cooperation with vocational schools.

In another sense, of course, the open shop contractor is even more dependent on training than his unionized counterpart. He does not have access to a pool of labor which can often be augmented by a telephone call from a business agent to outlying communities. Although open shop contractors can and do cooperate by lending and trading workers, this process lacks the central brokerage agent which unions supply. Open shop hiring halls or referral agencies are still in their formative stage. Thus, for the open shop contractor, a training program, which can issue a steady flow of competent craftsmen, is the most reliable source of skilled labor, and the largest open shop contractors recognize this fact.

It can be said with some confidence that investment in formal training by open shop contractors will, in large measure, be a function of their employer organizations. Except for the giants of the industry, these contractors are, for the most part, small even by the standards of the atomistic construction industry. The underwriting of a modern training program can be prohibitively expensive for the individual contractor, especially with the danger that the trained worker will leave to take a job elsewhere. By pooling his resources with others similarly situated, he can benefit from the economies of scale associated with most undertakings.

It is thus no accident that as the ABC has grown, its commitment to training has materially increased. As new ABC chapters are formed, older ones initiate new training programs. As open shop contractors have gained influence within the AGC, training programs to benefit them have been developed. Organizations like the NAPHCC, in which open shop contractors have become a major force, have initiated open shop training programs. The all-open shop IEC has always stressed training.

The dominant form of training and development in the open shop sector remains on-the-job training, but off-the-job formal instruction is becoming widespread and increasing faster than the Business Roundtable study indicated. Moreover, the significance of the helper and of on-the-job training are not properly credited in the Business Roundtable report. Nevertheless, the Business Roundtable deserves enormous credit for highlighting the need for more open shop training. As a result of its efforts, the ABC and other organizations have formed the National Open Shop Training Trust. In turn, these associations have asked the large users that are members of the Business Roundtable to finance the trust so that training can be underwritten. Since these same users contribute substantially to union-management training programs whenever they engage a union contractor, open shop associations maintain that they are only requesting parity treatment. The future of open shop training is thus dependent upon the users as well as upon open shop associations and contractors.

Regardless of the success of the training trust, open shop training seems certain to expand. The needs of the contractors require this. In particular, the expansion of the open shop in the higher skilled trades ensure it. The result will not only be more training, but more effective training as well, based upon task instead of time requirements and geared to the needs of construction now and in the future, rather than upon either artificial journeyman-apprentice ratios or the use of apprentice trainees to do work that a helper might better perform while learning on the job.

CHAPTER XI

Wages, Benefits, and Hours

Employee compensation in open shop construction is a very difficult subject to analyze, for there are no regular statistical series distinguishing the union and nonunion sectors. The most comprehensive wage and benefit data apply to union construction. These data were published annually and updated quarterly by the U.S. Bureau of Labor Statistics (BLS) until discontinued in 1981, and included extensive information on wages, hours, and benefits by craft and market area.[1] The figures, broken down by industry division, were published monthly, but they did not include any occupational detail nor did they separate the union and nonunion sectors. After the Bureau discontinued these efforts, the Construction Labor Research Council (CLRS) continued to publish excellent data, and Construction Data Services, Inc., published handbooks in 1983 and 1984 which list wage and benefit rates of the organized crafts throughout the United States.

Over the years, the BLS has also made studies of union and nonunion construction wages in selected cities, and various individual researchers and associations have added to this information. We shall utilize all these studies in this chapter, but, of course, they have been made only intermittently and cover only a segment of open shop construction. Using also the data which have been generated by the Industrial Research Unit, we shall also present a montage from scattered and disparate sources, but it will permit some reasonably confident conclusions to be drawn with respect to both wage levels and wage structures in open shop construction. The biggest gap, however, will be in regard to the behavior of wages over time; we shall be able to say little about the determinants of wage movements in the open shop, especially as they might relate to settlements in the unionized sector.

The various sources of information available on employee compensation suggest three firm conclusions:

---

[1] U.S. Bureau of Labor Statistics, *Union Wages and Hours: Building Trades* (Washington, D.C.: Government Printing Office, published annually until 1982).

1. Average hourly wage rates in open shop construction are significantly lower than comparable union scales. It is sometimes asserted that the nonunion contractor must pay at or near the union rate in order to recruit the labor he needs, but the facts manifestly show otherwise.
2. Wages in open shop construction are appreciably more dispersed than in the union sector. Since most union workers are paid the negotiated wage, it is not surprising that there is more diversity of wages, both among and within employing units, for any given occupational group in the open shop sector. Somewhat less predictable perhaps is the observation that there also is a wider spread of wages between occupations, especially between the skilled and unskilled trades.
3. Supplementary employee benefits in the open shop are, on average, less common, lower than those in the union sector, and even lower than those in the labor force generally. In other words, straight-time pay (or even gross pay) accounts for a larger proportion of total compensation in nonunion construction. This conclusion does not appear to have been significantly affected by the development of various benefit plans by open shop contractors' associations, but then such plans are still in their infancy.

In this chapter we will summarize the results of several studies of the last twenty years which throw light on the level and structure of open shop wages and their relation to union scales. We will also compare some of these findings with those of a comprehensive survey of construction wages conducted by the BLS nearly four decades ago, for such a comparison may indicate whether the anatomy of wages in the industry has roots in the past.

THE BUREAU OF LABOR STATISTICS STUDY OF 1936

As noted in earlier chapters, this study covered over 13,000 projects by 5,450 contractors in 105 cities with populations of 10,000 or more. These contractors employed about 186,000 workers.[2] The study was limited to urban building construction, and therefore, the findings do not represent conditions in rural areas or in other subdivisions of the industry.

Table XI-1 summarizes the major findings of the study. In the interest of saving space, only eleven of the thirty-one occupational

---

[2] Edward P. Sanford, "Wage Rates and Hours of Labor in the Building Trades," *Monthly Labor Review*, Vol. 45 (August 1937), pp. 281–300.

TABLE XI-1

Average Hourly Union and Nonunion Wages in Selected Occupations in the Building Trades by Geographic Divisions, 1936

	United States			New England			Middle Atlantic			E.N. Central			W.N. Central		
Occupation	U[a]	NU[b]	Ratio	U	NU	Ratio	U	NU	Ratio	U	NU	Ratio	U	NU	Ratio
Bricklayers	$1.35	$0.97	.72	$1.25	$1.03	.82	$1.45	$0.94	.65	$1.34	$1.03	.77	$1.33	$1.02	.77
Carpenters	1.15	.81	.70	1.07	.80	.75	1.24	.85	.69	1.22	.80	.66	1.10	.77	.70
Cement finishers	1.22	.78	.64	1.25	.90	.72	1.30	.80	.62	1.21	.79	.65	1.18	.83	.70
Electricians	1.36	.86	.63	1.17	.90	.77	1.52	.89	.59	1.34	1.02	.76	1.26	.83	.66
Painters	1.09	.70	.64	1.00	.69	.69	1.14	.75	.66	1.25	.73	.58	1.11	.70	.63
Plumbers	1.28	.95	.74	1.21	.93	.77	1.36	.99	.73	1.31	1.00	.76	1.28	.92	.72
Sheet metal workers	1.20	.82	.68	1.25	.80	.64	1.30	1.01	.78	1.19	.82	.69	1.20	.85	.71
Structural-iron workers	1.40	1.11	.79	1.24	.76	.61	1.39	1.30	.94	1.35	1.09	.81	1.35	.88	.65
Hod carriers	.85	.58	.68	.76	.56	.74	.90	.60	.67	.80	.65	.81	.86	.56	.65
Helpers[c]	.78	.49	.68	.72	.51	.71	.80	.57	.71	.87	.52	.60	.75	.46	.61
Laborers	.63	.42	.67	.67	.49	.73	.57	.48	.84	.70	.49	.70	.74	.44	.59

TABLE XI-1 (continued)

Occupation	S. Atlantic			E.S. Central			W.S. Central			Mountain			Pacific		
	U	NU	Ratio	U	NU	Ratio	U	NU	Ratio	U	NU	Ratio	U	NU	Ratio
Bricklayers	$1.29	$0.95	.74	$1.21	$1.01	.83	$1.21	$0.89	.74	$1.32	$0.92	.70	$1.41	$0.93	.66
Carpenters	1.15	.76	.66	.90	.71	.79	.98	.75	.77	1.19	.80	.67	1.09	.92	.84
Cement finishers	1.24	.64	.52	1.01	.65	.64	1.01	.81	.80	1.17	.93	.79	1.16	1.01	.87
Electricians	1.36	.79	.58	1.04	.78	.75	1.12	.87	.78	1.21	.92	.76	1.24	.83	.67
Painters	1.05	.67	.64	.92	.65	.71	.96	.64	.67	1.16	.77	.66	.98	.80	.82
Plumbers	1.25	.87	.70	1.15	.93	.81	1.25	.99	.79	1.28	1.00	.78	1.21	1.05	.87
Sheet metal workers	1.25	.77	.62	.94	.77	.82	1.25	.83	.66	1.14	.71	.62	1.10	.80	.73
Structural-iron workers	1.57	.82	.52	1.12	.91	.81	1.01	.97	.96	1.22	1.01	.83	1.29	1.07	.83
Hod carriers	.60	.46	.77	.58	.43	.74	.55	.40	.73	.86	.61	.71	1.04	.61	.59
Helpers[c]	.59	.44	.75	.47	.41	.87	.46	.43	.93	.67	.51	.76	.70	.55	.79
Laborers	.56	.37	.66	.35	.34	.97	.36	.36	1.00	.51	.44	.86	.65	.57	.88

Source: Edward P. Sanford, "Wage Rates and Hours of Labor in the Building Trades," *Monthly Labor Review*, Vol. 45 (August, 1937), pp. 281-300.

[a] Union.

[b] Nonunion.

[c] Not elsewhere classified.

groups covered are included in the table. In the aggregate, it appears that union wages were approximately one-half as high as nonunion wages. Unfortunately, however, the detailed breakdowns do not reveal any systematic pattern of relationships across geographic areas. For example, such highly unionized areas as the Middle Atlantic, Pacific, and East North Central do not regularly show wider or narrower wage differentials than such sparsely unionized areas as the East South Central and West South Central. Nor are the data unambiguous when viewed across occupations. Relatively high differentials appear for such heavily organized occupations as electricians and cement finishers, as well as for occupations with large nonunion components, such as painters, helpers, and laborers. On the whole, however, there is a slight tendency for the most tightly unionized occupations to depict lower union-nonunion differentials. This tendency becomes significantly sharper if electricians—who were then the most unionized and have the highest differential—are excluded from the table. As we shall see, however, this relationship does not appear in the findings of later studies, which tend to reveal wider differentials where unions are strong.

BUREAU OF LABOR STATISTICS STUDIES 1965–1980

Since 1965, the BLS has published eight studies dealing specifically with construction wages which make some distinction between the union and open shop sectors. Three of these studies cover each of the three major divisions of the industry: general building contractors, heavy and highway contractors, and special trades contractors. The data were obtained directly from employers. The fourth study covers the entire workforce, and construction is reported as one unit. The data for that study were obtained from workers through supplementary questions asked during the March 1971 Current Population Survey. Similar studies were made in 1977 and 1980. In 1971–1972, 1973, and 1977, the BLS studied compensation in construction in large metropolitan areas, breaking down the results by crafts, by industry sectors, and by union and nonunion employees. All these studies are reviewed in this section.

General Building Contractors (1965)[3]

This study was based on a sample of 409 companies employing about 69,000 workers. A distinction was drawn between employers

[3] U.S. Bureau of Labor Statistics, *Compensation in the Construction Industry,* Bulletin 1656 (Washington, D.C.: Government Printing Office, 1970), Appendix A.

with a majority of workers covered by bargaining agreements and those with none or a minority of workers covered by such agreements.[4] Based on this study, open shop workers were found to average less than 63 percent of the total compensation of union employees: $3.10 compared to $4.94. There is no breakdown, however, by craft or geographic region, nor is there an estimate of the relative annual earnings of the two groups.

The study provided further insights into the compensation picture of the open shop worker. Most of his remuneration took the form of wages: over 89 percent of total compensation was straight-time pay. The comparable figure for union workers was 84 percent. The most conspicuous difference was in the area of private welfare plans, particularly health insurance and pensions. Clearly, the benefit programs recently initiated by various open shop contractors' associations have responded to a real gap in the compensation practices of individual employers.

Finally, the study demonstrates that there is a considerable range of wage payments within the open shop sector. The nonunion employer at the 75th percentile offered a wage more than one-half as great as his counterpart at the 25th percentile. In other words, the middle half of the open shop sample ranged from $2.15 an hour to $3.26. The comparable figures for the union sector were $3.90 to $5.06, or a difference of less than one-third. The numbers here suggest considerably greater flexibility for open shop contractors in wage setting.

Heavy Construction Industry (1971)[5]

The heavy construction study covered 1,375 firms with a total payroll of about 159,000 manual workers. These numbers represented about 9 percent of all establishments and 22 percent of all workers in the industry division. A little more than one-half of all workers were employed in union establishments.

This study paints much the same picture as the previous study with one exception: it found the spread between union and open shop compensation to be even larger. The results are shown in Table XI-2. Open shop workers averaged only about 52 percent of total compensation paid in the union sector, with nonunion highway and

[4]All information on union and nonunion differences is contained in a single table. See *ibid.*, Table A-3, pp. 85–86.

[5]U.S. Bureau of Labor Statistics, *Employee Compensation and Payroll Hours, Heavy Construction Industry, 1971,* Report 428 (Washington, D.C.: Government Printing Office, 1974).

TABLE XI-2
Construction Worker Compensation in Union and Nonunion Establishments Heavy Construction, General Contractors, 1971

Compensation Item	Heavy Construction		Heavy Construction except Highway		Highway	
	Union	Non-union	Union	Non-union	Union	Non-union
Total	$8.06	$4.13	$8.17	$4.35	$7.92	$3.88
Pay for work time	6.61	3.62	6.73	3.80	6.49	3.41
Straight-time pay	6.30	3.42	6.44	3.62	6.15	3.19
Premium pay	.30	.20	.24	.17	.35	.23
Pay for leave time	.19	.05	.20	.06	.17	.04
Retirement	.62	.21	.62	.22	.63	.19
Life and health insurance	.52	.18	.54	.21	.49	.16
Unemployment benefits	.10	.05	.10	.05	.11	.05
Other	.02	.02	.02	.02	.02	.03

Source: U.S. Bureau of Labor Statistics, *Employee Compensation and Payroll Hours, Heavy Construction Industry, 1971,* Report 428 (Washington, D.C.: Government Printing Office, 1974), Tables 1.6, 2.11, and 2.12.
Note: Because of rounding, sums of individual items may not equal totals.

street contractors paying less than one-half of the comparable union rates. The differences are slightly less when only pay for work time is considered, since a larger proportion of compensation in the union sector takes the form of fringe benefits. For the division as a whole, work-time pay is 82 percent in the union sector and nearly 88 percent in the open shop sector.

As the authors of the study conclude, the compensation "gap results from a combination of factors including differences in skill and occupational mix of the workforce, type of work, and geographic location, as well as the influence of unions on compensation levels."[6] Unfortunately, the study offers no clues as to the degree of importance of these various factors, nor does it examine the effect of intermittent employment on differences in compensation between open shop and union workers.

It is not clear, however, whether these data represent an industrywide change between 1965 and 1971 or a historical difference between the building and heavy/highway segments of the industry. Given the explosive wage settlements in union construction between 1969 and 1971, it is a plausible inference that the union-nonunion

[6] *Ibid.*, p. 2.

differential widened appreciably throughout the industry between the time of the two studies.

Special Trades Contractors (1969)[7]

This study was based on a sample of 41,000 workers in 823 establishments, or less than one percent of all firms and about 2.5 percent of all employees in the industry division. The report contains summary material on the specialty contractors as a whole, with separate detail on each of its three largest specialty contractor components: plumbing and heating, electrical, and masonry. As Table XI-3 shows, the overall findings are much the same as those of the previous studies: open shop workers average 57 percent of the union rates. If, however, one looks at the three reports chronologically, there is an unmistakable trend toward a wider union-nonunion differential and a growing emphasis on fringe benefits in both sectors. The table also shows a striking difference between plumbing and heating contractors and electrical contractors, on the one hand, and masonry contractors on the other; open shop masons appear to fare much better in relation to their unionized counterparts. Although the difference may be in part attributable to sampling error (the sample included only 823 firms in all specialties), it may also reflect variations in the extent of union representation among the various specialties. In addition, open shop establishments paid out 89 percent of total compensation in the form of pay for working time, compared to 85 percent in unionized companies. Again, however, none of these relationships is controlled for firm size or location, so that the determinants of the wage structure cannot be readily discerned.

Construction Industry (1970)[8]

This study reports on the employment and earnings profile of the United States labor force in 1970. Unfortunately, most of the earnings data are presented in terms of "year-round full-time wage and salary workers," so that the relative effect of intermittent employment in the union and nonunion sectors cannot be readily meas-

[7] U.S. Bureau of Labor Statistics, *Employee Compensation and Payroll Hours, Construction—Special Trade Contractors, 1969,* Report 413 (Washington, D.C.: Government Printing Office, 1972).

[8] U.S. Bureau of Labor Statistics, *Selected Earnings and Demographic Characteristics of Union Members, 1970,* Report 417 (Washington, D.C.: Government Printing Office, 1972).

TABLE XI-3
Construction Worker Compensation by Bargaining Agreement Coverage Construction—Special Trade Contractors, 1969

Compensation Item	Special Trades		Plumbing, Heating Air Conditioning		Electrical		Masonry	
	Union	Non-union	Union	Non-union	Union	Non-union	Union	Non-union
Total	$6.54	$3.71	$6.82	$3.62	$6.87	$3.51	$6.07	$4.28
Pay for work time	5.59	3.30	5.77	3.19	5.95	3.16	5.22	3.86
Straight-time pay	5.41	3.20	5.60	3.09	5.67	3.05	5.13	3.78
Premium pay	.18	.10	.17	.10	.28	.11	.09	.07
Pay for leave time	.12	.06	.14	.08	.16	.07	.06	.04
Retirement	.43	.16	.49	.16	.43	.16	.41	.18
Life and health insurance	.33	.13	.35	.13	.26	.08	.30	.13
Unemployment benefits	.06	.04	.06	.03	.05	.03	.07	.05
Other	.01	.03	.01	.02	.01	.01	.01	.01

Source: U.S. Bureau of Labor Statistics, *Employee Compensation and Payroll Hours, Construction—Special Trade Contractors, 1969,* Report 413 (Washington, D.C.: Government Printing Office, 1972), Tables 2, A-2, B-2, and C-2.
Note: Because of rounding, sums of individual items may not equal totals.

ured.[9] The data on year-round full-time workers, however, may be regarded as reasonably accurate reflections of wage rates.

Table XI-4 summarizes the pertinent results of the study. Again, the average wage of union workers was 50 percent greater than that of open shop workers. The occupational and regional detail, not available in the establishment studies, offers further insight into the wage structure of the industry. Specifically, there is a marked difference in the union-nonunion ratio between skilled and unskilled workers and between the South and the rest of the nation. Since it is clear that laborers and the South both have lower levels

[9] One exception is a table of median earnings by industry and region which includes data on all workers. Such a statistic is problematic with reference to construction, however, because summer employees in construction work more hours than in any other industry and thus distort the averages. There is some evidence that these summer workers are more numerous in the open shop sector. This would render the union-open shop comparison misleading. See Howard G. Foster, *Manpower in Homebuilding: A Preliminary Study,* Manpower and Human Resources Studies, No. 3 (Philadelphia: Industrial Research Unit, The Wharton School, University of Pennsylvania, 1974), and Chapter IX, above. For the record, the figures show substantially higher differentials for all workers than for year-round full-time workers. See *Selected Earnings, 1970,* Table 10. For more on this point, see Chapter IX, above.

TABLE XI-4
Median Earnings of Male Year-Round Full-Time Wage and Salary Workers in Construction by Occupation and Region, 1970

Industry, Occupation Region	Union	Nonunion	Ratio
Construction industry	$11,244	$7,619	.68
Occupation: carpenters[a]	10,274	6,897	.67
Other craftsmen[a]	11,223	7,826	.70
Laborers[a]	8,693	5,128	.59
Region: Northeast	11,226	9,185	.82
North Central	11,472	8,554	.75
South	10,089	6,346	.63
West	11,884	9,629	.81

Source: U.S. Bureau of Labor Statistics, *Selected Earnings and Demographic Characteristics of Union Members, 1970,* Report 417 (Washington, D.C.: Government Printing Office, 1972), Tables 6, 7, and 10.
[a] Includes construction workers employed outside the construction industry.

of unionization in construction, it would appear that the presence of a sizable open shop component tends to hold down nonunion wages relative to union scales.

Finally, the study again demonstrates the greater dispersion of wages among open shops. Nearly 57 percent of male year-round workers in open shops earned either less than $7,000 or more than $15,000 in 1970, while 43 percent earned between those amounts. By contrast, only 29 percent of the union workers were at the extreme of the distribution, while 71 percent earned between $7,000 and $15,000. Put another way, less than one-half of the nonunion workers fell within 30 percent of the median for that group, whereas the comparable proportion for union workers was about three-fifths.

Compensation in Construction in Metropolitan Areas (1971–1972)

In April and May 1971, the BLS conducted a pilot study of wages and benefits in five labor market areas: Biloxi, Chicago, Des Moines, Houston, and Washington. Based on this pilot survey, the bureau "launched its first occupational wage survey of union and nonunion construction, in September 1972, in 21 areas. The construction surveys resulting are now part of the BLS regular industry wage survey program."[10] A number of such surveys followed but the last one was taken in 1976–1977. Each of these is briefly reviewed here

[10] U.S. Bureau of Labor Statistics, *Summary Tabulations, Earnings and Supplementary Benefits in the Construction Industries—A Pilot Study, April–May 1971* (Washington, D.C.: Government Printing Office, 1974), p. 1.

beginning with the first in 1972. Although the pilot study was limited to employees in contract construction, the first study in this series also included operative builders, those who erect buildings—mostly residential—for sale on their own account. Most of the major crafts were also included in this study.[11]

Open shop firms, unfortunately, were underrepresented in the sample of firms surveyed, since a lesser proportion of smaller establishments than larger companies were included. As a result, the nonunion data base for some trades as a whole, and for other trades in several of the labor market areas, were not sufficiently large to meet BLS publication standards.[12] Thus, the study offers no comparative data for bricklayers, pipefitters, roofers, and structural-iron workers, and only limited data for cement masons, electricians, plumbers, sheet metal workers, and equipment operators. Similarly, in heavily unionized areas like Chicago, San Francisco, Portland, and St. Louis, the BLS was unable to generate comparative wage rates for any of the trades surveyed. Despite the spottiness of the data, however, certain conclusions emerge, the thrust of which is largely consistent with the earlier studies described above. Average hourly earnings of workers covered by the survey are summarized in Table XI-5.

Table XI-5 depicts the familiar wage spread in different regions of the country, with the relatively low rates in the South as the most prominent feature. The table also reveals, however, a somewhat smaller range of union-nonunion differentials than those observed in earlier studies. It is especially interesting to compare these figures with those of the 1936 BLS survey (see Table XI-1). In both cases, a union rate of about 1.5 times the nonunion rate appears to be the norm. A further noteworthy observation is that the differential shows no systematic variation by level of skill. Laborers, for example, appear to exhibit differentials of much the same magnitude as carpenters, electricians, and plumbers, but the absence of small firms from the sample could account for this.

Union and open shop wages differ not only in respect to their levels, but also their relative dispersion. Union rates tend to be fixed at a single figure, with only minor variations around it. Even

---

[11] U.S. Bureau of Labor Statistics, *Industry Wage Survey: Contract Construction, September 1972,* Bulletin 1853 (Washington, D.C.: Government Printing Office, 1975). This report was also summarized in Martin E. Personick, "Union and Nonunion Pay Patterns in Construction," *Monthly Labor Review,* Vol. 97 (August 1974), pp. 71–75.

[12] In some cases, mainly in southern cities, the data base for unionized workers was also too small for comparative analysis.

TABLE XI-5
Average Hourly Earnings of Workers in Construction Industries by Selected Occupations and Areas September 1972

Occupation and union status	Northeast					South					
	Boston	Buffalo	Hartford	New York[1]	Phila-delphia	Atlanta	Biloxi[2]	Dallas	Memphis	Miami	Wash-ington
Journeymen											
Bricklayers—total[3]	$8.13	$8.66	$8.56	$8.41	$8.41	$7.77	$5.23	$7.16	....	$8.01	$7.87
Union	840	8.66	8.71	8.41	8.72	7.80	......	7.38	....	8.10	8.32
Carpenters—total	7.91	7.83	7.08	8.41	8.02	6.12	5.19	5.82	$5.72	7.77	6.49
Union	8.09	8.11	8.12	8.58	8.65	7.40	6.08	6.62	6.85	7.94	7.76
Nonunion	6.46	5.03	4.41	7.49	5.24	5.03	4.23	4.91	4.21	5.75	5.27
Cement masons—total	8.23	8.89	8.76	7.89	7.16	6.53	5.11	4.87	5.49	7.46	6.00
Union	8.63	9.04	8.76	8.09	7.52	6.95	6.02	6.50	6.57	7.71	7.52
Nonunion	......	....	....	.....	5.37	3.61	4.29	4.41	4.88	5.58	5.18
Electricians—total	8.08	9.71	7.75	8.48	8.35	8.20	6.05	5.40	6.74	8.33	7.83
Union	8.58	9.71	8.72	8.49	9.30	8.67	6.37	7.40	6.83	8.50	8.72
Nonunion	5.07	....	5.32	.....	5.86	....	4.15	4.49	....	7.00	5.64
Pipefitters—total[3]	9.01	9.30	8.31	8.07	8.69	7.57	......	6.88	7.21	9.16	8.59
Union	9.01	9.30	8.79	8.07	8.98	7.76	.....	7.20	7.17	9.16	8.87
Plumbers—total	8.23	9.14	6.84	8.11	7.31	7.34	4.50	5.62	7.18	8.39	6.89
Union	9.35	9.20	8.65	8.43	8.98	....	......	......	7.44	9.16	8.76

Nonunion	7.11	......	5.52	5.40	5.36	......	3.99	5.09	......	6.23	5.36
Roofers—total	8.00	......	8.01	7.46	......	5.39	......	4.21	......	6.61	6.24
Union	8.00	......	8.01	8.30	......	5.65	......	......	......	7.77	......
Sheet-metal workers—total	7.85	8.50	8.70	9.81	9.16	4.61	5.88	6.06	6.87	8.67	6.73
Union	8.68	8.50	8.70	10.33	9.70	......	......	......	7.22	9.20	8.05
Nonunion	......	......	......	......	5.68	3.91	......	3.81	......	......	4.99
Structural-iron workers—total[3]	8.13	8.71	9.30	9.25	8.60	6.80	......	......	......	8.25	8.28
Equipment Operators											
Back-hoe operators—total	8.11	8.58	7.20	9.16	8.67	4.90	5.07	3.58	4.15	6.25	6.20
Union	8.71	8.59	8.01	9.14	9.37	7.03	......	......	6.08	7.22	6.91
Nonunion	......	......	6.60	9.54	......	4.10	3.98	3.29	3.80	......	5.75
Bulldozer operators—total	7.96	6.26	6.24	8.77	8.39	4.86	4.69	3.54	3.96	5.49	5.87
Union	8.16	8.53	6.83	8.98	8.39	......	6.55	......	5.79	6.03	7.01
Nonunion	......	......	5.62	......	......	3.85	3.71	3.54	3.46	4.11	4.93
Truckdrivers—total	6.08	7.36	5.06	6.61	5.41	3.06	2.58	2.76	3.11	3.39	4.05
Union	......	7.19	5.61	6.61	5.59	......	......	......	......	......	4.29
Nonunion	......	......	4.11	......	4.04	2.97	2.34	2.76	2.76	3.21	3.68
Helpers and Laborers											
Bricklayers' helpers—total[3]	6.47	......	6.25	7.42	6.17	......	......	4.04	......	......	4.37
Carpenters' helpers—total[3]	3.66	4.92	3.59	3.84	3.63	3.80	2.68	3.91	......	4.21	4.02
Construction laborers—total	6.14	6.38	5.73	6.86	5.81	3.54	2.80	3.25	3.21	5.31	4.33
Union	6.36	6.43	6.39	7.04	6.11	4.49	3.92	4.64	3.97	5.74	5.66
Nonunion	5.04	......	4.57	4.97	3.74	3.00	2.52	2.62	2.42	3.47	3.47
Electricians' helpers—total[3]	......	......	......	......	3.62	......	2.77	2.87	......	4.21	3.39
Plumbers' helpers—total[3]	3.24	......	2.81	3.58	......	......	2.72	3.10	......	3.23	4.44

TABLE XI-5 (continued)

Occupation and union status	North Central						West			
	Chicago	Des Moines	Indian-apolis	Kansas City	Minne-apolis	St. Louis	Denver	Los Angeles[2]	Portland	San Francisco-Oakland
Journeymen										
Bricklayers—total[3]	$8.95	....	$7.87	$7.73	$7.61	$7.81	$8.21	$7.78	....	$8.50
Union	8.95	....	8.55	7.73	7.61	7.81	8.25	7.78	....	8.50
Carpenters—total	8.32	$6.18	7.69	8.00	7.12	7.79	6.29	6.76	$6.78	8.10
Union	8.32	7.01	8.17	8.00	7.13	7.79	6.57	6.76	6.78	8.10
Nonunion	....	....	5.77	....	....	.....	4.81	....	....	....
Cement masons—total	8.88	6.80	6.82	8.14	7.66	7.24	6.43	6.16	6.81	7.18
Union	8.89	6.80	7.07	8.14	7.66	7.24	6.43	6.16	6.81	7.18
Nonunion	8.60	....	....	....	....	.....	6.41	....	....	....
Electricians—total	8.98	7.47	7.97	7.81	7.99	7.85	7.64	9.09	7.50	8.03
Union	9.05	7.47	8.20	7.87	8.00	7.85	8.04	9.09	7.50	8.03
Nonunion	6.63	....	....	....	....	.....	5.68	....	....	....
Pipefitters—total[3]	8.98	....	7.15	8.62	7.35	7.81	7.70	8.31	6.61	8.43
Union	8.98	....	7.90	8.62	7.35	7.81	7.70	8.31	6.61	8.43
Plumbers—total	8.69	7.05	5.80	8.71	7.34	8.18	7.50	8.19	6.61	8.29
Union	8.75	7.09	8.15	8.71	7.32	8.18	7.70	8.34	6.61	8.29
Nonunion	....	....	4.59	....	....	.....	....	7.35	....	....
Roofers—total	8.62	....	....	....	....	.....	....	7.02	....	7.92
Union	8.62	....	....	....	....	.....	....	7.09	....	7.92
Sheet-metal workers—total	8.54	7.50	6.10	8.70	7.78	7.95	8.17	8.80	6.83	8.11
Union	8.54	7.50	7.93	8.70	7.78	7.95	8.17	8.80	6.83	8.11
Nonunion	....	....	5.14	....	....	.....	....	....	....	....
Structural-iron workers—total[3]	9.30	6.98	8.25	8.50	....	8.05	7.25	8.58	7.31	8.57
Equipment Operators										
Back-hoe operators—total	8.68	5.91	7.01	8.50	7.57	7.86	5.64	7.87	6.86	8.59
Union	8.68	6.28	7.48	8.50	7.57	8.08	5.73	7.87	....	8.59

Nonunion	...	...	...	...	...	...	5.50	...	...	...
Bulldozer operators—total	8.52	5.94	7.29	8.48	7.38	8.06	5.56	7.89	...	8.27
Union	8.52	6.13	7.57	8.48	7.38	8.06	5.59	7.89	...	8.27
Nonunion	...	...	...	...	...	...	...	...	...	...
Truckdrivers—total	5.98	5.10	4.94	7.07	6.53	6.67	4.65	6.34	6.18	6.89
Union	6.01	5.10	4.95	7.07	6.56	6.67	4.93	6.34	...	6.89
Nonunion	...	...	...	...	...	...	...	...	...	...
Helpers and Laborers										
Bricklayers' helpers—total[3]	...	...	...	5.93	6.05	7.23	4.74	6.00	...	...
Carpenters' helpers—total[3]	...	...	...	...	...	...	4.19	...	...	...
Construction laborers—total	6.21	5.12	4.82	6.24	5.91	6.88	4.23	5.50	5.01	5.47
Union	6.22	5.73	5.51	6.23	5.93	6.89	4.36	5.50	5.10	5.47
Nonunion	5.85	3.27	3.74	...	5.46	...	3.41	...	...	...
Electricians' helpers—total[3]	...	...	...	...	...	...	3.31	...	...	...
Plumbers' helpers—total[3]	...	4.94	...	5.78	...	7.32	3.71	...	...	...

Source: Reproduced from Martin E. Personick, "Union and Nonunion Pay Patterns in Construction," *Monthly Labor Review,* Vol. 97 (August 1974), pp. 72–73.

[1] The survey reference month was October 1972.

[2] Shortened terms for Biloxi-Gulfport and Pascagoula area and combined SMSA's of Los Angeles-Long Beach and Anaheim-Santa Ana-Garden Grove area.

[3] Insufficient published data to warrant separate presentation of union and/or nonunion rates.

NOTE: Dashes indicate no data reported or data that do not meet publication criteria. Average hourly earnings exclude premium pay for overtime and hazardous work and for work on weekends, holidays, and late shifts. Zone rates (usually based on distance between local union headquarters and the construction site) are included in straight-time rates for purposes of this survey.

The survey covered establishments employing 8 workers or more and engaged primarily in construction, i.e., building construction by general contractors; construction other than building by general contractors; construction by selected special trades contractors; and construction by operative builders, those building for sale on their own account. (Industry Groups 15, 16, part of 17, and 656 as defined in the 1967 edition of the Standard Industrial Classification Manual, prepared by the U.S. Office of Management and Budget.) Specifically excluded were special trades contractors primarily engaged in painting, paper hanging, and decorating; plastering and lathing; terrazzo, tile, marble, and mosaic work; floor laying; water well drilling; ornamental iron work; glass and glazing work; and special trades contractors not classified separately in the Manual.

All areas studied except Biloxi were Standard Metropolitan Statistical Areas, as defined by the Office of Management and Budget through November 1971.

these variations, moreover, tend to represent differences among local unions within a metropolitan area, rather than among workers within a single employing unit. Nonunion rates, on the other hand, vary widely within a given area and even within the same employing unit.

The September 1973 Survey—Seventeen Areas[13]

This study closely followed the format of the 1972 study, although it covered four fewer areas.[14] The union-nonunion wage differentials within areas remained substantial, typically ranging from 35 to 55 percent for carpenters, 40 to 60 percent for cement masons and plumbers, and 45 to 70 percent for construction laborers. As in the case of the 1972 survey, the union-nonunion differential also reflected differentials between the residential sector, which was found (as noted in Chapter III) to be heavily nonunion, and the commercial sector, which in 1972 and 1973 was much more unionized. Moreover, the nonunion areas surveyed were concentrated in southern and border states, and therefore, the union-nonunion comparisons also reflect a comparison of the lower-wage South with the higher-wage cities of the Northeast, Midwest, and Pacific states. Comparing just union or nonunion residential and commercial earnings, however, the differences between the two sectors are not so large. Table XI-6 makes such a comparison for carpenters and construction laborers in four southern and border cities—the only areas in which enough data on open shop operations were gathered for comparison purposes. The differentials between residential and commercial earnings are larger in all cities for the union sector than for the nonunion one, perhaps because the open shop commercial work involved smaller jobs closer to areas where residential work was under way.

The 1973 survey did not develop sufficient nonunion data for most crafts in the bulk of the cities surveyed to meet the BLS publication standards, because small firms (under eight employees) were excluded from the sample. A large portion of the nonunion data collected applied to residential construction. Nevertheless, the same trends of earnings dispersion in the nonunion sector, and earnings concentration in the union one were observed, as well as the aforementioned union-nonunion differentials.

---

[13] U.S. Bureau of Labor Statistics, *Industry Wage Survey: Contract Construction, September 1973,* Bulletin 1911 (Washington, D.C.: Government Printing Office, 1976).

[14] The city areas covered were Atlanta, Baltimore, Boston, Chicago, Cleveland, Dallas, Denver, Detroit, Los Angeles-Long Beach, Miami, Nassau-Suffolk, New York; Newark, New Jersey; New York City, Philadelphia, San Francisco-Oakland, Seattle-Everett, and Washington, D.C.

TABLE XI-6

Workers and Average Hourly Earnings by Union Status and Type of Construction Selected Occupations and Areas, September 1973

Type of Construction	Carpenters				Construction laborers			
	Atlanta	Baltimore	Dallas	Washington	Atlanta	Baltimore	Dallas	Washington
	Union and Nonunion combined							
Commercial:								
Workers	2,528	1,636	1,857	3,130	5,232	3,308	2,547	6,341
Earnings	$6.52	$7.14	$6.93	$7.44	$4.30	$4.67	$4.59	$5.49
Residential, under 5 stories:								
Workers	767	388	798	2,315	610	720	894	3,237
Earnings	$5.33	$5.18	$5.04	$5.70	$3.27	$3.44	$2.82	$3.74
Commercial as a percent of residential:								
Earnings	122	138	138	131	131	136	163	147
	Nonunion							
Commercial:								
Workers	1,412	936	530	1,108	2,280	1,960	808	2,156
Earnings	$5.63	$6.57	$5.78	$6.22	$3.39	$4.00	$3.56	$4.07
Residential, under 5 stories:								
Workers	704	388	792	2,315	610	720	882	2,987
Earnings	$5.12	$5.18	$5.04	$5.70	$3.27	$3.44	$2.81	$3.65
Commercial as a percent of residential:								
Earnings	110	127	115	109	104	116	127	112

Source: U.S. Bureau of Labor Statistics, *Industry Wage Survey: Contract Construction September 1973,* Bulletin 1911 (Washington, D.C.: Government Printing Office, 1976), p. 6.

Seventeen Large Metropolitan Areas, 1977[15]

This 1977 survey covered contract construction firms and operative builders (as defined in previous surveys) employing eight workers or more, in the same seventeen cities as the 1973 survey. Again, this had the effect of eliminating small companies, especially those in residential construction, who are most likely to be nonunion. Specialty contractors employed 55 percent of the nonsupervisory construction workers covered by the survey; heavy construction contractors, 24 percent; and general building contractors and operative builders combined, 21 percent. Seventeen different areas and twenty-eight occupational classifications were examined in this survey. Of these areas, only Baltimore, Washington, Denver-Boulder, and the South could be classified as open shop areas, except in residential construction. Residential was also then heavily unionized in New York, Philadelphia, Newark, Chicago, Cleveland, Detroit, San Francisco-Oakland, Seattle-Everett, and Los Angeles-Long Beach. A comparison of the degree of unionization found in cities surveyed by the BLS in 1973 and 1977 is found in Table IV-1. It shows a considerable decline in union strength during this period.

As in all previous studies, unionized workers enjoyed a substantial, although widely varying wage advantage over their nonunion counterparts. The wage advantage of union workers averaged from 22 percent for carpenters and plumbers to 89 percent for truck drivers in all areas. For carpenters and laborers—the two largest trades studied—union wages exceeded nonunion wages by 51 percent and 83 percent, respectively. Since the nonunion population was heavily concentrated in the South, this differential was obviously affected by regional considerations. Moreover, the differentials also reflect the pay practices of the various construction sectors. Surprisingly, however, the low-rise residential sector was not always the one with the lowest earnings, especially if the comparison is confined to the union and nonunion segments.

Table XI-7 sets forth union earnings as a percentage of nonunion earnings for ten crafts in the areas where sufficient information was available. Table XI-8 shows average hourly earnings by union status and type of construction for five trades in the seventeen areas. Not shown in Table XI-8 is the usual wide dispersion of open shop rates as compared with the concentration of union rates.

---

[15] U.S. Bureau of Labor Statistics, *Compensation in the Construction Industries in Large Metropolitan Areas,* Report No. 61 (Springfield, Va.: National Technical Information Service, 1980).

TABLE XI-7
Unionized Workers' Wages as a Percent of Nonunion Workers' Wages, Selected Crafts Seventeen Large Metropolitan Areas March 1977

Trade	Union Pay as a Percent of Nonunion Pay: Median	Union Pay as a Percent of Nonunion Pay: Range Among Areas Included
Bricklayers (6 areas)	126	111-138
Carpenters (14 areas)	144	110-153
Cement masons (8 areas)	144	129-181
Electricians (12 areas)	156	131-176
Painters (13 areas)	142	111-188
Plumbers (11 areas)	150	136-183
Roofers (11 areas)	141	124-196
Sheet metal workers (12 areas)	154	132-192
Truckdrivers (9 areas)	142	98-155
Construction laborers (17 areas)	143	115-174

Source: U.S. Bureau of Labor Statistics, *Compensation in the Construction Industries in Large Metropolitan Areas,* Report No. 610 (Springfield, Va.: National Technical Information Service, 1980), p. 4.

Characteristics of Union Represented Workers, 1977 and 1980

In 1977 and 1980, special questions inserted in the CPS sought information regarding labor union representation and the wages of union-represented and nonrepresented workers. The results for construction workers are set forth in part in Table XI-9. These data show unionized construction workers receiving weekly earnings approximately 50 percent higher than nonunion workers. The dispersion, not shown in Table XI-9, was quite different for the two groups. For example, the modal weekly range for organized construction workers in 1977 was $350 or more; for 1980, it was $400 or more. The 1977 modal range for the nonunion construction workers was $150-$199 and the 1980 range, $250-$299. In both years, the union population was much more concentrated in one or two ranges than was the nonunion one. For example, in 1977, 65.8 percent of the unionized construction workers had earnings of $300 per week or more; in 1980, 52.2 percent earned $400 or more, and 63.5 percent, $350 or more. In 1977, 18.3 percent of nonunionized construction workers earned between $100-$149 per week; 23.5 percent between $150-$199; 22.5 percent between $200-$249; 11.8 percent between $250-$259, and 12.6 percent between $300-$349. For 1980, the dispersion of nonunion earnings was even wider: 18.5

TABLE XI-8
Number of Workers and Average Hourly Earnings in Selected Crafts, Construction Industry, by Industry Segment Seventeen Large Metropolitan Areas[a]
March 1977

Craft and Industry Sector	Union Workers		Nonunion Workers	
	No. of Workers[b]	Average Hourly Earnings	No. of Workers	Average Hourly Earnings
		$		$
Carpenters	35,680	10.27	12,436	6.82
Commercial	20,252	10.12	6,583	7.07
Residential (5 stories or more)	1,137	10.35	526	5.96
Residential (under 5 stories)	9,833	10.55	4,565	6.56
Street and highway	682	10.11	325	6.03
Other heavy[c]	3,776	10.36	437	7.30
Electricians	26,990	11.20	5,055	6.46
Commercial	25,164	11.20	2,916	7.02
Residential (5 stories or more)	99	10.91	167	8.54
Residential (under 5 stories)	1,727	11.22	1,972	6.22

Plumbers	13,650	10.84	7,781	6.90
Commercial	9,620	10.86	2,327	7.02
Residential (5 stories or more)	517	10.19	385	5.96
Residential (under 5 stories)	3,513	10.86	5,069	6.92
Sheet metal workers	12,178	10.98	3,608	6.31
Commercial	11,016	10.96	1,870	6.56
Residential (under 5 stories)	1,105	11.19	1,555	6.08
Construction laborers	50,482	7.96	24,326	4.35
Commercial	20,646	7.93	9,226	4.54
Residential (5 stories or more)	989	8.12	346	3.47
Residential (under 5 stories)	3,847	8.12	6,372	4.27
Street and highway	10,409	7.84	3,801	3.85
Other heavy[c]	14,591	8.03	4,581	4.59

Source: U.S. Bureau of Labor Statistics, *Compensation in the Construction Industries in Large Metropolitan Areas,* Report 610 (Springfield, Va.: National Technical Information Service, 1980), Tables 2 and 3.

[a] For metropolitan areas involved, see Text Footnote 14.

[b] Totals in a craft include those in segments not presented in table because of small numbers.

[c] Definition of "heavy" includes industrial construction as defined in Chapter V.

TABLE XI-9
Average Weekly Earnings for Full-Time Wage and Salary Workers Construction Industry, by Union Representation May 1977 and May 1980

	Average Weekly Earnings[a]					
	1977			1980		
	Union	Nonunion	Earnings Ratio[b]	Union	Nonunion	Earnings Ratio[b]
	$	$		$	$	
Construction industry	343	230	1.49	405	284	1.42
Carpenters	336	217	1.55	—[d]	252	—[d]
Construction craft workers except carpenters[c]	338	222	1.52	409	278	1.47

Source: U.S. Bureau of Labor Statistics, *Earnings and Other Characteristics of Organized Workers, May 1977,* Report 556 (Washington, D.C.: Government Printing Office, 1979), Tables 10 and 11; *Earnings and Other Characteristics of Organized Workers, May 1980,* Bulletin 2105 (Washington, D.C.: Government Printing Office, 1981), Tables 10 and 11.

[a] Mean.

[b] Ratio of earnings of workers covered by labor-management agreements to those not covered by agreements.

[c] May include non-construction workers.

[d] Base too small for calculation.

percent had weekly earnings either between $150-$199 or between $200-$249, 19.8 between $250-$299, and 15.5 percent earned $400 or more.[16] Thus, these studies clearly confirm the very different range dispersion of union and nonunion construction wages and earnings, as well as the tendency of union shop rates to exceed open shop rates by 50 percent or more. They also indicate some widening of the union-nonunion differential since 1972, but this may be attributable to the absence of small companies from the samples.

BLS Studies—Final Comments

The various studies of the BLS confirm significant differences between union and open shop wage structures in the construction industry. In the cities studied, very often the union rates were 50

[16] U.S. Bureau of Labor Statistics, *Earnings and Other Characteristics of Organized Workers, May 1977,* Report 556 (Washington, D.C.: Government Printing Office, 1979), p. 32, Table 11; and *ibid., Earnings and Other Characteristics of Organized Workers, May 1980,* Bulletin 2105 (Washington, D.C.: Government Printing Office, 1981), p. 31, Table 11.

percent or more above the open shop rates. Part of this differential is attributable to work in different sectors of the industry; a considerable amount may well be the result of different job structures, particularly in the wide use of various grades of helpers by the open shop and the quite successful attempts of unions to eliminate the helper category. Most important, however, is the fact that open shop contractors pay their workers on the basis of their ability and performance, while union contractors pay their workers a single rate regardless of their qualifications, experience, or performance. All the BLS studies showed a high concentration of union rates within each craft and a very wide dispersion of open shop rates. Not being covered by union personnel or pay practices, the open shop contractors can not only deploy personnel in the most advantageous manner, but they can pay personnel in accordance with their qualifications and position. The greater flexibility of open shop contractors in labor deployment and wage setting lowers the cost of building open shop and, consequently, provides open shop contractors with an enormous advantage in competing for work.

Another finding from these studies is that the differential in pay between open shop and unionized workers is very substantial at the unskilled level. Since World War II the Laborers' union has been able to narrow differentials between its members and craftsmen. Open shop contractors not only pay laborers considerably less than do their unionized counterparts, but they utilize laborers for a wide variety of jobs for which the union sector must use journeymen—for example, handling and unloading materials. Again, the savings for the open shop segment are substantial.

NONGOVERNMENTAL STUDIES

Individual scholars have also made a number of studies comparing union and open shop wage rates in construction.[17] The most recent and thorough of these studies is found in an unpublished paper by Professors Jeffrey M. Perloff and Robin C. Sickles.[18] Using an elaborate and carefully constructed econometric model and applying it to male craftsmen who listed their industry as construction in the May 1973, 1974, and 1975 CPS's, they concluded that the mean union wage rate exceeded the mean nonunion rate by 80.6 percent, that nonunion employees worked longer hours by 4.7 percent, and

[17] See Appendix E for a bibliography of these studies.

[18] Jeffrey M. Perloff and Robin C. Sickles, "FIML Estimation of Union Wage, Hours, and Earnings Differentials in the Construction Industry: A Nonlinear Limited Dependent Variable Approach," unpublished manuscript, March 1983.

that the mean earnings of union workers exceeded those of nonunion workers by 60.3 percent.[19] There was considerable variation among demographic subgroups. Experience and education were found to raise nonunion wages relatively and "simulations indicate that the union wage markup was smaller for non-whites than for whites."[20] There were also regional variations and large craft wage differences.

The Perloff-Sickles study found larger differentials than have previous investigations, which have been summarized by these authors in Table XI-10. Perloff and Sickles attribute these differences to their more advanced econometric techniques and to their ability to examine subgroups within their sample and account for the effect of "experience, race and other observable characteristics."[21] Moreover, by examining hours as well as wages, Perloff and Sickles were able to distinguish the earnings differential ratio from the wage differential and to point up the effect of longer hours worked by the open shop sector, which will be discussed later in this chapter.

Analysis and Update of Perloff-Sickles Results

The Perloff-Sickles estimate of a 60 percent wage and earnings differential between open shop and unionized construction workers may well have been too low for the period 1976 to 1984, even if one takes into consideration the many union concessions made between 1981 and 1984. For reasons that are not clear, these authors, writing in early 1983, used data that were eight to ten years old rather than more recent figures. If, however, they had been able to utilize later and more complete information, as presented herein above, they might have had even more surprising results. For during the late 1970s, union construction wages skyrocketed, as Figure XI-1 demonstrates.

Their paper also pays scant attention to the growing strength and policies of open shop contractors during the 1973-1975 period. The authors state, in relation to changes in the wage rate for this period: "Newspaper articles at the time suggested that nonunion workers were making inroads into the construction industry during this period and suggested that this competition had more of an effect than did cyclical fluctuations."[22] Our first edition, however, systematically researched this point for the period covered by their

---

[19] *Ibid.*, p. 27, Table 5.
[20] *Ibid.*, p. 23.
[21] *Ibid.*, pp. 26–31.
[22] See *ibid.*, p. 24.

data and found that, faced with growing competition from open shop contractors, union wage levels have actually risen dramatically.

Figure XI-1 presents data gathered by the Construction Labor Research Board (CLRB), showing that over this decade wages and fringes in the unionized portion of the construction industry increased about 125 percent for the country as a whole, almost 100 percent even for the very low unionized Southeast, and considerably more in other sectors, including the South Central, where the open shop is also very strong, and the Northwest, where the open shop has recently made great gains. Despite a rate increase of 4.6 percent in 1983, the smallest since the mid-1960s, increases during the five-year period prior to January 1, 1984, amounted to an average of $6.25 per hour or 47.5 percent, bringing the average hourly rate in the United States to $19.35.[23] Increases for all crafts during this five-year period were close to the 47.5 percent national average, with the highest paid trades receiving the largest dollar increases.

Since it has undoubtedly been the inflation of union wage costs which has led to the great expansion of open shop construction during the 1970s and early 1980s, it is reasonable to conclude that during this period, and particularly since 1975, the last year for which data were utilized in the Perloff-Sickles paper, the wage differential between open shop and union shop workers widened more rapidly. This is because, as we shall demonstrate in a later section of this chapter, open shop contractors generally pay for fewer fringe benefits and keep the costs of those that they do provide lower.

Perhaps more important than wage rates when considering costs are the deployment factors discussed in Chapter IX. An open shop contractor can readily pay the top union rate—which he rarely must do—and still have a huge cost advantage by utilizing one journeyman at that rate, plus a helper and two laborers at one-fourth to one-half that rate, while his union competitor is forced to employ four persons at the journeyman's rate. Given the wage, fringe, and deployment differentials, plus restrictive union work rules which limit productivity, claims, such as those of Professor Allen (discussed in Chapter II), that union construction is less expensive must be viewed with enormous skepticism.

Finally, the Perloff-Sickles finding that nonunion craftsmen work longer hours than union craftsmen and that this, in turn, must

---

[23] "1984 Construction Labor Rate Trends and Outlook," *Construction Labor Research Council,* February 1984, p. 2.

TABLE XI-10
Male Construction Craftsmen
Union Wage Differentials in Various Studies

Study	Estimation technique	Data	Time period	Occupation	Percentage markup		
					White	Nonwhite	All
Lewis based on Sobotka	OLS[f] and judgment	Macro—large cities	1939	Skilled; unskilled			25
BLS Bulletin	Survey	21 large SMSAs	1972	Carpenters			35–50
				Electricians and plumbers		55–70	
Parnes et al.	OLS	Parnes	1969–70	Craftsmen	53		
Ashenfelter	OLS	CPS—all industries	1967	Craftsmen	30.5	41.1	
			1973	Craftsmen	42.1	45.4	
			1975	Craftsmen	40.4	21.5	
Lee	Consistent: union endogenous	SEO	1967	All			27.8
Allen	OLS	CPS	1973–1975	All			53.7[a]
Bourdon	OLS	Parnes—older men	1969	Craftsmen			67.1[b]
			1969	Craftsmen			77.3[c]
			1971	Craftsmen			67.6[c]

Freeman and Medoff	OLS	CPS	1973–1975	Craftsmen	37[d]
				Craftsmen	50[e]
Moore and Raisian	OLS	Income dynamics panel	1967	All	57.0
			1968	All	61.3
			1969	All	41.7
			1970	All	41.8
			1971	All	48.1
			1972	All	37.3
			1973	All	42.2
			1974	All	50.8
			1975	All	60.6
			1976	All	35.7
			1977	All	50.6

Source: Jeffrey M. Perloff and Robin C. Sickles, "FIML Estimation of Union Wage, Hours, and Earnings Differentials in the Construction Industry: A Nonlinear Limited Dependent Variable Approach", unpublished manuscript, March 1983.

Note: For citation to studies, see Appendix E.

[a] This specification does not include a race variable.

[b] This specification uses an experience variable but not an occupational training variable.

[c] This specification uses an occupational training variable but not an experience variable.

[d] Assumes that 20 percent of the market was unionized.

[e] Assumes that 80 percent of the market was unionized.

[f] Ordinary least squares.

FIGURE XI-1
Percent Increase in Wage Plus Fringe Rate
United States and Regions
January 1, 1974-January 1, 1984

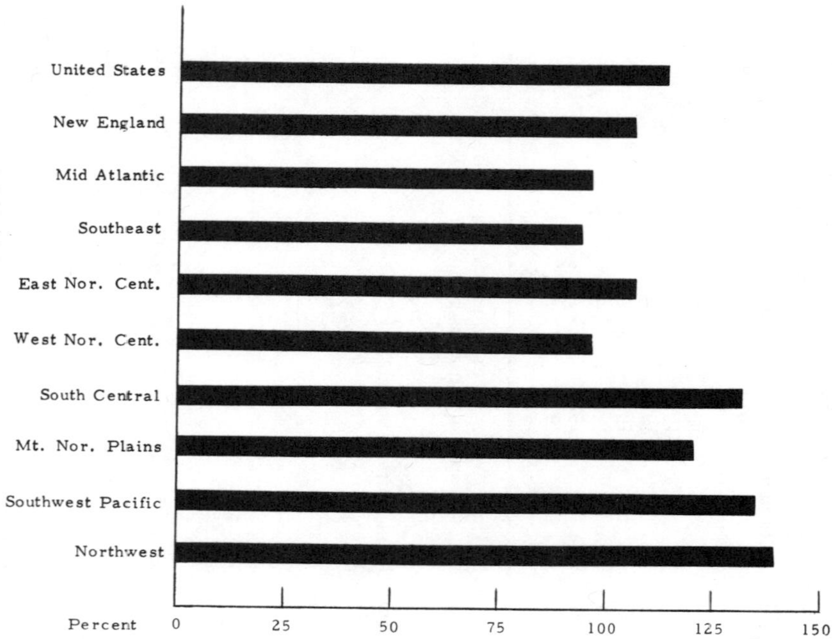

Source: "1984 Construction Labor Rate Trends and Outlook," *Construction Labor Research Council*, February 1984, p. 2.

narrow the earning gap over time, is quite important. Analysis of this aspect is reserved for later in this chapter.

The Personnel Administration Services Study

Personnel Administration Services, Inc., (PAS) is an Ann Arbor, Michigan consulting firm specializing in personnel administration and human resources development. In early 1984, PAS sent a mail questionnaire to members of the ABC requesting information on wages and benefits and received replies from 248 companies, employing about 30,000 workers.[24] This represents a response rate of about one percent. The replies were heavily weighted by six companies headquartered in the Southeast.

Table XI-11 shows the average wage rates found by the PAS study for carpenters, electricians, pipefitters, and laborers, as well as the

[24] *1984 Merit Shop Wage and Benefit Survey* (Ann Arbor, Mich.: Personnel Administration Services, 1984).

TABLE XI-11
Weighted Average[a] Hourly Wage Rates of Open Shop Journeymen and Laborers by Craft and Construction Sector First Quarter, 1984

Construction Sector[b]	All Crafts Combined[c]	Carpenters	Electricians	Pipefitters	General Laborers
	$	$	$	$	$
Commercial	10.08	10.20	10.69	12.67	6.51
Industrial	10.21	10.48	10.94	12.63	6.32
Institutional	8.66	9.19	9.25	10.45	6.35
Residential-1	8.79	8.66	9.64	10.66	5.84
Residential-2	8.62	9.88	8.90	8.58	6.22
Heavy	10.74	11.05	12.00	12.93	6.89
Highway	9.24	9.47	—	—	6.93
Municipal	8.71	8.59	9.20	11.30	5.94
All Sectors	10.10	10.34	10.92	12.63	6.57

Source: *1984 Merit Shop Wage and Benefit Survey* (Ann Arbor, Mich.: Personnel Administration Services, 1984).

[a] Weighted average is defined as the average wage reported by each company times the number of employees in the craft which is then totaled and divided by the total number of employees.

[b] PAS study defines residential-1 as under four stories, residential-2 as four stories or more; no other sectoral definitions given.

[c] The PAS study covers forty-five journeyman and/or foreman categories but, of course, each sector has a different number of categories and the definitions do not in all cases agree with unionized craft definitions.

weighted average for all crafts, by type of construction. The lack of a union rate in the same study, of course, precludes an exact union-open shop comparison. A rough comparison, however, can be made by utilizing union construction data from the CLRC, which is much more complete, and of course, much easier to obtain. Table XI-12 compares the PAS findings for open shop trades with those of the CLRC for the same period. The CLRC rates, like those of PAS, do not include fringe benefits and are therefore lower than the combined wages and fringe figures used in Figure XI-1.

Table XI-12 shows that open shop rates are approximately 70.7 percent to 60.6 percent of union shop skilled trade rates and 54.5 percent of laborers' rates. Again, therefore, we find that the skilled-unskilled wage differential is greater among open shop than among union shop contractors. As we shall also see below, the expensive union fringe benefit costs add to this differential. Moreover, Table XI-12 undoubtedly underestimates the union-open shop differential because the returns to the PAS questionnaire were undoubtedly from the larger ABC firms, who are likely to pay higher wages than

TABLE XI-12
Comparison of Open Shop and Union Shop Weighted Average Hourly Wage Rates All and Four Crafts, All Sectors First Quarter, 1984[a]

Craft	Union	Open Shop	Percent Open Shop Rate of Union Rate
	$	$	%
Carpenters	15.67	10.34	55.0
Electricians	18.03	10.92	60.6
Pipefitters	17.87	12.63	70.7
Laborers	12.06	6.57	54.5
All Crafts	15.65[b]	10.10[c]	64.5

Source: Open Shop, Table XI-11; Union Shop, Construction Labor Research Council.
[a] See text for limits of data for comparative purposes.
[b] Sixteen standard crafts.
[c] Forty-five crafts, but majority are subdivisions of sixteen union crafts and/or helper classifications.

their smaller competitors. The PAS survey found in this regard that the larger the project, the higher the wages paid by open shop contractors. Moreover, open shop contractors performing work primarily on a fixed or firm price were found by PAS to be paying lower wages than those who worked under cost plus arrangements.[25]

Table XI-11 affirms what the BLS found in its large metropolitan area surveys—not too great a spread between wages by sector among open shop builders. It is believed, however, that the absence of small homebuilders, who are more likely to be members of the National Association of Home Builders (NAHB) than of the ABC, account for these results. Such contractors are likely to pay less than larger builders. Regardless of these limitations, the PAS survey adds to our knowledge of open shop wages. Since it also obtained information on fringe benefits and overtime practices, we shall once again utilize it in the later sections of this chapter.

INDUSTRIAL RESEARCH UNIT SURVEYS

The Industrial Research Unit's 1974 and 1982 surveys of both union and open shop contractors solicited wage information some-

[25] *Ibid.* A good summary of this report is found in *Cockshaw's Construction Labor News + Opinion,* Vol. 13 (March 1984), No. 4.

what differently from the studies just described. The 1974 survey attempted to find out whether open shop contractors paid different wages to employees in the same trade. By 1984 this point was no longer speculation but, in fact, had been well documented by numerous BLS studies. Thus, the 1982 survey focused on probing the actual differential between union and open shop wages. We shall thus discuss the two surveys separately in this section.

The 1974 Questionnaire

The 1974 questionnaire requested that the respondents report both the highest and lowest wage rates paid to employees in several occupational categories. The purpose of this approach was to determine the nature of wage structures within and among employing units. This approach also enabled us to compare the results of this study with those of more comprehensive federal studies.

Table XI-13 shows the median wage rates reported by union and nonunion respondents in the three trades most commonly employed by firms in our sample—carpenter, equipment operator, and laborer. The differences between union and nonunion rates were of the same magnitude as those found in other studies. As before, it is difficult to discern a clear pattern of variations among geographic areas or trades, although the skilled-unskilled differential is an appreciably higher percentage among the open shop firms. Still, the union-nonunion differentials are not systematically higher or lower in the skilled trades or in the more heavily unionized regions

TABLE XI-13
Median Wages of Highest Paid and Lowest Paid Workers by Union Status and Trade, 1974

	Union			Nonunion		
Trade	Number Reporting	High	Low	Number Reporting	High	Low
Bricklayer	221	$8.75	$8.45	174	$6.53	$5.21
Carpenter	405	8.36	7.87	398	5.74	4.37
Electrician	165	9.51	9.07	151	6.10	4.38
Equipment operator	397	8.79	7.86	342	5.25	3.94
Laborer	519	6.54	5.97	518	3.99	2.83
Painter	99	7.63	7.18	98	4.87	3.71
Plumber	155	9.38	9.10	109	6.28	4.80
Sheet metal worker	108	8.94	8.75	92	5.64	3.85

Source: Industrial Research Unit Questionnaire, 1974.

of the country. Both union and nonunion wages, however, are highest in the northeastern and east central states, and lowest in the southern and mountain states.

To emphasize the differing internal wage structures of union and open shop contractors, Table XI-14 presents the median wages reported by respondents for both their highest paid and lowest paid workers in eight trades. Since the high and low figures relate to the same group of employers, they are not seriously affected by differences prevailing between industry divisions (for example, residential and commercial construction) or between geographic areas. Among the higher paid workers, median wages in the union sector average about 54 percent higher than those in the nonunion sector for the eight trades. Among lower paid workers, however, wages in the union sector are nearly double the open shop average. This higher differential for lower paid workers, particularly construction laborers, has, of course, been found in most previous surveys, including the BLS studies discussed above.

In our first edition of *Open Shop Construction*, we showed the greater dispersion of wages within a single trade and region in the nonunion segment by means of the coefficient of variation. In view of the data already presented here on this point it is not necessary to reproduce this table again, but merely to reiterate that the 1974 questionnaire clearly supported this distinction between the unionized and open shop wage structures.[26]

The 1982 Questionnaire

The 1982 questionnaire aimed its inquiries at the relative position of open shop wage rates to the union scale. Table XI-15 shows the results from the question as to whether the wage level which the contractor respondents' *average* skilled craftsmen received was equal to, less than, or more than the union rate. Not surprisingly, the greater number of unionized respondents reported that the average craftsmen received the union rate, whereas the open shop contractors reported that the average craftsmen were paid below union scale.

The next question asked for the same differentiation for the *most* skilled craftsmen. There was same upward movement of the re-

---

[26] For these calculations, see Table XI-8 of the original edition. Although the equipment operator trade appears to have a dispersed rate schedule, this is merely a result of the policy of the Operating Engineers' International Union which demands a separate rate for each type of machine in order to push up wages by charging more for operating larger equipment.

TABLE XI-14

Median Wage Rates Reported by Contractors

Three Trades, by Region and Union Status, 1974

Trades	New England	Middle Atlantic	E.N. Central	W.N. Central	South Atlantic	E.S. Central	W.S. Central	Mountain	Pacific
Union firms									
Carpenter-high	$9.00	$ 8.94	$9.03	$8.05	$8.08	$7.75	$7.28	$8.26	$7.77
Carpenter-low	8.75	8.60	8.63	7.50	7.75	7.38	6.70	7.31	7.50
Eq. op.[b]-high	8.76	10.06	9.20	8.38	7.76	7.50	6.94	7.32	8.86
Eq. op.-low	8.67	8.57	8.31	7.59	6.76	6.00	6.25	7.12	8.01
Laborer-high	7.00	7.18	7.08	6.46	5.59	5.64	4.90	5.71	6.33
Laborer-low	6.60	6.73	6.65	5.96	4.80	5.00	4.25	5.53	6.00
Nonunion firms									
Carpenter-high	5.94	6.29	5.91	4.98	5.15	5.75	5.79	5.76	[a]
Carpenter-low	4.66	4.81	4.68	3.83	4.10	3.75	3.95	5.00	[a]
Eq. op.-high	5.76	5.82	5.90	5.21	4.83	5.35	4.85	6.08	[a]
Eq. op.-low	4.82	4.55	4.91	3.86	3.59	4.21	3.46	5.75	[a]
Laborer-high	4.59	4.44	4.93	3.60	3.59	3.47	3.44	3.95	[a]
Laborer-low	3.34	3.38	3.36	2.64	2.68	2.62	2.55	2.85	[a]

Source: Industrial Research Unit Questionnaire, 1974.

[a] Responses too few to permit calculation of median.

[b] Equipment operator.

TABLE XI-15
Percent Distribution of the Wages Received by Average Skilled Craftsmen Compared with Union Rate, 1982

Trade	Union[a]				Nonunion[b]			
	Less	Equal	More	Do Not Know	Less	Equal	More	Do Not Know
Bricklayer	5.0	26.9	1.3	7.6	15.0	9.4	0.8	13.9
Carpenter	7.2	39.0	2.1	5.5	24.1	13.5	2.0	10.4
Equipment operator	6.8	38.0	1.8	5.7	22.0	12.8	2.3	13.1
Electrician	5.8	22.3	1.2	7.3	21.1	10.5	1.2	14.1
Laborer	9.4	46.6	2.1	4.6	34.8	14.7	2.6	10.1
Painter	6.8	26.6	1.6	7.4	19.4	7.4	1.4	14.3
Plumber	6.3	36.9	1.3	6.1	21.2	12.5	1.3	14.7
Sheetmetal	6.5	33.0	1.5	6.1	23.7	11.7	1.1	15.4
Steamfitter	5.4	36.3	1.5	6.9	16.0	9.5	0.9	16.7

Source: Industrial Research Unit Questionnaire, 1982
[a] N = 1828
[b] N = 2022

sponses, as shown in Table XI-16, for unionized contractors—that is, a slightly larger percentage of the unionized contractors paid above the union scale for the most skilled craftsmen, but the union scale was the mode by only a small percentage less for the most skilled craftsmen than it was for the average craftsmen.

For the open shop respondents, the results of the question concerning the most skilled craftsmen are quite different than that for the average craftsmen. Indeed, in all trades, the open shop respondents paid their most skilled craftsmen either the union rate or more in a majority of cases, whereas in no trade was that true for average craftsmen. Once again, this points to the fact that open shop contractors tend to pay rates dependent upon individual qualifications, whereas unionized contractors are most likely to pay the union rate to anyone classified as a journeyman.

Tables XI-15 and XI-16 also provide evidence in support of the claim, often voiced by open shop contractors in our interviews, that they frequently paid the union scale or better to highly skilled craftsmen, but gained the advantage by paying average and less skilled personnel lower rates. Again this demonstrates how more economical deployment of labor and wage scales provide material cost advantages to open shop contractors.

A third question on the 1982 questionnaire asked open shop respondents to estimate the percent of their hourly workers whose

TABLE XI-16
Percent Distribution of the Wages Received by Most Skilled Craftsmen Compared with Union Rate, 1982

Trade	Union[a]				Nonunion[b]			
	Less	Equal	More	Do Not Know	Less	Equal	More	Do Not Know
Bricklayer	2.6	21.4	4.4	7.5	8.3	11.1	2.6	13.9
Carpenter	3.9	29.0	13.2	4.7	12.1	17.3	7.9	10.6
Equipment operator	3.6	33.1	6.3	4.9	12.7	15.0	5.8	13.5
Electrician	2.4	18.7	4.0	7.1	12.4	12.1	4.5	14.2
Laborer	5.1	39.3	9.6	3.8	22.4	18.4	6.5	9.8
Painter	2.8	21.9	6.2	7.4	11.8	9.1	3.2	15.0
Plumber	3.1	28.2	9.1	6.0	11.9	14.7	4.8	15.1
Sheetmetal	3.1	25.7	8.1	6.2	14.1	14.4	4.7	15.6
Steamfitter	2.7	26.9	10.4	6.6	9.0	11.1	3.2	16.8

Source: Industrial Research Unit Questionnaire, 1982.
[a] N = 1828
[b] N = 2022

wage was as high or higher than the union scale. Table XI-17 shows that 58.8 percent of the respondents reported that 25 percent or less of their hourly workers were paid the union rate or more, and another 13.2 percent estimated that 26 to 50 percent of their hourly workers were paid at that rate. Only 20.5 percent of those surveyed estimated that 51 percent or more of their hourly workers were paid at a rate equal to the union scale. These results, of course, confirm our general findings of a substantial union-open shop wage differential.

FRINGE BENEFITS

The differentials between union and nonunion compensation in the construction industry are greater than the wage differentials examined in the previous sections. Unionized workers receive much higher average fringe benefits than do open shop employees. All studies demonstrate this very clearly. Open shop workers, on the other hand, obtain a larger proportion of their remuneration in direct wage payments.

Before analyzing union and open shop benefits, two cautions must be emphasized. First, the real differentials are sometimes greater than those which appear on the surface because both unionized and open shop employers must pay taxes on government-mandated ben-

TABLE XI-17
Open Shop Respondents' Estimates of Their Workers with a Comparable or Higher Hourly Wage Than Union Scale, 1982

Total	Number and Percentage of Workers							
	0-25		26-50		51-75		76-100	
	No.	Percent	No.	Percent	No.	Percent	No.	Percent
2022	1189	58.8	266	13.2	213	10.5	203	10.0

Source: Industrial Research Unit Questionnaire, 1982.

efits, including old-age (social security), unemployment, and workers' compensation. And second, open shop fringe benefits do not include such items as funds for apprenticeship training, industry advancement, or substantial administrative costs, all of which are placed into union funds whether or not they are utilized to the degree negotiated. In addition, of course, comparisons are difficult, as in the case of wages, because of the paucity of data on the open shop sector. Finally, the BLS surveys of benefit plans excluded those for which employees paid the whole cost. This probably reduced the incidence of open shop health insurance plans reported.

Bureau of Labor Statistics Studies

All BLS studies thus far examined confirm that union contractors provide greater fringe benefit coverage and incur higher fringe benefit costs than their open shop counterparts. Thus, Table XI-18 shows that the BLS studies issued during 1965–1971 found that the open shop paid 3.4 to 5.7 percentage points less in fringe benefits, but this is understated for the benefits include legally required expenditures for social security, unemployment insurance, and workers' compensation. The data below (from Table XI-18) summarize the percentages of employer expenditures for voluntary, private benefits in each of the industry divisions:

General Building		Special Trades		Heavy/Highway	
Union	Nonunion	Union	Nonunion	Union	Nonunion
5.8	1.9	8.1	3.5	10.9	3.6

In its 1972 study, the BLS found employer contributions to funds providing pensions and health and life insurance "applied to almost

TABLE XI-18
Construction Worker Compensation by Union Status and Type of Expenditure 1965–1971

	General Building		Heavy/Highway		Special Trades	
Expenditure	Union	Nonunion	Union	Nonunion	Union	Nonunion
Total	100.0	100.0	100.0	100.0	100.0	100.0
Pay for work time	86.7	90.7	85.4	88.8	82.0	87.7
Straight-time	84.4	89.4	82.6	86.2	78.2	82.8
Premium pay	2.3	1.3	2.8	2.6	3.7	4.8
Pay for leave time	1.0	.7	1.8	1.5	2.3	1.1
Vacations	.1	.3	.1	.9	.3	.6
Holidays	.2	.2	.1	.6	.4	.4
Payment to funds	.7	.2	1.6	*	1.6	*
Retirement programs	4.4	3.2	6.6	4.4	7.7	5.0
Social security	2.4	2.8	3.3	4.0	3.2	4.1
Private plans	2.0	.4	3.3	.4	4.6	.9
Life/health benefit programs	5.3	3.1	5.0	3.5	6.4	4.4
Life/health insurance	2.3	.6	2.8	.8	3.7	1.0
Sick leave	*	*	*	.1	*	.1
Workers' compensation	3.0	2.5	2.1	2.6	2.6	3.3
Unemployment benefit programs	2.1	2.0	1.0	1.0	1.3	1.2
Unemployment insurance	2.1	2.0	.9	1.0	1.3	1.1
Severance pay	*	*	*	*	*	*
Supplemental unemployment benefit funds	*	*	*	*	*	*
Nonproduction bonuses	.5	.2	.2	.7	.3	.5

Source: U.S. Bureau of Labor Statistics, *Compensation in the Construction Industry,* Bulletin 1656, Table A-3; *Employee Compensation and Payroll Hours, Heavy Construction Industry, 1971,* Report 428, Table 1.6; *Employee Compensation and Payroll Hours, Construction—Special Trade Contractors, 1969,* Report 413, Table 2, (Washington, D.C.: U.S. Government Printing Office, 1970, 1974, and 1972 respectively).
Note: Because of rounding, sums of individual items may not equal totals.
* Less than 0.05 percent.

all union workers studied," with employer contributions to separate vacation funds "less prevalent," and employer contributions to funds combining these benefits covering only about one-third of the areas studied.[27] The data for nonunion worker fringe benefit coverage was considerably less well documented. It did show, however,

[27] *Industry Wage Survey: Contract Construction, September 1972,* pp. 10–11.

that most nonunion workers received paid holidays, and that "paid vacation not provided by union funds applied to at least some workers in most of the occupations and areas." Nevertheless, the proportion of workers covered by benefits of all types was decidedly below that of nonunion employees in a given area. Thus "only one-half of the nonunion carpenters and one-fifth of the nonunion construction laborers in Washington were in firms with formal vacation provisions.... [Approximately] 40 to 50 percent of the carpenters in Atlanta, Biloxi, Dallas, and Memphis... were nonunion but only 7 to 16 percent were covered by health or insurance plans." Retirement plan coverage was much more scarce for nonunion workers.[28]

By 1973, the BLS reported that employer contributions to pension and insurance funds "applied to almost all union workers studied," and that additional contributions to several other type benefit funds or combination of all benefit funds were less prevalent. Again the study found that nonunion benefit plan coverage was less complete, but the data do indicate that open shop fringe benefits were growing. Thus, most nonunion workers received paid holidays, but a smaller proportion—often less than one-half—received paid vacations. Health and life insurance programs were found for open shop contractor employees in all areas studied, but worker coverage by such plans often fell well below that for nonunion workers in an area. Thus, two-thirds of the carpenters in Atlanta were nonunion, but only one-fourth had life and health insurance coverage. In Dallas, almost two-thirds of the sheet metal workers were nonunion but less than 10 percent were covered by these benefits.[29]

In the BLS's last survey of this type, conducted in 1976–1977, it found again that employer contributions to funds for insurance and pension plans applied to the overwhelming majority of unionized workers, and that contributions to vacation funds were somewhat less prevalent. Contributions for apprenticeship, industry advancement, and other such matters were also widespread in the unionized portion.[30]

For the open shop segment, by 1977 the incidence of health and insurance plans was generally equal to or higher than that for vacation or holiday plans. Open shop health plans were then more likely to include surgical, medical, major medical, and hospitalization benefits than sickness and accident, sick leave, and life in-

---

[28] *Ibid.*, pp. 12–13.
[29] *Industry Wage Survey: Contract Construction, September 1973*, pp. 7–11.
[30] *Compensation in the Construction Industries in Large Metropolitan Areas*, p. 5.

surance benefits.[31] Open shop retirement plans, as shown in Table X-19, were applicable to only one-third of the workforce covered by the survey, except in the case of a few highly skilled craftsmen—for example, plumbers and electricians—in a few industry sectors. Given the fact that the BLS report excluded the small companies, who are likely to provide the least benefits, by 1977 open shop employers were far behind their union shop competitors in providing fringe benefits and, as a result, were able to operate at a considerable cost advantage.

The Industrial Research Unit Questionnaire, 1982

The 1982 Industrial Research Unit questionnaire produced some surprising results indicating that fringe benefit payments of open shop contractors were becoming more extensive, but were differently oriented than those of unionized contractors. Table XI-20 shows that health insurance is still more extensively provided under union contracts than under open shop contracts. Life insurance, however, was provided for all or most employees by 61.7 percent of the open shop respondents and by 57.1 percent of the union shop respondents; paid vacations were given by 60.2 percent of the open shop employers for all or most employees but only by 46.4 percent of the union shop employers; and paid holidays or sick leave were granted by 53.3 percent of the open shop contractors, but only by 28.2 percent of the unionized contractors. Finally, the open shop contractors were far more likely to provide bonus or profit-sharing plans; 50.6 percent had such plans for all or most of their employees, as compared to only 16.2 percent of the unionized contractors. On the other hand, only 15.8 percent of the open shop respondents participated "in any funded benefit plan such as those administered by an employer." Pension plans for unionized employees are, of course, very common.

What these responses reflect is not only the growing willingness of open shop contractors to provide fringe benefits, but also the manner in which their different labor deployment policies direct such benefits. Life insurance, paid vacations, and holidays are benefits that are more applicable to permanent employees, not those who work for various contractors within a labor market. Construction unions stress portable pensions, overtime for holidays if worked, and vacation funds rather than paid vacation time. Open shop contractors have long found bonus and profit-sharing plans to be an

[31] *Ibid.*, pp. 5–6.

TABLE XI-19

Percent of Workers in Establishments Providing Specified Benefits Not Provided by Union Funds, Selected Trades, Commercial and Residential Construction Seventeen Metropolitan Areas,[a] March 1977

Occupation and Sector	Percent of Workers in Establishments Providing Selected Benefits									
	Paid Holiday	Paid Vacation	Life Insurance	Hospitalization	Surgical	Medical	Major Medical	Sickness and Accident or Sick Leave	Retirement Plans	No Insurance or Retirement Plans
Carpenters	16	15	21	22	22	21	21	18	12	76
Commercial	14	13	19	19	19	19	19	17	12	79
Residential-1[b]	5	31	10	14	11	11	7	8	6	86
Residential-2[c]	24	19	25	28	28	27	27	20	12	70
Highway	17	18	37	37	37	37	34	17	7	63
Other Heavy[d]	9	13	18	17	17	12	16	16	13	81
Electricians	29	19	18	18	18	17	17	14	9	80
Commercial	24	12	10	12	12	11	11	9	6	87
Residential-1	66	66	18	41	41	41	41	9	—	48
Residential-2	57	58	57	54	54	51	51	45	29	39

Plumbers	27	37	30	33	33	31	31	18	15	63
Commercial	15	18	13	12	12	10	10	10	7	83
Residential-1	33	33	22	31	31	31	14	6	3	67
Residential-2	39	59	50	56	56	54	54	27	24	41
Sheet Metal Workers	22	24	20	23	23	22	22	12	14	76
Commercial	15	16	14	16	16	15	15	5	9	84
Residential-1	84	84	80	80	80	80	80	2	—	20
Residential-2	59	60	52	62	62	60	60	52	43	38
Laborers	18	16	24	25	25	24	22	18	13	72
Commercial	14	15	18	18	18	18	18	16	11	80
Residential-1	15	11	12	14	14	14	9	4	1	86
Residential-2	28	20	28	33	31	26	26	26	21	64
Highway	9	9	24	28	28	27	17	8	5	72
Other Heavy	23	21	30	30	30	28	28	25	17	67

Source: U.S. Bureau of Labor Statistics, *Compensation in the Construction Industries in Large Metropolitan Areas,* Report 610 (Springfield, Va: National Technical Information Service, 1980), Tables 6 and 7.

[a] For areas covered, see Text footnote 14, above.

[b] Residential-1 covers five stories and above.

[c] Residential-2 covers four stories and below.

[d] Includes heavy and industrial as defined in Chapters V and VI.

TABLE XI-20

Respondents' Estimates of Coverage of Selected Fringe Benefits by Union Status, 1982

Benefit	Respondents' Estimates of Coverage									
	All		Most		Few		None		DK/NR[a]	
	No. of Employers	Percent	No. of Employers	Percent	No. of Employers	Percent	No. of Employers	Percent	No. of Employers	Percent
Health insurance										
Union	1,368	74.8	218	11.9	66	3.6	55	3.0	121	6.6
Open shop	1,074	53.1	426	21.1	226	11.2	216	10.7	80	4.0
Life insurance										
Union	824	45.1	226	12.0	219	12.0	298	16.3	261	14.3
Open shop	898	44.4	350	17.3	249	12.3	367	18.2	158	7.8
Paid vacation										
Union	660	36.1	189	10.3	410	22.4	337	18.4	232	12.7
Open shop	930	46.0	287	14.2	379	18.7	308	15.2	118	5.8
Paid holiday or sick leave										
Union	374	20.5	140	7.7	476	26.0	544	29.8	294	16.1
Open shop	847	41.9	231	11.4	371	18.3	407	20.1	166	8.2
Bonus or profit sharing										
Union	183	10.0	113	6.2	524	28.7	695	38.0	313	17.1
Open shop	706	34.9	317	15.7	461	22.8	404	20.0	134	6.6

Source: Industrial Research Unit Questionnaire, 1982.
N = Union, 1828; Open Shop, 2022.
[a] DK/NR = Do not know or no reply.

especially effective benefit. The Industrial Research Unit 1974 questionnaire reported that 42 percent of the open shop respondents, as compared with 15 percent of the unionized respondents, had profit-sharing plans, and the 1982 questionnaire saw the margin on this benefit widen between the open and unionized segments, with 34.9 percent of the open shop respondents but only 10 percent of the unionized respondents providing such plans to all employees. Such plans are utilized to encourage loyalty and productivity and can be particularly effective in small companies. The BLS did not include profit sharing in its questionnaires unless the benefits are paid in the form of a pension, in which case it was reported as pension coverage. Since open shop contractors, according to our field research, generally provide direct annual, or occasionally more frequent, bonuses with profit-sharing funds, this benefit did not appear in the BLS studies.

Associated Builders and Contractors' Plans

Recognizing that the absence of fringe benefits can work as a disadvantage to open shop firms in attracting and retaining employees, the ABC, AGC, and other associations have adopted fringe benefit programs which they make available to their members. Through the ABC Insurance Trust, established in 1957, the ABC offers dental, life, and medical insurance plans with many options. Approximately, 90,000 persons were enrolled in these ABC plans as of 1984. The plans are fully portable from one ABC contractor to another who is covered.[32]

Figure XI-2 shows fringe benefit coverage provided by ABC contractors for their employees as of 1977, based on an Opinion Research Corporation survey. Holidays and vacations were provided by about 70 percent of ABC contractors, health and accident insurance by 80 percent, and life insurance by 62 percent. ABC subcontractors were a bit more willing to provide these benefits than were the general contractors.

Of special interest is the extent to which ABC contractors provided pensions in 1977. Figure XI-2 shows that 20 percent of ABC contractors provided pension plans, and another 9 percent had plans to which employees contributed. The 1977 BLS study of large metropolitan areas included only employer-contributed plans, and the data therein indicated that the ABC contractors were more generous in this regard than those surveyed by the BLS (see Table XI-19).

---

[32] Data from ABC.

FIGURE XI-2
Survey of Contractor and Subcontractor Members Associated Builders and Contractors, 1977

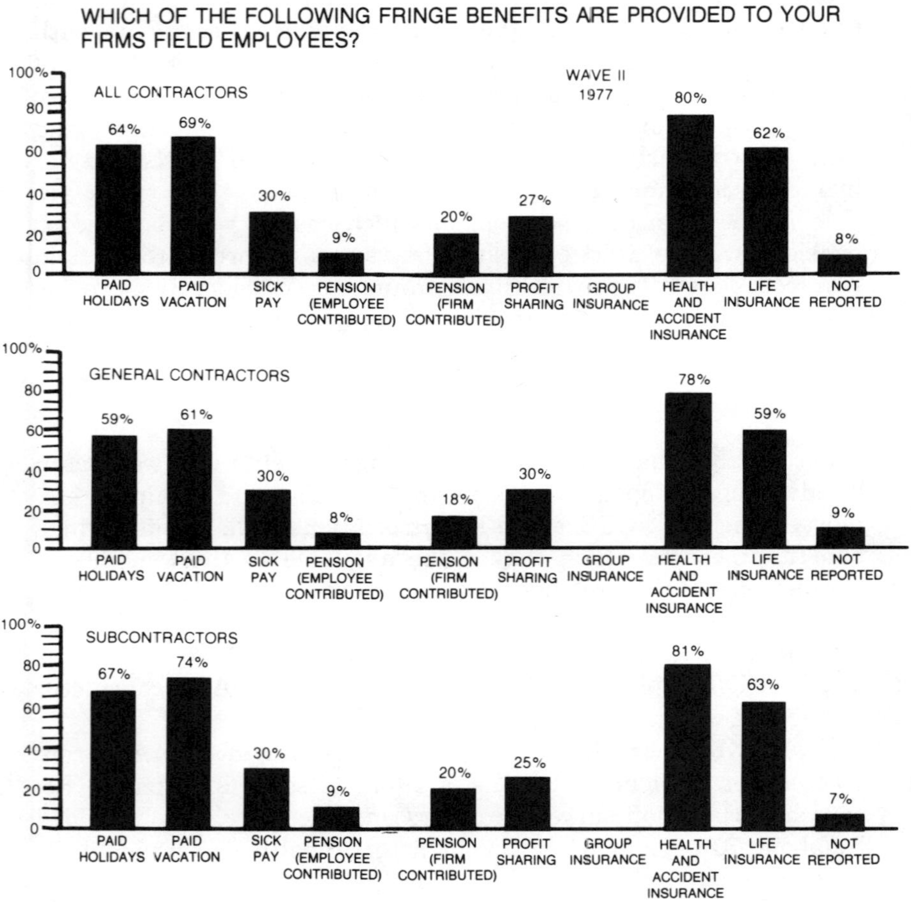

Source: Opinion Research Corporation, Associated Builders and Contractors' Trend Study II (Princeton, N.J. 1977).

Moreover, an additional 27 percent of the ABC contractors provided profit sharing, most of which was undoubtedly paid in bonuses, but some of which probably included a pension feature.

Further light on ABC contractors' policies relating to fringe benefits can be derived from the 1982 Industrial Research Unit questionnaire. Table XI-21 shows that a majority of respondents provided health insurance, life insurance, and paid holidays or sick leave to

TABLE XI-21
Associated Builders and Contractors Respondents' Estimates of Coverage of Selected Fringe Benefits, 1982

	Respondents' Estimates of Coverage									
	All		Most		Few		None		DK / NR[a]	
Benefit	No. of Employers	Percent	No. of Employers	Percent	No. of Employers	Percent	No. of Employers	Percent	No. of Employers	Percent
Health insurance	596	60.4	185	18.8	101	10.3	66	6.7	34	3.5
Life insurance	496	50.5	151	15.4	117	11.9	146	14.9	72	7.3
Paid vacation	479	48.8	131	13.3	174	17.7	144	14.7	54	5.5
Paid holiday or sick leave	446	45.4	91	9.3	182	18.5	191	19.5	72	7.3
Bonus or profit sharing	362	36.9	129	13.1	223	22.7	205	20.9	63	6.4

Source: Industrial Research Unit Questionnaire, 1982.
N = 982.
[a] Do not know or no reply.

all or most employees, while just 50 percent gave bonuses or profit sharing. In addition, 27.2 percent participated in funded benefit plans, "such as those administered by an employers' association." The responses to the 1982 questionnaire clearly indicate that ABC members have expanded their benefit coverage to equal or even surpass that of open shop contractors in general, as represented in Table XI-20. The fact that more than one-half of the ABC respondents provide bonus or profit-sharing plans is indicative of the importance which open shop contractors attach to benefits. ABC members, like all open shop contractors, continue to lag in providing pension plans.

Associated General Contractors' Plans

In 1972, the AGC established the National Employee Benefit Trust (NEBT) which provides a complete array of surgical, hospitalization, and medical benefits as well as group life insurance to AGC members and chapters not covered by union-funded plans. As of January 1984, twenty AGC chapters were participating in the program. In addition, a survey conducted by the national AGC office in late 1983 found that fourteen of the sixty-seven chapters responding had established group programs for medical and related insurance.[33]

The 1982 Industrial Research Unit questionnaire permits a comparison of the benefit structure of unionized and open shop AGC contractors. For reasons that are not clear, a smaller percentage of AGC open shop respondents provided benefits in all categories to all or most of their employees than did open shop respondents in general or the ABC respondents. The open shop AGC contractors, unlike those in general, did not have a greater percentage in the "all" or "most" categories for life insurance, and were about even on paid vacations. Like other open shop respondents, the AGC contractors were more likely to have paid vacation plans and considerably more likely to have profit-sharing plans for all or most employees. Like other open shop respondents, however, they also lagged behind on pension plan coverage; only 20.2 percent participated in association-funded pension plans.

Single-Company and Other Benefit Plans

The large open shop concerns such as Brown & Root, Daniel International, H. B. Zachry, BE&K, Austin, and others provide their

[33] Letter to the author from Dennis M. Bradshaw, Executive Director, AGC, January 23, 1984.

TABLE XI-22

Associated General Contractors Respondents' Estimates of Coverage of Selected Fringe Benefits, by Union Status, 1982

	Respondents' Estimates of Coverage									
	All		Most		Few		None		DK / NR[a]	
Benefit	No. of Employers	Percent	No. of Employers	Percent	No. of Employers	Percent	No. of Employers	Percent	No. of Employers	Percent
Health insurance										
Union	615	73.5	113	13.5	29	3.5	29	3.5	51	6.1
Open shop	191	36.1	145	27.4	96	18.1	77	14.6	20	3.8
Life insurance										
Union	318	38.0	126	15.1	129	15.4	156	18.6	108	12.9
Open shop	155	29.3	119	22.5	103	19.5	115	21.7	37	7.0
Paid vacation										
Union	257	30.7	100	11.9	227	27.1	168	20.1	85	10.2
Open shop	138	26.1	92	17.4	154	29.1	111	21.0	34	6.4
Paid holiday or sick leave										
Union	139	16.6	62	7.4	250	29.9	278	33.2	108	12.9
Open shop	115	21.7	89	16.8	141	26.7	137	25.9	47	8.9
Bonus or profit sharing										
Union	63	7.5	50	6.0	277	33.1	340	40.6	107	12.8
Open shop	123	23.3	102	19.3	150	28.4	122	23.1	32	6.0

Source: Industrial Research Unit Questionnaire, 1982.
N = Union, 837; Open Shop, 529.
[a] Do not know or no reply.

TABLE XI-23
Wage Rates, Benefits, and Benefits as a Proportion of Total Hourly Wages and Benefits Building Trades, 1970–1980

Year	Wage rate	Employer benefit payment[1]	Wage rate-plus-benefit payment	Benefit payment as percent of wage rate-plus-benefit payment
	All building trades			
1970	$6.18	$0.80	$6.97	11.5
1971	6.88	.95	7.82	12.1
1972	7.27	1.06	8.34	12.7
1973	7.62	1.21	8.83	13.7
1974	8.14	1.45	9.59	15.1
1975	8.88	1.68	10.56	15.9
1976	9.48	1.95	11.43	17.1
1977	10.02	[1]2.26	[1]12.28	[1]18.4
1978	10.61	2.46	13.07	18.8
1979	11.32	2.66	13.98	19.0
1980	12.21	2.83	15.05	18.8
	Journeymen trades			
1970	$6.54	$0.84	$7.39	11.4
1971	7.28	.99	8.27	12.0
1972	7.69	1.10	8.79	12.5
1973	8.02	1.26	9.28	13.6
1974	8.55	1.50	10.05	14.9
1975	9.32	1.74	11.06	15.7
1976	9.92	2.02	11.94	16.9
1977	10.44	[1]2.37	[1]12.81	[1]18.5
1978	11.05	2.56	13.60	18.8
1979	11.81	2.78	14.59	19.1
1980	12.72	2.97	15.69	18.9
	Helper and laborer classifications			
1970	$4.86	$0.61	$5.47	11.2
1971	5.43	.78	6.22	12.5
1972	5.68	.92	6.61	13.9
1973	6.06	1.02	7.08	14.4
1974	6.53	1.25	7.79	16.0
1975	7.06	1.44	8.51	16.9
1976	7.54	1.64	9.18	17.9
1977	8.03	[1]1.77	[1]9.79	[1]18.1
1978	8.54	2.00	10.54	19.0
1979	9.15	2.11	11.26	18.7
1980	9.80	2.22	12.02	18.5

Source: U.S. Bureau of Labor Statistics, *Union Wages and Benefits: Building Trades, July 1, 1980,* Bulletin No. 2091 (Washington, D.C.: Government Printing Office, 1981), p. 3.

[1] Beginning in July 1977, benefits studied were expanded to include payments for supplemental unemployment benefits, savings funds, and paid holidays in addition to contributions for insurance, pension, and vacation plans. Had the July 1977 benefit

own benefit plans. These vary considerably in scope, including life and medical insurance, holidays, and vacations, and for the most part, pensions as well. The cost of these plans is usually less than that of union-funded plans is an advantage to open shop contractors in competing for work.

As we noted in Chapter VIII, the National Western Life Insurance Company had developed a benefits program for open shop contractors and, by early 1984, had sold it to 900 open shop concerns. There are likely to be more such companies in the future, for the information set forth above clearly indicates that coverage of benefits for employees of open shop contractors is increasing. Moreover, as already noted, other associations, for example, the Independent Electrical Contractors (IEC), have group plans for its members.

Thus, open shop benefit plans, particularly multiemployer ones, are likely to increase. Their broader scope, lower costs, and greater administrative convenience are attractive. Fringe benefits still lag in the open shop sector, particularly pensions, but the differential between the union and open shop sectors in fringe benefit coverage has narrowed considerably since 1975. Costs, as the next section demonstrates, remain far apart.

Benefit Costs

Table XI-23 shows hourly wage and fringe benefits and employer benefit payments as a percentage of wage rates plus benefit payments from 1970 to 1980. The latter year is the last for which such data were published by the BLS. A CLRC study found that, as of January 1, 1984, the weighted average hourly union wage and fringe rate in the industry was $19.35; the wage rate alone, $15.36; and the benefit payment, $3.99.[34] Therefore, the benefit payment as a percentage of wage rate plus benefits, on January 1, 1984 was 20 percent.

[34] These data were kindly supplied by Robert M. Gasperow of the CLRC.

payments been limited to insurance, pension, and vacation plans, the 1977 figures for employer benefit payments would have been $2.18 for all building trades, $2.29 for journeymen, and $1.71 for helpers and laborers; the corresponding wage-plus-benefit totals would have been $12.20, $12.73, and $9.74, respectively. The proportion of the three-benefit package to the total wage-plus-benefit average would have been 17.9 percent of all trades and for journeymen and 17.6 for helpers and laborers.
Note: Because of rounding, sums of items may not equal totals.

There are no comparable data for open shop benefits. The PAS study, as already noted, is skewed toward the Southeast and heavily weighted by six companies. It found that fringe benefits paid by the employer varied from 2.3 percent to 12.5 percent of the pay package, depending upon the industry sector and trade involved.[35] The weighted average hourly wage rate in the PAS study was $10.10, making the fringe rate anywhere from $0.23 to $1.26 per hour.

The experience in the Exxon refinery case, discussed in Chapter V, lends some credence to the PAS data, at least for the Southwest. Information made available to the author showed that fringe benefit payments required for union contractors exceeded the average for Brown & Root, Daniel, and Zachry by approximately $2.80 per hour. Subtracting that figure from the $3.99 weighted average of the CLRC, we get a rate for the three big open shop contractors of $1.19, which is in the upper range of the PAS study findings. One would, of course, expect these three contractors to be on the high end of the open shop range. When, however, the unions sought to win the Lower Colorado River Authority power plant contract by reducing costs an additional $2.80 per hour, Zachry was able to underbid the unionized contractors, in part because its benefit costs are below the average of the three that bid on the Exxon project.

We conclude, therefore, that benefit payments are a serious cost disadvantage to unionized contractors. Because contractors are eager to avoid withdrawal liability problems, these costs, as discussed in Chapter VIII, are likely to continue to remain high.

HOURS, OVERTIME, AND RESTRAINTS

The forty-hour week prevails in construction as in most other industries. According to the BLS survey of unionized construction in 1980: "The number of weekly hours required before overtime premiums take effect has changed very little since 1967. . . . Just over seven-eighths of the union members were covered by contracts specifying forty hours of work before overtime premiums are paid; most of the remaining members were under contracts calling for straight-time pay for the first thirty-five hours worked per week."[36] Open shop contractors work a forty-hour week probably without exception. The Fair Labor Standards Act (FLSA) of 1938 requires payment of time and one-half the wage rate for hours worked over forty in any one week.

---

[35] See *1984 Merit Shop Survey.*

[36] U.S. Bureau of Labor Statistics, *Union Wages and Benefits: Building Trades July 1, 1980,* Bulletin 2091 (Washington, D.C.: Government Printing Office, 1981), p. 4.

On other overtime features, the open shop and unionized contractors differ substantially. Union contracts have typically required doubletime pay for work on Sundays and holidays regardless of the weekly hours worked. Open shop contractors are more likely to grant holidays to employees, but usually pay either straight time or time and one-half when work is scheduled on holidays or Sundays. Union contracts also usually require at least time and one-half, but mostly, doubletime for Saturday work regardless of weekly hours worked, whereas open shop contractors only pay overtime for Saturdays if the weekly hours worked exceed forty.[37]

Union contracts require premium pay in numerous other situations. For example, if the scheduled hours are for 8 a.m. till 4 p.m. and the workers are requested to begin work at 7 a.m. for a particular reason on a specified day, union contracts require that the pre- or post-scheduled time be paid usually at doubletime, but at least at time and one-half regardless of the weekly or even daily hours worked.[38] Some union contracts prohibit shift work or require premiums of doubletime or time and one-half; others provide for eight hours' pay for evening and night shifts. Again, the open shop does not pay such extras in the limited number of shift operations which occur in construction.

A significant advantage to the open shop is found in daily overtime practices. Union contracts have typically required premiums after eight hours in one day, more often at doubletime than time and one-half. Open shop contractors, except on government jobs do not pay overtime after eight hours in a day except where the weekly hours exceed forty. This is not only a cost advantage in itself, but it also increases flexibility. For example, if cement has to be poured for twelve straight hours, it can be done without paying overtime provided the employee receives compensatory time off that week. More important, however, the requirement to pay daily overtime inhibits the use of the four-day, ten-hour schedule which open shop contractors have found advantageous.

Reporting Pay and Ten-Hour Shifts

Reporting pay for time not worked is another significant feature of union contracts. Nearly all construction agreements have required from one- to eight-hours' reporting pay (often four hours) if

---

[37] A good description of union premium pay requirements is found in *Industry Wage Survey: Contract Construction, 1973,* p. 7 and Tables 52–107. Although these data are over ten years old, the premiums were well entrenched by then and did not weaken till the giveback period of the early 1980s.

[38] *Ibid.*

the worker reports and there is no work, unless he has been notified beforehand. Open shop contractors do not maintain this practice, but instead strive to keep the workforce working whenever possible. Thus, when the weather is inclement, union workers are laid off without pay while open shop workers remain on the job. A four-day, ten-hour shift adds to the ability of open shop contractors to maintain work in inclement weather without increasing costs.

In Chapters IV and V, we noted that the adamant refusal of the president of the International Brotherhood of Electrical Workers (IBEW) to consider waiving overtime premiums after eight hours in one day was materially disadvantageous to unionized contractors in competition with open shop contractors and caused IBEW and other unionized tradesmen to lose work. Working four ten-hour days has a number of advantages both to employees and contractors. First of all, it reduces workers' travel expenses, which is especially advantageous to those workers who travel a considerable distance to the job (a not uncommon occurrence in the construction industry). When inclement weather prohibits work during one or more days during a week, the four-day, ten-hour schedule allows the day or days to be made up without overtime pay by using the fifth weekday and/or Saturday and Sunday. This provides steady pay for the workers' and hastens completion of the project, thereby reducing costs for the contractor and the owner. It is viable, however, in ordinary circumstances only if daily overtime pay is not required, and thus, most unionized contractors cannot use this system.

Walsh-Healey Act

The basic wage and hour law—the FLSA—requires time and one-half premium pay only after forty hours per week. It is silent on daily overtime requirements. A little-known law, the Public Contracts Act of 1936, known as the Walsh-Healey Act, however, requires time and one-half pay after eight hours' work per day on all federal government contracts which total $10,000 or more. Attempts to modify this provision, even to permit four-day, ten-hour work schedules, have failed because of the opposition of the AFL-CIO, particularly the building trades unions.[39] Where union contracts are the prevailing practice pursuant to Davis-Bacon or state prevailing wage laws, overtime after eight hours would also obviously be required for government-financed construction. This is, of course,

---

[39] "Senate Tables Amendment That Would Authorize Four Tens Work Schedule," *Construction Labor Report,* Vol. 28 (July 14, 1982), p. 682.

another reason why government-financed projects usually mean higher costs.

Scheduled Overtime

To make up for their lower rates and to push jobs to faster completion—a significant saving in itself on interest rates, cash flow, etc.—open shop contractors often offer their workforce significant overtime opportunities. At one time large users also pushed their unionized contractors to work overtime but have found that this practice was counterproductive. In the early 1970s, therefore, the Business Roundtable, the organization of large corporations and construction users, encouraged its members to permit overtime only on a spot basis and to avoid scheduled overtime. The Roundtable published a study in 1974 which concluded that:

> Placing field construction operations of a project on a *scheduled* overtime basis is disruptive to the economy of the affected area, magnifies any apparent labor shortage, reduces labor productivity, and creates excessive inflation of construction labor cost with no material benefit in schedule.[40]

In 1980, the Roundtable reexamined this issue and came to the same conclusion.[41] It found that work fatigue and holding back on production to ensure overtime offset productivity gains from the extra hours; that overtime was often scheduled to attract labor to an area, but instead induced a whole area to adopt an overtime schedule which, in effect, does not reduce job hopping nor increase the supply of manpower; and that such scheduled overtime increased the demands of unions for higher rates. Consequently, the main impact of overtime was inflationary, rather than hastening the completion of a job. Even prior to the publication of the 1974 study, the efforts of the Roundtable had significantly reduced scheduled overtime on major industrial projects.

As a result of givebacks, the CLRC reported in August, 1984, that about 11 percent of the contracts surveyed in mid-1984, as compared with about 4 percent in 1980, provided for four ten-hour days without daily overtime. According to the CLRC:

> Contracts with 4-10's language were previously concentrated in the south central and southeast regions. There are now at least a few

[40] "Effect of Scheduled Overtime on Construction Projects," in *Coming to Grips with Some Major Problems in the Construction Industry,* a Business Roundtable Report (New York: The Roundtable, 1974), p. 1. Italics in original.

[41] *Scheduled Overtime Effect on Construction Projects,* Report C-2, Construction Industry Cost Effectiveness Project (New York: The Business Roundtable, 1980), p. 1.

contracts with these clauses in each region and they have increased to a third of the agreements in the southeast. Almost half the boilermaker agreements permit this alternative, but virtually none of the electricians.

The increase in Saturday make-up provisions has been widespread. Regionally, the mountain / northern plains region had been, and still is, the region with the highest incidence of these clauses—greater than 20 percent. The increase in Saturday make-up classes has been most rapid in the west north central region.[42]

Annual Hours

Because of the increased amount of pay for time not worked (e.g., paid vacations and holidays) the average workweek in American industry over a yearly period has declined considerably since World War II, although the forty-hour week has remained the norm. For this reason, both companies and scholars are looking at annual rather than weekly hours of work as a more significant criterion from which to judge costs.

The most thorough study of this phenomenon found, not surprisingly, that annual hours of work in nearly all industries declined between 1947 and 1979 to the extent that many were actually on, or close to, a four-day workweek if averaged on an annual basis. The decline in the construction industry during this period, however, was significantly less than in other industries. This slower rate of decline is partially attributable to the simultaneous growth of open shop construction, since nonunion construction employees tend to work longer hours than their union counterparts.[43]

Hours and Overtime—Final Comment

Union policies on hours of work are basically designed to prevent overtime work or scheduling work after regular work hours. The union theory is that this will spread the work and expand employment and union membership, or at least greatly increase the income of union members because of expensive premium pay. In fact, of course, the net effect is to cause unemployment among union members by making the costs of unionized contractors noncompetitive so that less projects are built and / or those that are built are increasingly done by open shop contractors.

---

[42] "Terms and Conditions Reference Guide. Work Schedules and Pay Provisions," *Construction Labor Research Council,* August 14, 1984, p. 9.

[43] Theresa Diss Greis, *The Decline of Annual Hours Worked in the United States Since 1947,* Manpower and Human Resources Studies No. 10 (Philadelphia: Industrial Research Unit, The Wharton School, University of Pennsylvania, 1984), pp. 81–82.

The givebacks in the early 1980s featured issues of hours and pay for time not worked in construction as in many industries. Construction unions in many parts of the country agreed to time and one-half instead of doubletime for overtime work, modified or abolished call-in pay, and in a few cases, even agreed to a four-day, ten-hour schedule without daily overtime. Nevertheless, the basically different philosophy regarding hours worked and the cost advantages resulting therefrom for the open shop contractors remain, even if in some areas they are lessened by the union realization of the effect of costs on competitiveness and employment. The employees of open shop contractors who accept lower overtime premiums, or none, and less pay for time not worked, or none, are more likely to be paid when weather is inclement or to work a schedule that permits forty hours' pay regardless of bad weather.

PREVAILING WAGE LAWS

In several previous chapters, reference has been made to prevailing wage laws, particularly the Davis-Bacon Act of 1931 and the state "Little Davis-Bacon" laws. Particularly as the Davis-Bacon Act was administered prior to 1984, and as many of the state laws are still operated, they have acted to offset the open shop's cost advantages in public works. Because these laws have been the subject of other Wharton Industrial Research Unit studies they will not be examined in depth here.[44] As already noted in several previous chapters, the impact of these laws has been to make the union rate the prevailing rate whether or not wage surveys actually found the union rate to be prevailing in a particular area, and to eliminate the use of helpers on government-financed jobs, thereby making government buildings the most expensive to construct.[45] Modifications in state laws and the already noted new regulations governing

---

[44] The original IRU study of the Davis-Bacon Act, now out of print, has been the basis for much of the discussion of the Act's inequities. See Armand J. Thieblot, Jr., *The Davis-Bacon Act,* Labor Relations and Public Policy Series, Report No. 10 (Philadelphia: Industrial Research Unit, The Wharton School, University of Pennsylvania, 1975). Dr. Thieblot is completing a revision of this work, which will be published in late 1984 or early 1985 and will also incorporate studies of the state "Little Davis-Bacon" laws and the federal Service Contract Act. An analysis of the state laws has been prepublished as "Prevailing Wage Laws of the States," *Government Union Review,* Vol. 4 (Fall 1983), pp. 1–65. The study of the Service Contract Act, by Beverly H. Burns, Esq., has been published under the title "The Service Contract Act: A Cure Where There is No Disease?" *Villanova Law Review,* Vol. 29 (January 1984), in press.

[45] "Public Buildings Top Cost List," *Engineering News-Record,* February 16, 1984, p. 14. This article reports on a study by the Construction Resources Analysis group, University of Tennessee.

Davis-Bacon wage and worker complement setting should ease this situation somewhat, but the fact remains that these laws all too frequently force open shop builders to adopt the costly practices of their unionized competitors instead of being able to pass on to the users and the public the benefits of their more economical deployment and wage practices.

CONCLUSION

The foregoing account of compensation in open shop construction provides several insights into the nature of the competitive threat posed by nonunion contractors to the building trades unions. A primary element, of course, is the large differential in average wages between the two sectors. Although, as we have seen, the precise magnitude of this differential is difficult to pinpoint, an estimate of 59 to 60 percent is reasonably consistent with the available evidence. Thus, if labor is assumed to be 24 percent of the total cost of a project, then other things being equal, the open shop contractor's total cost of production will be at least 8 percent lower as a result of the wage difference alone.

The wage differential itself appears to have two elements. First, even the top workers in open shop construction are very often paid less than the union scale. The situation appears to vary, however, since open shop contractors sometimes pay the union rate or higher for a top-rated craftsman.

Second, open shop contractors pay an appreciably broad range of wages to journeymen and helpers of differing levels of skill and experience, whereas the union contractor is usually bound to pay all workers who fall under a given trade classification the same wage. The open shop firm, on the other hand, can and often does, especially on private sector work, pay different rates even within trades. The result is a wage distinction between the fully qualified craftsman and the narrowly trained specialist, one which the union employer cannot make. Thus, average wages in the open shop are lower than in the union shop, even though open shop contractors pay their top craftsmen at the union scale in order to retain them as employees.

The wage differential alone does not account for the total difference in worker compensation between the two sectors. Studies have consistently shown that open shop workers receive a greater proportion of total employment expenditures "in the envelope" than do their organized counterparts. This factor, of course, adds to the labor cost advantage enjoyed by open shop firms, as does the absence of daily premium requirements, call-in pay, and other additions to compensation. None of this, moreover, takes into account possible differences in worker productivity or job components to which, according to many open shop spokesmen, are attributable the most significant labor cost advantage for their contractors.

For the analyst, however, the relative absence of systematic longitudinal data on open shops constitutes the major impediment to a broader understanding of wage determination in construction. Despite the BLS studies and others, knowledge is fragmentary. Nevertheless, the data available all indicate that the union-open shop differential has not lessened over the years but has probably increased somewhat. Moreover, the sharp decline in union wage increases during the early 1980s and the widespread wage-and-rule concessions during this period, all point to the direct effect of open shop compensation practices on union policy.

Third, it is obvious that open shop wage practices are directly responsive to market as well as economic realities. In contrast, until major open shop expansion and unemployment brought the market to bear, union wages tended in many areas to rise dramatically despite the operation of these two forces. The dispersion of wage rates in open shop construction, both among and within employing units, suggests a rough correspondence between wages and marginal product.

The tendency for differentials to be relatively narrow in tightly organized localities, and wider where open shop activity is more substantial, again intimates an accommodation of nonunion wages to the dictates of labor supply. Thus, an important determinant of the union-nonunion differential is the extent to which the building trades are able to raise their members' wages above market-determined levels. Given the 50 to 60 percent union-open shop wage differential, this has been considerable. The experience of the early 1980s, however, suggests that the vast growth of the open shop has become a greater restraint on union wage aggrandizement than heretofore. This will vary in the future, of course, because union power will exert itself when employment is high and open shop competition in an area construction sector is weak. Nevertheless, the growth of the open shop, the increased size and capability of open shop contractors, and the greater willingness of users to build open shop may well point to considerably greater restraint on construction union wage inflation in the 1980s than was experienced in the 1970s.

CHAPTER XII

Minority and Female Employment

The construction industry has always been a key source of employment for minorities. This is especially true in the South, where the black craftsman has played a key role since the days of slavery. On the other hand, the construction industry has also featured some of the most invidious discrimination against minorities. Black craftsmen from the South have been denied work and/or union membership in the North, and until recently training was closed to blacks in many areas of the North and South. Since the 1960s, however, federal policies and the rise of the open shop have both contributed to improving the employment situation of blacks and other minorities. In addition to exploring in depth the role of blacks in the construction industry, this chapter will pay special attention to the increasing participation of Hispanics and women.

Within this context, the relative utilization of minorities by union and open shop contractors, like most other aspects of open shop activity, has not heretofore been explored except in the original edition of this work.[1] As in that volume, the conclusions set forth in this chapter must be considered tentative since the data were collected in conjunction with research on other aspects of open shop construction rather than for an in-depth study of minority employment.

BLACK EMPLOYMENT IN CONSTRUCTION SINCE 1890

Table XII-1 shows the number and proportion of blacks in the construction industry generally and in seven crafts in the United States, and in ten southern states, from 1890 to 1940.[2] Some very

---

[1] The Industrial Research Unit was given a grant in 1973 by the U.S. Department of Labor, Employment and Training Administration, to study the status of minorities in construction. Before much was accomplished, then Secretary of Labor Brennan, who before and after his short tenure in Washington was president of the New York City Building and Construction Trades Council, ordered that the grant be withdrawn. No such study has been made since.

[2] Discussion of the period 1890–1940 is based upon Herbert R. Northrup, *Organized Labor and the Negro* (New York: Kraus Reprint Co., 1971), Chapters I and II.

TABLE XII-1
Number and Proportion of Blacks in Selected Occupations of the Building Trades in the United States and the South,[a] 1890–1940

Occupation and Area	1890			1900			1910		
	All Workers	Blacks	Percent Blacks	All Workers	Blacks	Percent Blacks	All Workers	Blacks	Percent Blacks
Carpenters									
United States	611,482	22,318	3.6	602,741	22,435	3.7	682,490	29,039	4.3
South	70,327	18,017	25.6	78,728	17,241	22.5	104,606	24,624	23.2
Painters									
United States	219,912	4,386	2.0	277,990	5,934	2.1	273,441	8,035	2.9
South	13,105	2,911	22.2	17,763	3,821	21.5	22,212	5,628	25.3
Bricklayers									
United States	158,518	9,647	6.1	161,048	14,457	9.0	160,151	12,014	7.5
South	14,628	6,857	47.0	14,855	7,933	53.4	16,125	8,817	54.7
Plasterers and cement finishers									
United States	39,002	4,006	10.3	35,706	3,754	10.5	47,682	6,175	13.0
South	2,408	1,261	52.5	2,964	1,841	62.1	3,329	2,213	66.5
Plumbers									
United States	56,607	616	1.1	97,884	1,197	1.2	119,596	1,990	1.7
South	n.a.	n.a.	n.a.	4,375	781	17.9	8,822	1,366	15.5
Electricians									
United States	n.a.	n.a.	n.a.	n.a.	n.a.	n.a.	47,024	293	0.6
South	n.a.	n.a.	n.a.	n.a.	n.a.	n.a.	8,056	250	2.9
Total six crafts									
United States	n.a.	n.a.	n.a.	n.a.	n.a.	n.a.	1,330,384	57,546	4.3
South	n.a.	n.a.	n.a.	n.a.	n.a.	n.a.	163,150	42,900	26.3

	1920			1930			1940[c]		
	All Workers	Blacks	Percent Blacks	All Workers	Blacks	Percent Blacks	All Workers	Blacks	Percent Blacks
Carpenters									
United States	887,379	34,217	3.9	929,426	32,413	3.5	690,526	25,427	3.9
South	125,913	25,774	20.5	125,714	20,699	16.5	129,779	17,785	13.7
Painters									
United States	248,497	8,026	3.2	430,105	15,677	3.6	439,472	16,841	3.8
South	23,067	5,228	22.6	40,184	6,830	17.0	52,194	7,578	14.5
Bricklayers[b]									
United States	131,264	10,606	8.1	170,903	11,701	6.9	135,013	8,159	6.0
South	13,634	7,317	53.6	17,386	7,716	44.4	19,228	6,050	31.5
Plasterers and cement finishers									
United States	45,876	7,079	15.4	85,480	13,465	15.8	68,750	10,431	15.2
South	4,105	2,619	63.8	10,325	6,317	61.2	10,515	5,729	54.5
Plumbers									
United States	206,718	3,516	1.7	237,814	4,729	2.0	198,477	4,299	2.2
South	15,287	2,002	13.1	20,340	2,445	12.0	22,210	2,461	11.1
Electricians									
United States	212,964	1,342	0.6	280,317	1,913	0.7	217,075	1,555	0.7
South	15,393	393	2.6	23,659	391	1.7	29,119	443	1.5
Total six crafts									
United States	1,732,698	64,786	3.7	2,134,045	79,898	3.7	1,749,312	66,712	3.8
South	197,399	43,333	22.0	237,608	44,398	18.7	263,045	40,046	15.2

Reproduced from Herbert R. Northrup, *Organized Labor and the Negro* (New York: Kraus Reprint Company, 1971), pp. 18–19.
Source: U.S. Census of Occupations, 1890–1940.
[a] Includes states of Alabama, Arkansas, Florida, Georgia, Louisiana, Mississippi, North Carolina, South Carolina, Tennessee, and Virginia.
[b] Includes stone cutters in 1940.
[c] Includes totals of "employed" and "experienced workers seeking work" groups, which are roughly comparable to "gainful worker" definition of previous enumerations.
n.a. = not available.

clear trends emerge from these data. First, there was a high concentration of black craftsmen in the South, where as late as 1940, more than three-quarters of our black population dwelled. Second, there was a steady erosion of the Negro position in the South in nearly all crafts except plasterers and cement finishers over the fifty-year period, without any significant gains nationwide. Third, there was a wide discrepancy between black representation in older crafts, such as carpentry, painting, and the trowel trades of bricklaying, plastering, and cement finishing, and newer crafts such as electrical and plumbing work. This differentiation is even more stark if one understands that the bulk of black plumbers in the South were sewer workers, while the newer work was largely reserved for white workers.

Unions played a substantial role in these developments. Craft consciousness came to be associated with race consciousness making it simple to keep blacks out of certain trades. Whereas in the South blacks were admitted to unions on a nondiscriminatory basis, as in the trowel trades, or given "separate but equal" local unions, as in the Carpenters or Painters unions, these black members found themselves unwelcome and were denied work in the border and northern states. The Electrical and Plumbers unions excluded blacks on a national scale by tacit consent, as did the Bridge, Structural, and Ornamental Iron Workers. The Sheet Metal Workers, long one of the most discriminatory unions in the country, offered blacks membership in "auxiliary unions" under the supervision of white locals, or excluded them altogether.

Licensing agencies also played a key role in eliminating the competition of blacks in the electrical and plumbing trades in the North and South. Unions were usually represented on the boards of these agencies, and blacks thus found it very difficult, if not impossible, to pass tests and to receive certification in these trades. If they did, they found that inspectors, who were often unionists, would find the work unsatisfactory and refuse to pass or to certify it.

Even though unions were small in the South, their role was significant. They were represented on licensing boards and as inspectors. Construction union members, traditionally of the white working class, had historically been in competition with blacks. Because blacks could not vote, whites were a force beyond their size.

The discrimination was communitywide. Segregated vocational schools did not teach blacks the trades they needed in order to compete. Contractors did not train them, and they were denied admission to apprentice programs. In short, they lost ground between 1890 and 1940 in the older crafts and never developed strength in the newer ones. Significantly, those blacks who did succeed in breaching the racial barriers outside of the trowel trades largely did so in the nonunion sector, particularly in residential and small commercial construction. Moreover, the open shop builder welcomed this source of cheaper labor. The racial policies of the craft unions thus fostered an alliance between Southern black workers and open shop contractors.

The Great Depression to 1964

The antagonism between craft unions and blacks became more severe during the Great Depression. Recognizing the need to aid unemployed blacks, special programs were inaugurated by the Roosevelt administration to provide a share of public works and relief employment for black construction workers. Quotas were established and pressure applied to contractors and unions to effectuate this program. The result was some improvement in black employment, and some increase in black union membership. In many areas, however, blacks did not receive their share of work in various crafts, and when they did, they were often required to obtain costly work permits and were not admitted to the unions.[3]

The relief construction of the 1930s, followed by the construction boom of the World War II years greatly enhanced union strength. Again the federal government moved to aid black workers and again, it had limited success. The same period saw a major migration of blacks from the South to the urban areas of the North, Midwest, and West. Black construction workers again found that discrimination by unions was often more severe in the North than in the South.

Table XII-2 shows the number and proportion of blacks in the construction industry and in the same seven crafts listed in Table XII-1 for the United States and ten southern states, for the period from 1950 to 1980. This table also includes data for construction laborers. The data indicate that in the United States, the proportion of black workers in all seven crafts rose from 4.8 percent in 1950 to 6.9 percent in 1970, but then declined to 6.6 percent in 1980. Of special interest have been the slow but steady gains throughout these thirty years in the electrical and plumbing trades, while the

---

[3] *Ibid.*, Chapter II; Robert C. Weaver, *Negro Labor, A National Problem* (New York: Harcourt, Brace & Co., 1946); and Marc W. Kruman, "Quotas for Blacks: The Public Works Administration and the Black Construction Worker," *Labor History*, Vol. 16 (Winter 1975), pp. 37–51.

TABLE XII-2

Number and Proportion of Blacks in Selected Occupations of the Building Trades in the United States and the South,[a] 1950–1980

Occupation and Area	1950			1960			1970			1980		
	All Male Workers	Blacks	Percent Blacks	All Male Workers	Blacks	Percent Blacks	All Male Workers	Blacks	Percent Blacks	All Male Workers	Blacks	Percent Blacks
Carpenters												
United States	907,728	34,582	3.8	816,195	35,830	4.4	831,363	43,530	5.2	1,275,666	62,958	4.9
South	195,564	22,565	11.5	187,150	22,513	12.0	195,327	24,726	12.7	271,669	28,718	10.6
Painters, glaziers, paperhangers												
United States	409,947	21,953	5.4	386,463	26,289	6.8	323,175	29,065	9.0	412,185	27,520	6.7
South	70,536	10,229	14.5	77,736	11,020	14.2	73,052	11,818	16.2	89,958	14,331	15.9
Masons, tile setters, stone cutters												
United States	172,876	18,003	10.4	191,169	21,738	11.4	191,432	29,083	15.2	234,868	31,915	13.6
South	33,055	11,256	34.1	42,503	13,385	31.5	50,315	17,589	35.0	54,188	16,618	30.7

Plasterers and cement finishers												
United States	89,112	17,227	19.3	86,678	18,772	21.7	92,950	25,457	27.4	102,858	21,429	20.8
South	16,327	9,167	56.1	17,592	9,801	55.7	21,833	13,197	60.4	20,084	10,097	50.3
Plumbers and pipefitters												
United States	275,892	8,290	3.0	303,541	10,120	3.3	377,769	17,371	4.6	483,423	31,742	6.6
South	40,383	4,409	10.9	48,600	5,157	10.6	69,358	7,488	10.8	99,480	13,146	13.2
Electricians												
United States	307,013	3,236	1.1	334,732	4,978	1.5	459,843	13,408	2.9	594,781	28,019	4.7
South	44,066	887	2.0	54,233	1,148	2.1	88,704	3,085	3.5	128,481	8,765	6.8
Total six crafts												
United States	2,162,568	103,291	4.8	2,118,778	117,727	5.6	2,276,532	157,914	6.9	3,103,781	203,583	6.6
South	399,931	58,513	14.6	427,814	63,024	14.7	498,589	77,903	15.6	663,860	91,675	13.8
Laborers												
United States	649,341	164,357	25.3	608,541	157,724	25.9	589,906	130,903	22.2	833,937	143,554	17.2
South	131,197	69,843	53.2	133,420	73,393	55.0	148,784	67,103	45.1	177,926	62,231	35.0

Source: U.S. Census of Population:

1950: Vol. II, Characteristics of the Population, United States Summary, Table 128, State Volumes, Table 77.
1960: Vol. I, Characteristics of the Population, United States Summary, Table 205, State Volumes, Table 122.
1970: PC(1)-D, Detailed Characteristics, United States Summary, Table 225, State Volumes, Table 173.
1980: PC80-S1-8, *Detailed Occupation and Years of School Completed by Age, for the Civilian Labor Force by Sex, Race, and Spanish Origin,* Table 1. PC80-1-D, *Detailed Population Characteristics,* State Volumes, Table 219.

[a] Includes states of Alabama, Arkansas, Florida, Georgia, Louisiana, Mississippi, North Carolina, South Carolina, Tennessee, and Virginia.

proportion of laborers has declined. The proportion of blacks in crafts with traditional representation oscillated throughout the thirty-year period, declining between 1970 and 1980.

The data in Table XII-2 also show that the decline in the proportion of black craftsmen in the South was arrested after 1940, and some small gains were made until 1970, after which a fall off occurred. Nevertheless, the ratio achieved during the period of stabilization in the South does not substantially exceed the 1940 ratio, the lowest proportion of blacks during the fifty years prior thereto. Moreover, black representation in crafts other than the trowel trades and painting continues to be substantially below their labor force ratio, both nationwide and in the South. In contrast, their share of the construction laborer category remained disproportionately high throughout the thirty-year period.[4]

During the post–World War II period until the passage of the Civil Rights Act in 1964, the construction unions' racial policies were attacked by a number of northern civil rights groups and state fair-employment agencies; but although there was much agitation, change appears to have been slight.[5] Black membership in electrical, plumbing, and sheet metal workers' unions remained low throughout the country. Meanwhile, as the number of open shop firms in commercial and residential construction grew, black craftsmen increasingly found work in these sectors of the industry.[6]

The Period Since 1964

Since the passage of the Civil Rights Act in 1964, the federal government has attempted to expand black participation in skilled construction jobs by a variety of approaches: encouragement and support of apprenticeship and training; enforced and voluntary plans to expand employment in various cities; contractual provisions to accomplish the same goal; and litigation to end discriminatory practices. The results have broken down barriers, opened up training and apprenticeship programs, and scored a number of local successes.[7] Change has indeed occurred, but it has moved slowly,

---

[4] There was, however, a curious decline in the proportion of black construction laborers in the South between 1960 and 1970, two possible explanations for which include migration and upgrading.

[5] For the period of the 1950s, see Ray Marshall, *The Negro and Organized Labor* (New York: John Wiley & Sons, Inc., 1965), especially Chapters 5, 6, and 8.

[6] These comments derive from numerous interviews throughout the country, as well as forty years of observation of these trends by the author.

[7] For analyses of these various programs, see Stephen A. Schneider, "Apprenticeship Outreach Program," in Charles R. Perry et al., *The Impact of Government*

as the data in Table XII-2 indicate. Blacks still remain considerably underrepresented in the skilled crafts in proportion to their ratio in the population and labor force.

Until recently, governmental programs designed to expand black employment were concentrated principally in union construction. This is particularly the case with apprentice outreach programs, the goal of which is to attract minorities into union construction. Yet this study indicates that the open shop sectors of the industry are expanding the most rapidly.

A fundamental problem which minorities in the unionized sector face is the very structure of the union organization. Union craft lines institutionalize the status quo, making it difficult for a laborer to gain experience in other crafts. The craft unions' successful drive to curtail the work of helpers has also prevented minorities from gaining trade skills. Unlike the open shop, where a worker may progress from laborer to helper and thus pick up craft knowledge, this opportunity in union shop construction has been seriously diminished both by the virtual elimination of the helper and by the existence of separate unions for laborers and individual crafts. Since minorities are disproportionately concentrated in the laborer classification, craft union lines act as a barrier preventing their advancement. The absence of such barriers in open shop construction is a major advantage for minorities.

Numerous court cases involving discrimination by union shop contractors attest to the difficulties minorities have had in gaining entrance to, or fair employment under, union rules, regulations, and policies.[8] This extensive litigation has resulted in some improvement, but without any fundamental change in the union organization structure, the problem is likely to remain. Given the strong craft consciousness and the desire to limit the labor market, which derives directly from that consciousness, blacks and other minorities are likely to remain underrepresented in the skilled

---

Manpower Programs, Manpower and Human Resources Studies, No. 4 (Philadelphia: Industrial Research Unit, The Wharton School, University of Pennsylvania, 1975), Chapter 10; and Richard L. Rowan and Lester Rubin, *Opening the Skilled Construction Trades to Blacks: A Study of the Washington and Indianapolis Plans for Minority Employment,* Labor Relations and Public Policy Series, Report No. 7 (Philadelphia: Industrial Research Unit, The Wharton School, University of Pennsylvania, 1972).

[8] The cases may be found in any legal service, e.g., the *CCH Reports* or the Bureau of National Affairs, Inc., *Fair Employment Reports.* As this was being written in April 1984, an important case involving discrimination by the electrical workers in Chicago was finally settled. See Ridgeway IBEW Local 134 and Chicago Chapter NECA, No. 74-C-3045 (N.D. Ill. 1984). For details see *Cockshaw's Construction Labor News + Opinion,* Vol. 14, No. 4 (April 1984), pp. 1–4. The case involves years of invidious discrimination which took ten years of litigation to settle.

construction trades, especially the electrical, plumbing, and mechanical trades.

In more recent years, civil rights measures have concentrated on promoting work for minority contractors in the hopes that this would spur increased minority employment and business development. This has involved state and federal governments, requiring that a portion of government contracts and subcontracts be given to companies owned and operated by minorities. These provisions for Minority Business Enterprises (MBEs) have been bitterly fought by the Associated General Contractors (AGC) and other contractor groups on various legal and practical grounds, including the limited number of qualified minority contractors. There is, for example, considerable evidence that some minority contractors have been set up by major firms in order to meet government subcontracting requirements and are not really owned or operated by the persons acting as principals. In one case, such a firm was disqualified as a minority enterprise, but this was at the instigation of the construction unions who were fighting the major contractor involved, an open shop firm which had garnered much of the formerly unionized work in Oregon.[9]

Whatever the drawbacks of minority set asides, they have created work and jobs for minorities. Moreover, officials of the Associated Builders and Contractors (ABC) believe that the accent on minority contractors and workers aids the open shop. They claim that most minority contractors operate open shop, and therefore, their work adds to the open shop share.[10] The extent to which this is accurate has not been thoroughly investigated.

HISPANIC AND FEMALE CONSTRUCTION EMPLOYMENT

Hispanics, primarily Mexican-Americans, have recently increased their share in the construction labor force in a number of states. In addition, this once all-male industry has been prodded by government equal employment programs to recognize that females can become proficient at crafts if given the opportunity. This section reviews the employment status of these two groups.

Hispanic Employment

Table XII-3 shows the employment of Hispanics in construction in the United States and in a five state area—Texas, New Mexico,

[9] "Feds Decertify Oregon MBE," *Engineering-News Record,* April 12, 1984, p. 63.

[10] Interview, Miami Beach, Florida, February 15, 1984.

TABLE XII-3
Number and Proportion of Hispanics in Selected Occupations of the Building Trades in the United States and Five Selected States[a]
1970 and 1980

Occupation and Area	1970 All Male Workers	1970 Hispanics	1970 Percent Hispanics	1980 All Male Workers	1980 Hispanics	1980 Percent Hispanics
Carpenters						
United States	831,363	31,273	3.8	1,275,666	71,587	5.6
Five states	136,463	23,034	16.9	235,905	43,239	18.3
Painters and paperhangers						
United States	323,175	17,721	5.5	412,185	39,544	9.6
Five states	64,985	12,539	19.3	93,105	25,076	26.9
Masons and tile setters						
United States	191,432	6,832	3.6	234,868	15,190	6.5
Five states	22,456	5,114	22.8	36,394	10,334	28.4
Plasterers and cement finishers						
United States	92,950	8,976	9.7	102,858	14,649	14.2
Five states	24,043	7,698	32.0	31,754	11,525	36.3
Plumbers and pipefitters						
United States	377,769	11,247	3.0	483,423	25,242	5.2
Five states	62,213	8,578	13.8	99,755	17,747	17.8
Electricians						
United States	459,843	12,282	2.7	594,781	22,934	3.9
Five states	74,061	8,211	11.1	112,465	14,069	12.5
Total six crafts						
United States	2,276,532	88,331	3.9	3,103,781	189,146	6.1
Five states	384,221	65,174	17.0	609,378	121,990	20.0
Laborers						
United States	589,906	44,449	7.5	833,937	89,239	10.7
Five states	109,915	36,759	33.4	168,762	63,228	37.5

Source: U.S. Census of Population:
1970: Detailed Characteristics, PC(1)-D1, U.S. Summary, Table 225; State Vols., Table 173.
1980: *Detailed Occupation and Years of School Completed by Age for the Civilian Labor Force by Sex, Race, and Spanish Origin,* PC80-S1-8, Table 1; *Detailed Population Characteristics,* PC80-1-D, State Vols., Table 219.

[a] Texas, Mexico, Arizona, California, and Colorado.

Arizona, California, and Colorado—for 1970 and 1980. Over the ten-year period, Hispanics made substantial gains, increasing their share of work in the United States in the construction crafts from 3.9 percent to 6.1 percent and showing increased representation in each craft group, as well among laborers. Overall, Hispanics have their largest share of employment in plastering and cement finishing, and their smallest in electrical and plumbing work. In the five states, however, they have a respectable share of this work also, as well as 20 percent of that in all six crafts combined.

Undoubtedly, the mitigation of discrimination is one reason for these Hispanic gains, but the major reason is probably the tremendous expansion of population, labor force, and construction work in the Southwest, California, and the Mountain states where most Mexican-Americans reside. As the data in Table XII-3 further demonstrate, Mexican-Americans have greatly increased their participation in construction in these states since 1970. Moreover, given the large number of illegal aliens, it is likely that the number shown by the census data is considerably understated.

Contractors interviewed in these five states were high in their praise of Mexican-American workers, stating that they worked hard and well. Many expressed preference for them over black and white workers. Our impression is that, in these states, they compose a larger proportion of the open shop than the unionized workforce. Some may prefer to avoid formal hiring halls because of questions of their legal status; others have experienced discrimination by unions or prefer an attachment to a contractor rather than to a union. It is, therefore, probably not coincidental that the expansion of Mexican-American participation in key crafts of the construction industry occurred during a period in which open shop contractors won sizable increases in their share of work in the states in which the Mexican-American population is concentrated.

Female Employment

Table XII-4 shows female construction employment in 1970 and 1980 for the same occupations which were examined in prior tables. It will be immediately seen that although women have begun to enter the industry, their numbers remain few, even after some small gains in the 1970s. The exception is among painters, paperhangers, and glaziers. Here female employment rose to 6.1 percent by 1980. This often less strenuous, inside work appears to be a niche that women have carved out in the industry. Minority women, not shown in Table XII-4, represent an even smaller portion of the workforce, considerably below that of their male counterparts, at less than 1 percent of all occupations including laborers. Construction remains a male bastion, although women have a foot in the door. No evidence of greater female participation in unionized or open shop concerns was uncovered during these investigations.

TABLE XII-4
Total Employed Females in Selected Occupations of the Building Trades, United States, 1970 and 1980

	1970			1980		
Occupation	Total Workers	Female	Percent Female	Total Workers	Female	Percent Female
Carpenters	852,134	11,110	1.3	1,295,891	20.225	1.6
Painters, paperhangers, and glaziers	365,834	14,735	4.0	473,245	29,036	6.1
Masons, tile setters, and granite cutters	196,710	2,475	1.3	237,783	2,915	1.2
Plasterers and cement finishers	97,799	1,569	1.6	104,254	1,396	1.3
Plumbers and pipefitters	384,359	4,048	1.0	489,644	6,221	1.3
Electricians	480,615	8,283	1.7	606,952	12,171	2.0
Total six crafts	2,377,451	42,200	1.8	3,201,548	71,964	2.2
Laborers	556,773	9,206	1.6	861,223	27,286	3.2

Source: U.S. Census of Population, 1970, PC(2)-7A, *Occupational Characteristics,* Table 39.
U.S. Census of Population, 1980, PC80-S1-8, *Detailed Occupation and Years of School Completed by Age, for the Civilian Labor Force by Sex, Race, and Spanish Origin,* Table 1.

RELATIVE MINORITY AND FEMALE EMPLOYMENT— OPEN AND UNION SHOP

It was a common, although not universal, view among persons contacted in our field surveys both in 1974 and 1982 that open shop contractors employ a larger portion of minorities in skilled work than do their unionized counterparts. There are, however, no definitive data on this subject. One reason, of course, is that open shop contractors are concentrated in the South where the tradition of black craftsmen in trowel and other trades remains strong, and where slightly more than one-half of the black population and a sizable proportion of the Hispanic population dwells. Another reason is undoubtedly that minorities, having experienced discrimination from the construction unions, have tended to seek work where those unions cannot influence employment decisions.

As we noted in Chapter II, however, there are also contrary factors. The most significant is that open shop construction is strongest outside central cities, while the bulk of blacks live in major metropolitan centers. Consequently, this often reduces the potential for minority employment, particularly in residential construction.

Bureau of Labor Statistics Studies

The U.S. Bureau of Labor Statistics (BLS) has made a number of studies of the demographic characteristics of union members based upon questions asked in the Current Population Survey of the U.S. Bureau of the Census. In 1970, it found that blacks had a higher labor union participation rate than whites in nearly every industry within each region. Construction, however, was an exception. About 9 percent of union construction workers in the sample were black, but blacks represented 10 percent of the industry's total wage and salary employment. In the South, where the highest percentage of blacks are found in the construction industry, 23.2 percent of the whites were union members, but only 17.4 percent of blacks. For construction craftsmen nationwide, the relative membership was even more disparate—56.2 percent for whites, 36.8 percent for blacks. Even among laborers, the union participation rates of blacks, at 29 percent, trailed that of whites, at 30.2 percent.[11]

A similar study conducted in May 1977 again found that minorities comprised 11.6 percent of the civilian labor force but 14.2 percent of the labor union membership.[12] This report found that minorities represented 8.8 percent of all employed construction workers and the same percentage of those represented by unions. It also reported that 51.9 percent of construction craft workers except carpenters were union represented, but only 40.3 percent of minorities in this category were so represented. On the other hand, 37.3 percent of all construction laborers were union represented in contrast to 39.6 percent of minority ones.[13] These data thus show that by 1977, minority construction laborers were more than proportionately union represented, but minority craftsmen less than proportionately unionized, again pointing up the historic craft union-minority antagonism in the construction industry and the tendency for minority craftsmen to work open shop.

The 1977 BLS report showed that Hispanic workers compose 4.3 percent of all construction workers and 4.0 percent of those union

---

[11] U.S. Bureau of Labor Statistics, *Selected Earnings and Demographic Characteristics of Union Members, 1970,* Report 417 (Washington, D.C.: Government Printing Office, 1972), pp. 3, 5, and Table 1, p. 6, and Table 8, p. 21.

[12] U.S. Bureau of Labor Statistics, *Earnings and Other Characteristics of Organized Workers, May 1977,* Report 556 (Washington, D.C.: Government Printing Office, 1976), p. 1. Blacks comprised 90 percent of all minorities.

[13] *Ibid.,* Table 1, p. 8, and Table 16, p. 49.

represented. In comparison, Hispanics compose 4.5 percent of employed wage and salary workers for all industries and 4.9 percent of those represented by labor organizations. Hispanics comprised 4.5 percent of carpenters and 5.8 percent of construction craft workers except carpenters, but only 3.9 percent and 5.1 percent, respectively, of the unionized workers in these categories. Surprisingly, Hispanics composed 7 percent of construction laborers but only 6.6 percent of those union represented.[14] Hispanics, like other minorities, compose a lower percentage of unionized construction workers than of construction workers industrywide. The opposite, however, is true of Hispanics and other minorities in the labor force in general, where their share of union membership exceeds their proportion in the workforce.

The BLS conducted a final study of this nature in May 1980. It found again that blacks and other minorities composed 11.7 percent of the labor force but constituted 14.9 percent of union members. Thirty-two percent of black employees, and 29 percent of Hispanics, were union represented, but only 24.9 percent of whites.[15]

This survey reported that in 1980 minorities constituted 9 percent of construction workers and 9.4 percent of that industry's union-represented workers. For carpenters, minorities were 7.4 percent of the labor force, but only 5.2 percent of union-represented workers. Similarly, 10.8 percent of construction craft workers except carpenters were minorities, but minorities were only 8.5 percent of those union represented.[16]

The results for Hispanics follow a similar pattern. This report found that they composed 5.4 percent of the labor force, but 6.1 percent of those union represented. In construction, however, Hispanics accounted for 6.4 percent of the labor force and 6 percent of those union represented. For carpenters, Hispanics were found to be 6.4 percent of the labor force, but only 3.9 percent of those union represented; for construction craft workers except carpenters, the figures were 6.2 percent of the labor force, and 4.8 percent of those union represented. In 1980, Hispanics composed 10.8 percent of construction laborers and 12.0 percent of those union represented, a significant increase over the figures for 1970.[17]

The BLS statistics indicate that between 1970 and 1980 minorities remained underrepresented in construction labor unions, thus sug-

[14] *Ibid.,* Tables 17 and 18, pp. 50–51.

[15] U.S. Bureau of Labor Statistics, *Earnings and other Characteristics of Organized Workers, May 1980,* Bulletin 2105 (Washington, D.C.: Government Printing Office, 1981), p. 1.

[16] *Ibid.,* Tables 16 and 17, pp. 65–66.

[17] *Ibid.,* Tables 18 and 19, pp. 67–68.

gesting that open shop projects employ a greater proportion of minority workers.

The Industrial Research Unit Questionnaires

In 1974 and 1982, contractors were asked to estimate the level of minority employment in open vs. union shops. Tables XII-5 and XII-6 summarize the replies. The 1982 questionnaire included females as well as minorities. Since, however, it was directed both to unionized and open shop contractors, and since female employment in the industry is such a small percentage of the total, the data in the two tables are reasonably comparable.

The results for the two questionnaires are rather different. Only 20 percent of the unionized contractors responding in 1982, as compared with 25.7 percent in 1974, felt that open shop contractors employed a higher proportion of minorities than did union contractors. Forty-one percent in 1982, as compared with 24 percent in 1974, felt the ratios were about the same, and 28.6 percent in 1982, as compared with only 9.7 percent in 1974, believed that the open shop had a smaller share of minorities. For the open shop respondents, 34 percent in 1982, in contrast to 42.3 percent in 1974, believed that the open shop proportion of minorities was greater. In 1982, 42.4 percent viewed the minority percentage as about the

TABLE XII-5
Percent Distribution of Respondents' Estimates of Minority Employment in Open Shop vs. Union Construction by Region and Union Status, 1974

	Union				Nonunion			
Region	Total	Greater	Less	Same	Total	Greater	Less	Same
Not reported	89	25.8	13.5	33.7	83	53.0	6.0	21.7
New England	27	29.6	11.1	40.7	63	42.9	4.8	30.2
Middle Atlantic	125	31.2	16.8	28.8	134	52.2	10.4	31.3
E.N. Central	147	27.9	14.3	28.6	75	49.3	4.0	36.0
W.N. Central	60	18.3	11.7	41.7	45	44.4	6.7	33.3
S. Atlantic	68	47.1	10.3	23.5	220	55.0	5.9	25.5
E.S. Central	32	31.3	9.4	43.8	45	51.1	2.2	28.9
W.S. Central	45	57.8	4.4	26.7	92	71.7	2.2	17.4
Mountain	52	38.5	9.6	32.7	21	57.1	4.8	19.0
Pacific	133	35.3	12.0	27.8	6	50.0	0.0	50.0
Total	778	25.7	9.7	24.0	784	42.3	4.5	21.3

Source: Industrial Research Unit Questionnaire, 1974.

TABLE XII-6
Percent Distribution of Respondents' Estimates of Minority and Female Employment in Open Shop vs. Union Construction by Region and Union Status, 1982

	Union				Nonunion			
Region	Total	Greater	Less	Same	Total	Greater	Less	Same
Not reported	44	18.2	38.6	27.3	49	42.9	16.3	36.7
New England	62	24.2	40.3	24.2	111	31.5	22.5	32.4
Middle Atlantic	275	21.8	31.3	36.7	228	28.5	22.4	39.0
E.N. Central	474	15.8	30.8	41.8	205	26.3	22.9	44.4
W.N. Central	220	26.8	20.5	45.5	157	36.3	14.0	45.2
S. Atlantic	109	25.7	28.4	37.6	550	34.0	12.2	45.8
E.S. Central	69	20.3	23.2	46.4	111	26.1	18.0	48.6
W.S. Central	105	33.3	20.0	40.0	284	46.8	12.0	37.0
Mountain	108	20.4	21.3	53.7	184	34.2	13.6	48.4
Pacific	362	13.5	31.2	42.8	143	30.8	23.1	36.4
Total	1,828	20.0	28.6	41.2	2,022	34.0	16.4	42.4

Source: Industrial Research Unit Questionnaire, 1982.

same, while in 1974 only 21.3 percent felt this way. In 1982, 6.4 percent believed that the open shop had a smaller percentage of minorities than the union shop, as compared with 4.5 percent in 1974.

Both questionnaires also requested contractor respondents to estimate the percentage of black and Hispanic workers in their labor force. The results are set forth by region and union status in Tables XII-7 through XII-10. Unfortunately, the percentage distributions in the two tables are somewhat different, but by combining categories, they can be compared. Looking first at Tables XII-7 and XII-8, we find that in 1974, 24.7 of the unionized contractors and 32.1 of the open shop contractors reported that they had no blacks. In 1982, these percentages rose to 40.3 percent of the unionized respondents and an astounding 51 percent of the open shop ones. Interestingly, both in 1974 and 1982, in the higher proportional categories open shop contractors had a larger share of blacks than did unionized contractors. In 1974, the open shop ratio exceeded the union shop ratio in all categories above 26 percent and was less in those below. In 1982, open shop contractors had a higher percentage of blacks in categories above 21 percent and a lower percentage in categories below.

The results for Hispanic workers, as set forth in Tables XII-9 and XII-10, are similar. In the lower percentage categories, the open

TABLE XII-7
Percent Distribution of Black Workers in Current Workforce by Region and Union Status, 1974

Region	Total	0	1–5	6–10	11–25	26–50	51–97	98–100
Union								
Not reported	89	24.7	20.2	9.0	11.5	24.2	2.2	0
New England	27	40.7	25.9	11.1	14.8	0	3.7	0
Middle Atlantic	125	31.2	28.0	14.4	12.8	5.6	3.2	0
E.N. Central	147	32.7	21.8	17.0	19.7	4.1	2.0	0
W.N. Central	60	33.3	35.0	11.7	10.0	3.3	0	0
S. Atlantic	68	16.2	16.2	10.3	20.6	25.0	7.4	0
E.S. Central	32	9.4	28.1	15.6	18.8	9.4	9.4	0
W.S. Central	45	13.3	13.3	22.2	24.4	20.0	2.2	0
Mountain	52	48.1	30.8	7.7	3.8	3.8	0	0
Pacific	133	46.6	23.3	12.8	9.0	0.8	0	0
Total	778	24.7	18.6	10.4	11.3	6.2	1.9	0
Nonunion								
Not reported	83	31.3	18.1	7.2	13.3	14.5	13.3	0
New England	63	71.4	12.7	6.3	1.6	3.2	1.6	0
Middle Atlantic	134	57.5	12.7	9.0	6.0	6.7	1.5	0.7
E.N. Central	75	68.0	8.0	8.0	9.3	1.3	2.7	0
W.N. Central	45	44.4	26.7	13.3	6.7	0	0	0
S. Atlantic	220	21.4	11.8	12.3	22.3	19.1	8.2	0
E.S. Central	45	17.8	24.4	17.8	11.1	24.4	2.2	0
W.S. Central	92	30.4	16.3	12.0	15.2	18.5	5.4	0
Mountain	21	76.2	19.0	0	0	4.8	0	0
Pacific	6	50.0	16.7	16.7	0	0	0	0
Total	784	32.1	11.5	8.1	9.8	9.5	4.0	1.0

Source: Industrial Research Unit Questionnaire, 1974.

shop contractors have a smaller proportion of Hispanics than do the unionized contractors. In the upper categories, the opposite is true. Variations because of regional concentration are more pronounced in the case of Hispanic workers than in the case of blacks.

In 1982, but not in 1974, an identical question was asked concerning female employees. The results are found in Table XII-11. Female employment is obviously small in both the open shop and the union sectors, and there seems little difference between the two groups.

While the thrust of the Industrial Research Unit interviews was that minorities were a higher portion of the labor force in open shop than in union shop concerns, the questionnaire results do not support this. The lack of skilled-unskilled breakdowns, the overrepresentation of large firms, and the greater likelihood that union firms have government contracts requiring higher minority obli-

TABLE XII-8
Percent Distribution of Black Workers in Current Workforce by Region and Union Status, 1982

	Union									Nonunion								
Region	Number	NR/DK[a]	0–10	11–20	21–30	31–40	41–50	51–75	76–100	Number	NR/DK[a]	0–10	11–20	21–30	31–40	41–50	51–75	76–100
Not Reported	44	6.8	50.0	27.3	9.1	6.8	0.0	0.0	0.0	49	2.0	42.9	26.5	10.2	4.1	12.2	2.0	0.0
New England	62	1.6	21.0	56.5	16.1	3.2	1.6	0.0	0.0	111	0.9	76.6	17.1	2.7	1.8	0.0	0.0	0.9
Middle Atlantic	275	2.9	28.4	42.5	14.9	5.8	4.0	1.1	0.4	228	3.1	58.8	24.6	6.6	3.5	1.8	0.9	0.9
E.N. Central	474	3.6	39.0	41.8	12.0	1.3	1.9	0.0	0.4	205	2.4	70.7	19.0	5.9	0.5	1.5	0.0	0.0
W.N. Central	220	2.3	55.5	33.6	6.8	1.8	0.0	0.0	0.0	157	1.3	77.1	17.2	3.2	0.0	0.6	0.6	0.0
S. Atlantic	109	1.8	11.9	34.9	21.1	14.7	10.1	4.6	0.9	550	3.1	23.3	27.8	16.9	9.3	12.4	5.5	1.8
E.S. Central	69	2.9	27.5	37.7	11.6	7.2	10.1	1.4	1.4	111	3.6	27.9	35.1	8.1	8.1	8.1	7.2	1.8
W.S. Central	105	2.9	28.6	38.1	11.4	5.7	6.7	6.7	0.0	284	2.1	40.5	32.0	12.0	3.5	3.9	4.6	1.4
Mountain	108	0.0	66.7	27.8	4.6	0.9	0.0	0.0	0.0	184	1.6	80.4	15.8	2.2	0.0	0.0	0.0	0.0
Pacific	362	2.2	50.6	37.8	6.1	2.2	0.8	0.3	0.0	143	4.9	72.0	20.3	2.8	0.0	0.0	0.0	0.0
Total	1,828	2.7	40.3	38.7	10.8	3.7	2.7	0.9	0.3	2,022	2.6	51.0	24.5	9.1	4.1	5.0	2.7	0.9

Source: Industrial Research Unit Questionnaire, 1982.

[a] Not reported or do not know.

TABLE XII-9
Percent Distribution of Hispanic Workers in Current Workforce by Region and Union Status, 1974

Region	Total	0	1–5	6–10	11–25	26–50	51–97	98–100
Union								
Not reported	89	59.6	18.0	1.1	5.6	3.4	0	0
New England	27	74.1	14.8	3.7	3.7	0	0	0
Middle Atlantic	125	76.0	16.0	0.8	0.8	1.6	0	0
E.N. Central	147	72.1	17.7	3.4	1.4	1.4	1.4	0
W.N. Central	60	63.3	28.3	0	1.7	0	0	0
S. Atlantic	68	73.5	8.8	5.9	4.4	2.9	0	0
E.S. Central	32	90.6	0	0	0	0	0	0
W.S. Central	45	51.1	13.3	11.1	4.4	11.1	4.4	0
Mountain	52	34.6	19.2	7.7	17.3	15.4	0	0
Pacific	113	39.8	11.3	15.8	18.8	5.3	1.5	0
Total	778	48.5	12.0	4.2	4.9	2.9	.6	0
Nonunion								
Not reported	83	69.9	13.3	4.8	3.6	3.6	2.4	0
New England	63	79.4	11.1	4.8	1.6	0	0	0
Middle Atlantic	134	82.8	6.7	2.2	1.5	0	0.7	0
E.N. Central	75	84.0	8.0	5.3	0	0	0	0
W.N. Central	45	62.2	20.0	6.7	2.2	0	0	0
S. Atlantic	220	75.9	11.4	2.3	2.7	0.5	0.9	1.4
E.S. Central	45	91.1	6.7	0	0	0	0	0
W.S. Central	92	41.3	8.7	10.9	12.0	13.0	12.0	0
Mountain	21	57.1	19.0	9.5	4.8	9.5	0	0
Pacific	6	33.3	33.3	16.7	0	0	0	0
Total	784	57.0	8.4	3.5	2.5	1.8	1.6	.3

Source: Industrial Research Unit Questionnaire, 1974.

gations may well skew our questionnaire results. Moreover, the concentration of union firms in cities and of open shop firms outside these areas affects the availability of minorities. Finally, the data discussed below show a very high representation of minorities in highway work, not surprising in view of the lower skill content of this sector's work. Nevertheless, regardless of reason, it must be stated that our questionnaires do not yield a definitive answer on relative minority employment in open shop and union shop construction.

A final question on both questionnaires elicited answers in regard to participation of contractors in programs designed to upgrade minority workers. It should be noted that our questionnaires were not sent to members of the National Association of Home Builders, the bulk of whom are nonunion, and therefore did not encompass

Table XII-10
Percent Distribution of Hispanic Workers in Current Workforce by Region and Union Status, 1982

	Union									Nonunion								
Region	Number	NR/DK[a]	0–10	11–20	21–30	31–40	41–50	51–75	76–100	Number	NR/DK[a]	0–10	11–20	21–30	31–40	41–50	51–75	76–100
Not reported	44	6.8	54.5	25.0	0.0	11.4	2.3	0.0	0.0	49	2.0	71.4	18.4	4.1	0.0	2.0	2.0	0.0
New England	62	1.6	50.0	45.2	3.2	0.0	0.0	0.0	0.0	111	0.9	81.1	0.0	0.0	0.0	0.0	0.0	0.0
Middle Atlantic	275	1.8	54.2	36.4	4.0	0.7	2.2	0.4	0.4	228	3.5	76.8	15.8	2.2	1.3	0.4	0.0	0.0
E.N. Central	474	3.4	64.8	30.0	1.7	0.2	0.0	0.0	0.0	205	3.4	82.0	11.7	2.0	0.5	0.0	0.5	0.0
W.N. Central	220	2.3	68.6	28.2	0.9	0.0	0.0	0.0	0.0	157	1.3	79.0	16.6	2.5	0.6	0.0	0.0	0.0
S. Atlantic	109	1.8	47.7	36.7	6.4	0.9	0.9	0.9	4.6	550	2.4	66.4	24.9	2.5	0.5	1.1	0.4	1.8
E.S. Central	69	2.9	87.0	10.1	0.0	0.0	0.0	0.0	0.0	111	4.5	85.6	8.1	1.8	0.0	0.0	0.0	0.0
W.S. Central	105	3.8	34.3	35.2	12.4	4.8	3.8	4.8	1.0	284	1.4	31.0	24.6	9.5	7.7	10.2	6.3	9.2
Mountain	108	0.0	22.2	33.3	20.4	10.2	3.7	7.4	2.8	184	1.6	28.8	39.7	8.2	6.5	7.6	4.3	3.3
Pacific	362	1.9	37.3	24.3	15.2	10.8	5.5	3.3	1.7	143	4.9	44.8	27.3	13.3	4.9	2.1	1.4	1.4
Total	1,828	2.5	53.0	30.1	6.6	3.5	2.0	1.5	0.9	2,022	2.5	62.2	21.9	4.5	2.4	2.7	1.6	2.2

Source: Industrial Research Unit Questionnaire, 1982.

[a] Not reported or do not know.

TABLE XII-11
Percent Distribution of Female Workers in Current Workforce by Region and Union Status, 1982

	Union									Nonunion								
Region	Number	NR/DK[a]	0–10	11–20	21–30	31–40	41–50	51–75	76–100	Number	NR/DK[a]	0–10	11–20	21–30	31–40	41–50	51–75	76–100
Not reported	44	6.8	47.7	38.6	6.8	0	0	0	0	49	2.0	22.4	67.3	8.2	0	0	0	0
New England	62	1.6	35.5	58.1	4.8	0	0	0	0	111	0.9	38.7	46.8	9.0	3.6	0.9	0	0
Middle Atlantic	275	1.8	32.0	62.2	4.0	0	0	0	0	228	3.1	26.8	54.4	13.2	2.6	0	0	0
E.N. Central	474	3.0	34.4	56.0	5.3	0.6	0.8	0	0	205	0.5	36.1	50.7	9.8	0.5	2.4	0	0
W.N. Central	220	1.8	33.2	58.2	5.5	1.4	0	0	0	157	1.3	30.5	54.8	10.8	1.3	0	0.6	0.6
S. Atlantic	109	2.8	30.2	58.8	5.5	2.6	0	0	0	550	2.0	26.5	61.3	7.8	1.6	0.5	0.2	0
E.S. Central	69	2.9	40.6	47.8	5.8	0	1.4	1.4	0	111	2.7	33.3	59.5	3.6	0.9	0	0	0
W.S. Central	105	2.9	25.7	61.0	9.5	0.9	0	0	0	284	2.1	26.1	56.3	10.6	4.6	0.4	0	0
Mountain	108	0	27.8	64.8	6.5	0.9	0	0	0	184	1.1	3.8	53.8	9.8	1.6	2.2	0	0.5
Pacific	362	1.9	37.3	53.6	4.4	2.2	0.6	0	0	143	2.8	25.9	44.8	16.1	5.6	3.4	0.7	0.7
Total	1,828	2.3	33.9	57.0	5.3	1.0	0.4	0	0	2,022	1.9	29.1	55.6	9.8	2.3	0.9	0.1	0.1

Source: Industrial Research Unit Questionnaire, 1982.
[a] Not reported or do not know.

TABLE XII-12
Percent Distribution of Participation in Local Programs to Upgrade Minority Workers, by Region and Union Status, 1974

Region	Union		Nonunion	
	Yes	No	Yes	No
Not reported	64.0	32.6	27.7	60.2
New England	55.6	44.4	30.2	66.7
Middle Atlantic	67.2	29.6	33.6	58.2
E.N. Central	63.3	31.3	36.0	58.7
W.N. Central	61.7	33.3	51.1	35.6
S. Atlantic	57.4	41.2	40.5	53.2
E.S. Central	71.9	21.9	31.1	62.2
W.S. Central	46.7	49.9	52.2	42.4
Mountain	82.7	72.2	38.1	57.1
Pacific	15.4	23.3	50.0	50.0
Total	65.3	30.8	38.1	54.7

Source: Industrial Research Unit Questionnaire, 1974.

the participants in NAHB plans, which strongly stress minority upgrading. Tables XII-12 and XII-13 show the responses by union status and region. It should come as no surprise to see that a larger portion of union firms are participating than are open shop firms. In Chapter X we noted that union firms are more likely than open shop firms to be involved in formal training programs. Such programs usually carry stipulations (not necessarily practiced) that require special efforts to attract minorities. Construction unions

TABLE XII-13
Percent Distribution of Participation in Local Programs to Upgrade Minority or Female Workers by Region and Union Status, 1982

Region	Union			Nonunion		
	Total	Yes	No	Total	Yes	No
Not reported	44	43.2	47.7	49	42.9	53.1
New England	62	51.6	45.2	111	27.9	67.6
Middle Atlantic	275	54.9	41.5	228	31.1	65.8
E.N. Central	474	58.4	36.9	205	31.2	65.9
W.N. Central	220	54.1	44.1	157	33.1	65.6
S.Atlantic	109	49.5	47.7	550	39.5	57.3
E.S. Central	69	53.6	43.5	111	31.5	65.8
W.S. Central	105	44.8	51.4	284	38.0	58.8
Mountain	108	63.9	36.1	184	34.8	63.0
Pacific	362	54.7	42.0	143	28.0	67.8
Total	1,828	54.9	41.7	2,022	34.8	62.2

Source: Industrial Research Unit Questionnaire, 1982.

have frequently negotiated with the Department of Labor for contracts to train minorities.

Department of Labor voluntary plans to increase minority participation in skilled trades have historically been confined in practice to the unionized sector. The Department of Labor has often gone to great lengths to ensure the exclusion of open shop firms from such programs. Open shop builders in Indianapolis attempted in vain to participate in a city-sponsored minority training program. They were unable to receive credit under the plan for the placement of minorities with open shop firms.[18] In Charlotte, North Carolina, where only four of the city's major builders were union firms (and even these firms maintained doublebreasted open shop affiliates), the Department of Labor signed a plan agreement with these four builders, excluding all others. A request by the Carolinas AGC for a more representative plan was denied.

At best, the Department of Labor has refused to certify plans in which unions did not participate. For example, in Dallas, the local AGC Chapter drew up a plan with union and minority representatives, but it was never effectuated. The Department of Labor would not approve it because some unions would not sign the agreement.

Tables XII-12 and XII-13 also show that open shop contractor participation in minority upgrading plans is greater where open shop associations are strongest, as in the southern regions. As we noted in Chapter X, the ABC and the open shop chapters of the AGC clearly stress minority recruitment in their training activities. They realize the significance of minorities to their present and future labor supply and emphasize this to their members. It was also clear from our interviews that the largest open shop contractors are also very cognizant of this fact. Many of these contractors are southern based and have depended upon blacks and/or Hispanics as a key labor element both for skilled and unskilled work since the inception of their companies. It, therefore, would seem likely that open shop contractor participation in training minorities will continue to expand. As noted earlier, the fight over minority business set asides has created bitterness and could probably have diminished interest in furthering minority employment programs.

EEOC Data

The U.S. Equal Employment Opportunity Commission (EEOC) requires companies that employ 100 or more full-time persons to

---

[18] See Rowan and Rubin, *Opening the Skilled Construction Trades*, Ch. VIII.

submit data concerning their employees by race, sex, and job category. Thus, the data supplied to the EEOC applies only to a limited number of contractors. Nevertheless, they do shed light on minority participation in the larger firms which are, of course, more likely to be unionized, except in the southern region, and whose employees compose between one-fourth and one-third of the industry's employees.

Tables XII-14 and XII-15 show the EEOC data for 1970 and 1981. The data indicate some change in minority representation over the eleven-year period. In the skilled trades, the black ratio increased from 6.4 percent in 1970 to 7.3 percent in 1981, and the Hispanic ratio from 3.1 percent to 5.3 percent. Overall, the black percentage declined from 11.9 percent to 9.4 percent, due almost entirely to a decline in the proportion of unskilled black workers. The overall proportion of Hispanics rose from 4.4 percent to 6.3 percent. Other minority and female participation remained minute. In comparison, the census reported that blacks composed 6.6 percent of the construction civilian workforce in 1980 and Hispanics 5.9 percent.

The EEOC data indicate that larger firms tend to employ a greater percentage of minorities than does the industry generally, but not necessarily a larger percentage of minority skilled workers. The changes from 1970 to 1981 may be attributable to increased utilization of skilled minorities, or to a different mix of companies employing 100 persons or more. They may, for example, be the result of a larger representation in 1981 of southern-based open shop concerns, which we found in Chapters IV and V to have grown substantially in this period, and which have often had a large percentage of black craftsmen. The decline in the percentage of black laborers during this period could mean many things, including different company mix or the upgrading of laborers to helper or craft jobs.

EEOC Referral Union Data. Besides requiring the submission of data from companies, the EEOC also mandates that referral unions, those which act as employment agents, supply membership information. These include unions in the maritime, amusement, printing, and other industries in addition to construction, but construction unions are by far most numerous among referral unions. Unfortunately, the union data are even more flawed for use in the construction industry than data collected from firms. The data are limited to unions with 100 or more members, which exempts a large portion of local construction unions from the need to report; and the EEOC has never made any serious effort to police this rule in order to ensure that local unions which are required to submit data

TABLE XII-14
Construction Employment by Race, Sex, and Occupational Group[a]
1970

Job Categories	Total	Male: White	Male: Black	Male: Hispanic	Male: Asian/ Pacific Islander	Male: Amer. Indian/ Alaskan Native	Female: White	Female: Black	Female: Hispanic	Female: Asian/ Pacific Islander	Female: Amer. Indian/ Alaskan Native
	Full Time Employees										
Officials/managers	44,760	42,531	576	404	142	173	896	24	5	2	7
		(95.0)	(1.3)	(0.9)	(0.3)	(0.4)	(2.0)	(0.1)	(0.0)	(0.0)	(0.0)
Professionals	35,734	32,640	261	694	1,179	71	801	35	19	30	4
		(91.3)	(0.7)	(1.9)	(3.3)	(0.2)	(2.2)	(0.1)	(0.1)	(0.1)	(0.0)
Technicians	19,847	17,188	526	692	339	85	920	36	32	25	4
		(86.6)	(2.7)	(3.5)	(1.7)	(0.4)	(4.6)	(0.2)	(0.2)	(0.1)	(0.0)
Sales workers	6,061	5,311	40	86	20	7	522	12	54	6	3
		(87.6)	(0.7)	(1.4)	(0.3)	(0.1)	(8.6)	(0.2)	(0.9)	(0.1)	(0.0)

Office/clerical	45,171	14,648 (32.4)	638 (1.4)	428 (0.9)	221 (0.5)	46 (0.1)	27,280 (60.4)	903 (2.0)	581 (1.3)	275 (0.6)	151 (0.3)
Workers											
Skilled	226,195	201,675 (89.2)	14,416 (6.4)	7,069 (3.1)	431 (0.2)	1,680 (0.7)	804 (0.4)	48 (0.0)	20 (0.0)	39 (0.0)	13 (0.0)
Semi-skilled	104,551	76,940 (73.6)	15,351 (14.7)	6,505 (6.2)	122 (0.1)	807 (0.8)	4,284 (4.1)	308 (0.3)	201 (0.2)	7 (0.0)	26 (0.0)
Unskilled	104,951	54,468 (51.9)	37,847 (36.1)	10,564 (10.1)	110 (0.1)	1,065 (1.0)	589 (0.6)	241 (0.2)	51 (0.0)	1 (0.0)	15 (0.0)
Service workers	29,856	11,431 (38.3)	3,625 (12.1)	573 (1.9)	270 (0.9)	54 (0.2)	10,848 (36.3)	2,629 (8.8)	371 (1.2)	42 (0.1)	13 (0.0)
Total	617,126	456,832 (74.0)	73,280 (11.9)	27,015 (4.4)	2,834 (0.5)	3,988 (0.6)	46,944 (7.6)	4,236 (0.7)	1,334 (0.2)	427 (0.1)	236 (0.0)

Source: U.S. Equal Employment Opportunity Commission

[a] Building construction—general contractors and operative builders (SIC 15) 1970

Construction other than building—general contractors (SIC 16) 1970

Construction—special trade contractors (SIC 17) 1970

TABLE XII-15
Construction Employment by Race, Sex, and Occupational Group[a]
1981

Job Categories	Total	Full Time Employees									
		Male					Female				
		White	Black	Hispanic	Asian/ Pacific Islander	Amer. Indian/ Alaskan Native	White	Black	Hispanic	Asian/ Pacific Islander	Amer. Indian/ Alaskan Native
Officials/managers	53,914	48,066 (89.2)	1,088 (2.0)	1,120 (2.1)	399 (0.7)	260 (0.5)	2,754 (5.1)	108 (0.2)	74 (0.1)	28 (0.1)	17 (0.0)
Professionals	34,832	29,062 (83.4)	617 (1.8)	648 (1.9)	1,161 (3.3)	97 (0.3)	2,889 (8.3)	160 (0.5)	71 (0.2)	112 (0.3)	15 (0.0)
Technicians	21,857	16,713 (76.5)	649 (3.0)	748 (3.4)	453 (2.1)	67 (0.3)	2,874 (13.1)	180 (0.8)	78 (0.4)	82 (0.4)	13 (0.1)

Sales workers	8,142	6,163 (75.7)	89 (1.1)	107 (1.3)	31 (0.4)	17 (0.2)	1,639 (20.1)	42 (0.5)	33 (0.4)	15 (0.2)	6 (0.1)
Office/clerical	45,052	9,089 (20.2)	564 (1.3)	363 (0.8)	150 (0.3)	47 (0.1)	31,498 (69.9)	1,588 (3.5)	1,209 (2.7)	358 (0.8)	186 (0.4)
Workers											
Skilled	215,152	182,472 (84.8)	15,790 (7.3)	11,461 (5.3)	855 (0.4)	2,143 (1.0)	2,013 (0.9)	256 (0.1)	117 (0.1)	26 (0.0)	19 (0.0)
Semi-skilled	90,603	61,388 (67.8)	12,821 (14.2)	7,519 (8.3)	551 (0.6)	790 (0.9)	5,250 (5.8)	1,152 (1.3)	866 (1.0)	194 (0.2)	72 (0.1)
Unskilled	74,910	39,451 (52.7)	18,840 (25.2)	11,807 (15.8)	331 (0.4)	643 (0.9)	2,567 (3.4)	839 (1.1)	328 (0.4)	60 (0.1)	44 (0.1)
Service workers	13,808	6,722 (48.7)	1,752 (12.7)	1,234 (8.9)	189 (1.4)	50 (0.4)	2,483 (18.0)	769 (5.6)	464 (3.4)	101 (0.7)	44 (0.3)
Total	558,270	399,126 (71.5)	52,210 (9.4)	35,007 (6.3)	4,120 (0.7)	4,114 (0.7)	53,967 (9.7)	5,094 (0.9)	3,240 (0.6)	976 (0.2)	416 (0.1)

Source: U.S. Equal Employment Opportunity Commission.
[a] Building construction—general contractors and operative builders (SIC 15) 1981
Construction other than building—general contractors (SIC 16) 1981
Construction—special trade contractors (SIC 17) 1981

do so. Some states did not even submit data for certain crafts. Nevertheless, the data set forth in Table XII-16 for 1979 do provide some further information concerning union racial policies.

It will be immediately seen that the data in Table XII-16 do not differ materially from the census data. Blacks and Hispanics are underrepresented in the electrical and plumbing trades in all regions, better represented among union trowel tradesmen and carpenters, and overrepresented among laborers. There are, however, a few noteworthy exceptions, such as electrical workers in the Middle Atlantic region. There blacks compose 7.7 percent and Hispanics, 8 percent of the union membership. This is attributable to the effective work of the state EEOCs in the three states involved, all of which antedate the federal Civil Rights Act of 1964 and have taken an active role in opening up such opportunities. Obviously, greater success in electrical work has been achieved in this region than in plumbing, where the reported union membership, as shown in Table XII-16, was 3.3 percent black and 0.5 percent Hispanic.

Hispanic workers also showed a high rate of union representation in the electrical craft in the Mountain Region, 13.3 percent, but only 4 percent in the West South Central and 6.9 percent in the Pacific. (California alone had a 9.8 percent level of Hispanic membership in the electrical unions, while Texas had only 6.5 percent.) In all these areas, the Plumbers' union minority membership was about one-half that of the electrical workers. One must also, therefore, conclude that the larger local unions of the electrical workers have accommodated the equal employment opportunity demands in a superior manner to those of the Plumbers' union.[19] Women and other minority representation was minuscule in all categories and regions.

In the southern regions minority union membership in the large local unions, except for laborers, is obviously far from satisfactory, even for the trowel trades. The reasons are clearly twofold. One is the historic union discrimination which, of course, has never been confined to the South. The other is the fact that the South is heavily open shop and most southern black craftsmen are not unionized. Thus the data in Table XII-16 show, for example, a higher proportion of black carpenters in the Mountain region, which has the lowest black population of any region, than in any of the three Southern regions, where blacks are proportionately most numerous.

[19] For the official names of these unions, see Table I-2.

TABLE XII-16

Membership in Construction Union Referral Bargaining Units of 100 or More, by Race, Sex, Occupational Group, and Region, 1979

Regions and Trade Categories	No. of Unions	All Members	Male					Female				
			White	Black	Hispanic	Asian/ Pacific Islander	Amer. Ind./ Alaskan Native	White	Black	Hispanic	Asian/ Pacific Islander	Amer. Ind./ Alaskan Native
New England												
Elec. workers (IBEW)	12	2,958	2,837	65	34	3	7	8	4	0	0	0
			(95.9)	(2.2)	(1.1)	(0.1)	(0.2)	(0.3)	(0.1)	(0.0)	(0.0)	(0.0)
Plumbers	18	4,026	3,910	70	17	3	18	8	0	0	0	0
			(97.1)	(1.7)	(0.4)	(0.1)	(0.4)	(0.2)	(0.0)	(0.0)	(0.0)	(0.0)
Trowel & misc. trades	19	6,348	6,067	126	61	10	24	45	1	7	0	7
			(95.6)	(2.0)	(1.0)	(0.2)	(0.4)	(0.7)	(0.0)	(0.1)	(0.0)	(0.1)
Carpenters	16	5,867	5,621	105	48	10	23	45	1	7	0	7
			(95.8)	(1.8)	(0.8)	(0.2)	(0.4)	(0.8)	(0.0)	(0.1)	(0.0)	(0.1)
Laborers	28	14,435	10,929	2,387	887	16	25	87	99	4	0	1
			(75.7)	(16.5)	(6.1)	(0.1)	(0.2)	(0.6)	(0.7)	(0.0)	(0.0)	(0.0)
Middle Atlantic												
Elec. workers (IBEW)	51	47,772	35,833	3,679	3,824	186	170	1,564	1,013	1,502	1	0
			(75.0)	(7.7)	(8.0)	(0.4)	(0.4)	(3.3)	(2.1)	(3.1)	(0.0)	(0.0)
Plumbers	34	22,312	21,330	739	122	7	78	23	13	0	0	0
			(95.6)	(3.3)	(0.5)	(0.0)	(0.3)	(0.1)	(0.1)	(0.0)	(0.0)	(0.0)
Trowel & misc. trades	125	59,220	52,374	4,285	1,968	76	260	150	83	17	5	2
			(88.4)	(7.2)	(3.3)	(0.1)	(0.4)	(0.3)	(0.1)	(0.0)	(0.0)	(0.0)
Carpenters	74	35,296	30,998	2,545	1,409	66	100	114	41	16	5	2
			(87.8)	(7.2)	(4.0)	(0.2)	(0.3)	(0.3)	(0.1)	(0.0)	(0.0)	(0.0)
Laborers	71	47,494	33,667	10,871	1,842	29	215	729	128	12	0	1
			(70.9)	(22.9)	(3.9)	(0.1)	(0.5)	(1.5)	(0.3)	(0.0)	(0.0)	(0.0)

TABLE XII-16 (continued)

Regions and Trade Categories	No. of Unions	All Members	Male White	Male Black	Male Hispanic	Male Asian/ Pacific Islander	Male Amer. Ind./ Alaskan Native	Female White	Female Black	Female Hispanic	Female Asian/ Pacific Islander	Female Amer. Ind./ Alaskan Native
E.N. Central												
Elec. workers (IBEW)	71	44,227	41,764	1,810	382	63	222	159	25	1	0	1
			(94.4)	(3.6)	(0.9)	(0.1)	(0.5)	(0.4)	(0.1)	(0.0)	(0.0)	(0.0)
Plumbers	71	45,966	43,320	1,662	612	78	176	97	18	2	0	1
			(94.2)	(3.6)	(1.3)	(0.2)	(0.4)	(0.2)	(0.0)	(0.0)	(0.0)	(0.0)
Trowel & misc. trades	136	71,187	67,533	2,181	629	47	318	424	27	27	0	1
			(94.9)	(3.1)	(0.9)	(0.1)	(0.4)	(0.6)	(0.0)	(0.0)	(0.0)	(0.0)
Carpenters	100	43,870	42,337	628	301	19	157	385	17	25	0	1
			(96.5)	(1.4)	(0.7)	(0.0)	(0.4)	(0.9)	(0.0)	(0.1)	(0.0)	(0.0)
Laborers	94	43,175	35,449	5,608	1,077	16	232	634	134	10	9	6
			(82.1)	(13.0)	(2.5)	(0.0)	(0.5)	(1.5)	(0.3)	(0.0)	(0.0)	(0.0)
W.N. Central												
Elec. workers (IBEW)	25	13,391	12,908	245	97	12	70	56	3	0	0	0
			(96.4)	(1.8)	(0.7)	(0.1)	(0.5)	(0.4)	(0.0)	(0.0)	(0.0)	(0.0)
Plumbers	37	13,801	13,170	262	101	14	225	28	0	0	0	1
			(95.4)	(1.9)	(0.7)	(0.1)	(1.6)	(0.2)	(0.0)	(0.0)	(0.0)	(0.0)
Trowel & misc. trades	36	17,034	16,511	138	149	6	133	93	0	0	0	4
			(96.9)	(0.8)	(0.9)	(0.0)	(0.8)	(0.5)	(0.0)	(0.0)	(0.0)	(0.0)
Carpenters	33	15,021	14,557	108	125	6	128	93	0	0	0	4
			(96.9)	(0.7)	(0.8)	(0.0)	(0.9)	(0.6)	(0.0)	(0.0)	(0.0)	(0.0)
Laborers	33	22,359	17,471	3,497	815	15	185	304	44	4	0	24
			(78.1)	(15.6)	(3.6)	(0.1)	(0.8)	(1.4)	(0.2)	(0.0)	(0.0)	(0.1)

S. Atlantic												
Elec. workers (IBEW)	31	15,535	14,632	376	338	22	42	80	32	12	0	1
			(94.2)	(2.4)	(2.2)	(0.1)	(0.3)	(0.5)	(0.2)	(0.1)	(0.0)	(0.0)
Plumbers	33	16,826	15,809	626	264	10	35	46	26	3	0	7
			(94.0)	(3.7)	(1.6)	(0.1)	(0.2)	(0.3)	(0.2)	(0.0)	(0.0)	(0.0)
Trowel & misc. trades	43	16,572	15,101	974	336	5	89	48	17	1	0	1
			(91.1)	(5.9)	(2.0)	(0.0)	(0.5)	(0.3)	(0.1)	(0.0)	(0.0)	(0.0)
Carpenters	37	13,067	12,170	575	196	4	66	40	15	0	0	1
			(93.1)	(4.4)	(1.5)	(0.0)	(0.5)	(0.3)	(0.1)	(0.0)	(0.0)	(0.0)
Laborers	25	20,220	10,361	7,714	674	20	34	491	921	5	0	0
			(51.2)	(38.2)	(3.3)	(0.1)	(0.2)	(2.4)	(4.6)	(0.0)	(0.0)	(0.0)
E.S. Central												
Elec. workers (IBEW)	19	11,926	11,187	526	11	6	12	134	47	1	1	1
			(93.8)	(4.4)	(0.1)	(0.1)	(0.1)	(1.1)	(0.4)	(0.0)	(0.0)	(0.0)
Plumbers	22	13,840	13,047	544	25	8	28	100	88	0	0	0
			(94.3)	(3.9)	(0.2)	(0.1)	(0.2)	(0.7)	(0.6)	(0.0)	(0.0)	(0.0)
Trowel & misc. trades	37	18,347	17,242	768	13	0	34	171	119	0	0	0
			(94.0)	(4.2)	(0.1)	(0.0)	(0.2)	(0.9)	(0.6)	(0.0)	(0.0)	(0.0)
Carpenters	34	17,871	16,787	751	13	0	34	169	117	0	0	0
			(93.9)	(4.2)	(0.1)	(0.0)	(0.2)	(0.9)	(0.7)	(0.0)	(0.0)	(0.0)
Laborers	17	12,868	7,192	4,539	11	0	0	549	576	1	0	0
			(55.9)	(35.3)	(0.1)	(0.0)	(0.0)	(4.3)	(4.5)	(0.0)	(0.0)	(0.0)
W.S. Central												
Elec. workers (IBEW)	36	16,213	14,301	828	651	16	257	132	25	2	1	0
			(88.2)	(5.1)	(4.0)	(0.1)	(1.6)	(0.8)	(0.2)	(0.0)	(0.0)	(0.0)
Plumbers	32	31,124	27,466	422	1,495	329	1,263	113	18	18	0	0
			(88.2)	(1.4)	(4.8)	(1.1)	(4.1)	(0.4)	(0.1)	(0.1)	(0.0)	(0.0)
Trowel & misc. trades	69	38,167	31,142	3,311	2,952	98	393	187	64	14	1	5
			(81.6)	(8.7)	(7.7)	(0.3)	(1.0)	(0.5)	(0.2)	(0.0)	(0.0)	(0.0)
Carpenters	53	34,361	29,359	1,791	2,480	96	385	182	48	14	1	5
			(85.4)	(5.2)	(7.2)	(0.3)	(1.1)	(0.5)	(0.1)	(0.0)	(0.0)	(0.0)
Laborers	23	19,560	3,774	12,337	2,388	12	34	341	636	26	5	7
			(19.3)	(63.1)	(12.2)	(0.1)	(0.2)	(1.7)	(3.3)	(0.1)	(0.0)	(0.0)

TABLE XII-16 (continued)

Regions and Trade Categories	No. of Unions	All Mem-bers	Male					Female				
			White	Black	His-panic	Asian/ Pacific Islander	Amer. Ind./ Alaskan Native	White	Black	His-panic	Asian/ Pacific Islander	Amer. Ind./ Alaskan Native
Mountain												
Elec. workers (IBEW)	12	6,919	5,442	176	918	37	214	70	16	21	10	15
			(78.7)	(2.5)	(13.3)	(0.5)	(3.1)	(1.0)	(0.2)	(0.3)	(0.1)	(0.2)
Plumbers	13	5,080	4,638	53	232	19	119	16	0	1	1	1
			(91.3)	(1.0)	(4.6)	(0.4)	(2.3)	(0.3)	(0.0)	(0.0)	(0.0)	(0.0)
Trowel & misc. trades	33	18,066	14,316	1,265	1,800	29	461	161	3	25	0	6
			(79.2)	(7.0)	(10.0)	(0.2)	(2.6)	(0.9)	(0.0)	(0.1)	(0.0)	(0.0)
Carpenters	33	18,066	14,316	1,265	1,800	29	461	161	3	25	0	6
			(79.2)	(7.0)	(10.0)	(0.2)	(2.6)	(0.9)	(0.0)	(0.1)	(0.0)	(0.0)
Laborers	3	1,165	1,017	17	76	1	15	34	2	3	0	0
			(87.3)	(1.5)	(6.5)	(0.1)	(1.3)	(2.9)	(0.2)	(0.3)	(0.0)	(0.0)
Pacific												
Elec. workers (IBEW)	39	36,645	31,405	1,190	2,526	574	445	402	33	47	14	9
			(85.7)	(3.2)	(6.9)	(1.6)	(1.2)	(1.1)	(0.1)	(0.1)	(0.0)	(0.0)
Plumbers	39	36,924	32,244	992	2,511	460	507	184	11	14	0	1
			(87.3)	(2.7)	(6.8)	(1.2)	(1.4)	(0.5)	(0.0)	(0.0)	(0.0)	(0.0)
Trowel & misc. trades	169	147,694	115,153	6,358	20,770	1,119	2,772	1,030	135	268	44	45
			(78.0)	(4.3)	(14.1)	(0.8)	(1.9)	(0.7)	(0.1)	(0.2)	(0.0)	(0.0)
Carpenters	134	123,841	98,483	4,962	16,972	848	1,518	669	99	248	31	11
			(79.5)	(4.0)	(13.7)	(0.7)	(1.2)	(0.5)	(0.1)	(0.2)	(0.0)	(0.0)
Laborers	54	64,684	27,337	12,275	22,227	268	701	1,123	335	391	4	23
			(42.3)	(19.0)	(34.4)	(0.4)	(1.1)	(1.7)	(0.5)	(0.6)	(0.0)	(0.0)

Source: U.S. Equal Employment Opportunity Commission.

THE HIGHWAY SECTOR

The highway sector of the industry, as we noted in Chapter VI, has several unique features including a less skilled labor force, different craft requirements, and a greater dependence upon federal funding. Since highway builders are major users of minority labor and the census provides an unusually sound data base for this industry, we devote a special section of this chapter to minority employment in this sector.

The Census Data

Table XII-17 shows that 7.4 percent of the basic equipment operators in road construction in the United States in 1970 and 6.6 percent in 1980 were black. In contrast to this decline, Hispanics increased their share from 3.9 percent in 1970 to 4.6 percent in 1980. In the ten southern states, the percentage of blacks remained virtually constant during the decade at nearly 18 percent. The percentage in Texas also remained stable at approximately 9 percent. In California, the black percentage declined from 3 percent to 2.1 percent. Although the data do not show this, it is very clear from observations and discussions with contractors that blacks have a high proportion of labor jobs in highway construction in all areas where the black population is significant. Black representation in the skilled crafts other than equipment operators is higher in this sector of the industry than in other construction sectors.

Hispanics play an increasing role in this sector. Their share in Texas jumped from 17.2 percent in 1970 to 24.6 percent in 1980, and in California from 14.4 percent in 1970 to 17.8 percent over this decade.

Blacks thus have a far greater share of this work in the open shop states of the South than in the rest of the country, and Hispanics have there become a major portion of the workforce. In California, the strong Operating Engineers' union has historically been very inhospitable to blacks, but as a result of EEOC instituted legislation, discussed below, it has sanctioned the utilization of Mexican-Americans on a large scale.

The Federal Highway Administration

As discussed in Chapter VI, where the data were utilized to estimate the extent of the open shop in the highway sector, the Federal Highway Administration prior to 1982 periodically collected detailed data on employment by region, occupation, race, and sex on

TABLE XII-17

Males Employed as Paving, Surfacing, and Tamping Equipment Operators, Machine Operators, Construction, Excavating, and Loading Machine Operators, Grader, Dozer, and Scraper Operators by Race and Geographic Area, 1970 and 1980 [a]

Region	1970					1980				
	All Male Workers	Blacks	Percent Black	Hispanics	Percent Hispanics	All Male Workers	Blacks	Percent Black	Hispanics	Percent Hispanic
United States	176,698	13,076	7.4	6,891	3.9	175,630	11,608	6.6	8,150	4.6
South[b]	43,539	7,679	17.6	327	0.8	46,386	8,201	17.7	612	1.3
Texas	13,294	1,157	8.7	2,287	17.2	16,162	1,434	8.9	3,985	24.6
California	14,254	428	3.0	2,052	14.4	9,437	195	2.1	1,675	17.8

Source: U.S. Census of Population, 1970, PC(1)D, *Detailed Characteristics,* U.S. Summary, Table 225, State Vols., Table 173; U.S. Census of Population, 1980, PC80-1-D, *Detailed Population Characteristics,* U.S. Summary, Table 323, State Vols., Tables 217 & 219, and PC80-S1-8, *Detailed Occupation and Years of School Completed by Age, for the Civilian Labor Force, by Sex, Race, and Spanish Origin,* Table 1.

[a] 1970 data from corresponding table in original edition revised to conform to 1980 data.

[b] Alabama, Arkansas, Florida, Georgia, Louisiana, Mississippi, North Carolina, South Carolina, Tennessee, and Virginia.

all highway projects for which federal funds are allocated. Table XII-18 shows these data for 1974 and Table XII-19 for 1981, the last year for which they were compiled.

In comparing the data in the two tables, it must be understood that they reflect the situation at two specific points in time. Thus the inception or termination of work at various locations before the date of data collection can materially alter minority or female representation. Nevertheless, some reasonably significant comparisons may be made.

In 1974, minorities composed 21 percent of all employees on these projects. By 1981, this percentage had risen to 23 percent. In 1974, 50 percent of all minorities were unskilled or semiskilled laborers. By 1981, that percentage had declined slightly to 48 percent. In 1974, minorities represented 4 percent of the officials and managers category, 5 percent of the supervisors, 11 percent of the foremen (probably including working foremen and straw bosses), and 9 percent of the clerical jobs. In 1981, the minority share of officials and managers had risen to 10 percent, undoubtedly reflecting the results of the minority set aside program, which is very strong in all projects administered by agencies of the U.S. Department of Transportation, including the Federal Highway Administration. Minority supervisors and foremen also increased their share in 1981, but more modestly, rising to 7 percent and 13 percent, respectively. Minority clerical workers remained at 9 percent.

Table XII-20 shows that several skilled and semiskilled trades had an increased percentage in 1981 over 1974. In both years, the minority percentage in the skilled crafts of the highway sector far exceeds those found in building construction. The numbers are small in such trades as electrical, plumbing and pipefitting, but it is rare indeed to have over 10 percent minorities in these trades in any situation. Here the percentages are two and one-half to three times greater. The reasons for this greater minority presence are, of course, first, the strong federal pressure and insistence on a fair share of minority employment and, second, the historically greater presence of minorities in highway work.

The regional breakdowns, which we utilized in Chapter VI to provide clues to the extent of open shop growth, also provide estimates of comparative union and open shop minority utilization. The key job is the equipment operator. The highest percentage of minorities in this category, 35 percent in 1974 and 40 percent in 1981, is found in Region 6, dominated by the open shop area of Texas. In 1974, Region 9, with 23 percent minorities, ranked second in minority utilization. This region includes the highly unionized states

TABLE XII-18

Statistical Summary of Employment Data for All Federal-Aid Highway Projects Compiled as of July 31, 1974

Region[a]	Number OF Projects	Total Dollar Value of Projects	Officials (Managers)			Supervisors			Foremen			Clerical			Equipment Operators			Mechanics			Truck Drivers			Iron Workers & Welders			Carpenters			Cement Masons (Finishers)		
			Total Employment	Minorities	Percent	Total Employment	Minorities	Percent	Total Employment	Minorities	Percent	Total Employment	Minorities	Percent	Total Employment	Minorities	Percent	Total Employment	Minorities	Percent	Total Employment	Minorities	Percent	Total Employment	Minorities	Percent	Total Employment	Minorities	Percent	Total Employment	Minorities	Percent
1	325	1,157,881,761	139	7	5	436	16	4	302	83	27	247	22	9	2650	118	4	433	21	5	1728	90	5	591	63	11	1025	87	8	221	37	17
3	837	1,998,940,718	235	7	3	724	14	2	1807	113	6	355	20	6	1469	452	10	530	16	3	2284	564	25	437	71	16	1786	203	11	732	308	42
4	804	1,901,493,212	214	3	1	771	25	3	1905	204	11	281	12	4	6163	1341	22	630	71	11	2260	694	31	409	143	35	1647	446	27	993	767	77
5	1064	2,018,499,729	286	9	3	862	23	3	1975	111	6	365	23	6	8773	712	8	493	22	4	4013	423	11	864	87	10	1787	154	9	890	251	28
6	588	1,257,020,752	187	7	4	546	60	11	1213	263	2	268	37	14	4184	1447	35	430	92	21	1618	717	44	355	232	65	1170	583	50	779	618	79
7	410	414,665,396	107	2	2	340	10	3	577	20	3	95	4	4	2288	108	5	265	8	3	1055	72	7	288	43	15	510	53	10	226	56	25
8	492	1,207,482,492	117	1	1	349	18	5	594	34	6	139	7	5	3010	206	7	517	31	6	1993	109	5	63	3	5	355	49	14	162	34	21
9	242	954,288,634	245	26	11	257	30	12	915	241	26	259	63	24	2106	477	23	230	42	18	1057	197	19	383	162	42	782	219	28	265	175	66
10	246	472,797,915	84	7	8	173	10	6	328	23	7	138	10	7	1969	141	7	236	15	6	1114	72	6	179	28	16	293	16	5	106	36	34
Total	5008	11,383,070,609	1814	69	4	4458	206	5	9616	1092	11	2147	198	9	32,612	5002	15	3764	318	8	17,122	2938	17	3569	832	23	9355	1810	19	4374	2282	52

Region [a]	Electricians			Pipe-Fitters Plumbers			Painters			Laborers Semi-Skilled			Laborers Unskilled			Apprentices			On-The-Job Trainees			Total Number of Employees	Total Number of Minorities	Percent of Minorities to Total Number of Employees	Total Number of Minority Semi-Skilled and Unskilled Laborers	Percent of Minority Simi-Skilled & Unskilled Laborers to Total Number of Employees	Percent of Minority Semi-Skilled & Unskilled Laborers to Total Number of Minorities
	Total Employment	Minorities	Percent	Total Employment	Minorities	Percent	Total Employment	Minorities	Percent	Total Employment	Minorities	Percent	Total Employment	Minorities	Percent	Total Employment	Minorities	Percent	Total Employment	Minorities	Percent						
1	174	22	13	9	1	11	98	17	17	2821	591	21	2025	228	11	155	55	35	341	182	53	13,395	1,403	10	819	6	58
3	261	53	20	41	5	12	137	23	17	1938	845	44	4594	1234	27	207	39	19	482	241	50	18,019	3,928	21	2079	12	53
4	123	16	13	82	53	85	87	11	13	2890	1227	42	6303	2654	42	71	15	21	596	320	54	25,425	7,667	31	3881		51
5	267	27	10	30	6	20	81	10	12	3381	808	24	6414	1349	21	662	230	35	544	353	65	31,687	4,015	13	2157	7	54
6	63	9	14	27	20	74	28	12	43	2600	1547	60	4018	2466	61	46	22	48	331	191	58	17,863	8,110	45	4013	22	49
7	86	3	3	23	1	4	39	5	13	1059	155	15	1479	178	12	69	35	51	256	116	45	8,762	718	8	333	4	46
8	45	2	4	8	1	13	5	0	0	1040	187	18	1893	286	15	133	41	31	149	85	57	10,572	968	9	473	4	49
9	148	43	29	24	11	49	23	3	13	1475	877	59	924	545	59	461	251	54	10	5	50	9,564	3,111	33	1422	15	46
10	175	12	7	97	9	9	11	0	0	801	88	11	690	93	13	127	37	29	115	43	37	6,636	560	8	181	3	32
Total	1342	187	14	341	107	31	509	81	16	18,005	6325	35	28,340	9033	32	1931	725	38	2824	1536	54	141,923	30,480	21	15,358	11	50

Source: Office of Civil Rights/Federal Highway Administration.

[a] Regions are defined by the Federal Highway Administration as follows (there is no Region 2):

1. Connecticut, Maine, Massachusetts, New Hampshire, New Jersey, New York, Rhode Island, and Vermont.
3. Delaware, District of Columbia, Maryland, Pennsylvania, Virginia, and West Virginia.
4. Alabama, Florida, Georgia, Kentucky, Mississippi, North Carolina, South Carolina, and Tennessee.
5. Illinois, Indiana, Michigan, Minnesota, Ohio, and Wisconsin.
6. Arkansas, Louisiana, New Mexico, Oklahoma, and Texas.
7. Iowa, Kansas, Missouri, and Nebraska.
8. Colorado, Montana, North Dakota, South Dakota, Utah, and Wyoming.
9. Arizona, California, and Nevada.
10. Alaska, Idaho, Oregon, and Washington.

Table XII-19
Statistical Summary of Employment Data for All Federal-Aid Highway Projects Compiled as of July 31, 1981

Region [a]	Number of Projects	Total Value of Projects	Officials (Managers)			Supervisors			Foremen			Clerical			Equipment Operators		
			Total Employment	Minorities	Percent	Total Employment	Minorities	Percent	Total Employment	Minorities	Percent	Total Employment	Minorities	Percent	Total Employment	Minorities	Percent
1	505	$ 1,357,964,660	255	23	9	445	38	9	1,045	116	11	246	20	8	2,272	266	12
3	508	$ 2,573,881,905	308	27	9	614	35	6	1,323	113	9	244	20	8	2,936	364	12
4	970	$ 2,279,290,417	332	34	10	844	58	7	1,617	252	16	375	27	7	4,936	1,283	26
5	906	$58,904,050,948	651	60	9	1,007	26	3	1,470	92	6	672	53	8	5,394	470	9
6	549	$ 1,919,115,327	284	27	10	673	88	13	1,109	307	28	329	56	17	3,417	1,383	40
7	338	$ 445,389,894	165	5	3	272	8	3	517	23	4	188	7	4	2,188	132	6
8	293	$ 450,908,604	89	6	7	250	6	2	430	34	8	88	3	3	1,841	108	6
9	1582	$ 795,138,842	185	46	25	186	34	18	454	117	26	184	30	16	1,203	275	23
10	207	$ 679,250,405	169	9	5	198	12	6	429	36	8	193	14	7	1,278	126	10
Total	5858	$69,404,991,002	2,438	237	10	4,489	305	7	8,394	1,090	13	2,519	230	9	25,465	4,407	17

Region [a]	Mechanics			Truck Drivers			Iron Workers & Welders			Carpenters			Cement Masons (Finishers)			Electricians		
	Total Employment	Minorities	Percent	Total Employment	Minorities	Percent	Total Employment	Minorities	Percent	Total Employment	Minorities	Percent	Total Employment	Minorities	Percent	Total Employment	Minorities	Percent
1	196	10	5	1,345	171	13	410	73	18	781	102	13	193	28	15	198	33	17
3	322	9	3	1,781	363	20	380	51	13	1,329	212	16	523	226	43	144	20	14
4	503	62	12	2,162	691	32	272	108	40	1,362	426	31	714	512	72	71	2	3
5	305	16	5	3,416	301	9	1,098	85	8	1,326	122	9	912	194	21	704	75	11
6	312	92	29	1,339	616	46	292	185	63	1,148	656	57	726	620	85	67	14	21
7	172	2	1	894	57	6	396	67	17	615	48	8	298	74	25	34	1	3
8	314	18	6	1,310	76	6	73	9	12	189	26	14	141	29	21	61	0	0
9	174	32	18	684	163	24	137	53	39	259	76	29	164	77	47	59	14	24
10	154	21	14	923	77	8	185	15	8	405	41	10	171	29	17	172	9	5
Total	2,452	262	11	13,854	2,515	18	3,243	646	20	7,414	1,709	23	3,842	1,789	47	1,510	168	11

TABLE XII-19 (continued)

Region [a]	Pipe-Fitters Plumbers			Painters			Laborers Semi-Skilled			Laborers Unskilled			Apprentices			On-the-Job Trainees		
	Total Employment	Minorities	Percent	Total Employment	Minorities	Percent	Total Employment	Minorities	Percent	Total Employment	Minorities	Percent	Total Employment	Minorities	Percent	Total Employment	Minorities	Percent
1	22	4	18	67	12	18	3,742	755	20	1,312	249	19	105	35	33	204	81	40
3	59	12	20	65	12	18	3,101	655	21	2,780	715	26	134	40	30	254	119	47
4	72	46	64	26	4	15	2,433	1,143	47	4,778	2,075	43	46	11	24	338	170	50
5	33	3	9	117	37	32	4,753	1,045	22	3,980	701	18	519	154	30	294	114	39
6	26	13	50	54	15	28	2,224	1,420	64	3,380	2,287	68	43	19	44	419	226	54
7	6	0	0	54	2	4	1,278	229	18	1,481	232	16	143	35	24	103	47	46
8	3	1	33	20	1	5	889	121	14	886	103	12	110	20	18	95	33	35
9	22	7	32	113	28	25	1,100	544	49	417	170	41	216	78	36	3	1	33
10	67	2	3	118	13	11	1,140	156	14	298	40	13	220	41	19	66	9	14
Total	310	88	28	634	124	20	20,660	6,068	29	19,312	6,572	34	1,536	433	28	1,776	800	43

Region [a]	Total Employees	Total Minorities	Percent of Minorities to Total Employees	Total Minority Laborers	Percent of Minority Laborers to Total Employees	Percent of Minority Laborers to Total Minorities	Total Women
1	12,529	1,900	15	1,004	8	53	564
3	15,909	2,834	18	1,370	9	48	740
4	20,497	6,723	33	3,218	16	48	879
5	25,838	3,280	13	1,746	7	53	1,350
6	15,380	7,779	51	3,707	24	48	764
7	8,558	887	10	461	5	52	446
8	6,584	541	8	224	3	41	548
9	5,341	1,666	31	714	13	43	337
10	5,900	600	10	196	3	33	486
Total	116,536	26,210	23	12,640	11	48	6,114

Source: See Table XII-18.
[a] For definitions of regions, see Table XII-18.

TABLE XII-20
Percent of Minorities in Skilled and Semiskilled Occupations
Federal-Aid Highway Projects
1974 and 1981

Occupation	Percent Minorities 1974	1981
Equipment operators	15	17
Mechanics	8	11
Truck drivers	17	18
Iron workers and welders	23	20
Carpenters	19	23
Cement masons (finishers)	52	47
Electricians	14	11
Plumbers and pipefitters	31	28
Painters	16	20

Source: Tables XII-18 and XII-19.

of California and Nevada. In 1981, however, Region 4, which includes the almost completely open shop Southeast, had a 26 percent share of minorities, which exceeded Region 9's 23 percent.

As we shall note below, Local 3 of the Operating Engineers, which dominates the California-Nevada area, has been under court order to open up jobs to minorities. It appears from these data that the decree has had an effect. In general, however, the open shop areas, which of course include a high percentage of blacks and Mexican-Americans, clearly lead in utilization of minority equipment operators.

In the craft jobs, Region 4, the Southeast, and Region 6, the Southwest, have a decided edge in nearly every category of minority utilization, except electricians in Region 4. Both the Southwest and California benefit in this regard from the training received by Mexican immigrants in Mexico. In the Southeast, and in the states of Texas, Louisiana, and Oklahoma, blacks have a craft tradition, but much of the training must be done by contractors.

The percentage of apprentices and on-the-job trainees in the highway sector augurs well for future work for minorities in this sector. Interestingly, the open shop-dominated Region 6 leads in both categories.

The number and percentage of women on federal-aided highway projects remains, as one would expect, quite small. In 1981, only 6,114 (.05 percent of the total) were women. Despite federal pressure, it could be some years before women compose even a few percent of the highway sector's labor force.

It is apparent from these data that minorities, but not women, play a major role in highway construction. Although the proportion of minorities in the highly unionized Region 9 is quite high, it is clear that the largely open shop areas of the Southwest and Southeast lead the country in utilization of minorities. We thus conclude that open shop contractors appear to have a decided lead in minority utilization in the highway sector.

The Policies of the Operating Engineers

The key union in the highway sector is the Operating Engineers International Union. Its policies on racial employment were inhospitable to minority workers for many years. Policies varied considerably, however, depending upon the local union involved. Thus Professor Mangum reported in 1964:

> The question of a color bar in the membership requirement of the operating engineers' union arose only at one convention. The issue was debated in 1910 when a southern local advocated a "whites only" clause. The color bar was rejected on pragmatic grounds. The union was then essentially a stationary engineer's organization with no control of the labor supply. The Negro engineers could not be excluded from employment. Therefore it was considered preferable to include them in order to control them. Negro members of stationary locals are not uncommon today, but they are rare in hoisting and portable locals. The matter has remained in local hands, and there is no doubt that considerable discrimination exists. Until racial discrimination became a widespread issue in the nation and in the labor movement, the international union developed no policy in the matter. As the issue has grown hotter, international officers have begun to warn local unions of possible problems from external sources in the future if discrimination is not ended. Beyond this, the international has taken no action.[20]

Since 1964, when Title VII of the Civil Rights Act became law, a number of Operating Engineers local unions, under national leadership, have encouraged minority membership through training programs and contracts for training with the U.S. Department of Labor. This has resulted in opening up previously closed ranks in a number of areas, such as Pittsburgh,[21] Chicago,[22] and several other locations. On the other hand, it has been necessary for the government or for individuals to file suits to end discriminatory admission

---

[20] Garth L. Mangum, *The Operating Engineers: The Economic History of a Trade Union* (Cambridge: Harvard University Press, Wertheim Publications in Industrial Relations, 1964), pp. 232-33.

[21] See Irwin Dubinsky, *Reform in Trade Union Discrimination in the Construction Industry: Operation Dig and Its Legacy* (New York: Praeger Publishers, 1973).

[22] Preliminary research in 1974 by the Industrial Reseaerch Unit on the "Chicago Plan" indicated that the Operating Engineers local there was one of the few in Chicago which attempted seriously to meet its obligations.

and work practices in California, Oregon, Philadelphia, Ohio, Kentucky, and other areas. In Philadelphia, the Commonwealth of Pennsylvania obtained an injunction to prevent Local 542 and some of its members and agents from "pursuing a course of conduct designed to intimidate, harass and preclude" blacks from pursuing a law suit against that local.[23] In California, the U.S. Department of Justice obtained a consent decree which ended the practice of confining blacks and Mexican-Americans to the less desirable jobs, refusing to give them credit for past experience, and denying them apprenticeship opportunities.[24] In 1975, Local 18 of the Operating Engineers and the Ohio Operating Engineers Apprenticeship Fund agreed not to exclude persons from the fund's apprenticeship training program because of race or color; to admit at least 90 of 154 prior applicants; to give black trainees who previously applied for apprenticeship the opportunity to transfer into the apprenticeship program without any loss in rate of pay or credit for prior training experience; and after 90 black prior applicants had been admitted, to select an additional 600 minorities among the next 1,200 applicants. The decree, which remained in effect until its provisions were filled, covered apprenticeship selection in all but three counties in Ohio and five Kentucky counties as well.[25]

A fundamental complaint against the policies and training programs of the Operating Engineer is that too often minorities end up as oilers or are confined to other lower paying jobs. On the other hand, the absence of craft jurisdictional lines in the open sector paves the way for a laborer to be upgraded to an equipment operator or truck driver, to move between the latter two jobs and to gain the capability to handle increasingly large equipment. Unquestionably, in highway construction, as in other sectors of the industry, the rigidity of craft jurisdictional lines acts as a barrier to advancement. Since blacks and other minorities are overly concentrated in the laborer classification, the result is to retard minority improvement.

CONCLUSION

The historic attitudes and policies of the craft unions, the barriers which craft organization sets up against upward mobility, the large

---

[23] *Commonwealth of Pennsylvania v. Local Union No. 542, International Union of Operating Engineers,* 347 F. Supp. 268 (E.D. PA, 1972); 648 Fc.2d 922 (Third Cir. 1981).

[24] *United States v. Local Union No. 3, International Union of Operating Engineers,* Fair Empl. Prac. Cas. 984 (1973).

[25] "Three Recent EEOC Consent Decrees with Employers and Union Set New Hiring and Selection Procedures," *Daily Labor Report,* No. 76 (April 18, 1975), p. A-15.

number of civil rights cases involving construction unions, and the lack of major success in many programs attempting to expand minority employment in construction all point to a greater problem in the union shop sector.

Bourdon and Levitt believe that any conclusion that minority utilization is greater in open shop than union shop firms is unwarranted, because minority participation in joint (union-management) apprentice programs is higher than in unilateral (open shop) programs which are approved by the U.S. Department of Labor.[26] In support of this contention, they presented data for 1976 showing a higher dropout rate for open shop than for unionized apprenticeship plans operating under the auspices of the U.S. Bureau of Apprenticeship and Training.[27] In contrast, in Chapter X, we presented data which showed that the dropout rate for both groups was about the same.

In fact, the Bourdon and Levitt apprenticeship data are anything but definitive in regard to relative minority employment. As they note, apprenticeship does not automatically create journeymen, although it is clearly a step in the right direction. But apprenticeship excludes the many other avenues of training, and particularly the major routes by which open shop workers learn craft jobs. The percentage of minorities in apprenticeship programs does not necessarily reflect the percentage in training or employed, particularly in the open shop segment.

Bourdon and Levitt also, apparently unknowingly, present evidence damaging to their position. They note that "the highest levels of minority participation [in U.S. Bureau of Apprenticeship and Training Programs] were in occupations such as cement masons, painters and roofers, while relatively lower proportions were . . . in trades such as plumbing, carpentry, and sheetmetal work."[28] In addition, they should have noted that higher proportions of minority apprentices are found in the trowel trades, and much fewer in electrical work. In short, the Bourdon and Levitt data emphasize the historic patterns of acceptance and discrimination, although they conclude that "whatever difficulties exist are common to both union and open-shop sectors of the industry and cannot be attributed to the influence of building trades unions alone."[29]

Such a statement ignores historical evidence and confuses the

---

[26] Clinton C. Bourdon and Raymond E. Levitt, *Union and Open-Shop Construction,* (Lexington, Mass.: D.C. Heath and Company, Lexington Books, 1980), pp. 69–73.

[27] Bourdon and Levitt, *Union and Open Shop Construction,* Tables 4-9, p. 73.

[28] *Ibid.*

[29] *Ibid.*

issue. Ample scholarly research has already been cited concerning the invidious discrimination historically practiced by the electrical, plumbing, sheetmetal, and other unions in construction.[30] No fair finding could charge that discrimination in the industry was, or is, attributable "to the influence of building trades unions alone." What is quite true and easily verifiable is the historically consistent patterns.

At the same time it is very clear that underutilization of minorities in the electrical, plumbing, and mechanical areas exists also in the open shop sector. Years of discrimination and lack of training have taken their toll. There are also open shop contractors who do not practice equal opportunity, and some who have run afoul of equal employment requirements. Nevertheless, it is equally clear that the union craft structure inhibits upward movement. Even in the absence of overt discrimination, this is a disadvantage for minorities, who tend to have less education, and training than white workers. The ability to work as a laborer, upgrade to a helper, and learn a trade at one's own pace, is a great advantage to minorities under many circumstances. The rigidities of the craft structure and its training system, union hostility toward the helper classification, and the traditional construction union-minority bitterness all tend to buttress the tentative conclusion, for which obviously more research is needed, that the open shop segment is both more hospitable as a whole to minority employment and structured to be more capable of dealing with the situation.

---

[30] See *supra,* notes 2, 3, 5, and 7.

PART FOUR

Conclusion

CHAPTER XIII

Conclusion

A brief review of the major points covered by this study will suggest some important trends to be expected in the future of open shop construction.

EXTENT OF OPEN SHOP CONSTRUCTION

In mid-1984, all the evidence indicates that open shop construction accounts for about 70 percent of construction volume in the United States, a substantial increase over the 50 to 60 percent open shop control estimated in 1975 in the first edition of this book. Open shop gains have taken place in every industry sector and in every region. Where open shop contractors were dominant, they have increased their share; where unionized contractors once had clear control, open shop contractors have, in a remarkable number of sectors and areas, wrested a share—sometimes a majority share—and often with surprising rapidity.

By industry sector, homebuilding is the most open shop, as it has always been. In mid-1984, single-family homes are being built about 95 percent open shop. Nearly all of the once unionized areas (except a few city centers where few such homes are built) either have become open shop or have some open shop residential construction. Multiple, particularly high-rise, dwelling construction, which as late as 1975 was estimated at 50 percent unionized, by mid-1984 is approximately 80 to 85 percent open shop, with unionized structures again confined largely to the inner cores of major cities. The open shop gains in high-rise apartment construction are especially important because builders who can construct high-rise apartments can also build high-rise office buildings. High-rise apartments thus provide a major access into commercial construction for open shop builders. Overall, the open shop segment probably controls 90 percent of all residential construction, which in turn accounts for about 40 percent of all construction.

It is in commercial construction that the open shop builders have made their greatest gains since the mid-1960s; in this sector, also, lies the greatest potential for future gains by open shop builders.

Again, the smaller the buildings and the farther from city centers, the greater the open shop penetration. The inflation of wages, benefits, and other costs has accelerated the search for economical alternatives to union construction. The open shop builder is one such alternative to which more and more users have turned; the result has been larger and larger facilities built open shop. The unionization of the commercial sector may be at its lowest ebb since World War II, with the open shop accounting for 55 to 60 percent of all volume and a much larger percentage of the number of projects.

Industrial construction remains the union stronghold. Yet larger and larger structures are being built open shop. Only three or four open shop contractors were capable of huge projects, such as public utility generating facilities or very large factories, in 1975. In mid-1984 ten or twelve open shop contractors can take on such work. In addition, most of the largest unionized contractors have doublebreasted open shop entities that can do these jobs. Given the large number of small structures built by open shop contractors, and their strength in process industry facility construction and maintenance, it is likely that open shop contractors have captured 35 to 40 percent of the industrial construction volume. This percentage could increase with the decline in the construction of nuclear power facilities, other public utilities and the huge "megaprojects" where unions retain a clear majority of work.

The highway and heavy construction sector of the industry has seen considerable gains by open shops since 1975, and is likely to see more under the new regulations of the Davis-Bacon Act, except in states which administer "little Davis-Bacon" laws in a manner that supports union power. Open shop penetration is probably in excess of 50 percent, although the data are somewhat conflicting.

Location and regional factors have played a key role in the expansion of the open shop. Union strength is greatest in the centers of major cities and weakens as construction moves to suburbia, to exurbia, and to rural areas. The older cities, particularly, have been losing population and business activity; this in turn has provided work for open shop builders at the expense of unionized builders in those cities.

The South, despite a few union enclaves, remains predominantly open shop. This is where the largest open shop builders grew and prospered, and here most of them are headquartered.

The Middle Atlantic states, the midwestern states, and the Pacific region are the most highly unionized, but such areas as the Pacific Northwest, upstate New York, central Pennsylvania, central Ohio, and large parts of Indiana have seen tremendous open shop gains.

The open shop segment is concentrated in the areas where industry is growing most rapidly. Moreover, the decline of unionism generally has reduced incentives to users to build union, and in some cases has even encouraged them to build open shop in order to reduce union influence in the areas of their facilities.

In the 1975 edition of this book it was found that open shop contractors were weakened by the lack of highly skilled nonunion subcontractors. This is no longer true. The majority of the members of the Associated Builders and Contractors (ABC) are now subcontractors, mostly in the electrical, plumbing, and pipefitting trades. Data presented in Chapter IV show a decided shift in the electrical and mechanical trades from unionized toward open shop operation which enhances the capability of open shop construction.

LABOR MARKET FACTORS

Open shop contractors hire, deploy, train, and compensate labor quite differently than do their unionized counterparts. The open shop sector places greater emphasis on job stability, diversity in job assignments, informal training, and annual earnings than on hourly wages. Open shop contractors gain by not having to concentrate their hiring through union hiring halls, but suffer from the lack of centralized employment sources. As a result, a number of open shop groups are establishing employment referral systems.

Flexibility in deploying workers gives open shop contractors a major advantage. Employees are used beyond the limits of craft concentration, no unnecessary workers need be employed, and skilled personnel are not used for unskilled and semiskilled work. These factors affecting productivity were repeatedly stressed in interviews seeking to determine why the open shop sector had achieved gains in recent years.

Most training in the open shop sector is of the informal, on-the-job type. As the open shop segment has expanded, however, so has its interest in more formal training programs. The large open shop contractors have developed training programs which emphasize upward movement with accomplishment rather than with time, and the open shop associations have developed traditional apprentice programs, which are growing but which remain a small part of open shop training. Other programs, for example the ABC's "Wheels of Learning," emphasize task training. The ABC, the open shop portion of the Associated General Contractors (AGC), and many other associations now sponsor formal open shop training programs. Because of the strength of the highly skilled specialty crafts in the open shop segment, more formal open shop training is obviously needed. At the same time, the flexible deployment system, with several grades of helpers and no craft barriers confining workers to laboring jobs, provides tremendous opportunities for on-the-job training and upgrading.

Although some open shop craftsmen receive hourly wages equal to or even greater than the union, most open shop construction workers are paid considerably less per hour. Open shop employers pay a range of rates depending upon the worker's service and competence; construction unions require a single rate for a classification, and the union rate is usually above the open shop one for the same classification. Pay for time not worked is significant in the unionized sector and is generally absent in the open shop one. Fringe benefits are much more important in the unionized than in the open shop sector. Most open shop contractors do not provide pension plans; profit sharing, however, is much more prevalent in the open shop segment. The open shop claims that by flexible working hours, work during inclement weather, and steady employment, its employees approximate or even exceed the annual earnings of unionized employees. On the basis of the data we have gathered, open shop wages are as much as 60 percent lower and benefits less common and generous than are unionized ones. Moreover, it is likely that the wage rate spread between the top groups has grown since 1975 and that this has been a key factor in the expansion of the open shop during this period.

Work hours also differ between the open and unionized groups. Open shop builders pay smaller and fewer overtime premiums, use varied workweeks and weekends to make up for inclement weather delays, and freely alter daily starting and closing times to expedite work and to reduce costs. Such flexibility is often made prohibitively expensive by union rules.

There has been no definite analysis of minority employment in the construction industry sectors, nor is there decisive evidence that open shop builders employ a larger percentage of minorities than do unionized ones. The data we gathered do, for the most part, point in this direction. The long record of discriminatory treatment of minorities by many construction unions, and the resulting community of interest between minorities and open shop builders help account for these findings. In addition, minorities are concentrated in many areas where the open shop contractors are the strongest, especially the South and southwest. Discrimination is certainly not absent in the open shop group, but the open shop nevertheless seems to be more hospitable to minority employment.

THE FUTURE OF OPEN SHOP CONSTRUCTION

The expansion of open shop construction since the mid-1960s is the result of the operation of market forces. The inflation of labor costs—including wages, fringes, and work rule inhibitions of productivity—undoubtedly created users' interest in less costly alternatives to unionized construction. Open shop builders were prepared to take advantage of their opportunities by years of development in suburban, exurban, and rural areas, and in the South and other regions where the construction unions are weak. Economic forces since the mid-1960s and early 1970s proved strong enough to overcome institutional and government restraints, such as past relationships of builders and users, political support of unionism, and union-favoring prevailing wage laws. The interesting question is whether these trends will continue with open shop construction expanding at the same or a greater pace.

In Chapter VIII, we saw how the construction unions responded to open shop competition by economic means, by political means, by use of the legal weapons of conflict, and by violence. We noted that the economic response varied, but that it by no means met the competitive threat in many areas, although the givebacks and restraints by the construction unions in the early 1980s have been the most significant since World War II. It is also not likely that union picketing or secondary boycotts will halt open shop expansion, since much such activity is illegal and can be controlled. Unions are finding that pension withdrawal liability, which they sponsored to keep companies from withdrawing from bargaining units, is a double-edged sword, for it keeps contractors from agreeing to go union. The new union drive against doublebreasted companies is also of questionable legality.

Union violence continues to break out, but it will not prevent the expansion of open shop construction, although it is quite clear that violence and the lack of even-handed police protection in many areas inhibit the use of open shop builders. Nevertheless, through the ABC and other organizations, the open shop contractors now are well prepared to fight violence and to compel law enforcement agencies to enforce the law, making violence less productive.

Open shop construction, as emphasized many times in this volume, has expanded because of market forces. The unions have priced themselves out of competition. They are therefore compelled to do one of two things: to become competitive by reducing wages, doing away with uneconomic work rules and other productivity restrictions, or to abolish, or modify, the craft union structure so that uneconomic labor deployment and artificial work restrictions are eliminated. Attempts to become economically more competitive have not been successful because the craft structure has remained sacrosanct. The construction unions, like the AFL-CIO generally, are therefore attempting to make economics irrelevant through politics. The unions are exerting maximum effort to elect former Vice-President Walter Mondale as President of the United States, and have prepared a legislative program which has AFL-CIO approval. (Similar programs have been or are being prepared at the state level.) This program includes:

1. Return of the U.S. Department of Labor and National Labor Relations Board to union control as during the Carter Administration.
2. Reinstitution of the basic rules and administration of the Davis-Bacon Act as prior to the changes initiated by Secretary of Labor Donovan.
3. Legislation banning doublebreasted companies by providing that "employee" within the meaning of the National Labor Relations (Taft-Hartley) Act is defined to "include all the employees under a contractor's control. Should the contractor decide to spin off a secondary operation, the employees of the operation would legally be members of existing bargaining units and covered by their contract."[2]
4. Passage of "common situs" legislation in order to permit boycotts of all building work when one craft is affected by a dispute. This would abolish the current legislation which provides that a construction union, although free to strike and to picket a firm with which it has a dispute, cannot legally picket or pressure any other firm on the site which is neutral in the particular dispute. This change would enable any one of the construction unions having a dispute with one of several firms on a construction site to picket indiscriminately and to shut down the total project. A common situs bill was enacted by Congress in 1976, but vetoed by President Ford; a similar bill was defeated in the House of Representatives in 1977.
5. Obtaining the right to utilize pension funds with little restraint for union purposes.[1]

There are other union political demands which, like the above, are designed to offset the economic advantages of open shop construction.[3] Their enactment would increase construction costs and

---

[1] This section is based on a study of construction union election and policy literature, 1982–84.

[2] "IUOE's Turner Scores Erosion of Legal Rights in Convention Speech," *Construction Labor Report,* Vol. 30 (April 18, 1984), p. 160. The quotation is from a speech by J.C. Turner, president of the Operating Engineers.

[3] One cannot help noting that if unionized construction had the advantage which Professors Steven Allen, Richard Freeman, and James Medoff claim, there would seem to be no need for political action of this nature.

could depress, not expand, the employment of union workers. If adopted, these demands would also adversely affect open shop contractors and workers.

Nevertheless, the economic advantages of open shop construction are so great that political action can inhibit but not destroy it. Open shop construction will probably continue to expand, although that expansion may slow, as the areas still under union control are likely to be the most difficult to dislodge. If, on the other hand, open shop construction is restricted by legislation granting more power to the already powerful construction unions, it would seem to follow that construction wages and costs will rise because a competitive restraint has been removed.

The first edition of this book concluded with the assertion that the future of open shop construction is in the hands of the builders. Open shop contractors and their organizations clearly met that challenge after 1975, and prospered accordingly. They nevertheless still have much work to do. They must develop the managerial and financial capability to manage larger jobs; they must develop new mechanisms for recruiting labor; they must develop more extensive training programs, so that they have a skilled workforce available; they must develop wage and benefit programs that meet their growing status and maintain their labor forces; and, although they probably lead the unionized sector in minority employment, they must take advantage of the opportunity for labor presented by minorities and expend more effort in affirmative action training and development. Now, after success, the challenge is still greater, and the future is more than ever in the hands of the open shop builders and their organizations.

APPENDICES

Appendix A

[Note: The interview schedule was developed for the 1975 study. It was utilized again for this work in order to facilitate comparisons. The main change made was to ask the same questions about women as are set forth in Part D for minorities.]

INTERVIEW SCHEDULE

These questions are designed for direct face-to-face administration to nonunion contractors and (particularly) spokesmen for nonunion contractors associations. They are purposely broad and open-ended to allow for maximum probing.

A. *Extent and Pattern of Nonunion Construction Activity*

1. What proportion of the work in your area is performed by nonunion contractors?
2. To what extent does the amount of nonunion work vary by industrial sector (industrial, commercial, heavy, and highway, residential, etc.)?
3. To what extent does it vary by size of contractor?
4. To what extent does it vary by functional specialty?
5. Has there ever been a history of more extensive union penetration in those fields which presently have a large nonunion component?
6. If so, what caused the unions to lose strength in those fields?
7. To what extent are nonunion contractors in your area represented by employer associations? (Get details on which *kinds* of contractors are represented by which associations.)
8. What labor relations functions, if any, do these associations perform? (e.g., gathering and disseminating wage information, sponsoring training, acting as intermediary in job-man matching, etc.)
9. To what extent do contractors operate on *both* a union and nonunion basis? In which fields?

10. To what extent do nonunion contractors operate with unionized subcontractors? In which fields are the subcontractors most likely to be unionized?

11. To what extent have union contractors gone nonunion in recent years? In any particular fields?

12. To what extent do nonunion workers migrate to the union sector? To what extent do union workers migrate to the nonunion sector. What are the reasons, if any, for these movements? Are they more prevalent for some trades than others?

B. *Labor Market Practices (Other than training)*

1. To what extent do each of the following provide a source of trained or partially trained workers for nonunion contractors:

   a) Vocational Schools

   b) Military

   c) Union construction

   d) Other industries (farming, manufacturing, etc.)

   e) Other (probe)

2. How might these sources vary with different trades?

3. What procedures or agencies are used by nonunion contractors to find the workers they need? Do they vary by trade? By size of contractor?

4. To what extent are craftsmen used to perform functions outside their primary craft (e.g., a carpenter laying brick)?

5. In your experience, what trades are most closely related so that an individual worker is likely to be able to perform in more than one?

6. About what proportion of nonunion craftsmen start out as laborers or helpers on construction jobs and work their way up? Is this process more prevalent in some trades than others?

7. During the summer months, about what proportion of nonunion workers are part-time (e.g., students)? Does this vary by trade?

8. How extensive is the practice of using students on a part-time basis while school is in session? Does this vary by trade?

9. If a contractor is unable to find the workers he needs, what adjustments is he most likely to make?

10. What is the normal seasonal ratio of employment from peak to trough? How does this vary by functional area?

11. For how many weeks might a regular employee of a nonunion contractor expect to be laid off during the slow season in a "normal" year? How does this vary by functional area?

12. Does the process of allocating and dividing work among employees (especially skilled vs. unskilled) differ during slow and busy periods? If so, how so? Does the practice vary by trade?

13. About what proportion of the nonunion work force would you characterize as "highly skilled"? To what extent does this proportion change with the tightness of the labor market? Could you try to describe the mechanics of this change, if any.

C. *Training and Upgrading*

1. Are there any industry-wide training programs, sponsored either locally or by a national organization, in your area?

2. If so, please describe the program. (Probe—information on nature of program, trades covered, administration, admission criteria, number of trainees, funding, post-training experience, etc.)

3. Do any individual contractors have their own programs? If so, describe.

4. Are there any formal arrangements between contractors (singly or as a group) and local vocational schools or institutes? If so, describe.

5. Apart from formal programs, what *specific* practices do contractors have to generate skilled workers? (If response is "OJT", try to learn what OJT entails—i.e., does

it include anything *outside* the normal production process?) Do these practices differ by type of contractor?

6. In your experience, are workers with formal training likely to reach a given skill level slower, faster, or in about the same time as one trained on the job? If there is a difference (either way), to what do you attribute this difference?
7. Do the training backgrounds of key people (e.g., foremen, supervisors, or contractors) tend to differ from those of rank and file craftsmen? Do these differences vary by trade?
8. Does the rate of which workers are upgraded internally vary with the pace of work at the time? Is the experience different for different trades?
9. Do you have (or know of) a *formal* upgrading mechanism? In other words, does a worker who moves up receive any explicit recognition that his status has changed?
10. Are the training practices of the large contractors different from the smaller ones?

D. *Minorities**

1. About what proportion of the nonunion construction work force in your area is black? Hispanic? Other minority?
2. Would this proportion vary by trade? If so, how?
3. In your estimation, is the level of minority participation greater than, less than, or about the same as that in the unionized sector?
4. If there is a difference, to what is it attributable?
5. Is there, in your area, any local program designed to increase minority participation in construction generally or nonunion construction in particular?
6. If so, do you have an opinion on how the program has worked? If it has worked well, why and by what standard? If not, why not?

* Women were asked the same question during the 1982–1983 interviews.

7. Do you feel that minority participation in your industry should be increased? If not, why not? If so, what in your judgment are the factors—and their relative importance—which have kept minority participation low up to now?
8. To your knowledge, are there any (or many) minority *contractors* in your area? If so, do they tend to be concentrated in any particular trades?
9. Do minority contractors face any special problems not typically experienced by white contractors?
10. Do you feel there should be any action taken by government or the industry to encourage minorities to become contractors? If not, why not? If so, what form might this action take?

E. *Employee Compensation*

1. Could you estimate the *range* of wage rates paid in your area to each of the following trades:

   a. carpenter
   b. bricklayer
   c. electrician
   d. equipment operator
   e. cement finisher
   f. plumber/pipefitter
   g. sheet metal worker
   h. roofer
   i. ironworker (rod)
   j. ironworker (structural)
   k. painter
   l. plasterer
   m. tile setter
   n. laborer

2. Are there any multi-employer benefit plans in your area? In which trades or among which kinds of contractors?
3. Could you estimate the prevalence and level of each of the following benefits in nonunion construction in your area:

   a. health insurance
   b. life insurance
   c. pension
   d. paid vacation
   e. paid holidays
   f. Christmas bonus
   g. profit-sharing
   h. severance pay
   i. SUB pay
   j. sick leave
   k. other (specify)

4. Do these benefits vary by trade or type of contractor?
5. Is overtime paid, and at what level, for:
    a) over 8 hrs/day?
    b) weekends?
    c) over 40 hrs/wk?
6. How prevalent is the "helper" classification with a pay scale between that of journeyman and laborer? Does it differ by trade?
7. About what proportion of the work performed by non-union contractors is subject to Davis-Bacon type regulations?
8. What is the impact of Davis-Bacon type regulations on the pay scales *and* pay structure of the industry?
9. About what proportion of hourly-paid nonunion workers earn at or above the union rate in their respective trades? Would this proportion vary by trade?
10. To the extent that nonunion wages are, on average, lower than those in the union sector, what in your judgment allows this differential to be maintained?

DAVIS-BACON QUESTIONS

1. Have you ever bid on a job where prevailing wages (according to the Davis-Bacon) applied?
2. If so, were you successful? If not was the DB requirement significant?
3. Were wage rates representative of the area rates?
4. How much could your bid have been reduced if the DB wages didn't have to be paid?
5. Have you any comments on how the DB Act effects:
    1) the cost structure of your bid,
    2) work force morale
    3) other?

6. 1) Unions, when they are a part of a "hometown plan", have the affirmative action requirement of the DB Act fulfilled. How do open shop firms comply with these requirements?
   2) Have DB requirements caused minority hiring problems?
   3) Have you had difficulty in reporting your compliance with DB requirements?

SUBCONTRACTING QUESTIONS

1. Do open shop employers in your area ever pay their employees by some form of piece work? If so, what type of employer is most likely to do so? How common is the practice?
2. Are there many subcontractors who do most or all of their work with a single contractor? If so, what type of subcontractor is most likely to do so?
3. Do contractors ever set up their own employees to be subcontractors? If so, what kinds of subcontractors are most likely to be so established? Why do contractors do this?
4. Does the general contractor typically supply materials for any of his subs? If so, which ones?
5. What kinds of work is the general contractor most likely to do with his own employees, and what kinds is he most likely to contract out?
6. *Note to interviewer*: If answers to previous questions seem to reflect a certain pattern of relationships with one group of subs (e.g., those doing basic structural work) and a different pattern with another group (e.g., those doing electromechanical or finishing work), try to find out *why* these patterns exist.

Appendix B

UNIVERSITY of PENNSYLVANIA

PHILADELPHIA 19174

The Wharton School

INDUSTRIAL RESEARCH UNIT
VANCE HALL — THIRD FLOOR
3733 SPRUCE STREET
HERBERT R. NORTHRUP, *Director*

OPEN SHOP CONSTRUCTION QUESTIONNAIRE

This questionnaire, prepared by the Industrial Research Unit of the University of Pennsylvania, concerns open shop construction. It is being sent to several thousand firms representing a cross-section of the construction industry. Whether you operate union or open shop, whether you are in favor of or opposed to unions or to open shop construction, your response is essential to the accuracy of this survey. The Associated General Contractors and the Associated Builders and Contractors are cooperating in this study.

The questionnaire is in three parts. The first asks questions about your company, the second asks about the area in which you operate, and the third about your labor market activities. We are interested in your views whether you work open or closed shop.

It should not take you more than thirty minutes to complete the survey. When completed, please staple it so the return address is on the outside, and mail. Postage is prepaid.

If you would like a statistical compilation of the finished survey, check here ____ and we shall send you one. These statistics will become part of a major study of all aspects of open shop construction. If you prefer to remain anonymous, you can still get a copy of the survey results by sending a separate letter.

We shall greatly appreciate it if you return your completed questionnaire within two weeks or as soon as possible.

Thank you.

Herbert R. Northrup

SECTION ONE
YOUR FIRM

1. What is the name and address of your firm?

____

____

____

____

(Omit this question if anonymity is desired)

SECTION ONE -- YOUR FIRM

2. Are you primarily a general contractor _____ or a subcontractor _____?

3. If you are a subcontractor, what kind of work do you specialize in? (e.g., flooring, electrical, concrete, etc.) ______________________.

4. What was the approximate volume of your business in 1973 (or last fiscal year)? $______________________.

5. Approximately what proportion of your work last year was in each of the following industry divisions?

Residential (single family and low rise)	_____%
Residential (high rise)	_____%
Commercial (stores, offices, etc.)	_____%
Industrial (factories, power plants, etc.)	_____%
Highway (streets, roads, interstates, etc.)	_____%
Heavy (bridges, dams, sewers, etc.)	_____%
Other (please specify)______________	_____%
TOTAL	100 %

6. Which of the following trades do you normally have on your payroll excluding subcontractors? (Please check all appropriate spaces.)

____bricklayer	____electrician	____plumber
____carpenter	____ironworker (structural)	____painter
____cement mason	____ironworker (rod)	____sheet metal worker
____equipment operator	____laborer	____steamfitter

7. Approximately what has been the average seasonal high number of non-supervisory workers on your payroll over the past three years? ____________ The seasonal low? ____________

8. Is the firm signatory to any union agreements, either individually or through a contractors association? _____ Yes _____ No

9. If "yes," which union(s)? ______________ ______________

10. Do you operate on a double-breasted basis? _____ Yes _____ No

11. If you are a general contractor, please list below the four trades you subcontract most frequently and indicate by a check (✓) how often you use nonunion subs.

	Always	Often	Occasionally	Never
a.____________				
b.____________				
c.____________				
d.____________				

12. If the answers to all of the above were "never," are there any trades for which you use nonunion subs at least occasionally? What are they?

________________ ________________

13. If you are a <u>subcontractor</u>, please check below how frequently you do work for unionized and open shop general contractors, respectively:

	Always	Often	Occasionally	Never
For union contractors				
For open shop				

SECTION TWO -- YOUR AREA

1. The table below refers to the amount of construction performed on an <u>open shop</u> basis in your geographic area. Please try to estimate wherever you can the percentage of open shop work in each type of construction indicated. Please insert a check (√) where appropriate:

	Percent Open Shop					
Type of Construction	0-20	21-40	41-60	61-80	81-100	Don't Know
Residential (single and low rise)						
Residential (high rise)						
Commercial (offices, etc)						
Industrial (factories, power plants, etc.)						
Highway (streets, etc.)						
Heavy (bridges, etc.)						
All construction (average)						

2. Are any of the percentages you indicated in the previous question different from what you would have said a few years ago? ___ Yes ___ No

3. If "yes," would you please indicate by an (X) in the above table the percentages that would have been appropriate then. (Example: if single-family residential is 81-100% open shop now and was 61-80% open shop a few years ago, please put a check (√) in the 81-100 column and an (X) in the 61-80 column.)

4. Regardless of the general level of open shop construction, are there any specialty areas (e.g., masonry, electrical, heating, etc.) which are significantly more tightly unionized than others? ___ Yes ___ No

5. If "yes," which ones? ________________ ________________

6. Have you ever directly experienced violence to person or property in connection with a labor dispute? ___ Yes ___ No

SECTION THREE -- LABOR MARKET ACTIVITIES

1. In the table below, please place checks (√) to show how often you use each of the indicated methods in seeking to hire new workers:

Method	Frequency Used			
	Always	Often	Occasionally	Never
Ask current employees to recommend someone				
Ask other employers to recommend someone				
Contact workers you have employed in the past				
Use union hiring hall				
Run a newspaper ad				
Contact workers who have approached you for a job in the past				
Use the state employment service				
Contact local vocational schools				
Hire "off the street"				

2. If you are an open shop employer, about what proportion of your workers were once (or are currently) union members? ________%

3. In the table below, please place checks (√) to show how often the workers you hire come to you with prior training from each of the sources indicated:

Source of Training	Frequency			
	Always	Often	Occasionally	Never
Vocational school				
Military (Corps of Engineers, CB's, etc.)				
Union apprenticeship				
Open shop apprenticeship				
Apprenticeships in other industries (farming, manufacturing, etc.)				
On the job with prior employer				
Other: ________				

4. Do you participate in any local training programs administered by an employer's association or by a union-management apprenticeship committee? ____ Yes ____ No

5. If "yes," how many of your employees are enrolled? ________

6. Do you provide any of your workers with <u>off the job</u> training or instruction on your own? ___ Yes ___ No

7. If "yes," please indicate the following:
   a. Number of employees receiving such training now ______
   b. Number who have already received such training ______
   c. Is instruction given by your own people ______ or through a vocational school ______?
   d. How long does training usually last (in weeks)? ______
   e. How long are the classes (hours per week)? ______

8. Please use this space to describe any training efforts you have undertaken which are not suggested in questions 4 through 7.

____

____

____

9. Do you typically put on supplementary summer help (college students, etc.) ___ Yes ___ No

10. If "yes," about how many did you employ in summer of 1973? ______

11. Have you had to lay off any workers in the past three years because of <u>seasonal</u> slowdowns? ___ Yes ___ No

12. If "yes," about how many were laid off in winter 1973-74? ______

13. Do you participate in any local programs to upgrade the employment opportunities of minority workers? ___ Yes ___ No

14. If "yes," who runs this program? ______

15. About what proportion of your current workforce is black? ______% Spanish speaking? ______% Indian? ______%

16. In your estimation, is the level of minority employment in open shop construction greater than ______, less than ______, or about the same as ______ that in union construction? (Please check.)

17. In the table below, please indicate the range of wage rates your workers receive: ($ per hour)

Trade	Wage Rates		
	Low	High	Foreman
Bricklayer			
Carpenter			
Equipment Operator			
Electrician			
Laborer			
Painter			
Plumber			
Sheet Metal Worker			
Steamfitter			

18. If you are an open shop employer, please estimate the percentage of your workers whose hourly wage is as high as or higher than the union scale. ____________________%

19. In the table below, please show by a check (✓) the approximate number of your workers who receive each of the fringe benefits indicated:

Benefit	Number of Workers Receiving: All	Most	Few	None
Health insurance				
Life insurance				
Paid vacation				
Paid holidays				
Paid sick leave				
Bonus or profit-sharing				
Other: ____________				

20. If you are an open shop employer, do you participate in any funded benefit plans such as those administered by an employers' association? ____ Yes ____ No

21. If "yes," who administers the fund? ______________________

UNIVERSITY of PENNSYLVANIA
PHILADELPHIA 19104

The Wharton School

INDUSTRIAL RESEARCH UNIT
VANCE HALL / CS. 3733 SPRUCE STREET
HERBERT R. NORTHRUP, *Director*

April 23, 1982

OPEN SHOP QUESTIONNAIRE

This questionnaire, prepared by the Wharton School's Industrial Research Unit, University of Pennsylvania, is part of the current revision of the 1975 study, Open Shop Construction. All replies will be in confidence and no companies responding will be identified.

This questionnaire, which is almost identical to the one which we sent out in April of 1975, is being mailed to several thousand firms representing a cross-section of the construction industry. Whether you operate union or open shop, whether you are in favor of or opposed to unions or to open shop construction, your response is essential to the accuracy of this survey. The Associated Builders and Contractors, the Associated General Contractors, the American Road and Transportation Builders Association, and many other specialty contractor organizations are cooperating in this study.

The questionnaire is in three parts. The first asks questions about your company, the second asks about the area in which you are headquartered or do most of your business, and the third about your labor market activities. We are interested in your views whether you work open or closed shop, or operate double-breasted.

It should not take more than thirty minutes to complete the survey. When completed, please remove the adhesive cover strip, press together, and drop it in any mail box. Postage is prepaid.

These statistics will become part of our major study of all aspects of open shop construction. The results will be available in early 1983. If you would like to order a copy of the final study, Open Shop Construction Revisited, please fill in your name and address where indicated at the end of the survey. Again, all replies will be in confidence.

We shall greatly appreciate it if you return your completed questionnaire within two weeks or as soon as possible.

Thank you for your cooperation.

Herbert R. Northrup
Professor of Industry

EXAMPLE	**IMPORTANT DIRECTIONS FOR MARKING ANSWERS.**
Wrong.	Use black lead pencil only. (2½ or softer).
1	Make heavy black marks that fill the circle completely.
Right.	Erase clearly any answers you wish to change.
2	Make no stray marks on this answer sheet.

1. Please indicate your Zip Code here for statistical purposes.

0 1 2 3 4 5 6 7 8 9

2. Are you primarily A. A general contractor. B. A subcontractor. A B

3. If you are a subcontractor what kind of work do you specialize in

- A Bricklayer
- B Carpenter
- C Cement Mason
- D Equipment Oper.
- E Electrician
- F Iron Worker (Structural)
- G Iron Worker (Rod)
- H Laborer
- I Plumber
- J Painter
- K Sheetmetal
- L Steamfitters

4. What was the approximate dollar volume of your business in 1981 or last fiscal year

A B C D E F G H — A 0 to 500,000 B 500,000 to 1 million C 1 to 10 million D 10 to 20 million E 20 to 50 million **F 50 to 100 million G 100 to 500 million H Over 500 million**

5. Approximately what portion of your work last year was in each of the following industry divisions

A 0 to 10% B. 11 to 20% C 21 to 30% D. 31 to 40% E 41 to 50% F. 51 to 75% G 76 to 100%

A	B	C	D	E	F	G	
A	B	C	D	E	F	G	Residential (Single Family Low Rise)
A	B	C	D	E	F	G	Residential (High Rise)
A	B	C	D	E	F	G	Commercial (Stores, Offices, Etc.)
A	B	C	D	E	F	G	Industrial (Factories, Power Plants)
A	B	C	D	E	F	G	Highway (Streets, Roads, Interstates)
A	B	C	D	E	F	G	Heavy (Bridges, Dams, Sewers)
A	B	C	D	E	F	G	Other

6. Which of the following trades do you normally have on your payroll, excluding subcontractors? Please fill in all that apply

- A Bricklayer
- B Carpenter
- C Cement Mason
- D Equipment Oper.
- E Electrician
- F Iron Workers (Structural)
- G Iron Workers (Rod)
- H Laborer
- I Plumber
- J Painter
- K Sheetmetal
- L Steamfitters

7. Are you a member of a contractors' association in your home territory A. Yes B. No (A) (B)

If so, which one A. A.B.C. B. A.G.C. C. A.R.T.B.A. D. N.U.C.A. E. I.E.C. F. P.H.C.C. G. N.E.C.A. H. P.D.C.A. I. N.R.C.A. J. S.M.A.C.N.A. K. M.C.A.A. L. A.C.C.A. M. Other

(A) (B) (C) (D) (E) (F) (G) (H) (I) (J) (K) (L) (M)

8. Is your firm signatory to any union agreements either individually or through a contractors' association? A. Yes B. No (A) (B)

9. Is your firm signatory to any union agreements? Please fill in all that apply.

- A Bricklayer
- B Carpenter
- C Cement Mason
- D Equipment Oper.
- E Electrician
- F Iron Worker (Structural)
- G Iron Worker (Rod)
- H Laborer
- I Plumber
- J Painter
- K Sheetmetal
- L Steamfitters

10. Do you operate on a double-breasted basis. A. Yes B. No (A) (B)

11. Please indicate how often you subcontract the following trades.
A. Always B. Often C. Occasionally D. Never

A	B	C	D	
○	○	○	○	Bricklayer
○	○	○	○	Carpenter
○	○	○	○	Cement Mason
○	○	○	○	Equipment Oper.
○	○	○	○	Electrician
○	○	○	○	Iron Worker (Structural)
○	○	○	○	Iron Worker (Rod)
○	○	○	○	Laborer
○	○	○	○	Plumber
○	○	○	○	Painter
○	○	○	○	Sheetmetal
○	○	○	○	Steamfitters

12. How often do you use non-union subs.
    - A. Always
    - B. Often
    - C. Occasionally
    - D. Never

13. If you are a subcontractor, please indicate how often you do work for union and non-union general contractors.
    A. Always B. Often C. Occasionally D. Never
    - For Union Contractors
    - For Non-Union Contractors

Section Two — Your Area

1. Please indicate the amount of construction performed on an open shop basis in your geographic area. Please try to estimate the percentage of open shop work in each type of construction listed.
   A. 0 to 20% B. 21 to 40% C. 41 to 60% D. 61 to 80% E. 81 to 100%.
   - Residential (Single and Low Rise)
   - Residential (High Rise)
   - Commercial (Offices, Etc.)
   - Industrial (Factories, Power Plants, Etc.)
   - Highway (Streets, Etc.)
   - Heavy (Bridges, Etc.)
   - All Construction (Averages)

2. Are any of the percentages you indicated in the previous question different from what you would have said a year ago.
   Yes No
   - One year ago
   - Five years ago
   - Ten years ago

3. If your answer to question #2 was "Yes," indicate the percentages that would have been appropriate five years ago.
   A. 0 to 10% B. 11 to 20% C. 21 to 30% D. 31 to 40% E. 41 to 50% F. 51 to 75% G. 76 to 100%.
   - Residential (Single and Low Rise)
   - Residential (High Rise)
   - Commercial (Offices, Etc.)
   - Industrial (Factories, Power Plants, Etc.)
   - Highway (Streets, Etc.)
   - Heavy (Bridges, Etc.)
   - All Construction (Average)

4. Regardless of the general level of open shop construction, are there any special areas (Masonry, Elec. Heating, Etc.) which are more tightly unionized than others.
   - A. Bricklayer
   - B. Carpenter
   - C. Cement Mason
   - D. Equipment Oper.
   - E. Electrician
   - F. Iron Worker (Structural)
   - G. Iron Worker (Rod)
   - H. Laborer
   - I. Plumber
   - J. Painter
   - K. Sheetmetal
   - L. Steamfitters

5. Have you ever experienced violence to person or property in connection with a labor dispute.
   A. Yes B. No

Section Three — Labor Market Activities

1. Please indicate how often you use each of the following methods to hire new workers.
A Always B Often C Occasionally D Never
A B C D Ask current employees for recommendations
A B C D Ask other employees for recommendations
A B C D Contact past employees
A B C D Use of union hiring halls
A B C D Newspaper advertisement
A B C D Contact job applicants on file
A B C D State employment service
A B C D Contact vocational schools
A B C D Hire off the street

2. If you are an open shop employer what portion of your workers were once (or are currently) union members.
A 25% B 50% C. 75% D. 100% E. Don't know A B C D E

3. Please indicate how often the workers you hire come to you with prior training from each of the sources indicated.
A. Always B. Often C. Occasionally D. Never
A B C D Vocational School
A B C D Military (Corps of Engineers, Seabees, Etc.)
A B C D Union apprenticeship
A B C D Open shop apprenticeship
A B C D Apprenticeships in other industries (Farming, Manuf., Etc.)
A B C D On the job with prior employer
A B C D Other

4. Do you participate in any local training programs administered by an employers' association *or* by a union management apprenticeship committee. A Yes B No
A B

5. Do you provide any of your workers with off the job training or instruction of your own. A. Yes B. No
A B

6. Is instruction given by: A Your own people B Vocational schools A B

7. How long does the training usually last. A. Up to 4 weeks B. 4 to 8 weeks C. 8 to 16 weeks D. Over 16 weeks
A B C D

8. How many hours per week do classes usually last. A. Up to 5 hrs. B. 5 to 10 hrs. C. 10 to 15 hrs. D. 15 to 20 hrs.
A B C D

9. Do you typically put on supplementary summer help. (College students, Etc.) A. Yes B. No
A B

10. Have you laid off any workers in the last 3 years due to seasonal slowdowns. A. Yes B. No
A B

11. Do you participate in any local programs to upgrade the employment opportunities of minority or female employees.
A. Yes B. No
A B

12. About what percentage of your current hourly work force is Black. A. 0% B. 1 to 10% C. 11 to 20% D. 21 to 30%
E. 31 to 50% F. 51 to 75% G. 76 to 100%
A B C D E F G

13. About what percentage of your current hourly work force is Spanish-speaking. A. 0% B. 1 to 10% C. 11 to 20% D. 21 to 30%
E. 31 to 50% F. 51 to 75% G. 76 to 100%
A B C D E F G

Section Three — Labor Market Activities (Continued)

14. About what percentage of your current hourly work force is Indian.

A B C D E F G A. 0% B. 1 to 10% C. 11 to 20% D. 21 to 30% E. 31 to 50% F. 51 to 75% G. 76 to 100%

15. About what percentage of your current hourly work force is female.

A B C D E F G A. 0% B. 1 to 10% C. 11 to 20% D. 21 to 30% E. 31 to 50% F. 51 to 75% G. 76 to 100%

16. In your estimation is the level of minority or female employment in open shop construction

A B C A. Greater than B. Less than C. About the same as that in union construction.

17a. Please indicate whether the wage level your *average* skilled craftsmen receive is A. Less than union rate B. Equal to union rate C. More than union rate D. Don't know

A B C D Bricklayer
A B C D Carpenter
A B C D Equip. Operator
A B C D Electrician
A B C D Laborer
A B C D Painter
A B C D Plumber
A B C D Sheetmetal
A B C D Steamfitter

17b. Please indicate whether the wage level your *most* skilled craftsmen receive is A. Less than union rate B. Equal to union rate C. More than union rate D. Don't know

A B C D Bricklayer
A B C D Carpenter
A B C D Equip. Operator
A B C D Electrician
A B C D Laborer
A B C D Painter
A B C D Plumber
A B C D Sheetmetal
A B C D Steamfitter

18. If you are an open shop employer please estimate the percentage of your workers whose hourly wage is as high as or higher than union scale. A. 0 to 25% B. 26 to 50% C. 51 to 75% D. 76 to 100%.

A B C D

19. Please show approximately how many of your hourly workers receive each of the fringe benefits listed A. All B. Most C. Few D. None

A B C D Health insurance
A B C D Life insurance
A B C D Paid vacation
A B C D Paid holidays or sick leave
A B C D Bonus or profit sharing

20. If you are an open shop employer, do you participate in any funded benefit plans such as those administered by an employers' association. A. Yes B. No

A B

Section Four

Please indicate your answer by filling in the correct numbers. (Ex. if your answer was 2,315, you would fill in the (2) in column one, the (3) in column two, the (1) in column three and the (5) in column four. If your answer was for example 12 you would leave the first two columns blank and fill in only columns three & four. If your answer was 1, you would fill in only column 4.)

1. Approximately what has been the average seasonal high number of nonsupervisory workers on your payroll over the past three years.

1	2	3	4
0	0	0	0
1	1	1	1
2	2	2	2
3	3	3	3
4	4	4	4
5	5	5	5
6	6	6	6
7	7	7	7
8	8	8	8
9	9	9	9

2. If you provide off the job training for your employees, please indicate how many of your employees have received this training, or who are currently in the process of receiving this training.

1	2	3	4
0	0	0	0
1	1	1	1
2	2	2	2
3	3	3	3
4	4	4	4
5	5	5	5
6	6	6	6
7	7	7	7
8	8	8	8
9	9	9	9

3. Did you hire supplementary help last summer.

A. Yes	B. No
A	B

4. If you have laid off workers in the past 3 years due to seasonal slowdowns, please indicate how many were laid off during the winter of 1981.

1	2	3	4
0	0	0	0
1	1	1	1
2	2	2	2
3	3	3	3
4	4	4	4
5	5	5	5
6	6	6	6
7	7	7	7
8	8	8	8
9	9	9	9

Appendix C

PROS AND CONS OF OPERATING UNION AND NONUNION DUAL SHOPS

BY A. SAMUEL COOK, ESQ.

Reproduced from A. Samuel Cook, *The McGraw-Hill Construction Business Handbook* (New York: McGraw-Hill, 1984) by permission of the author and publisher. An earlier version was serialized in the *Builder and Contractor,* which also gave permission to republish it here. N. Peter Lareau, Esq., Laurence J. Merwin, Esq., Phyllis Cole Friedman, Esq., and Richard Markey, Esq., assisted the principal author in developing this work.

A. Samuel Cook, a native Baltimorean, is a graduate of Princeton University and the Law School of the University of Maryland. After his admission to the Maryland Bar, he served on the staff of the Solicitor to the U.S. Department of Labor, and on the staff of the General Counsel, National Labor Relations Board, in Washington, D.C. He then became Industrial Relations Manager to the Davison Chemical Division of W. R. Grace & Co., followed by his appointment as Labor Counsel to the parent W. R. Grace & Co. on a national and international basis. Today Mr. Cook is an author, lecturer, and senior partner in the Baltimore and Washington, D.C. law firm of Venable, Baetjer and Howard. He heads up his firm's Labor Section and enjoys a national practice as a consultant on labor relations, EEO and OSHA to business and industry and to State and local governments.

INTRODUCTION

The building construction trades unions are in a bind. They have priced themselves right out of the competitive marketplace, and their unemployment lines range from 15 percent to 50 percent of their membership. In short, the unions have performed a hysterectomy on the goose that laid the golden egg.

Customers of the construction industry—smarting under the burden of soaring prices, shrinking profits, a loss of public confidence, and uncertainty over the economy—have developed strong feelings of anxiety over building costs. After studying recent inflation figures and disturbing projections for the future, these cost-sensitive customers are now focusing attention anew on "merit shop" nonunion contractors throughout the country. Such nonunion operations currently garner more than half of the owner-users' business nationwide, up from only a 20 percent share of the market in the 1950s.

This trend has come about because unorganized employers can deliver impressive savings. Depending on the project, they may work for 15 to 40 percent lower labor costs than unionized operations—unhindered by labor agreements which restrict labor-saving devices and systems, add costly featherbedding work rules and practices, inflate wages and fringe benefits, impose job foremen controlled by the unions, force use of skilled personnel to perform unskilled work, create interunion jurisdictional disputes, require restrictive apprenticeship training ratios, and establish other productivity-inhibiting practices.

Lately, with so many construction users building merit or open shop, many unionized employers have begun exploring ways to recapture some of their lost market. Two methods are frequently considered. One option is for the contractor to offer the unions representing its employees two alternatives: either negotiate moderate, competitive wage scales and operating efficiencies through collective bargaining or take the consequences of an all-out confrontation for survival.

A company which chooses this option must be able to negotiate on its own behalf and not be required to abide by the terms and conditions which are negotiated jointly by a group of employers. Therefore, if the contractor has previously delegated its bargaining rights to an employer association, it must rescind the authority given to that association and give timely notice thereof to the union and the association.

An employer with freedom to negotiate independently may bargain for whatever terms and conditions it in good faith believes are necessary to operate its business competitively. The mere fact that this may entail a decrease in wages and fringe benefits or a modification of featherbedding work practices does not make such demands unlawful. Of course, in most cases union representatives will not agree to such a revised contract. Therefore, any contractor determined to embark on this course of action should be aware that in all probability the company will be facing a strike.

Concomitant with the union's right to strike, however, is the contractor's right to permanently replace the striking employees with new hires. If this should occur, the union may decide to abandon the workers and concede defeat. Alternatively, once the contract has expired, the newly hired employees are free to petition the National Labor Relations Board (NLRB or Board) to conduct an election to *decertify* the union as the workers' bargaining agent. It should be noted that if any such election were conducted, both the new employees and the replaced strikers would be eligible to vote.

The second and more popular method of competing with the union shop contractor is to establish what is commonly referred to as a "dual shop" or "double-breasted" operation. This concept refers to the creation of two distinct companies, one of which is party to a collective bargaining agreement and one which is not. In doing so, the contractor is free to work on a wider variety of projects. A broader market is thus available and, consequently, the business is more competitive. However, unions are obviously less enthusiastic than employers about these arrangements, and have attacked dual shop operations from several legal angles. The National Labor Relations Act (the Act), the Employee Retirement Income Security Act (ERISA), and antitrust laws have all been used to limit the conditions under which such dual shops may operate. The employer who seeks to operate dual shops must therefore follow stringent legal guidelines imposed by the NLRB and the courts to successfully establish and operate in both the union and merit shop sectors of the market.

When the employees of one part of a dual shop operation are represented by a labor organization, it is often asked whether, under the Act, the second company must also recognize that labor union and apply the terms and conditions of the existing collective bargaining agreement to its employees. The ultimate resolution of this issue turns on the degree of "separation" versus the degree of "integration" of operation of the two businesses, as evidenced by the facts and circumstances of each situation and the applicable law.

Of major concern to the contractor is the immediate financial risk of setting up a dual shop. If a dual shop is not properly established, the second firm could be held jointly liable retroactively for "underpaid" wages and "delinquent" payments to various labor union pension, health and welfare, and other fringe benefit funds under the parent company's collective bargaining agreement. The second firm could also face a Board order merging the employees of the second firm into the original company's labor contract, with all of its costly and restrictive work provisions. Thus, an employer operating or planning to operate as a dual shop must take steps to ensure that his two businesses are not viewed as a single firm.

SEPARATE OR SINGLE EMPLOYER

There are several theories used by the Board and the courts to extend the provisions of a collective bargaining agreement between a union and a contractor to a business which is related to the union contractor. The most common theory is the "single employer" principle, under which two apparently separate companies are considered to be one for labor relations purposes, in spite of the maintenance of two or more distinct operations. The second theory, which is sometimes discussed interchangeably with the single-employer principle, is the *alter ego* principle. Under this theory, the use of separate business entities is considered to be a sham established to circumvent or evade collective bargaining agreements with the union. Consequently, workers in both companies are viewed as employees of the unionized operation, and the labor agreements of that company are applied to everyone.

Many decisions by the NLRB as to whether two enterprises will be treated as one under the single employer principle have been made in the context of determining the Board's jurisdiction over such related businesses:

> In applying the present jurisdictional standards, the Board early reaffirmed the long-established practice of treating separate concerns which are closely related as being a single employer for the purpose of determining whether to assert jurisdiction.
>
> The question in such cases is whether the enterprises are sufficiently integrated to consider the business of both together in applying the jurisdictional standards.
>
> The principal factors which the Board weighs in deciding whether sufficient integration exists include the extent of:
>
> - Interrelation of operations;
> - Centralized control of labor relations;
> - Common management; and
> - Common ownership or financial control.
>
> No one of these factors has been held to be controlling, but the Board opinions have stressed the first three factors, which go to show "operational integration," particularly centralized control of labor relations. The Board has declined in several cases to find integration merely upon the basis of common ownership or financial control.[1]

The Board has applied these same four criteria in dual shop situations in determining whether a second, newly created nonunion company is subject to the collective bargaining agreement between the parent company and the union and must recognize that union as bargaining representative for its employees. Within this context, the union often claims that since the new nonunion company is not really a separate employer, the new company must recognize the union and apply the terms and conditions of the collective bargaining agreement between the union and the parent company.

The Board's use of these four factors has been approved by the federal circuit courts of appeal and by the Supreme Court of the United States. The Supreme Court has agreed with the Board that:

> The controlling criteria set out and elaborated in Board decisions [which examine whether nominally separate business entities are a single employer], are interrelation of operations, common management, centralized control of labor relations and common ownership.[2]

Years of litigation have led the Board and courts to place greater stress on certain of these factors and less emphasis on others. As the NLRB noted as early as 1956, it has emphasized those factors

[1] Twenty-first Annual Report of the NLRB (1956), pp. 14–15.

[2] *Radio & Television Broadcast Technicians Local Union 1264 v. Broadcast Service of Mobile,* 380 U.S. 255, 256 (1965).

showing operational integration, with special consideration given to control of labor relations.

The Board's current position on common ownership reflects the view that unless there are other indications of common control, the same corporation, person, or group can maintain 100 percent financial control of two companies engaging in the same general type of business in the same locality, *without* subjecting the second company to the terms of a collective bargaining agreement entered into by the first. Thus, as a general rule, common ownership is less important than the other factors.

While the NLRB has focused on common control of labor relations as strong evidence of operational integration, it has sometimes minimized even this factor when other evidence appeared more convincing. In one case, the Board refused to accord controlling significance to the fact that the two related companies had separate and distinct labor policies, holding that:

> the presence or absence of a common labor relations policy is not conclusive in determining whether separate legal entities constitute a "single employer."[3]

Almost every case decided by the Board and courts follows a pattern of considering the four criteria for finding single employer legal status. What must be constantly remembered, however, is that no case is ever presented to the Board or a court in a vacuum or on identical facts. Each instance in which several firms are alleged to constitute a "single employer" requires the NLRB to look at the totality of the circumstances.

Common Ownership

The question of a dual shop will seldom arise unless a person or group which owns a substantial portion of the parent company also has a similar interest in a nonunion firm. If there is no common ownership, it is unlikely that there will be such other evidence of the operational integration of several firms as to warrant a single-employer finding.

However, it is possible for the same corporation, person, or group to maintain 100 percent financial control over two companies in the same locality and same type of business without having the two firms considered a single employer, provided there is no further substantial evidence of integration between the two companies. As the Board noted in its 21st Annual Report, the degree of common

[3] *Canton Carp's, Inc.*, 125 NLRB 483, 484 (1959).

ownership or financial control is rarely absolutely controlling in decisions of whether to overlook corporate form to find that two or more nominally separate companies are a single legal entity.

Many kinds of common ownership structures have been considered by the Board and courts. Yet their decisions on the single-employer issue do not seem readily classifiable on the basis of *how* common ownership is actually carried out. Thus, the fact that two companies may be wholly owned subsidiaries of a parent company demonstrates common ownership as significantly as if a single individual owned all or substantially all outstanding stock in two companies.

Common ownership is not easily hidden, and questions that usually arise concern the degree of the ownership interrelation. In closely held businesses, family-owned or otherwise, it may be difficult to avoid the level of common ownership the Board has found sufficient to constitute a single employer.

Common Management

How and by whom top-level and intermediate management decisions are made in firms alleged to be a single employer is a factor which may demonstrate integration of operations. Common management has been inferred, despite an employer's claim that two businesses were separately managed, on evidence that a unionized contractor established a second company to do business nonunion, continued to solicit jobs as he had before forming a second company, and simply apportioned the work between the two companies as cost requirements dictated. Exercise of such control over these two companies was held to be incompatible with nonintegrated and separately managed enterprises.

It is not enough, however, that the potential for common management exists. An individual or group with substantial financial interests in two companies could invariably choose to exercise financial control to influence management decisions. But the actual exercise of common management is controlling, not the potential to do so. In other words, the possibility of financial interests exerting pressure on day-to-day management decisions will not, without other indicators of operational integration, lead to a finding that several businesses constitute a single employer.

Normally, proof as to active exercise of common management is found in the day-to-day operation of the business. Thus, if supervisors from one company perform similar functions for a second enterprise; if the chief managing officials are identical or possibly rearranged (the president of one company is vice president of the other, and vice versa); or if officials of one company ordinarily lack authority to undertake jobs and make major capital expenditures without permission from officials of another firm, active common management will be found to exist.

Control of Labor Relations

The NLRB has characterized this third factor—common control of labor relations—as "critical . . . in determining whether separate legal entities operate as a single employing enterprise."[4] Unfortunately, however, evidence that the Board considers relevant in a decision as to whether there is a common labor relations policy has not been firmly established.

The Board's concern has sometimes centered on the actual, active control of day-to-day labor problems as well as the creation of overall personnel policies. As is true with the test for common management, proof that the same individuals have the potential, by virtue of their financial interests in two companies, to enforce a common labor relations policy has not normally been given significance. The control of labor matters must be active rather than potential.

Thus, a company whose supervisors lack the power to hire, fire, and discipline without permission from the officers or of another business entity is unlikely to be seen as anything other than a single employer joined with the related firm. Similarly, the fact that a majority owner of two companies insists that he be consulted on such daily, routine labor relations decisions is also strong evidence that the two businesses have a common labor relations policy.

Another, much broader view of what constitutes common control over labor relations policy has been expressed recently. A federal appellate court has suggested that a contractor's mere choice to establish a second business as initially nonunion might constitute *per se* the common control of labor relations necessary for a finding of "single employer."[5] However, this view was only dictum, not the actual holding of the case.

The appellate court reasoned that the present company's establishment of a nonunion framework within which the second company was to operate constituted the touchstone for day-to-day decisions as to wages, hours, and working conditions, and was therefore significant to the single-employer issue.

[4] *Gerace Construction, Inc.,* 193 NLRB 645 (1971).

[5] *Local No. 627, International Union of Operating Engineers v. NLRB,* 518 F.2d 1040, 1046 (D.C. Cir. 1975), *modified,* 425 U.S. 800 (1976).

The appellate court's dictum has had some effect on the well-established rule that owners of a business have the legal right to own more than one firm without automatically and universally applying a collective bargaining agreement to every commonly owned enterprise. Subsequent cases indicate that the very decision to establish a dual shop may be used as evidence showing common control of labor relations.

However, the great weight of legal authority requires the NLRB to review the amount of control over day-to-day decisions on such matters as hires, discharges, transfers, wages, hours, and working conditions. A shareholder's direct involvement in these labor relations decisions of two firms in which he owns majority interests, going beyond a general concern for the companies' financial conditions, will likely cause the two firms to be viewed as possessing a common labor relations policy.

If a single individual is responsible for grievances, employee remuneration, and benefit programs for the two companies, or if the two firms offer the same benefit to employees, the Board is more likely to find that the two enterprises constitute a single employer.

The Board's "total circumstances" test allows it to minimize the importance of any one factor when it believes a single-employer finding is otherwise appropriate. However, although the existence of a common labor policy may be critical, this factor is but one of the four general factors weighed on a case-by-case basis. As one appellate court has stated, "even a very substantial qualitative degree of centralized control of labor relations does not in itself determine the joint employer issue."[6]

Interrelation of Operations

The NLRB and courts have ruled that several companies constitute a single employer when the facts demonstrate that their operations are so dependent on one another as to operate, in effect, as one single and continuous production or sales organization. Thus, outside the construction industry, if one firm performed the first four steps in a production process and then sold all of its partially completed goods to a commonly owned second firm for the completion of the manufacturing process, single-employer status will almost certainly be found. For example, a clothing manufacturing company that performs only cutting functions and sells all its completed piece goods to a commonly owned second company for com-

---

[6] *Pulitzer Publishing Co. v. NLRB*, 618 F.2d 1275 (8th Cir. 1980).

pletion of the clothing will demonstrate such integration of operations as has been required by the Board.

In the construction industry, the Board will test for interrelated operations in the same manner as it tests for common management. Where commonly owned contracting companies engage in similar types of construction, the method of operation and decision as to which jobs each company will bid may demonstrate operational integration.

Thus, the fact that one company—operating under a union contract—seeks only those types of jobs for which there is no significant nonunion competition, or that the common management has a policy that one firm will bid for smaller contracts while the other firm will bid on larger jobs, demonstrates some degree of interrelation of operations. Where one person sets these policies for both companies, there is a chance that they will be found to constitute a single legal entity.

Seemingly minor indicators of day-to-day operating integration may also be significant. Two companies' use of a single or adjoining office facility, share office clerical personnel, or use of one bookkeeper for both firms, have been cited by the Board as evidence of interrelation of operations. In construction industry cases, the Board has also said that the use of one company's tools by another company, without proportional reimbursement, supports operational integration, as does the shared use of mail, telephone, and payroll facilities, lines of credit, and common customers.

These and similar facts might most appropriately be labeled secondary, since they are not conclusive and do not necessarily point to integration. But such facts will almost always be thrown into the balance as "bootstrap" evidence supporting a Board decision. Although these secondary facts demonstrate integration only by inference, the Board may rule that such extensive, though minor, relationships would not exist in nonintegrated businesses.

On the other hand, interchange between companies of employees in the bargaining unit is a *major* indication of interrelationship of operations. If employees of one company with a union contract are ordered to perform work for another commonly owned company, but continue to be paid by their own company, the Board may find an operational integration significant enough to partially support a single-employer finding. The impact of such employee interchange and transfer will be somewhat blunted if each company pays a proportionate share of the salary of any employee of the related enterprise who performs work for it. Even then, however, if workers for two companies are frequently or routinely transferred back and forth, the inference of integration will be certain.

Where there is only a one-time transfer of some workers from one company to another at the time of the establishment of the second company, and no further employee interchange, a finding of a single employer is unlikely. In a 1971 case decided by the Board,[7] for example, the owners of Company 1 decided to set up a second firm. When Company 2 was initially staffed, there was a transfer of employees of Company 1 to Company 2. Although a continued transfer of employees back and forth between the two firms might have persuaded the Board to make a single-employer finding, the Board declined on the facts of the case to hold that the initial transfer of employees, standing alone, demonstrated the degree of integration required to make the two companies legally a single entity.

Representative "Single Employer" Decisions

The various factors in dual shop-contested cases are never reviewed in a vacuum. The *totality* of the particular facts of each case will prompt the NLRB or the courts to give varying weight to the four factors. Several cases involving double-breasted operations may help to reveal the Board's reasoning and illustrate the type of evidence that tips the balance in a dual-shop case.

In the 1971 *Gerace Construction Co.*[8] case, for example, the Board considered union charges that a unionized contractor, Gerace Company, unlawfully refused to apply its labor contract to Helger, a separately incorporated nonunion firm. Helger had been established with the majority owner of Gerace and his wife owning similar controlling stock. Three-fourths of the capital required to establish Helger originated with Gerace Company directors, but Helger had never borrowed any money from Gerace.

During the first five months of its existence, Helger operated under Gerace's close supervision and control, and Gerace and a Gerace Company official were two of the three Helger directors. Although he and his wife started with majority ownership of Helger stock, Gerace began severing his ties with Helger within a few months. He resigned from his positions with Helger and transferred his stock holdings to Helger's principal management officer. Fol-

---

[7] *Gerace Construction, Inc., supra* note 4.

[8] *Supra* note 4.

lowing this transfer, Gerace neither possessed nor exercised any further authority over Helger's operations.

Helger was incorporated to operate nonunion and was set up as a separate legal entity, with separate bank and payroll accounts. The primary reason for the decision to have Helger refrain from signing a prehire agreement with a union was Gerace's desire to obtain a particular construction job, for which a prerequisite was the giving of a guarantee that there would be no work stoppages.

Additionally, Gerace testified that for several years Gerace Company had been unable to win construction contracts for jobs smaller than $100,000 because of hiring costs associated with union jurisdictional requirements. Since Helger competed for jobs smaller than those Gerace normally undertook, it was understandable that Helger's average job was normally less than $50,000, while Gerace's was usually over $250,000. While Helger's top employment was 16 employees, Gerace employed 95 persons at its peak. Helger never hired any of Gerace's employees and there was no other evidence of employee interchange.

The two companies only rarely submitted joint job bids and normally did not bid on the same jobs. It appeared that Helger's manager and Gerace did consult on approximately one of every 10 bids submitted by Helger. Although Helger used a trailer owned by Gerace for its temporary office and used tools and equipment owned by Gerace, it paid proportionate rent for such use. Its offices were later moved to a location a mile and one-half from Gerace's. The companies maintained separate health, welfare, and workers' compensation insurance contracts.

The NLRB, reversing its administrative law judge, held that the two contractors did not constitute a single employer. It ruled that while the common ownership of the two firms had established the potential for common control, actual common control of labor relations had never been established. In addition, the Board was persuaded that Gerace's *gradual relinquishment* of control over Helger's operations to a separate manager was a significant indication that the two firms were now separate, even if they had not been separate at their founding.

Another decision involving a contractor's attempt to go dual shop was *J. J. Cook Construction Co.*[9] After considering the four main factors of common ownership, common management, interrelation of operations, and centralized control over labor relations, the Board

[9] *J. J. Cook Construction Co.,* 203 NLRB 41 (1973).

found that the two construction companies did constitute a single employer.

The two firms, one union and the other nonunion, were owned in the same proportion by the same three people. There could be no doubt of common ownership. The two men who jointly held 98 to 100 percent of the stock of each company's operations determined whether a particular job would be bid by the union or the nonunion company. Water and sewer contracts and residential work were normally bid by the nonunion company, while schools, churches, and other industrial buildings were normally bid by the union company.

The owners freely testified to the Board that this arrangement was used because of the structure of the construction industry. They noted that in those types of work which were typically done by "open-shop" contractors, a company forced to pay higher, union scale wages would be unable to compete.

In addition to this evidence of an interrelationship of operations, the Board noted that the two companies occupied adjoining office suites in a building held by their majority owner. Although the Board found little evidence that one company's clerical staff performed work for the other firm, it pointed out that these employers were paid by checks prepared by one computer located in the nonunion firm's offices. All of the supervisors were interchanged between the two companies, and under an apparently arbitrary decision, three-fourths were listed on the union company's rolls while the others were carried on the rolls of the nonunion company. Given this evidence, the Board pierced the "corporate veil" indicating the existence of two separate firms, and held that the two construction companies were in fact a single employer.

Undoubtedly the most important construction industry dual shop decision involves a company known as Peter Kiewit Sons' Co. A holding company (Kiewit, Incorporated) wholly owned two subsidiaries (Kiewit and South Prairie), both of which were distinct, separately incorporated legal entities. Each subsidiary performed general and highway construction. Kiewit had operated for many years as the only union highway construction contractor in Oklahoma; South Prairie had operated nonunion for many years in states other than Oklahoma. Kiewit's costs as a union shop made it less competitive in bidding for Oklahoma jobs. In order to compete with nonunion contractors in Oklahoma, Kiewit, Incorporated decided to activate South Prairie in that state. The union filed Section 8(a)(5) and (1) charges with the NLRB, seeking to have its contract with Kiewit apply to South Prairie.

The Board[10] concluded that since Kiewit and South Prairie constituted separate entities rather than a single employer, South Prairie faced no potential liability under the Kiewit collective bargaining agreement. The Board recognized the common ownership and the similarity in type of business, but held these factors to be not controlling. The Board found no significant interrelation of operations; the two companies maintained separate Oklahoma offices, bank accounts, telephone numbers, supervisory personnel, and office clerical staffs; the subsidiaries also submitted separate job bids to avoid competing with each other. The Board also found no common labor relations policy, since South Prairie's day-to-day policies were set by its president, while Kiewit, Incorporated officials determined labor policies for Kiewit.

On appeal,[11] a two-to-one majority of the Court of Appeals for the District of Columbia reversed the Board decision. In finding that Kiewit and South Prairie were a single employer, the court noted the common ownership of the two subsidiaries. The court majority also found evidence of the most important factor—centralized control of labor relations. As noted above, the court reasoned that Kiewit, Incorporated's original establishment of a nonunion "framework" for South Prairie in itself "constituted a very substantial qualitative degree of centralized control of labor relations." The court surprisingly equated this initial and potential framework of overall labor relations with the day-to-day decision on labor matters which had heretofore formed the touchstone of the test for a common labor relations policy. The court stated simply that the relationship between the two firms' frameworks would not be found in an arm's-length relationship existing between nonintegrated companies.

In addition, the appellate court disagreed with the Board's implicit holding that there was no common management or interrelation of operations. The court focused on a shift of numerous management personnel from Kiewit to South Prairie's Oklahoma management and supervisory staff, in addition to some evidence of rank-and-file interchange. These personnel changes were viewed as demonstrating "a reasonable likelihood" that Kiewit's employees would lose work at the expense of South Prairie. As further support for this position, the court pointed to the fact that after South Prairie's activation in Oklahoma, Kiewit's bidding activity (and its success rate) declined.

---

[10] *Peter Kiewit Sons' Co.,* 206 NLRB 562 (1973).

[11] *Local No. 627, International Union of Operating Engineers v. NLRB, supra,* n. 5.

Both the employer and the NLRB petitioned the Supreme Court for review of the case. The Board argued that the single employer finding was erroneous, and that even if the two firms were a single enterprise, the D.C. Circuit had usurped the role of the Board to ascertain whether the employees of both Kiewit and South Prairie possessed a sufficient community of interest to constitute an appropriate collective bargaining unit.

In a brief *per curiam* opinion, the Supreme Court[12] in part affirmed and in part vacated the D.C. Circuit decision. The Court left intact, without discussion, the Court of Appeals' reasoning and holding on the single-employer issue. However, the Supreme Court concluded that the lower court improperly reached and ruled upon the issue of whether employees of both firms should be included in one collective bargaining unit. The Court stated that this decision should always be made by the NLRB in the first instance. The Court observed that a determination that two firms constitute a single employer "in the construction industry . . . does not necessarily establish that an employer-wide unit is appropriate, as the factors which are relevant in identifying the breadth of an employer's operation are not conclusively determinative of the scope of an appropriate unit."[13] Accordingly, the Court vacated that part of the lower court judgment which held that the two firms' employees constituted the appropriate bargaining unit, and remanded the case to the Board for that determination.

In *Don Burgess Construction Corp.,*[14] the Ninth Circuit enforced a Board decision that a nonunion (formerly union, until the collective bargaining agreement expired) corporation (B.C.) operating as general contractor for light commercial construction, and a union partnership (V.B.), operating as general contractor and carpentry subcontractor, constituted a single employer. After reviewing precedents establishing the basic principles of the single employer determination, the court looked at each of the four factors in detail. The court found interrelated operations, based on the transfer of employees between the two entities, common supervision, use of the same tools, and the fact that nearly all of V.B.'s work came from B.C. Although V.B. and B.C. maintained separate payrolls, bank accounts, and tax returns, they used the same accountant.

The court found substantial common management in part because Don Burgess, the president of B.C., exercised "significant" control

[12] *South Prairie Construction Co. v. Local No. 627, International Union of Operating Engineers,* 425 U.S. 800 (1976).

[13] *Id.* at 805.

[14] *NLRB v. Don Burgess Construction Corp.,* 596 F.2d 378 (9th Cir. 1979).

over V.B. Moreover, V.B.'s California contractor's license designated Burgess as the individual "responsible for exercising such direct supervision and control . . . as is necessary to secure full compliance with the law." The record also revealed that Burgess made labor relations policy decisions for both firms—e.g., the decisions to form a nonunion corporation, to have V.B. employees work with the B.C. crew, and to sign union contracts for B.C. and V.B. Finally, the court found Burgess's 70 percent ownership of B.C. and 50 percent partnership interest in V.B. to be sufficient common ownership.

In another recent dual-shop case, *Edward J. White, Inc.,*[15] the Board found two contractors to be a single employer. White (union) and Repairs (mostly nonunion) were independent corporations operating as plumbing and heating contractors performing service and repair work. The Board found that White and Repairs performed essentially the same type of work, using the same facilities, tools, equipment, and often the same employees. The Board found substantial common ownership, common corporate officers and directors, and common day-to-day management. Common control of labor relations was inferred from a single individual's role in negotiation of the union contract with White, and in the decision to establish Repairs and determine the manner in which it would operate. The Board concluded that Repairs was created in order to avoid White's obligation to bargain with the union.

The Board and court decisions after *Peter Kiewit Sons' Co.* indicate that the law on the single-employer issue has changed little in recent years. The Board still examines the four indicators of integration—interrelation of operations, common management, centralized control of labor relations, and common ownership. Common ownership alone is not decisive in determining single-employer status. Centralized control of labor relations is probably the most critical factor, although a recent appellate decision[16] indicates that even this factor is not controlling. The major change in the past few years has been the Board's willingness to consider, as evidence of common control of labor relations, a company's decision to establish a dual shop.

In sum, the Board will consider the totality of the circumstances in determining whether two companies operating as a dual shop in fact are a single employer. Seemingly insignificant facts such as common use of telephones, buildings, mail facilities, tools, and clerical personnel may lead the Board to require the newly established nonunion company to recognize the union and apply the collective bargaining agreement.

---

[15] *Edward J. White, Inc.* 273 NLRB 1020 (1978).

[16] *Pultizer Publishing Co. v. NLRB, supra,* n. 6.

THE APPROPRIATE BARGAINING UNIT

The Kiewit Decision

The Supreme Court opinion in *Kiewit* established two prongs to the question of whether related companies may operate double-breasted. Even if the two firms are, in fact, a single employer, the Board must further determine whether the employees of the firms share a sufficient community of interest for the Board to include all in a single collective bargaining unit. Therefore, it is possible that although a unionized operation and nonunion operation are a single employer, the Board nonetheless will not require the non-union shop to adhere to the union contract.

This possibility became fact in a supplemental decision on the appropriate bargaining unit issued by the Board in the Kiewit case.[17] Although Kiewit and South Prairie constituted a single employer, the Board concluded that the employees of the two subsidiaries lacked community of interest. Since the employees of the two firms could not be joined together in a single bargaining unit, the Board refused to impose upon South Prairie the collective bargaining agreement executed between Kiewit and the union.

The *Kiewit* decisions thus clearly establish that the Board will examine two distinct issues in the single-employer context: first, whether the union and nonunion operations are so closely integrated as to constitute a single employer; and, second, whether employees of both firms should be included in one bargaining unit.

In addition, in the supplemental *Kiewit* decision the Board offered guidance on determining the appropriate unit:

> The following factors are particularly relevant: the bargaining history; the functional integration of operations; the differences in the types of work and the skills of employees; the extent of centralization of management and supervision, particularly in regard to labor relations, hiring, discipline, and control of day-to-day operations; and the extent of interchange and contact between the groups of employees.[18]

Representative Bargaining Unit Cases

In a recent decision involving the lumping of employees in a double-breasted operation into a single bargaining unit, *A-1 Fire Protection, Inc.,*[19] the Board found evidence that the union knew of

---

[17] *Peter Kiewit Sons' Co.,* 231 NLRB 76 (1977).
[18] *Id.* at 77.
[19] *A-1 Fire Protection, Inc.,* 250 NLRB 217 (1980)

the company's dual-shop operation and acquiesced in the existence of a nonunion enterprise. The company did not deceive the union or transfer work away from union employees to nonunion employees. The Board held:

> We recognize that the fact that CAS and A-1 constitute a single employer does not alone dictate that separate units are inappropriate (citing *Kiewit*). The Board and courts have acknowledged that in the construction industry a single employer may have one company to perform contracts under union conditions and another company to operate under non-union conditions, and the Board has refused in the past to include non-union company employees in the same unit with employers of the union company. Thus, if in fact there is a voluntary agreement on unit scope here, we will not disturb that agreement in this unfair labor practice proceeding, and we will not allow the Union to avoid the terms of that agreement by asserting an unfair labor practice in Respondent's refusal to enlarge the unit to include A-1 employees.[20]

One of the Board's recent bargaining unit decisions is *The Arundel Corporation.*[21] In that case, a large construction firm decided to put into operation a quarry located 37 miles from its headquarters. A dormant corporation was revived and given control over the quarry, so that the quarry could be operated nonunion. The Board concluded that the quarry employees could not be lumped into a single bargaining unit with employees at Arundel Corporation headquarters.

The Board found that the quarry employees warranted an independent unit. The quarry was a separate, autonomous operation, with the Corporation exercising limited oversight. There was almost no interchange of personnel, and the work performed at the quarry was different from that performed at other facilities of the corporation. The Board noted that "an employer may legitimately attempt to establish part of its business as a nonunion operation," but cautioned that a new nonunion enterprise should not be used as a device to drain off work to be performed by union employees under the collective bargaining agreement. The nonunion shop should be a "parallel and simultaneous operation," rather than a substitute for an existing unionized company.

A Second Defense

The Supreme Court's *Kiewit* opinion established two prongs to the determination of whether related companies may operate dou-

[20] *Id.* at 13–14.
[21] *The Arundel Corporation,* 252 NLRB 397 (1980).

ble-breasted. Even if the two operations are sufficiently integrated to be considered a single employer, the Board must further consider unit appropriateness factors. In theory, the *Kiewit* decision aids employers operating dual shops by adding an additional defense to unfair labor practice charges. *Kiewit* establishes that a dual operation may be a single employer, but that employees of both firms will not necessarily be lumped into a single bargaining unit.

However, in practice only rarely has the Board found a dual operation to be a single employer, and yet concluded that a single unit is inappropriate. Because many of the factors relevant in determining the single-employer question are also relevant to the unit question, the Board in most cases will leap both hurdles in a single bound.

THE ALTER EGO

In recent years, some court and Board decisions have sought to avoid *Kiewit's* requirement to analyze both the degree of affiliation between union and nonunion shops and the appropriateness of the bargaining unit before applying a labor agreement to the employees of both operations. It has been argued that one way to avoid the two-step analysis is to view the relationship of a nonunion shop as an *alter ego* (other self) of a union shop.

The *alter ego* doctrine, which most frequently is raised when one company succeeds another, holds that an employer may not evade its obligations under a labor agreement by merely changing the form of its business. Typically, the issue arises when one company transfers its assets to another and then "abandons" its original business. The successor company ends up with a different name, but the same management, operation, equipment, customers, and ownership as the former company. The claim may also arise when a company transfers work which once was performed by union workers to a nonunion affiliate. In both cases, the nonunion company claims that it is not bound by any labor agreement entered into by the union shop. Notwithstanding such a claim, the contract is applied to the employees because the business arrangement is viewed primarily as a disguised continuance of the former company, or an attempt to avoid the obligations of the collective bargaining agreement through a sham transaction or mere technical change in the union company's operation. The successor is deemed to be the *alter ego* of the former company and therefore it is also bound by the first company's labor obligations.

Questions of the single employer and the *alter ego* are sometimes treated as the same issue.[22] Nevertheless, more recent cases hold that they are separate principles which are distinguished primarily by a substantial continuity of the work force from one company to the other, and the employer's motivation to leave the first business to avoid dealing with the union.[23] The difference between the two principles is important because, at least according to some courts, the *alter ego* doctrine does not require a separate examination of the bargaining unit before the labor contract is applied to the *alter ego* company. The employer is thus denied his second defense before the Board if this dual-shop operation is involuntarily dismantled.[24]

As the Fifth Circuit has recently explained:

> The Board, in applying the single employer doctrine, makes an independent, careful investigation into whether the employees of two firms held to constitute a single-employer constitute an appropriate bargaining unit. Only where the employees do constitute an appropriate bargaining unit will the non-signatory firm be bound to the collective bargaining agreement entered into between the signatory firm and the union. However, when the Board makes a finding that a non-signatory employer is the *alter ego* of a signatory employer which has voluntarily agreed to recognize the union's representative status in a unit stipulated in the collective bargaining agreement, the Board generally will not reconsider the unit under the community of interests test, but will simply make a far more limited determination whether the stipulated unit is repugnant to any policy embodied in the NLRA.[25]

An additional problem that could be raised in the *alter ego* context is demonstrated in the Board's recent decision in *Milwaukee Spring Division & Illinois Coil Spring Co.*[26] In that case, the Board held a manufacturer guilty of an unfair labor practice because it transferred work from a union to a nonunion plant without the consent of the union which represented its one facility. Although the move was motivated by lawful economic reasons, and the company had bargained with the union regarding the decision to move and its effects on unionized workers, the Board found that the transfer violated the Act. The Board's remedy was to order the company to rescind the decision and to restore the status quo ante by returning the operation to the union facility. Both the decision and the remedy

---

[22] See, e.g., *Walter v. Cooke, Inc.*, 262 NLRB No. 74 (1981).

[23] See, e.g., *Carpenters Local Union No. 1846 v. Pratt-Farnsworth, Inc.*, 690 F.2d 489 (5th Cir. 1982); *Pinter Brothers, Inc.*, 263 NLRB No. 91 (1-82); *Air-Vac Industries, Inc.*, 259 NLRB No. 51 (1981).

[24] *Carpenters Local Union No. 1846 v. Pratt-Farnsworth, Inc., supra* n. 23.

[25] *Id.* at 508–09.

[26] *Milwaukee Spring Division & Illinois Coil Spring Co.*, 265 NLRB No. 28 (1982).

will undoubtedly be tested in the courts. Nevertheless, the case demonstrates the extreme penalties that could occur if the dual-shop operation is not properly approached from the beginning.

To protect against the loss of the right to require both a review of the degree of integration between union and nonunion operations *and* the appropriateness of the unit, contractors should continue to operate both the union and nonunion shops, if possible, and avoid using employees of the union shop to staff the nonunion operation. The risk of raising the *alter ego* issue will therefore be minimized and greater protection for the dual-shop operation will be established.

ANTITRUST PROBLEMS

In addition to direct attacks against dual shops under the single-employer and *alter-ego* principles, unions have used the antitrust laws in their attempts to prevent contractors from creating merit shop operations. However, courts have not yet clearly defined what, if any, antitrust implications exist in the context of dual-shop relationships.

In a recent decision, the United States Court of Appeals for the Ninth Circuit indicated that a conspiracy among construction industry employers to encourage the use of open-shop and double-breasted operations may violate federal antitrust laws. In *California State Council of Carpenters v. Associated General Contractors of California,*[27] the union charged that AGC violated the Sherman Act by promoting dual-shop operations in a seminar, in addition to other activities. The union alleged damages of $25 million, tripled to $75 million under the antitrust law. A two-to-one majority of the Ninth Circuit held that AGC was not exempt from antitrust liability, and ruled that the case should move forward to trial to determine whether AGC in fact violated the Sherman Act.

The court found that the union's complaint presented the reverse of the situation in *Connell Construction Co. v. Plumbers Local 100,*[28] a 1975 Supreme Court decision. In *Connell,* a union picketed a general contractor to compel it to do business exclusively with subcontractors who had signed contracts between the union and the general contractor. Reasoning that the union's activity would prevent more efficient nonunion subcontractors from effectively com-

---

[27] *California State Council of Carpenters v. Associated General Contractors of California,* 648 F.2d 527 (9th Cir. 1980).

[28] *Connell Construction Co. v. Plumbers & Steamfitters Local Union No. 100,* 421 U.S. 616 (1975).

peting, the Supreme Court held that the union was not exempt from antitrust liability.

The Ninth Circuit applied the rationale of *Connell* in concluding that a conspiracy among employers to persuade a stranger employer to operate dual shop might violate the Sherman Act. Such a conspiracy would coerce the stranger to use only nonunion subcontractors, thus preventing union subcontractors from effectively competing.

In a vigorous dissent, Judge Sneed argued that the threat of antitrust liability would lead an employer to divert all subcontracting to union firms. Thus, the Ninth Circuit's decision allows the union to achieve precisely what the Supreme Court in *Connell* prohibited—a restraint on trade by the elimination of competing nonunion subcontractors. Judge Sneed also argued that the union's sole injury (if any)—an impairment of its ability to represent a greater number of construction workers—was not protected by the Sherman Act.

The *California State Council of Carpenters* decision raises the ominous possibility of enormous antitrust liability for employers who meet together to discuss dual-shop operation. However, the case is only in its initial stage. The union must still prove the existence of a conspiracy and show an anticompetitive effect. Since the creation of a dual shop *increases* competition by introducing into the local market a nonunion company which operates at lower cost and offers lower prices to consumers, it seems unlikely that the AGC chapter will suffer heavy antitrust damages.

An important recent Fifth Circuit case has followed the rationale of the *California State Council of Carpenters* case, but has left open an important issue which may ultimately reduce the danger of antitrust liability against dual-shop contractors and their associations. In *Carpenters Local No. 1846 v. Pratt-Farnsworth, Inc.,*[29] unions and their affiliated health and welfare funds mounted an all-out assault on the dual-shop system of doing business in the New Orleans area. Allegations of wrongdoing were filed under a variety of legal theories, including antitrust. The court of appeals followed the Ninth Circuit's decision in *California State Council of Carpenters* and refused to dismiss the claims because of the labor exemptions to the antitrust laws. Furthermore, even though antitrust law deals exclusively with commercial competition (competition between two businesses as opposed to competition between unions and unorganized workers for jobs with an employer), the unions stated a cause

---

[29] *Supra* n. 23.

of action when they claimed that nonunion contractors had unlawfully joined together to drive union contractors out of business.

Although the Fifth Circuit held that the pleadings stated a claim for antitrust violations, it acknowledged that a preliminary issue regarding the standing of the unions to pursue those claims must be considered before the damages could be awarded. The parties had not raised the right of the unions to bring claims for the alleged anticompetitive activities and therefore the court would not rule on the issue. Nevertheless, the court recognized that the damages which were suffered as a result of the alleged conspiracy occurred only to the business competitors of the dual-shop operations; the unions had no such loss. The court stated:

> Although we have held that a cause of action exists under the antitrust laws because of restraints at the level of contractor services, we wish to clarify our analysis further by explicitly separating our views on the existence of a cause of action from the question of standing. . . . Our analysis has established that *Connell* seems to allow a cause of action against the defendants. However, that analysis does not necessarily point to the Unions or the Funds as the natural plaintiffs who are entitled to bring that cause of action. Rather, the most obvious candidates for standing are those most directly injured at the contractor level: the signatory contractors themselves.[30]

It thus appears that the Fifth Circuit is not inclined to entertain an antitrust claim by the unions unless they are able to establish that they are commercial competitors of the contractors. The issue of the unions' standing to raise the antitrust claims was therefore returned to the district court for further analysis.

POSSIBLE WITHDRAWAL LIABILITY UNDER THE MULTIEMPLOYER PENSION PLAN AMENDMENTS ACT OF 1980

Since most collective bargaining agreements in the construction industry require an employer to contribute to a jointly trusteed, multiemployer pension plan, the Multiemployer Pension Plan Amendments Act of 1980 (MEPPAA) may pose an added risk to existing as well as newly formed dual shops.

Contributions to multiemployer plans are generally fixed dollars amounts based upon hours worked. Prior to 1980, contributing employers, believing this was the limit of their liability, often paid little or no attention to whether the plans had sufficient assets to

---

[30] *Id.* at 535–36.

pay the generous benefits promised. In fact, many plans were underfunded.

A recent survey of 186 construction industry plans revealed that almost half had unfunded vested liabilities, with the total liability estimated to run as high as $3.57 billion. In other words, almost half of the multiemployer plans surveyed were insolvent. Nevertheless, as long as employers remain in a plan and continue contributions, the cash flow permits the continued payment of benefits despite a plan's insolvency. The benefits promised participants, however, would be in serious jeopardy if employers stopped or severely reduced their contributions, either because of changes in bargaining agreements or the full or partial cessation of their businesses.

To protect the benefits promised employees under multiemployer pension plans, Congress passed MEPPAA in 1980 to place the responsibility for any unfunded liabilities in such plans on contributing employers. MEPPAA thus provides that if a contributing employer stops or substantially reduces its contributions to a multiemployer plan, it will immediately be assessed a proportionate share of any unfunded liabilities. This proportionate share is known as "withdrawal liability" or "partial withdrawal availability [*sic*]" and is frequently very substantial. In many cases employers, who were unaware that they had been contributing to a multiemployer plan, sold or liquidated businesses ony to find an assessment of withdrawal liability in an amount greater than they had realized from the sale or liquidation of their businesses. One company with only nine covered employees received a bill for $68,000 after going out of business; another employer with a bargaining unit of approximately 150 employees was faced with withdrawal liability of approximately $2 million.

At the same time that Congress imposed this "withdrawal liability," it created a limited exception for the construction industry. This exception generally permits an employer contributing to a construction industry multiemployer pension plan to avoid the assessment of withdrawal liability or partial withdrawal liability when its contributions cease or substantially diminish because of the termination or substantial contraction of its business, as long as it does not shift its work to a nonunion operation. The construction industry exception for withdrawal liability, however, may be lost if a contractor has a dual-shop operation.

As discussed previously, in deciding when union and nonunion companies will be treated as a single firm, the NLRB has focused on the absence of: interrelated operations; centralized control of labor relations; and common management. For withdrawal liability purposes, however, MEPPAA adopts the common control test for a single employer found in the Internal Revenue Code.[31] This test focuses solely on percentage ownership interest without regard to the interrelation of operations, central control of labor relations or common management.[32]

The result of this dichotomy in definition may be that, although a nonunion company is insulated from the collective bargaining agreement of its sister company, it could cause its sister company to lose the construction industry exemption from withdrawal liability. This could happen because, in the construction industry, in order for an employer to qualify for the special exemption from withdrawal liability when it stops paying into a multiemployer plan, it must cease operations in the jurisdiction of the collective bargaining agreement.[33] However, if both the nonunion and union shops are under common ownership within the Internal Revenue Code definition, and the nonunionized shop continues operations, the "employer" will be deemed not to have ceased operations. The consequence of common ownership will then be that the union shop will not qualify for the construction industry exemption from withdrawal liability; withdrawal liability will be assessed; and it will be owed by the "employer" (consisting of both the union and nonunion shop).

A more complex problem is presented when the union shop temporarily ceases operations while the nonunion shop continues to operate. Whether withdrawal liability will be imposed under such circumstances depends upon how the Pension Benefit Guarantee Corporation and ultimately the courts construe the statutory phrase "ceases to have an obligation to contribute under the plan." If the phrase is construed to mean that whenever a union shop has no employees for whom it has a present obligation to contribute, it has ceased to have an obligation to contribute, then, in a double-breasted operation, full withdrawal liability could be assessed simply by the absence of current work for the union operation. This result could

[31] Under MEPPAA, for purposes of withdrawal liability, all employees of trades or businesses which are under common control are treated as employed by a single employer and all such trades and businesses as a single employer. Sec. 4001(c).

[32] MEPPAA prescribes regulations coextensive with the regulations for Section 424(c) of the Internal Revenue Code which treat employees employed by a business under common control as employed by a single employer. Generally, when a parent organization owns 80 percent of the subsidiary it will be deemed under common control; and when the same five or fewer persons own 50 percent of a brother-sister group of businesses they will be deemed under common control. Treas. Reg. § 11.414(c)-2.

[33] MEPPAA Sec. 4203(b)(2).

be disastrous for dual-shop operations, but it is too soon to tell whether such a strict construction will prevail.

For partial withdrawals, that is, prolonged business slowdowns that cause contributions to the plans to decline without the actual cessation of operations, the construction industry exception shields union contractors from liability if the contractor has not shifted its work from a union company to a nonunion operation.[34] In a dual-shop situation, because the "employer" could be deemed to be both operations if there is common ownership, the exemption could be lost if the union work decreases so that it is insubstantial when compared to total operations. In such a case, a partial withdrawal may be deemed to have occurred and partial withdrawal liability could be assessed.[35]

While the application of MEPPAA to dual-shop operations is still in its infancy, no matter how these rules are construed there is no question that double-breasted operation now face a greater jeopardy of withdrawal liability assessment than other union construction operations. For this reason, careful attention must be paid to determining the amount of withdrawal liability should it be assessed, as well as taking whatever steps are available to minimize it.

A DUAL SHOP CHECKLIST

The following checklist, though not exhaustive, should guide a contractor in his initial decision whether to establish dual shops:

1. **Act on lawful motivation**

This first and most important caveat should be considered carefully. An employer who has signed a collective bargaining agreement with a labor union commits an *unfair labor practice* if it establishes or purchases a new business intending to evade its legal obligation to bargain with that union or to avoid the terms and conditions of its contract with that union.

Any action which could be viewed by the NLRB as an effort to establish a nonunion company to siphon business away from a unionized construction company may lead to a Board order requiring not only that the new firm recognize the union as the collective bargaining agent of its employees, but that the new firm also adhere to the terms and conditions of the collective bargaining agreement

[34] MEPPAA Sec. 4208(d)(1).

[35] It should be noted, however, that even if there is a substantial shift in work, there will not be a partial withdrawal unless the ordinary tests for partial withdrawal are met (e.g., a 70 percent contribution decline).

with the union company, including retroactive payment of union scale wages and contributions to fringe benefit funds (pension, health, and welfare, etc.). If factually true, the corporate minutes of organizational meetings relating to creation of the new firm should express a "neutral" policy and a lack of anti-union "animus."

Where a second company is established, there is nothing to prevent a union from seeking to negotiate with the contractor concerning the wages and conditions of employment of workers of that second firm. While the contractor is under no legal obligation to bargain with the union until such time as the union can demonstrate that a majority of the second company's employees desire it to represent them, the contractor may nevertheless have an open mind in working out a realistic and practical agreement with the union. The mere fact that the contractor was willing to negotiate with the union could be significant evidence of a lack of anti-union animus on the part of the contractor.

If such negotiations are undertaken, there is nothing that requires the contractor to capitulate to the union's excessive demands. In fact, the contractor may legally insist on signing a contract only if the union is willing to agree to the more moderate and less restrictive terms and conditions which the contractor proposes.

Because unions desire that all members receive uniform wages and benefits, it is unlikely that a union would agree to any contract which deviated substantially from the "standard agreement" which it had executed with other area contractors, including its "favored nations" cause. However, if the union has had no previous dealings with a contractor, the union may agree to a contract providing lower wages and benefits than the standard agreement. By sacrificing wages and benefits in the short term, a union may try to seduce a presently unorganized employer into a union contract. Once it has its foot in the door, the union can begin to pressure the employer to accept the standard agreement.

2. **Make an audit of local construction conditions and labor climate**

Before going double-breasted, one should survey all business and economic conditions in the locality where the new firm will do business. Obviously, the new firm's competitive prospects and profit-making potential must be carefully considered. Some of the other questions that need to be answered include:

- Are the local unions in the area powerful and aggressive?
- Will the public attitude be friendly or hostile?
- Are the construction users prepared to award jobs to open-shop contractors?

- Is there adequate law enforcement against labor union coercion and violence?
- Is there an ample supply of skilled nonunion craftsmen and apprenticeship training programs?

3. **Review the geographical jurisdiction of your current collective bargaining agreements**

In most cases, collective bargaining agreements in the construction industry cover a limited geographical area equal to the local union's territorial jurisdiction. As such, the terms and conditions of a specific agreement apply only to employees working within that area. It may well be possible for a contractor to commence operations outside of the geographical area covered by its current collective bargaining agreements. In such a case, the new operation would not be bound by those agreements unless the contractor specifically agreed to extend the current union contract to cover another geographical area.

4. **Purchase an active, ongoing business**

As noted, financial ownership or control of a second business, standing alone, will not lead the Board to find that separate business entities are actually a single employer. A great degree of separation in management, operations, and labor policies may be obtained by purchasing an existing, nonunion construction company in the desired type of construction and allowing its executives to retain their positions and to make all day-to-day decisions on the newly acquired firm's operations. In such ideal circumstances, the only major interrelationship between the companies would be financial.

5. **Determine whether your union contracts contain restrictive "anti-dual shop" clauses**

Some unions include a provision in their standard labor agreements designed to prevent unionized employers from going dual shop. One version of the clause reads:

> If any employer covered by this agreement controls, operates, or has any financial interest in any other business within the trade and territorial jurisdiction of the union, such other business entity shall either have a signed agreement with the union or this agreement shall be interpreted as including such business entity under the term "employer" in this agreement.

A more sophisticated sample is the following language, extracted from a mechanical trades union contract with a mechanical contractors association:

> (1) In the event that the employer, a stockholder or stockholders of the employer whose stock constitutes a controlling interest of the stock of said employer, or an owner or owners of the employer who

> own a controlling interest in said employer in any form other than stock, own, on the effective date of this agreement, or acquire, at any time thereafter during the term of this agreement, a controlling interest in any other company engaged in the performance of all or any of the types of construction work covered by this agreement, whether said controlling interest is owned directly or indirectly, then said other company shall be bound and covered by and subject to each and every provision of this agreement as if already a signatory hereto, and said other company shall sign and adopt this agreement as its own collective bargaining agreement with the union immediately upon being requested to do so by the union.
>
> (2) The purpose of this article is to terminate, once and for all, the growing practice of employers subject to this agreement attempting to operate two companies, one subject to this agreement and one not subject to this agreement, for the obvious purpose of attempting to circumvent and escape from the obligations of this agreement, thereby jeopardizing the wages, hours, and other terms and conditions of employment of the employees covered by this agreement and making a mockery of collective bargaining relations between the union and the employers in the construction industry.
>
> (3) Under no circumstances shall this article be applied or enforced in a manner which, or be applied to circumstances which, would constitute an unfair labor practice by the employer or union under any provisions of the National Labor Relations Act, as amended.

Obviously an employer should carefully consider the full potential of such a clause when negotiating a labor contract. Any contractor contemplating dual shop operation should attempt to avoid entering into or remaining under a labor contract containing such a restrictive clause.

Collective bargaining agreements which contain clauses prohibiting dual-shop operation may violate federal or state antitrust laws. The contractor considering dual shop should discuss this possibility with legal counsel. In recent years, courts have more frequently reviewed provisions of collective bargaining agreements which restrain trade and reduce competition. Although organized labor is often exempt from the reach of the antitrust laws, the courts have not clearly defined the scope of this exemption.

A dispute over interpretation of these restrictive anti-dual shop clauses may be initially resolved through binding arbitration, provided the collective bargaining agreement at issue contains a broad arbitration provision. An arbitrator is not bound by the tests developed in the context of Board jurisdictional and unfair labor practice questions or by antitrust law, and his decision will almost surely be immune from any collateral attack in a federal court. Thus, arbitration may be an avenue for the contractor to escape from or modify a clause prohibiting dual-shop operation.

6. **Ascertain whether at the time the union was originally recognized or at any time thereafter it ever represented a majority of the contractor's employees in the bargaining unit**

Where a union has never represented a majority of the contractor's employees in an appropriate bargaining unit, it may not be an unfair labor practice for a contractor to fail to abide by any agreement it may have signed with the union. Consideration of this option should be approached with extreme caution and only after thorough review with legal counsel.

7. **Vest overall policy as well as day-to-day control of labor relations in the second firm's management**

No recommendation can be more strongly emphasized. The Board considers the existence of a common labor relations policy to be a critical factor in finding a single employer. If there is to be an ownership connection between two firms, the owners should select separate managers with authority to fix and administer separate labor relations policies and practices on a day-to-day basis. The more an owner involves himself in the daily labor relations of a second firm, the greater the risk that the Board will view the two firms as a single employer.

8. **Delegate absolute control over day-to-day operations to a general manager or superintendent who has no connection with the original business**

It is vital that the day-to-day operation of any construction company under common ownership with a unionized firm be as separate and distinct as possible. If one person or family owns majority holdings in the two companies, there may be strong practical considerations favoring installing a member of the family as manager of the daily affairs of the second business. Nevertheless, the better-advised owner will delegate complete authority over all normal business decisions to an independent outsider.

Extensive involvement of family members in a group of family-owned businesses has been viewed as evidence upon which to base a finding of a single employer. If such an intrafamily arrangement is chosen for a second company, notwithstanding the risks, every effort should be made to assure that any family member not responsible for the second business displays no more than an expected financial interest in how the business is doing in the long run.

9. **Avoid interchange of rank-and-file employees**

The easiest method of staffing a second, commonly owned construction company may appear to be the hiring or transferring of workers from the first company. But such a practice should be avoided: employee interchange is one of the critical factors considered by the NLRB in determining whether several nominally separate enterprises constitute a single employer.

Since the unionized contractor will likely obtain his workers through a hiring hall, while the nonunion company may use newspaper advertisements or word-of-mouth, there should be no need to interchange construction workers from one company to another. Since even a one-time, one-person interchange has been used as evidence that two companies should legally be viewed as one, if possible no employee on a unionized company's payroll should be employed by the nonunion firm.

10. **Engage in a different type of construction**

The establishment of a second nonunion company to engage in identical types of construction and in the same locality as a commonly owned unionized company runs a higher risk that the Board will find the second company's creation to be a subterfuge to draw business away from bargaining unit union members employed by the first company.

Moreover, if the two companies are engaged in different types of construction, the Board is less likely to lump both groups of employees into a single bargaining unit. If there exists a particular segment of the construction market which the unionized contractor has never attempted to penetrate, it would be wise for the second company to commence dual-shop operations in this new market.

11. **Avoid interlocking officers and boards; use strict accounting for salaries of officials who serve both companies**

When two or more firms are closely held, the owners may wish to use the same personnel in similar official capacities for each company. For instance, the president of Company 1 might serve as vice president of Company 2, while the same person might be treasurer of both firms, and the vice president of Company 1 might serve as president of Company 2 or as a member of its board of directors. If possible, interlocking officerships and boards should be avoided. Where no alternative is available, care should be taken to ensure that each officer or board member is paid separately by the two firms, in direct proportion to the work performed for each company. For example, if the person who serves as president of one company and vice president of another spends 75 percent of his time working as president and 25 percent of his time working as vice president, his compensation from the two companies should reflect this ratio.

12. **Avoid interchange of supervision**

The use of a unionized company's supervisors or bid estimators to staff any new or newly acquired company has been cited by the

NLRB as a factor supporting a finding of a single employer. If it can be avoided, supervisors should never be transferred directly from a unionized contractor to a prospective nonunion firm. If, after consulting counsel, it is decided that no practical alternative exists, a solution might be to *temporarily* transfer the supervisors to the new firm but to return them to the original firm's employ as soon as possible. Of course, during the time of the transfer, the supervisors should be carried on the new company's payroll and removed from their former company's payroll.

13. **Use separate office facilities**

A contractor with established office space will be tempted to operate a second company out of the same location. If possible, this should be avoided. Geographic separation of offices is more costly, but it will serve to further demonstrate the separateness of the two firms. Where it is not feasible to establish offices in separate spaces or a separate building, the second enterprise should at least pay its proportionate share of the rent (as well as rent on furniture, office machines, and other fixtures) to the original company. Once the second firm establishes itself financially, however, the move to a separate location should be made promptly.

14. **Use separate office clerical personnel**

Sharing of clerical personnel should also be avoided. However, if the two companies share clerical personnel at the same office location, billing and payroll records of all employees, including clerks, should be maintained separately for each firm. Clerical personnel should be proportionately compensated by each company for the time spent on that company's business.

15. **Share tools and equipment, if any, on an arms-length, businesslike basis**

When a contractor starts a second business, he will understandably try to reduce his initial capital investment by allowing the new business to use the expensive tools and equipment of the unionized company without cost. This should be avoided. Each construction company should ideally own all its tools and equipment, or at least lease at fair market value.

16. **Use separate payrolls**

Some businessmen attempt to set up a second, dual-shop enterprise but continue to pay employees of the new firm from the payroll of the old. The payrolls should be separated in as many ways and as much as possible. For example, use different banks and have different personnel sign the checks for the two companies. If computerized accounting is employed, use different companies for the different businesses; pay on different days. If the same bookkeeper or accountant must be used, ensure that this employee is compensated by the two firms in direct proportion to the time spent on the two payrolls.

17. **Do not refer customers from one business to the other**

Some evidence of common management may be found in one firm's consistent referral of customers to a sister company. The unionized company's management ideally should avoid any referral of customers to the nonunion company so that the construction user may obtain a lower price on similar work. Such actions may be held to imply that the two firms are linked more closely than by ordinary business dealings.

18. **Submit separate job bids**

Even if the new business engages in a completely different type of construction or contracting, it is not advisable for two commonly owned companies owned by legally separate enterprises to submit joint bids covering work to be done by the two firms. This could be viewed as evidence of integration of operations and less than an arms-length relationship between the companies. The effect of submitting separate bids on projects will be diminished if the same persons estimate work for both firms. Independent employer status may be more difficult to establish if there is an overlap in use of even these personnel.

19. **Use competitive bidding to award subcontracting work**

The Board will consider the companies less interrelated if work is *not normally subcontracted* to a company sharing some form of common ownership with the subcontracting firm. Use of competitive bids for all work will ensure that regardless of ownership of the companies, the award of subcontracting jobs will not be subject to the charge that the decision was based solely on family ties or common ownership.

20. **Establish an independent line of credit for the new company; avoid interchange of guarantees of performance bonds**

A sometime troublesome issue for firms seeking to go dual shop is the problem of establishing a line of credit for the new business. Any employer who is the sole majority owner of a unionized business should avoid personally guaranteeing the credit of or securing a performance bond for a second company which he also owns or controls. If possible, the new enterprise should obtain any financing (after the initial capitalization) from established financial institutions in its own right. In any event, this separate credit line should be established as soon as possible.

21. **Use separate bank accounts**

Integration of management is implied by the undifferentiated use of one bank account by two companies.

22. **Avoid insuring the two companies under a single policy**
Even the use of the same insurance policy for accident, workers' compensation, or health insurance, or pension benefits could be used as evidence supporting a finding of a single employer. If the two companies are separate and do not share common management, it would be reasonable to expect the new firm to shop around for its insurance rather than to rely on the same policy—or even the same company—which had insured the original business.

23. **Use different time clocks**
Where time clocks are used and two companies share offices or job sites, each company should provide its own clock for its personnel.

24. **Advertise separately**
Each firm should advertise separately to attract its own business.

25. **Differentiate vacation policy**
The two companies' policies on vacations should not be coordinated in any manner. Vacation policies should be delegated to persons as far down the line as possible. In no case should there be a policy of coordinating vacations granted employees of one firm with those for employees of the other firm.

26. **Use different telephone lines**
Where a new company maintains its offices in a different location than the established firm, there normally would not be a problem with telephones. Where two or more companies share office space or use adjoining offices, different telephone numbers and a different switchboard should be maintained. Shared use of telephones has been frequently cited by the Board as a fact suggesting that two firms are a single employer.

CONCLUSION: IS A DUAL SHOP WORTH THE RISK?

While the preceding discussion deals with broad legal guidelines for dual-shop operation, contractors must also look at harsh reality. If a local union hears about a new double-breasted setup, how will it react? Will there be slowdowns, work stoppages, or other forms of intimidation, coercion, and violence? To be sure, different unions will behave differently.

Some owners of dual shops have endured threats, property damage, and other harassment. Other contractors who have gone double-breasted report that the building trades unions have not bothered them at all. The union reaction in any particular situation may well depend upon, among other facts, the character and personalities of the particular union representative involved, the number of union tradesmen out of work, and the strength of the open-shop movement in the area where the companies operate.

Obviously the maintenance of dual shops is not a sure bet for all. Those who have made the switch to operate in both the union and nonunion sectors of the construction business, however, say enthusiastically that the increased business and versatility is well worth the risk, which for the most part has been minimal. These same contractors also warn against making a spur-of-the-moment decision and forming a dual shop too hastily. Without proper planning, effective implementation, and competent legal counsel, a dual shop is doomed to costly failure. Set up properly, however, a double-breasted operation might well be the key to many unionized contractors' survival.

Mr. Cook gratefully acknowledges the assistance of N. Peter Lareau, Esquire, Lawrence J. Merwin, Esquire, and Phyllis Cole Friedman, Esquire in the preparation of this article.

Appendix D

THE IMPACT OF ANTITRUST LAWS ON OPEN SHOP CONTRACTING

by

Edward B. Miller, Esq.
Senior Partner
Pope, Ballard, Shepard & Fowle, Ltd. (Chicago)
Former Chairman, National Labor Relations Board

Prepared September 1983

INTRODUCTION

In a 1925 decision in an antitrust case, the Supreme Court of the United States said:

> With the conflict between the policy of the "closed shop" and that of an "open shop," or with the "American plan," *per se,* we have nothing to do.[1]

But today, that broad declaration to the effect that the antitrust laws do not reach into conflicts between unions and employers on open shop issues can no longer be taken as gospel. Indeed, recent cases suggest that both unions and employers may expect increasing confrontations under the antitrust laws as each seeks to further its own interests with respect to the issue of whether construction work should be all-union, all-nonunion, or of a double-breasted variety.

May Unions Force Contractors to Keep All Jobs All-Union?

In *Connell Construction Co. v. Plumbers' Local 100,*[2] the local had insisted that Connell, a general contractor who had been utilizing both union and nonunion subcontractors, sign an agreement that

[1] Industrial Association v. United States, 268 U.S. 64, 76 (1925).
[2] 421 U.S. 616 (1975).

he would henceforth subcontract work only to plumbing contractors who were parties to a union contract with Local 100. After having one of his jobs shut down by the local (ironically, one on which he was using a union subcontractor), he signed the agreement and then filed his antitrust suit against the local. The Supreme Court held that the agreement Connell was forced to sign, at a time when he was not an employer of plumbing employees, ran afoul of the antitrust laws. The thrust of the Court's reasoning was that while, in general, labor disputes may be beyond the reach of the antitrust laws, employers and unions may not enter into agreements which exclude a class of competitors from a given market. Thus, agreements which exclude an entire class of nonunion subcontractors from the marketplace may be sufficiently anticompetitive to be more than just a simple "labor dispute."

While some thought that the *Connell* decision would be of great assistance to the cause of open shop contracting, it has been very narrowly construed and, as the law has developed, probably has only limited application. The Supreme Court held in *Woelke & Romero Framing, Inc. v. NLRB,*[3] that agreements similar to those in *Connell* are lawful if they are restricted to preventing union and nonunion workers from having to work side by side on a given job site. The Court also specifically found agreements lawful which, as in *Connell,* require the use of union subcontractors on any job site *if* those agreements are made "in the context of a collective bargaining relationship"—that is, if the general contractor employs union craftsmen himself and is a party to a contract covering them.

All of this, strangely enough, grows out of interpretations of Section 8(e) of the National Labor Relations Act (NLRA), even though the central issues are those of antitrust law. In the *Connell* case, the Supreme Court made clear that if the kind of subcontracting agreement there in issue was one permitted by what is called the "construction industry proviso to Section 8(e)" of the NLRA, then the matter would be one subject to the exclusive jurisdiction of the National Labor Relations Board (NLRB) and not subject to the antitrust laws. It found, in *Connell,* that the agreement was not protected by the NLRA's construction industry proviso because the general contractor did not employ any construction workers himself, but instead subcontracted all his work. Thus, the agreement there was not "in the context of a collective bargaining relationship."

The *Woelke & Romero* decision has opened a significant loophole in the ruling of *Connell.* A union which desires to have a general

---

[3] 456 U.S. 645 (1982).

contractor agree to use only union-signatory subcontractors now may find it not difficult to do so and still stay within the law. Most general contractors will, on some occasion, employ members of one or more crafts; generally most, if not all, of the craft unions in an area will be members of a local building trades council. If a union such as the Plumbers' Local involved in *Connell* does not have a collective bargaining relationship with the general contractor, it may nevertheless be able to get the local building trades council to act on its behalf and secure an agreement from the general contractor to use only contractors who have agreements with member unions of the building trades council. Such an agreement is probably lawful under *Woelke & Romero* if it is included in a collective bargaining agreement between the council and the general contractor, even if the contractor employs only a very few employees who are represented by perhaps only one of the unions forming a part of that council.

Another loophole in the *Connell* decision suggests that if the Plumbers' Local had asked for an agreement requiring only that the general contractor use all-union subcontractors on any job site on which any union employees were working, the Supreme Court would apparently sanction such an agreement.

An even larger loophole in *Connell* was found in *Donald Schriver, Inc. v. NLRB*.[4] In that case the circuit court approved a holding by a divided NLRB to the effect that a "Section 8(f) Contract" is "in the context of a collective bargaining relationship" and thus not within the scope of the *Connell* ruling.

A Section 8(f) contract is a prehire agreement pursuant to which the contractor usually agrees that any persons hired at the trade represented by the union will be required to become members of that union within seven days of the hiring. The contract also contains provisions requiring certain rates of pay, fringe benefits, and the like. In the *Schriver* case, the prehire contract was with the general contractor, and contained provisions prohibiting subcontracting except to a contractor who was a signatory to an agreement with the union. The NLRB found such an agreement valid and, unlike the agreement in *Connell,* to qualify for the construction industry exemption from Section 8(e) of the NLRA.

Thus, unions now have another readily available means to evade the holding in *Connell.* In *Connell,* the union went to the general contractor and secured from him an agreement to use only subcon-

[4] 635 F.2d 859 (D.C. Cir. 1980), *cert. denied,* 451 U.S. 976 (1981).

tractors who were signatory to contracts with the local building trades union. That was illegal. But if that same general contractor had agreed to pay workers hired within the jurisdiction of that local the prescribed union wages and required the employees to join the union within seven days, the union could then lawfully secure the very agreement condemned in *Connell*—namely, that the general contractor would utilize only subcontractors who had agreements with this same union. The NLRB and the D.C. Circuit Court agreed that, when included within this so-called Section 8(f) agreement, such a restriction on subcontracting was lawful, could be applied at any job site (not just those at which the local had members working), and could be "union specific"—that is, could require the general contractor not only to utilize union contractors, but also those contractors having an agreement with the specific locals.

The Board and the courts did find one provision of the Section 8(f) agreement in *Schriver* unlawful—one which permitted the union to resort to strikes and pickets if the general contractor violated his agreement to utilize only signatory subcontractors. The union, however, would be free to enforce the agreement by other means, such as court action.

A similar Section 8(f) agreement was found lawful in *A.L. Adams Const. Co. v. Georgia Power Co.*,[5] and the *Connell*-type antitrust allegations of the complaint were dismissed, even though the "general contractor" was in fact a power company. The court held that the power company was entitled to the construction industry exemption when it was serving as its own general contractor with respect to the construction of a power plant.

A long saga in Montgomery Country, Pennsylvania, appears to have ended with another loophole being found by the United States District Court in the *Connell* case, again relying heavily on the Section 8(f) exemption. *Altemose Construction Company v. Building and Construction Trades Council of Philadelphia and Vicinity*[6] was a case in which Altemose Construction Company; the Associated Builders and Contractors, Inc. (ABC); and the Chamber of Commerce of the United States of America all joined to try to stop building trades unions (and thereby, unionized building trades contractors) in the Philadelphia area from attempting to monopolize construction work by the use of violence. When Altemose obtained a general contract to build Valley Forge Plaza and hired the lowest-bidding subcontractors, whether or not they employed union labor, picketing

[5] 537 F.Supp. 168 (S.D. Ga. 1983).

[6] Civil Action No. 73-773 (E.D. Pa. 1983).

and extraordinary violence—referred to by a Pennsylvania trial judge as a "virtual military assault"—resulted in the "systematic decimation of the project. Damage was estimated at $300,000." (The violence is described in greater detail in Chapter VIII.)

When Altemose refused to employ only unionized labor, this and other Altemose job sites were picketed, at various times from 1973 through 1975. There was violence, and threats were made against nonunion subcontractors. Investors in the project were urged by the unions to withdraw their financial support; all union members in the area and the general public were urged to withdraw their funds from First Pennsylvania Bank, a major financial backer of the Valley Forge project. The list of secondary activity goes on and on.

Underlying this activity were the agreements between the Building and Construction Trades Council and the General Building Contractors Association (GBCA) of Philadelphia giving union members the right not to work with nonunion workers, and requiring GBCA-member contractors (a) to hire only union employees, (b) to employ them under the terms of the union agreements, and (c) never to contract with subcontractors not on the council's "fair contractor" list. General contractors regularly agreed to do business only with subcontractors who had entered into collective bargaining agreements with the appropriate craft union members of the council. The result of these well-enforced agreements has been that since 1966, 80 to 90 percent of the nonresidential construction work in Philadelphia and 40 to 75 percent in the suburbs has been performed by union contractors, union subcontractors, and all-union employees.

Altemose Construction Company, the ABC, and the U.S. Chamber of Commerce sued the Building and Construction Trades Council under the antitrust laws, alleging a conspiracy to restrain trade and to monopolize the construction market in a five-county geographic area. Count I of the complaint alleged an illegal combination of the labor defendants with union contractors and suppliers and lending institutions, whereas Count II alleged a conspiracy solely among the labor unions. The court wrote a lengthy opinion, which referred to almost every significant case ever decided under the antitrust laws and described much of the history of the development and application of those laws. The nub of the decision, so far as Count II was concerned, was that labor unions, when acting alone, enjoy an exemption from the antitrust laws, at least to the extent that their activities are confined to seeking traditional labor objectives. Hence, Count II of the complaint was dismissed. With respect to Count I, the court found that the agreements exacted by the Building and Construction Trades Council were lawful Section 8(f) (prehire) agreements as construed under *Connell.* Hence the court dismissed this count as well. The opinion also seems to have been considerably influenced by the availability of NLRB and Section 303 remedies, even though the Supreme Court in *Connell* had made reasonably clear that those remedies are *not* exclusive.

Although some of the opinion is less than crystal clear, the court also seems to regard the unions' "organizational interest" as justifying the agreements, thus once again resorting to judicial legislation on policy ground. Interestingly, too, this policy ground seems directly contrary to the language of the Supreme Court in *Connell* which indicates that an organizational objective, if achieved through "top-down" agreements forcing employees into a union mold, is *not* consistent with federal labor policy.

The net effects of the decision are to make even larger the many loopholes in *Connell* and to widen into an easily travelled road the little path first opened by *Woelke & Romero,* discussed earlier in this section. Unions have also adopted other techniques which are easily adaptable for the purpose of avoiding the antitrust vulnerability of a direct *Connell*-type, signatories-only approach.

In *Genser v. International Brotherhood of Electrical Workers Local 134,*[7] the owners of an electrical contracting business located outside Chicago sued the International Brotherhood of Electrical Workers (IBEW) and the Electrical Contractors Association of the City of Chicago, complaining that they were being kept out of the Chicago market by monopolistic arrangements which violated the antitrust laws. The plaintiffs complained, among other allegations, that out-of-the-city contractors could not get an "A" letter, which was required in order to hire apprentices and to be included on the union contractors' "fair list." It was also alleged that the union president was the chief electrical inspector for Chicago, and that he licensed "supervising electricians, thereby controlling the right to do business in Chicago."

The union's defense was that out-of-the-city firms which only temporarily operate in Cook County could not effectively participate in an apprentice training program, which lasts for several years. The union also said that including temporary, out-of-the-city contractors on a "fair list" would require too-frequent revisions to be practicable. Judge Getzendanner dismissed the complaint, saying essentially that the union, acting alone, could not be found guilty

---

[7] 522 F.Supp. 1153 (N.D. Ill. 981).

of a conspiracy, and that there were insufficient allegations to support a claim that the union had monopoly power over Chicago's electrical contracting.

Since the plaintiffs in *Genser* were, in fact, union contractors (although not "local" union contractors), the case does not directly involve an effort to keep open shop contractors at bay. Yet in an area like Chicago, where construction continues to be heavily dominated by union contractors, the devices described in this case would obviously be equally effective at keeping out any out-of-the-area open shop contractors who may seek to try to break into the Chicago market.

The potential impact of *Connell* has been further weakened by another case: *Larry V. Muko, Inc. v. Southwestern Pennsylvania Building and Construction Trades Council.*[8] There a nonunion general contractor, Muko, built a restaurant building for Long John Silver's Inc., a corporation operating fast-food seafood restaurants. After the restaurant was built, the local building and construction trades councils distributed handbills to customers, and this apparently had some adverse effect on patronage. Silver's then met with the building trades councils, as a result of which it agreed to use in the future only union contractors certified by the councils.

The Third Circuit had held, earlier in the proceedings, that an antitrust complaint filed by the nonunion contractor (Muko) against Silver's and the union was sufficient to set forth a cause of action. When the case was tried, however, the plaintiff won a battle or two, but lost the war. The jury found that there was, indeed, an agreement or combination to refuse to grant construction contracts to nonunion builders which had the effect of imposing a restraint on free competition beyond that which would flow from the elimination of competition based solely on wage rates and working conditions. The jury also found, however, in response to a specific question put to it by the judge, that the agreement was not "an unreasonable restraint of trade."

The plaintiff appealed from the judgment of dismissal which resulted from this verdict, but the Third Circuit turned down the appeal. The court found that Silver's goal was not an anticompetitive one, but was only to retain the good will of its customers. It also found that the union's goal was not anticompetitive, but instead was only the laudable goal of ensuring protection for the prevailing union wage. Neither of these purposes, said the court, suggested

[8] 109 L.R.R.M. 2469, 670 F.2d 421 (3rd Cir. 1982), *cert. denied,* October 1982, No. 81-1798, 111 L.R.R.M. 2608.

that either of the defendants was attempting to gain an advantage over Muko in an economic or competitive sense. The court also found that the agreement did not have the necessary effect of destroying Muko's business, since Muko could engage in construction other than building restaurants, and noted that the evidence showed that Muko's profits actually increased during the time period in question.

A very strong dissent espoused the view that the lower court should not even have permitted an inquiry into whether the restraint of trade was "reasonable." The dissent would have found that the facts evidenced a boycott of nonunion contractors, which can never be "reasonable" because it is a per se violation of the antitrust laws.

There has traditionally been a difference of opinion among the circuit courts as to whether, in labor cases, boycotts of the general nature involved here are subject to analysis on a "rule of reason" basis, or whether they constitute per se violations (as a dissent would have found in the *Muko* case). If the Supreme Court's denial of *certiorari* in *Muko* means that the "rule of reason" has become the law of the land in cases like *Connell* and *Muko,* it will be almost impossible to predict the outcome of this kind of antitrust suit. Who can predict whether any given jury will find an agreement to use only union contractors "reasonable" or "unreasonable"? Antitrust suits are costly and time-consuming for all involved. It seems doubtful that very many contractors, particularly small ones, would undertake the costly gamble of an antitrust suit if juries are to have as wide discretion as that permitted in *Muko.*

Yet in May 1982, a jury in California awarded over $3 million, trebled by the court to over $9 million, to a San Diego producer of television and radio commercials in an antitrust suit of the *Connell* variety. In *Tuesday Productions, Inc. v. American Federation of Television and Radio Actors,*[9] the antitrust evil was found in contracts between the AFTRA and a group of advertisers and advertising agencies. The offending contract, like that in *Connell,* required the use of only union-signatory production houses. The jury in this case apparently did not find the resulting restraint of nonunion competition "reasonable."

That *Connell* may still have some impact is also suggested by the case of *James Julian, Inc. v. Raytheon Co.*[10] There a plaintiff contractor, whose employees were not represented by building trades

[9] U.S.D.C. S. Calif. No. 78-0644, June 8, 1982, DAILY LAB. REP., No. 122, p. A-3.
[10] 499 F.Supp. 949 (D. Del. 1980); and 557 F.Supp. 1058 (D. Del. 1983).

unions, alleged that he was forced off a construction job at a solid-waste disposal plant being built for the Delaware Solid-Waste Authority. A group of building trades unions and contractors had met with representatives of the authority and objected to the participation in the project by the plaintiff or any other contractor whose employees were not represented by the building trades council or its member unions. When that meeting failed to result in the removal of the plaintiff contractor from the site, violent picketing occurred. This did force the plaintiff contractor off the job, because the general contractor canceled the plaintiff's subcontract. The district court held this complaint sufficient to allege violation of the antitrust laws. Thus, although there are major loopholes in *Connell,* there are nevertheless some circumstances in which it retains vitality.

Is *Connell,* then, a decision of great importance or a mere flash in the pan? It is too early, at this time, to offer a final answer as to how much practical application *Connell* will have, in the long run, to the open shop scene.

May the Antitrust Laws Be Invoked by Unions Against Employers to Stop the March to Doublebreasted Operations?

The Supreme Court, on narrow grounds, has held in *California State Council of Carpenters v. Associated General Contractors of California, Inc.,*[11] that a union cannot successfully turn antitrust guns on employers in an effort to stop the spread of doublebreasted operations. In this case, the Carpenters' Union had claimed that the Associated General Contractors (AGC) was attempting to undermine and weaken the collective bargaining relationship between the union and various contractors by encouraging and allegedly coercing contractors to maintain doublebreasted operations and to refuse to enter into collective bargaining relationships with the union. The complaint also alleged that the AGC was encouraging and coercing land owners to hire contractors and subcontractors who were not signatories to collective bargaining agreements.

The Court of Appeals for the Ninth Circuit refused to dismiss the complaint, holding it to have properly alleged an illegal conspiracy to boycott union contractors in violation of the antitrust laws. The court held that the pleadings appeared to present the flip side of *Connell*—i.e., an attempt to create a monopoly or a sheltered market for nonunion construction firms. The Supreme Court granted

---

[11] 698 F.2d 527 (9th Cir. 1981), revised, — U.S. —, 112 L.R.R.M. 2753 (1983).

certiorari and reversed the Ninth Circuit. Like *Connell,* the case attracted a good deal of attention; *amicus* briefs were filed with the United States Supreme Court by a number of employer organizations, including the United States Chamber of Commerce, and by union groups, including the AFL-CIO.

There were a number of potentially highly significant antitrust issues before the Court, including whether the open shop, union shop, and doublebreasted shop issues presented the same kind of "labor dispute" which is normally exempted from antitrust law application by Section 6 of the Clayton Act—at least when unions are the defendants. Another issue was whether the encouragement of doublebreasted contracting on the West Coast was actually procompetitive rather than anticompetitive, since it was breaking the monopoly previously held by the union shop over the West Coast construction business. The Supreme Court found it unnecessary to decide any of these issues and instead reversed, solely on the ground that the union did not have standing to sue in this kind of antitrust action.

Although the question of whether the union had standing to sue may be an important legal issue, it is on too narrow ground for the decision to offer much comfort to associations or groups of contactors seeking to advance the cause of doublebreasting. For although a *union* may not be able to bring action, what about the dyed-in-the-wool union shop *contractor* whose business is hurt by the advent of nonunion competition? Borrowing the gun the Carpenters' Union attempted to use in *California State Council of Carpenters,* can such a union contractor have found a valuable weapon? Because the Supreme Court failed to decide the key underlying issues, this question, so far as the Court is concerned, remains unanswered.

The Court of Appeals for the Fifth Circuit, however, has broadly hinted as to what its answer would be. In *Carpenters Local Union No. 1846 v. Pratt-Farnsworth,*[12] the court remanded a case very similar to the *California State Council of Carpenters* case, for further consideration of the issue which the Supreme Court found controlling in *California State Council of Carpenters*—i.e., whether the union had proper standing to sue. But it did so only after analyzing the other, perhaps more significant long-run, legal issues. And while the court had doubts as to the union's standing to sue, it commented that "the most obvious candidates for standing are those most directly injured at the contractors' services level: the signatory contractors themselves." A broader hint that a union contractor might successfully be able to sue would be hard to find.

---

[12] 690 F.2d 489 (5th Cir. 1982).

As to the other issues not reached by the Supreme Court in the *California State Council of Carpenters* case, the court rejected, out of hand, the contention that this was a "labor dispute" and that the employers were entitled to a statutory labor exemption under the Clayton and Norris-LaGuardia Acts. Relying on language from the Supreme Court's decision in *Allen Bradley Co. v. International Brotherhood of Electrical Workers,*[13] to the effect that "[a] business monopoly is no less such because a union participates," the court in *Pratt-Farnsworth* said: "To this we may add that it is no less such because a union does not participate."[14]

It also found that the nonstatutory labor exemption, which sometimes applies if the conduct complained of seems consistent with the congressional policy favoring collective bargaining under the NLRA, did not apply, because in the court's view this exemption is applicable only where the conduct grows out of collective bargaining agreements entered into with unions. Although the court's treatment of the availability of these exemptions to employers counters the decisions of some other courts and is open to considerable criticism, it is nevertheless a decision by a United States Court of Appeals, and cannot be ignored.

The court then went on to hold that although a concerted refusal by employers to deal with a building trades union did not give rise to a cause of action under the antitrust laws, a violation could result from a concerted refusal to deal with union subcontractors or with contractors who did not agree to create doublebreasted arrangements. The court concluded:

> Although a restraint at the level of the labor market itself does not state a claim of violation of the antitrust laws, a restraint at the level of subcontractor or contractor services does state such a claim.

Thus the door appears to have been opened, at least in the Fifth Circuit, perhaps not to the "flip side of the *Connell*" (which is what the Ninth Circuit called the *California State Council of Carpenters* fact situation), but at least to the flip side of *Muko*. If local building trades councils may agree with a restaurateur to use only *union* contractors if its goals seem "reasonable," what if a group of nonunion contractors persuades such a restaurateur to use only *open shop* contractors? If the Fifth Circuit decision becomes established law, such an agreement would be valid only if a jury would be willing to find the objectives of the open shop contractors and the restaurateur to be "reasonable." A jury in a strong union community may be very unwilling to do so.

---

[13] 325 U.S. 797 (1945).

[14] 690 F.2d at 531.

One must conclude, then, that there are real dangers inherent in cooperative action among open shop contractors, at least when such action is used to secure agreements by landowners or general contractors to make greater or exclusive use of open shop contractors. Unions may not be able to sue under these circumstances, but the possibility of suits by union contractors can no longer be lightly dismissed.

The Perils of Multiemployer Bargaining in the Organized Segment of the Industry

Open shop contractors may at least take some comfort in the fact that they are spared the perils which appear to lurk in some facets of the kind of multiemployer bargaining which commonly takes place in the organized sector of the construction industry. Association bargaining itself is not a combination vulnerable to antitrust attack, as was found in *California State Council of Carpenters v. Associated General Contractors of Calif., Inc.* opinion denying rehearing,[15] and *United States v. San Francisco Electrical Contractors Association.*[16] Lockouts and strike insurance plans have also been cleared of antitrust accusations as in *Clune v. Publishers Association of New York City,*[17] and *Kennedy v. Long Island Railroad Company.*[18] Yet some problems remain.

If multiemployer bargaining groups obtain commitments from the unions with which they deal to force high rate and fringe benefit packages on employers who may thereby be forced out of business, their intent may be found to be "predatory" and thus subject to antitrust attack.[19] Groups of employers also may not agree with a union or unions to fix prices or to monopolize a market, even if the union's motivation is a desire to secure a lawful union shop agreement.[20]

In *National Electrical Contractors Association, Inc. v. National Constructors Association,*[21] an employer association together with a union was found to be in violation of the antitrust laws by agreeing that all employers obtaining workmen through the union's hiring hall (with certain exceptions not here relevant) would be required

---

[15] — F.2d —, 107 L.R.R.M. 2724 (9th Cir. 1981).

[16] 57 F.Supp. 57 (N.D. Cal. 1944).

[17] 214 F.Supp. 520 (S.D. N.Y. 1963), *aff'd* on the opinion of the district judge, 314 F.2d 343 (2nd Cir. 1963).

[18] 319 F.2d 366 (2nd Cir. 1963), *cert. denied,* 375 U.S. 830 (1963).

[19] United Mine Workers v. Pennington, 381 U.S. 657 (1965); Smitty Baker Coal Co., Inc. v. United Mine Workers of America, 620 F.2d 416 (4th Cir. 1980), *cert. denied,* 449 U.S. 870 (1980).

[20] Allen Bradley Co. v. Local Union No. 3, IBEW, 325 U.S. 797 (1945).

[21] 678 F.2d 492 (4th Cir. 1982).

to contribute a specified percentage of payroll to an industry fund, regardless of whether or not they belonged to the employer association. The admitted purpose of this requirement was to eliminate the competitive advantage of nonmembers who paid no dues or service charges to the association. The agreement was found to be unlawful price fixing within the meaning of the *Allen Bradley* case.

Yet another hint of potential antitrust liability which could be of serious concern to unionized construction employers is found in *Frito-Lay, Inc. v. Retail Clerks' Union Local No. 7.*[22] In that case, the court remanded for trial a case which had been dismissed by the court below. The complaint dealt with a provision in a contract between the Retail Clerks' Union and a multiemployer bargaining group pursuant to which employees of all suppliers except bakeries and dairies were prohibited from entering retail grocery stores to stock merchandise. The exemption of bakery and dairy companies was inserted apparently to protect certain Teamsters' Union employees. The court's opinion, in large part, dealt with whether such an agreement was legal under the NLRA because it involved work preservation, or whether it might be illegal as attempted work acquisition.

Significantly, however, the court held that if on remand it should be found that the work was not "traditional Clerks' work" and the employer was limited too severely in his freedom to assign the work, the "potential liability under the antitrust laws must be considered."

This case has received surprisingly little attention in antitrust commentary. Its potential, however, cannot be ignored. It may mean that an all-embracing work jurisdiction clause (almost universal in construction agreements) can run afoul of antitrust laws if the employer does not have the right of control over the work. Since in construction an architect may prescribe prefabricated parts and a given construction union cannot show that the work has been traditionally performed by its constituency, then will unions and unionized contractor groups both run into antitrust problems if they attempt to use a broad work jurisdiction clause to exclude prefabricated parts?

Antitrust Implications of New Collective Bargaining Proscriptions on Doublebreasted Operations

Another development, not yet tested in the courts (so far as this author is aware), is the attempt by the building trades unions to

---

[22] 629 F.2d 653 (10th Cir. 1980).

outlaw doublebreasting via the collective bargaining route. Building trades unions are now seeking, and obtaining in a number of currently negotiated collective agreements, clauses requiring that if the employer directly or indirectly performs any work of the type covered by the agreement, either under its own name or under the name of another entity controlled by the employer, all such work is to be covered by the agreement. The clauses' definition of "direct or indirect" control would mean that if family members have a majority stock ownership in the nonunion company, it is indirectly controlled and thus would come under the union contract. Therefore, if a separate corporation is formed by a contractor's son or brother, which may engage in a separate kind of work—either by geographic area (e.g., suburban v. urban) or by function (e.g., residential v. commercial)—and even be separately managed by a different family member or an outsider, it would nevertheless be required to be covered by the contract. This is obviously a direct attack, through collective bargaining, on the practice of doublebreasting.

If the *Pennington* doctrine is construed as outlawing agreements to force collectively bargained terms on bargaining units not represented at the bargaining table, such agreements may violate the antitrust laws. There is a considerable difference of opinion among both courts and legal scholars as to whether "predatory intent" must be proved in order for the *Pennington* doctrine to apply. If predatory intent is required, this new type of agreement may escape the *Pennington* rule, because the unions will undoubtedly claim that it is not designed to force anyone out of business, but only to ensure the preservation of wage and benefit standards. There may also be some question as to whether *Pennington* applies if there is some nexus, such as familial ties, between the employer represented at the bargaining table and the other bargaining unit or units to be covered. On the other hand, employers will surely argue that the intent of these agreements is very similar to the intent found improper in both *Pennington* and *Connell*—i.e., to create a sheltered market and a business monopoly for such union contractors. It is too early to predict how vulnerable this new type of agreement may be to antitrust attack. Early rumblings already suggest, however, that they will soon be tested by litigation.

It also seems reasonably likely that insistence to impasse upon this new form of agreement may run afoul of the NLRA. A proposal to apply agreement terms to employees not in the certified or recognized bargaining unit is, under NLRB law, not a mandatory subject of bargaining. Insistence to impasse on nonmandatory demands is an unfair labor practice. The unions may argue that a double-breasted operation is not really two employers but is, in reality, only one. Whether common ownership and control results in two employers being treated as a single employer is, however, a legal issue and not, one would think, a mandatory subject of bargaining.

Because doublebreasting has become so common, there seems little doubt that this new clause will be tested at an early point, probably under both the federal labor and antitrust laws. Employers wishing to avoid being caught up in litigation under either or both sets of laws would do well to consult carefully with experienced counsel familiar with both labor and antitrust laws before agreeing to this most recent type of anti-doublebreasting union demand.

Thus, the organized segment of the industry must be concerned, in several important areas, about the impact of antitrust laws on its activity.

CONCLUSION

In summary, although both employers and unions paid relatively little attention to the antitrust laws for several decades while developing and implementing labor policy, recent decisions suggest that neither open shop contractors, unionized contractors, nor labor organizations can any longer afford blissful ignorance of this body of law. An ounce of preventive counsel may ward off the troubled ravages of serious and costly antitrust litigation.

Appendix E

SELECTED STUDIES OF UNION-NONUNION CONSTRUCTION CRAFTSMEN WAGE DIFFERENTIALS

Allen, Steven G. "Unionized Construction Workers Are More Productive." Unpublished manuscript, December 1980. (For a critique of Allen's work, see Chapter II.)

Ashenfelter, Orley. "Union Relations Wage Effects: New Evidence and a Survey of Their Implications for Wage Inflation." Mimeographed. Princeton, N.J.: Princeton University, 1976.

Bourden, Clinton C. and Levitt, Raymond E. *Union and Open-Shop Construction.* Lexington, Mass: D.C. Heath and Company, Lexington Books, 1980.

Freeman, Richard B. and Medoff, James L. "The Impact of the Percentage Organized on Union and Nonunion Differentials." *The Review of Economics and Statistics* 63 (1981): 561–72.

Lee, Lung-Fei. "Unionism and Wage Rates: A Simultaneous Equations Model with Quantitative and Limited Dependent Variables." *International Economic Review* 19 (1978): 415–34.

Lewis, H. Gregg. *Unionism and Relative Wages in the United States: An Empirical Inquiry.* Chicago: University of Chicago Press, 1963.

———. "Union Relative Wage Effects: A Survey of Macro Estimates." *Journal of Labor Economics* I (1983): 1–27.

Moore, W.J. and Raisen, J. "A Time Series Analysis of the Growth and Determinants of Union/Nonunion Relative Wage Effects, 1967–1977." U.S. Bureau of Labor Statistics. Working Paper No. 115, 1981.

Parnes, Herbert S.; Miljus, R.C.; Spitz, R.S.; and Associates. *Career Thresholds.* Manpower Administration, U.S. Department of Labor. Mimeograph No. 16. Washington, D.C.: Government Printing Office, 1975.

Sabotka, Stephen P. "The Influence of Unions on Wages and Earnings of Labor in the Construction Industry." Unpublished Ph.D. dissertation. Chicago: University of Chicago, 1952.

———. "Union Influence on Wages: The Construction Industry." *Journal of Political Economy* 61 (1953): 127–43.

Welch, Stephen W. "Union-Nonunion Construction Wage Differential." *Industrial Relations* 19 (1980): 152–62.

Index

Racial Policies of American Industry Series

1. *The Negro in the Automobile Industry,* by Herbert R. Northrup. 1968
2. *The Negro in the Aerospace Industry,* by Herbert R. Northrup. 1968
3. *The Negro in the Steel Industry,* by Richard L. Rowan. 1968
4. *The Negro in the Hotel Industry,* by Edward C. Koziara and Karen S. Koziara. 1968
5. *The Negro in the Petroleum Industry,* by Carl B. King and Howard W. Risher, Jr. 1969
6. *The Negro in the Rubber Tire Industry,* by Herbert R. Northrup and Alan B. Batchelder. 1969
7. *The Negro in the Chemical Industry,* by William Howard Quay, Jr. 1969
8. *The Negro in the Paper Industry,* by Herbert R. Northrup. 1969
9. *The Negro in the Banking Industry,* by Armand J. Thieblot, Jr. 1970
10. *The Negro in the Public Utility Industries,* by Bernard E. Anderson. 1970
11. *The Negro in the Insurance Industry,* by Linda P. Fletcher. 1970
12. *The Negro in the Meat Industry,* by Walter A. Fogel. 1970
13. *The Negro in the Tobacco Industry,* by Herbert R. Northrup. 1970
14. *The Negro in the Bituminous Coal Mining Industry,* by Darold T. Barnum. 1970
15. *The Negro in the Trucking Industry,* by Richard D. Leone. 1970
16. *The Negro in the Railroad Industry,* by Howard W. Risher, Jr. 1971
17. *The Negro in the Shipbuilding Industry,* by Lester Rubin. 1970
18. *The Negro in the Urban Transit Industry,* by Philip W. Jeffress. 1970
19. *The Negro in the Lumber Industry,* by John C. Howard. 1970
20. *The Negro in the Textile Industry,* by Richard L. Rowan. 1970
21. *The Negro in the Drug Manufacturing Industry,* by F. Marion Fletcher. 1970
22. *The Negro in the Department Store Industry,* by Charles R. Perry. 1971
23. *The Negro in the Air Transport Industry,* by Herbert R. Northrup et al. 1971
24. *The Negro in the Drugstore Industry,* by F. Marion Fletcher. 1971
25. *The Negro in the Supermarket Industry,* by Gordon F. Bloom and F. Marion Fletcher. 1972
26. *The Negro in the Farm Equipment and Construction Machinery Industry,* by Robert Ozanne. 1972
27. *The Negro in the Electrical Manufacturing Industry,* by Theodore V. Purcell and Daniel P. Mulvey. 1971
28. *The Negro in the Furniture Industry,* by William E. Fulmer. 1973
29. *The Negro in the Longshore Industry,* by Lester Rubin and William S. Swift. 1974
30. *The Negro in the Offshore Maritime Industry,* by William S. Swift. 1974
31. *The Negro in the Apparel Industry,* by Elaine Gale Wrong. 1974

Order from: Kraus Reprint Co., Route 100, Millwood, New York 10546

STUDIES OF NEGRO EMPLOYMENT

Vol. I. *Negro Employment in Basic Industry: A Study of Racial Policies in Six Industries (Automobile, Aerospace, Steel, Rubber Tire, Petroleum, and Chemicals),* by Herbert R. Northrup, Richard L. Rowan, et al. 1970.

Vol. II. *Negro Employment in Finance: A Study of Racial Policies in Banking and Insurance,* by Armand J. Thieblot, Jr., and Linda Pickthorne Fletcher. 1970.

Vol. III. *Negro Employment in Public Utilities: A Study of Racial Policies in the Electric Power, Gas, and Telephone Industries,* by Bernard E. Anderson. 1970.

Vol. IV. *Negro Employment in Southern Industry: A Study of Racial Policies in the Paper, Lumber, Tobacco, Coal Mining, and Textile Industries,* by Herbert R. Northrup, Richard L. Rowan, et al. 1971.

Vol. V. *Negro Employment in Land and Air Transport: A Study of Racial Policies in the Railroad, Airline, Trucking, and Urban Transit Industries,* by Herbert R. Northrup, Howard W. Risher, Jr., Richard D. Leone, and Philip W. Jeffress. 1971. $13.50*

Vol. VI. *Negro Employment in Retail Trade: A Study of Racial Policies in the Department Store, Drugstore, and Supermarket Industries,* by Gordon F. Bloom, F. Marion Fletcher, and Charles R. Perry. 1972.

Vol. VII. *Negro Employment in the Maritime Industries: A Study of Racial Policies in the Shipbuilding, Longshore, and Offshore Maritime Industries,* by Lester Rubin, William S. Swift, and Herbert R. Northrup. 1971.

Vol. VIII. *Black and Other Minority Participation in the All-Volunteer Navy and Marine Corps,* by Herbert R. Northrup, Steven M. DiAntonio, John A. Brinker, and Dale E. Daniel. 1979.

Order from University Microfilms, Inc.
Attn: Books Editorial Department
300 North Zeeb Road
Ann Arbor, Michigan 48106

* Order this book from the Industrial Research Unit, the Wharton School, University of Pennsylvania, Philadelphia, Pennsylvania 19104.